GAUCHER DISEASE

GAUCHER DISEASE

EDITED BY
ANTHONY H. FUTERMAN
ARI ZIMRAN

CRC Press
Taylor & Francis Group
Boca Raton London New York

CRC Press is an imprint of the
Taylor & Francis Group, an **informa** business
A TAYLOR & FRANCIS BOOK

CRC Press
Taylor & Francis Group
6000 Broken Sound Parkway NW, Suite 300
Boca Raton, FL 33487-2742

First issued in paperback 2019

© 2007 by Taylor & Francis Group, LLC
CRC Press is an imprint of Taylor & Francis Group, an Informa business

No claim to original U.S. Government works

ISBN-13: 978-0-8493-3816-8 (hbk)
ISBN-13: 978-0-367-39061-7 (pbk)

Library of Congress Cataloging-in-Publication Data

Gaucher disease / edited by Anthony H. Futerman and Ari Zimran.
    p. ; cm.
    Includes bibliographical references and index.
    ISBN 0-8493-3816-6 (alk. paper)
    1. Gaucher's disease. I. Futerman, Anthony H. II. Zimran, Ari.
    [DNLM: 1. Gaucher Disease. WD 205.5.L5 G265 2006]

RC632.G36G38 2006
616.3'997—dc22
                                       2006041804

Visit the Taylor & Francis Web site at
http://www.taylorandfrancis.com

and the CRC Press Web site at
http://www.crcpress.com

Preface

Gaucher disease is a multi-system disease that was first described by the French physician Philippe Gaucher in 1882. Since this first clinical description, there have been tremendous advances in our understanding of Gaucher disease pathology, progression, and management. It has been known for about 50 years that Gaucher disease belongs to a group of human inherited diseases classified as lysosomal storage disorders (LSDs), which are caused by the defective activity of one or more enzymes found in the lysosome. Gaucher disease is the most common LSD and often acts as a prototype for the others, inasmuch as many of the treatments now available for LSDs were first tested and their efficacy shown in Gaucher disease; this is true for enzyme replacement therapy (ERT), substrate reduction therapy (SRT), bone marrow therapy, and other new and emerging therapies, some of which are discussed in this book.

Part of the stimulus for writing this book at this time comes from the recent renewal of interest in the underlying biochemical mechanisms responsible for Gaucher disease pathology. In 1982, when the first book dedicated to Gaucher disease was published by Desnick, Gatt, and Grabowski (*Gaucher Disease: A Century of Delineation and Research*, Desnick, R.J., Gatt, S., and Grabowski, G.A., eds., New York, Alan R. Liss), little basic science was included and there was still no therapy for the disease. When a more recent book was published in 1997 (Gaucher's Disease. In *Clinical Haematology*, Zimran, A, ed., Balliere Tindall, London), ERT had become available, but SRT and pharmacological chaperones were not available as potential therapies, and the focus of the book was more on clinical aspects than on basic science. In contrast, the current book includes a number of chapters dealing with molecular biology, crystallography, cell biology, and animal models, and includes new and updated chapters about various clinical and diagnostic features, as well as new chapters on novel therapeutic approaches other than ERT, including SRT and chaperone therapy. It is fair to say that much greater progress has been made in the 23 years between 1982 and 2005 than was made during the 100 years between 1882 and 1982, reflecting the exponential advances of science and biotechnology during the recent period.

Another innovative feature of this book is the inclusion of a section on the ethical, societal, and medical considerations associated with Gaucher disease as a model for LSDs and orphan diseases in general, with chapters written by patient organizations, pharmaceutical companies, physicians, public health leaders, and ethicists. We believe that the inclusion of this section significantly adds to the relevance and importance of this book.

Finally, it has been a challenge to write an up-to-date book on Gaucher disease. We believe that we have covered the most recent advances and apologize if something has been inadvertently left out. We would like to thank all of the authors of chapters for their excellent contributions, Jessica Futerman for help with editing, and Israela Tishler for help with formatting. We trust that the current book will be of use to basic scientists, to clinicians, to pharmaceutical companies looking for new areas of research with therapeutic potential, and perhaps of most importance, to patients and their families who are seeking to understand more about the disease with which they have been afflicted.

The Editors

Anthony H. Futerman obtained his Ph.D. in 1987 and is currently a professor in the Department of Biological Chemistry at the Weizmann Institute of Science, Rehovot, Israel. His research work focuses on understanding the physiological roles of sphingolipids and glycolipids, and on the pathophysiological mechanisms by which glycolipid accumulation in lysosomal storage diseases (such as Gaucher disease) cause cell dysfunction. His work also focuses on attempts to provide alternative therapeutic options for Gaucher disease patients. He has published over 100 papers and frequently lectures at international meetings. He is the Joseph Meyerhoff Professor of Biochemistry at the Weizmann Institute of Science, a former member of the editorial boards of the *Journal of Biological Chemistry* and the *Journal of Neurochemistry*, and was chair of the 2006 Gordon Conference on Glycolipid and Sphingolipid Biology.

Ari Zimran graduated from the Hebrew University, Hadassah Medical School, Jerusalem, Israel in 1975, and completed his internship at Rambam Hospital, Haifa in 1976. He then served several years as a medical officer in the Israeli army, prior to completing his residency in internal medicine at Shaare Zedek Medical Center in Jerusalem in 1986. During three years of research fellowship at the Scripps Research Institute in La Jolla, under the mentorship of Professor Ernest Beutler, he gained interest in both molecular and clinical aspects of Gaucher disease. Upon return to Israel, he founded a referral center for patients with Gaucher disease, where over 600 patients have been followed. He participated in several clinical trials that led to market approval of new treatments for patients with Gaucher disease, both multi-center and single center studies. He has published more than 150 papers and edited two previous books — one on Gaucher disease and the other on lysosomal storage disorders.

Contributors

Johannes M.F.G. Aerts
Academic Medical Center
University of Amsterdam
Amsterdam, The Netherlands

Erik M. Akkerman
University of Amsterdam
Amsterdam, The Netherlands

Abby Alpert
University of Maryland
College Park, Maryland

David J. Begley
Kings College London
London, United Kingdom

Bruno Bembi
Pediatric Hospital Burlo Garofolo
Trieste, Italy

Ernest Beutler
The Scripps Research Institute
La Jolla, California

Roscoe O. Brady
National Institutes of Health
Bethesda, Maryland

Tanya Collin-Histed
European Gaucher Alliance
London, United Kingdom

Timothy M. Cox
University of Cambridge
Cambridge, United Kingdom

Robert J. Desnick
Mount Sinai School of Medicine of
  New York University
New York, New York

Deborah Elstein
Shaare Zedek Medical Center
Jerusalem, Israel

Paola A. Erba
University of Pisa Medical School
Pisa, Italy

Jian-Qiang Fan
Mount Sinai School of Medicine of
  New York University
New York, New York

Maria Fuller
Children, Youth, and Women's Health
  Service
Adelaide, Australia

Anthony H. Futerman
Weizmann Institute of Science
Rehovot, Israel

Alan M. Garber
Stanford University
Stanford, California

Dana P. Goldman
RAND Corporation
Santa Monica, California

Gregory A. Grabowski
Children's Hospital Research
  Foundation
University of Cincinnati
Cincinnati, Ohio

Carla E.M. Hollak
Department of Internal Medicine
Academic Medical Center
Amsterdam, The Netherlands

Silvia Locatelli Hoops
Kekulé-Institut für Organische Chemie
  und Biochemie der Universität
Bonn, Germany

John J. Hopwood
Children, Youth, and Women's Health
  Service
Adelaide, Australia

Kathleen S. Hruska
National Institute of Mental Health and
  National Human Genome Research
  Institute
Bethesda, Maryland

Avi Israeli
Ministry of Health
Jerusalem, Israel

Andrzej Kazimierczuk
Children's Hospital Research
  Foundation
University of Cincinnati
Cincinnati, Ohio

Thomas Kolter
Kekulé-Institut für Organische Chemie
  und Biochemie der Universität
Bonn, Germany

William Krivit
University of Minnesota School of
  Medicine
Minneapolis, Minnesota

Mary E. LaMarca
National Institute of Mental Health and
  National Human Genome Research
  Institute
Bethesda, Maryland

Robert E. Lee
Univeristy of Pittsburgh
Pittsburgh, Pennsylvania

Susan Lewis
European Gaucher Alliance
London, United Kingdom

Benjamin Liou
University of Cincinnati
Cincinnati, Ohio

Mario Maas
University of Amsterdam
Amsterdam, The Netherlands

Greg Macres
Children's Gaucher Research Fund
Sacramento, California

Jeremy Manuel
European Gaucher Alliance
London, United Kingdom

Giuliano Mariani
University of Pisa Medical School
Pisa, Italy

David P. Meeker
Genzyme Corporation,
Cambridge, Massachusetts

Peter J. Meikle
Children, Youth, and Women's Health
  Service
Adelaide, Australia

Pramod K. Mistry
Yale University School of Medicine
New Haven, Connecticut

Silvia Muro
University of Pennsylvania
Philadelphia, Pennsylvania

Gregory M. Pastores
New York University School of
  Medicine
New York, New York

Charles Peters
University of Missouri at Kansas City
  Medical School
Kansas City, Missouri

Frances M. Platt
University of Oxford
Oxford, United Kingdom

Lakshmanane Premkumar
Weizmann Institute of Science
Rehovot, Israel

Konrad Sandhoff
Kekulé-Institut für Organische Chemie
  und Biochemie der Universität
Bonn, Germany

Raphael Schiffmann
National Institutes of Health
Bethesda, Maryland

Edward H. Schuchman
Mount Sinai School of Medicine
New York, New York

James A. Shayman
University of Michigan Medical School
Ann Arbor, Michigan

Ellen Sidransky
National Institute of Mental Health and
  National Human Genome Research
  Institute
Bethesda, Maryland

Israel Silman
Weizmann Institute of Science
Rehovot, Israel

Avraham Steinberg
Shaare Zedek Medical Center
Jerusalem, Israel

Ying Sun
University of Cincinnati
Cincinnati, Ohio

Joel L. Sussman
Weizmann Institute of Science
Rehovot, Israel

Henri A. Termeer
Genzyme Corporation
Cambridge, Massachusetts

Ashok Vellodi
Great Ormond Street Hospital for Sick
  Children, NHS Trust
London, United Kingdom

Jill Waalen
The Scripps Research Institute
La Jolla, California

Kondi Wong
Wilford Hall Medical Center
Lackland Air Force Base, Texas

You-Hai Xu
University of Cincinnati
Cincinnati, Ohio

Ari Zimran
Gaucher Clinic
Shaare Zedek Medical Center
Jerusalem, Israel

Contents

CHAPTER 1

Introduction: Overview and Historical Perspective

Roscoe O. Brady

CONTENTS

ORIGINAL DESCRIPTIONS OF PATIENTS WITH GAUCHER DISEASE

Gaucher disease is named for the French dermatologist Philippe C.E. Gaucher who described a 23-year-old female patient in 1882 who had an enlarged spleen that he believed was due to an epithelioma.[1] He noted the presence of large, unusual cells in the patient's spleen. Reports of additional patients with similar presentations appeared shortly thereafter,[2-4] and the eponym "Gaucher's disease" was applied. In addition, the term "Gaucher cell" also became commonly used to specify the characteristic engorged cells in the organs of such patients. Brill suggested in 1901 that Gaucher disease was a familial disorder.[5] Involvement of the lymph nodes, the liver,

1

Figure 1.1   P. C. E. Gaucher, Paris, France. First description of a patient with Gaucher disease.

and the bone marrow was reported in 1904.[6] The first premortem diagnosis of a patient with the disorder was made by Brill and his co-workers in 1905.[7]

It was soon realized that the age at which the signs and symptoms associated with Gaucher disease become manifest varied considerably. Collier reported a patient in 1895 who was 6 years old.[2] Kraus[8] and Rusca[9] documented the disorder in infants. Neurologic impairment in an infant was first reported by Oberling and Woringer in 1927.[10] This phenotype eventually became known as the infantile or acute neuronopathic form. These patients are currently classified as having Type 2 Gaucher disease. Thirty-two years later, another discrete clinical presentation was described by Hillborg.[11] Neurologic signs in these patients appear during the preteen and early teen years. This phenotype has been called the juvenile or chronic neuronopathic form. These patients are now classified as Type 3 Gaucher disease.

The most prevalent phenotype is Type 1 Gaucher disease. Patients in this category were previously designated as the "adult" form of Gaucher disease without overt evidence of central nervous system involvement. In some of these individuals, systemic signs may appear as early as the first year of life. At the other extreme, the occurrence of Gaucher disease has been documented as late as the eighth and ninth decades in individuals who were only mildly affected, or even completely asymptomatic.[12] Clinical features of patients with Gaucher disease are described in

Figure 1.2 Left: Roscoe O. Brady, National Institutes of Health, Bethesda, Maryland, U.S. Demonstrated the enzymatic defect in Gaucher disease and developed diagnostic tests and enzyme replacement therapy for patients with this disorder. Right: Christian de Duve, Nobel Laureate, Belgium. Discovered lysosomes.

detail in Chapter 10 and Chapter 11. Pathological aspects are discussed in detail in Chapters 12 and 13.

ACCUMULATING MATERIALS

Glucocerebroside

In 1907, Marchand reported that a hyaline-like material was stored in the "so-called idiopathic splenomegaly" of the Gaucher type.[13] He erroneously believed that it was not a lipid because the material did not react with osmic acid. The accumulating substance was identified as a cerebroside by Lieb in 1924.[14] Cerebrosides consist of three components: two are lipids (sphingosine and fatty acid), and the third is a carbohydrate. The combination of sphingosine to which a long-chain fatty acid is linked by an amide bond to the nitrogen atom on carbon 2 of sphingosine is called ceramide. Galactocerebroside (galactosylceramide) had been known since the beginning of the 20th century to be the preponderant lipid of the brain on a weight basis. Lieb believed that this cerebroside accumulated in the organs and tissues of patients with Gaucher disease. However, the optical rotation of an aqueous solution of the sugar derived from the accumulating cerebroside was incompatible with its being galactose. In 1934, Aghion reported that the sugar moiety of the accumulating cerebroside in Gaucher patients was glucose rather than galactose.[15] This finding conclusively established that the principal accumulating lipid in patients with Gaucher disease is glucocerebroside (glucosylceramide). Glucocerebroside is synthesized by virtually every cell in the body. However, much of the burden of this

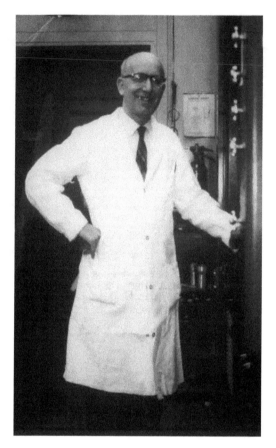

Figure 1.3   David Shapiro, Weizmann Institute of Science. Guided the synthesis of ^{14}C-labeled glucocerebroside used to identify the metabolic defect in Gaucher disease.

metabolite appears to arise from the catabolism of glycosphingolipids in the membranes of senescent red and white blood cells.[16]

Glucopsychosine (glucosylsphingosine)

Glucocerebroside that accumulates in the brain of patients with the neuronopathic forms of Gaucher disease appears to arise from the turnover of acidic sphingoglycolipids known as gangliosides. In addition to the accumulation of glucocerebroside in systemic organs and the central nervous system in patients with Types 2 and 3 Gaucher disease, augmented levels of glucopsychosine have also been demonstrated in the brain.[17,18] The enzymatic synthesis of glucopsychosine has been demonstrated in subcellular preparations of neonatal rat brain.[19] It is highly cytotoxic to cultured neuronal cells,[20] and it is likely to play a significant role in the pathogenesis of the neuronopathic forms of Gaucher disease.

Figure 1.4   Peter Pentchev, National Institutes of Health, Bethesda, Maryland, U.S. First purification of glucocerebrosidase from human placental tissue.

ELUCIDATION OF THE ENZYMATIC DEFECT IN GAUCHER DISEASE

Considerable speculation about the metabolic basis of Gaucher disease took place from 1934 until the early 1960s. Because of the possibility of an error of carbohydrate metabolism that resulted in substitution of glucose for galactose in the cerebrosides of patients with Gaucher disease, a galactose tolerance test was performed by Thannhauser in a patient with Type 1 Gaucher disease and found to be normal.[21] My colleagues and I began to investigate the cause of this disorder in 1956. The biochemical abnormality associated with Gaucher disease, or of any other lipid storage disorder, was not known. Among possible metabolic errors were the production of an abnormal substance or the synthesis of an excessive quantity of a normally occurring minor material. We began with an examination of the formation of glucocerebroside in slices of spleen tissue obtained from Gaucher patients who were undergoing splenectomy. We found that the pathway of cerebroside formation was normal and that the rate of formation of glucocerebroside was not accelerated. We postulated that the accumulation of glucocerebroside was due to a deficiency of an enzyme required for the catabolism of this lipid.[22] At the time, nothing was known about the catabolism of sphingolipids. To test this hypothesis, my colleagues and I chemically synthesized glucocerebroside labeled with 14C in the glucose, and in another preparation, in the fatty acid portion of this molecule. We discovered that all mammalian tissues contain the enzyme glucocerebrosidase that catalyzes the hydrolytic cleavage of glucose from glucocerebroside.[23] When we examined gluco-cerebrosidase activity in tissues from patients with Gaucher disease, we discovered that it was dramatically reduced from that in normal individuals.[24,25] The deficiency of glucocerebrosidase in Gaucher disease has been universally confirmed. This discovery provided pivotal understanding of the etiology of all of the sphingolipid storage disorders. Gaucher disease and other diseases in this category were shown to be lysosomal storage disorders.[26]

DIAGNOSTIC TESTS AND GENETIC COUNSELING

Practical benefit quickly followed the discovery of the enzymatic defect in Gaucher disease. My colleagues and I developed enzyme assay procedures to rapidly diagnose patients with these disorders. The determination was originally performed with white blood cells,[27] and was quickly extended to the use of cultured skin fibroblasts.[28] Further investigations revealed that the majority of the carriers of the Gaucher disease trait could be identified by such assays.[29] Procedures were then developed for the prenatal diagnosis of Gaucher disease.[30] These techniques are widely used for genetic counseling.

THERAPY

Enzyme Replacement

Soon after the discovery of the enzymatic defect in Gaucher disease, I began to consider what might be done to help patients with this and other metabolic storage disorders and postulated that enzyme replacement might be beneficial.[31] In order to examine this possibility, my colleagues and I undertook the purification of gluco-cerebrosidase from human placental tissue to reduce the possibility of sensitizing recipients to the exogenous protein.[32] When placental glucocerebrosidase was injected into two patients with this disorder, glucocerebroside that had accumulated in the liver was reduced significantly in both patients.[33] In addition, elevated gluco-cerebroside in the blood of the recipients returned to normal within 3 days following injection of the enzyme. The reduction of glucocerebroside in the circulation lasted many months.[34]

Greater quantities of glucocerebrosidase were required to remove the large amounts of glucocerebroside that accumulated in the organs of many patients with Gaucher disease. Therefore, my associates and I developed a large-scale procedure for the purification of placental glucocerebrosidase.[35] The enzyme is a glycoprotein with four oligosaccharide chains. It was discovered that it was necessary to modify these oligosaccharide moieties to target glucocerebrosidase to macrophages, the scavenger cells in the body in which glucocerebroside accumulates. Macrophages have a lectin on their surface that interacts avidly with glycoproteins with terminal molecules of mannose.[36] Targeting glucocerebrosidase to these cells was accomplished by the removal of oligosaccharides by the sequential action of three exoglycosidases.[37] Mannose-terminal glucocerebrosidase that is effectively targeted to macrophages was produced in this fashion.[38]

Dose-response and clinical efficacy trials were undertaken with macrophage-targeted glucocerebrosidase. All of the recipients experienced dramatic improvement.[39] Their anemia and thrombocytopenia improved greatly. Splenomegaly and hepatomegaly diminished significantly and skeletal benefit occurred in all of the patients. Before receiving the enzyme, many of the patients had severe bouts of pain due to infarctions in their spleens, livers, and bones. They became completely free from pain. No patient in the trial had an untoward reaction or became sensitized to the enzyme.[40] Enzyme replacement therapy for patients with Gaucher disease was approved by the U.S. Food and Drug Administration in 1991. It was subsequently approved in 50 additional countries.

Macrophage-targeted glucocerebrosidase was later produced recombinantly in Chinese hamster ovary cells. Recombinant glucocerebrosidase was shown to be as effective as the enzyme obtained from placenta,[41] and it was approved in the U.S. for the treatment of patients with Gaucher disease in 1994. More than 4500 patients throughout the world are currently receiving this life-saving therapy. The remarkable benefit of enzyme replacement in Gaucher disease appeared to indicate the potential of this strategy for the treatment of other human metabolic storage disorders.

Gene Therapy

Because successful bone-marrow transplantation can cure patients with type 1 Gaucher disease, it was considered reasonable to examine gene therapy in patients with this phenotype. In particular, the use of a retroviral vector containing the normal DNA sequence for glucocerebrosidase to transduce patients' stem and progenitor cells was believed to be a reasonable therapeutic undertaking.[42] Only temporary expression of the transgene could be detected in a patient with Gaucher disease following intravenous administration of the patient's autologous CD34+ cells that had been transduced and expanded *in vitro*.[43] It is likely that an insufficient number of cells had been transduced or there had been inadequate separation of transduced from nontransduced prior to administration to the recipient. Recently, improved transfuction efficiency has been demonstrated using HIV-based vectors.[44] This approach has been shown to cause the expression and secretion of glucocerebrosidase *in vitro* and *in vivo*.[45,46] If lentiviral vectors are found to be safe for human administration, it is anticipated that gene therapy trials in patients with Gaucher disease and other metabolic storage disorders will be undertaken with these agents.

THE FUTURE

In addition to enzyme replacement and gene therapy, several other treatment modalities are currently under consideration. One of them is known as substrate depletion therapy. In this approach, the enzymatic synthesis of glucocerebroside is blocked by an inhibitor of ceramide:UDP-glucoslytransferase. One of the principal substances employed in this regard is N-butyldeoxynojirimycin.[47] One wonders if a combination of enzyme replacement therapy with a small molecule inhibitor of glucocerebroside formation that can cross the blood-brain barrier will provide augmented benefit to patients with neuronopathic forms of Gaucher disease. A second novel strategy is called molecular chaperone therapy. Some mutations of genes coding for glucocerebrosidase cause mis-folding of nascent enzyme as it emerges from the endoplasmic reticulum. Mis-folded proteins are destroyed by intracellular quality control proteolytic enzymes before they reach the lysosome. Certain inhibitors of the catalytic activity of enzymes at less than maximal inhibitory levels can be used to maintain the proper configuration of a mutated enzyme and deliver it in an active form to the lysosome. Such substances are called molecular chaperones. One of them has already been shown to increase the activity of glucocerebrosidase containing the N370S mutation.[48] Gene editing is yet another potential strategy in which mutations in genes are corrected by approximation of an overlapping strand of normal DNA that induces excision and repair of the abnormal nucleotide. This approach has been reported to increase catalytic activity of mutated α-glucosidase in Pompe disease.[49] No reports have yet appeared concerning such correction of the mutated acid β-glucosidase in patients with Gaucher disease. Finally, one would hope that in the future the logistics of enzyme therapy that is now usually administered intravenously every 2 weeks to patients with Gaucher disease might be improved. In particular, an orally available method for its administration seems highly desirable. Rapid progress is being made in the development of micro-[50] and nanoparticles as delivery agents for proteins. It is anticipated that orally administered glucocerebrosidase will become available in the not too distant future.

ENVOI

Much beneficial information has been acquired in the course of investigating the etiology of Gaucher disease and the development of effective treatment for patients with this disorder. It is therefore entirely appropriate that a detailed examination of the multiple aspects of Gaucher disease be presented at this time. It is anticipated that lessons learned during the investigations on this metabolic disorder will prove beneficial for the management of patients with many other hereditary metabolic disorders.

REFERENCES

1. Gaucher, P.C.E., De l'épithélioma primitif de la rate, *Thèse de Paris*, 1882.

2. Collier, W.A., A case of enlarged spleen in a child aged six, *Tr. Path. Soc. Lond.*, 46, 148, 1895.
3. Picou, R. and Ramond, F., Splénomégalie primitive épithélioma primitif de la rate, *Arch. d'Med. Exper. et d'Anat. Path.*, 8, 168, 1896.
4. Bovaird, O.J.R., Primary splenomegaly-endothelial hyperplasia of the spleen-2 cases in children-autopsy and morphological examination in one, *Am. J. Med. Sci.*, 120, 377, 1900.
5. Brill, N.E., Primary splenomegaly with a report of three cases occurring in one family, *Am. J. Med. Sci.*, 121, 377, 1901.
6. Brill, N.E., A case of "splenomegalie primitive": with involvement of the haemopoietic organs, *Proc. NY Path. Soc.*, 4, 143, 1904.
7. Brill, N.E., Mandelbaum, F.S., and Libman, E., Primary splenomegaly — Gaucher type. Report on one of four cases occurring in a single generation in one family, *Am. J. Med. Sci.*, 129, 491, 1905.
8. Kraus, E.J., Zur Kenntnis der Splenomegalie Gaucher, insbesondere der Histogenese der Grobzellen Wucherung, *Z. Angew. Anat.*, 7, 186, 1920.
9. Rusca, C.L., Sul morbo del Gaucher, *Haematologica* (Pavia), 2, 441, 1921.
10. Oberling, C. and Woringer, P., La maladie de Gaucher chez le nourrisson, *Rev. Franc. de Pediat.*, 3, 475, 1927.
11. Hillborg, P.O., Morbus Gaucher: Norrbotten, *Nord. Med.*, 61, 303, 1959.
12. Brady, R.O., Glucosyl ceramide lipidosis: Gaucher's disease, in *The Metabolic Basis of Inherited Disease, IVth Edition*, Stanbury, J.B., Wyngaarden, J.B., and Fredrickson, D.S., Eds., McGraw-Hill, New York, 1978, 731.
13. Marchand, F., Über sogennante idiopathische Splenomegalie (Typus Gaucher), *Münchener Med. Wchnschr.*, 54, 1102, 1907.
14. Lieb, H., Cerebrosidespeicherung bei Splenomegalie Typus Gaucher, *Ztschr. Physiol.Chem.*, 140, 305, 1924.
15. Aghion, H., La maladie de Gaucher dans l'enfance, *Thèse, Paris,* 1934.
16. Kattlove, H.E. et al., Gaucher cells in chronic myelocytic leukemia: an acquired abnormality, *Blood*, 33, 379, 1969.
17. Nilsson, O. and Svennerholm, L., Accumulation of glucosylceramide and glucosylsphingosine (psychosine) in cerebrum and cerebellum in infantile and juvenile Gaucher disease, *J. Neurochem.*, 39, 709, 1982.
18. Orvisky, E. et al., Glucosylsphingosine accumulation in tissues from patients with Gaucher disease: correlation with phenotype and genotype, *Mol. Genet. Metab.*, 76, 262, 2002.
19. Curtino, J. A. and Caputto, R., Enzymic synthesis of glucosylsphingosine by rat brain microsomes, *Lipids*, 7, 525, 1972.
20. Schueler, U.H. et al., Toxicity of glucosylsphingosine (glucopsychosine) to cultured neuronal cells: a model system for assessing neuronal damage in Gaucher disease type 2 and 3, *Neurobiol. Dis.*, 14, 595, 2003.
21. Thannhauser, S.J., *Lipidoses. Diseases of Cellular Lipid Metabolism*. Christian, H.A., Ed., Oxford University Press, New York, 1950.
22. Trams, E.G. and Brady, R.O., Cerebroside synthesis in Gaucher's disease, *J. Clin. Invest.*, 39, 1546, 1960.
23. Brady, R.O., Kanfer, J., and Shapiro, D., The metabolism of glucocerebrosides. I. Purification and properties of a glucocerebroside-cleaving enzyme from spleen tissue, *J. Biol. Chem.*, 240, 39, 1965.

24. Brady, R.O., Kanfer, J.N., and Shapiro, D., The metabolism of glucocerebrosides. II. Evidence of an enzymatic deficiency in Gaucher's disease, *Biochem. Biophys. Res. Commun.*, 18, 221, 1965.

25. Brady, R.O., Kanfer, J.N., Bradley, R.M., and Shapiro, D., Demonstration of a deficiency of glucocerebroside-cleaving enzyme in Gaucher's disease, *J. Clin. Invest.*, 45, 1112, 1966.

26. Weinreb, N.J., Brady, R.O., and Tappel AL., The lysosomal localization of sphingolipid hydrolases, *Biochim. Biophys. Acta*, 159, 141, 1968.

27. Kampine, J.P. et al., The diagnosis of Gaucher's disease and Niemann-Pick disease using small samples of venous blood, *Science*, 155, 86, 1967.

28. Ho, M.W. et al., Adult Gaucher's disease: kindred studies and demonstration of a deficiency of acid β-glucosidase in cultured fibroblasts, *Am. J. Hum. Genet.*, 24, 37, 1972.

29. Brady, R.O., Johnson, W.G., and Uhlendorf, B.W., Identification of heterozygous carriers of lipid storage diseases, *Am. J. Med.*, 51, 423, 1972.

30. Schneider, E.L. et al., Infantile (Type II) Gaucher's disease: *in utero* diagnosis and fetal pathology, *J. Pediatr.*, 81, 1134, 1972.

31. Brady, R.O., The sphingolipidoses, *N. Engl. J. Med.*, 275, 312, 1966.

32. Pentchev, P.G. et al., Isolation and characterization of glucocerebrosidase from human placental tissue, *J. Biol. Chem.*, 248, 5256, 1973.

33. Brady, R.O. et al., Replacement therapy for inherited enzyme deficiency: use of purified glucocerebrosidase in Gaucher's disease, *N. Engl. J. Med.*, 291, 989, 1974.

34. Pentchev, P.G. et al., Replacement therapy for inherited enzyme deficiency: sustained-clearance of accumulated glucocerebroside in Gaucher's disease following infusion of purified glucocerebrosidase, *J. Mol. Med.*, 1, 73, 1975.

35. Furbish, F.S. et al., Enzyme replacement therapy in Gaucher's disease: large-scale purification of glucocerebrosidase suitable for human administration, *Proc. Natl. Acad. Sci. U.S.*, 74, 3560, 1977.

36. Stahl, P.D. et al., Evidence for receptor-mediated binding of glycoproteins, glycoconjugates, and lysosomal glycosidases by alveolar macrophages, *Proc. Natl. Acad. Sci. U.S.*, 75, 1599, 1978.

37. Furbish, F.S. et al., 1981, Uptake and distribution of placental glucocerebrosidase in rat hepatic cells and effects of sequential deglycosylation, *Biochim. Biophys. Acta*, 673, 425, 1981.

38. Brady, R.O. and Furbish, F.S., Enzyme replacement therapy: specific targeting of exogenous enzymes to storage cells, in *Membranes and Transport, Vol. 2*, Martonosi, A.N., Ed., Plenum Publishing Corp., New York, 1982, chap. 170.

39. Barton, N.W. et al., Replacement therapy for inherited enzyme deficiency — macrophage-targeted glucocerebrosidase for Gaucher's disease, *N. Engl. J. Med.*, 324, 1464, 1991.

40. Murray, G.J. et al., Gaucher's disease: lack of antibody response in 12 patients follow ingrepeated intravenous infusions of mannose-terminal glucocerebrosidase, *J. Immunol. Methods*, 137, 113, 1991.

41. Grabowski, G.A. et al., Enzyme therapy in Gaucher disease Type 1: Comparative efficacy of mannose-terminated glucocerebrosidase from natural and recombinant sources, *Ann. Int. Med.*, 122, 33, 1995.

42. Medin, J.A. et al., A bicistronic therapeutic retroviral vector enables sorting of transduced CD34+ cells and corrects the enzyme deficiency in cells from Gaucher patients, *Blood*, 87,1754, 1996.

43. Dunbar C.E. et al., Retroviral transfer of the glucocerebrosidase gene into CD34+ cells from patients with Gaucher disease: *in vivo* detection of transduced cells without myeloablation, *Hum. Gen. Ther.*, 9, 2629, 1998.
44. Mochizuki, H. et al., High-titer immunodeficiency virus type 1-based vector systems for gene delivery into nondividing cells, *J. Virol.*, 72, 8873, 1998.
45. Kim, E.Y. et al., Expression and secretion of human glucocerebrosidase mediated by recombinant lentivirus vectors *in vitro* and *in vivo*: implications for gene therapy of Gaucher disease, *Biochem. Biophys. Res. Commun.*, 318, 381, 2004.
46. Kim, E.Y., Long-term expression of the human glucocerebrosidase gene *in vivo* after transplantation of bone-marrow-derived cells transformed with a lentivirus vector, *J. Gene Med.*, 7, 878, 2005.
47. Cox, T. et al., Novel oral treatment of Gaucher's disease with N-butyldeoxynojirimycin (OGT 918) to decrease substrate biosynthesis, *Lancet*, 355, 1481, 2000.
48. Sawkar, A.R. et al., Chemical chaperones increase the cellular activity of N370S β-glucosidase: a therapeutic strategy for Gaucher disease, *Proc. Natl. Acad. Sci. USA*, 99, 5428, 2002.
49. Lu. I-L. et al., Correction/mutation of acid alpha-D-glucosidase gene by modified single-stranded oligonucleotides: *in vitro* and *in vivo* studies, *Gene Ther.*, 10, 910, 2003.
50. Kohane, D.S. et al., pH-triggered release of macromolecules from spray-dried polymethacrylate microparticles, *Pharm. Res.*, 20, 1553, 2003.

CHAPTER 2

Gaucher Disease: Molecular Biology and Genotype-Phenotype Correlations

Kathleen S. Hruska, Mary E. LaMarca, and Ellen Sidransky

CONTENTS

Almost two decades have passed since the gene for glucocerebrosidase (GBA), the enzyme deficient in Gaucher disease (MIM 230800, 230900 and 231000), was first localized to 1q21 [1] and then cloned and sequenced [2–4]. There were great expectations that the "molecular era" would provide the key to explaining the vast phenotypic differences encountered among patients with Gaucher disease. Yet while there have been vast advances in our understanding of the gene structure and locus [5], the identification of mutant alleles [6–8], the enzyme structure [9] and genotype-phenotype correlations [10–14], the precise mechanisms by which the mutant enzyme leads to the patient phenotype still remains a mystery.

THE GLUCOCEREBROSIDASE GENE (GBA)

The GBA gene contains 11 exons and 10 introns, covering 7.6 kb of sequence. It became immediately apparent that there is a homologous pseudogene (psGBA) located 16 kb downstream [4] (Figure 2.1). The pseudogene spans 5.7 kb with the same exon and intron number and structure as the functional gene, which is longer

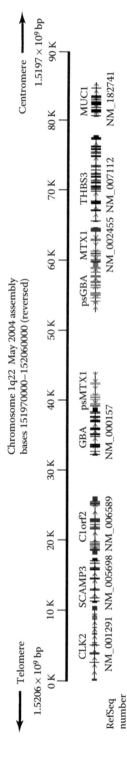

Figure 2.1 Glucocerebrosidase gene region on chromosome 1. The 85-kb region surrounding the glucocerebrosidase gene (GBA) contains seven genes and two pseudogenes. The known genes, their exonic structures, transcription direction and relative positions along chromosome 1q are shown as adapted from the UCSC Human Genome Browser, May 2004 assembly (http://genome.ucsc.edu). The pseudogenes for GBA and MTX1 have been added. Note that while GBA has been localized to 1q21 in cytogenetic studies, the location within the human genome sequence is 1q22. The GenBank RefSeq accession numbers for the mRNAs shown are given below the gene name and structure (http://www.ncbi.nlm.nih.gov/entrez/query.fcgi?CMD=search&DB=nucleotide). CLK2, CDC-like kinase 2; SCAMP3, secretory carrier membrane protein 3; C1orf2, chromosome 1 open reading frame 2 (cote1); GBA, glucocerebrosidase; psMTX1, metaxin 1 pseudogene; psGBA, glucocerebrosidase pseudogene; MTX1, metaxin 1; THBS3, thrombospondin 3; MUC1, mucin 1, transmembrane.

than psGBA because of several *Alu* insertions in intronic tracts. Despite the length differences, psGBA has maintained 96% sequence identity with the functional GBA gene. One important distinction between the two sequences is a 55-bp deletion in exon 9 of the pseudogene. The existence of a highly conserved pseudogene at the same locus is significant because many Gaucher mutations and groups of mutations result from recombination events between psGBA and GBA [15–19].

Sequencing of the region surrounding human glucocerebrosidase led to the discovery that the locus is particularly gene rich, with seven genes and two pseudo-genes within an 85-kb region of chromosome 1q [5,20] (Figure 2.1). Immediately downstream of the glucocerebrosidase pseudogene is the gene for metaxin (MTX1), which is transcribed convergently to GBA [20]. Metaxin, in turn, shares a bidirec-tional promoter with the nearby gene for thrombospondin 3 [21]. Metaxin, a protein that appears to play a role in the preprotein import complex in the outer mitochondrial membrane [22], also has a pseudogene that is located in the 16-kb region between GBA and psGBA. These two pseudogenes have resulted from a duplication of the region, and are present in primates, but not in other species [5]. Multi-species comparative sequence analyses [23] demonstrate conservation of the exonic regions of both MTX1 and GBA between human and nine mammalian species [24].

The glucocerebrosidase cDNA is approximately 2 kb in length. Northern blots have revealed several mRNA species that could be due to splice variants, alternate polyadenylation sites, or transcription of the pseudogene [25,26]. The GBA mRNA has two in-frame ATG translational start sites located in exons 1 and 2 [3]. Both start sites are efficiently translated, producing two polypeptides with different size signal peptides [27,28]. The protein synthesized from the first ATG (NP_000148) has a 39-residue leader, while the second ATG yields a 19-residue leader; both are processed to a 497-residue mature enzyme. The catalytic residues are p.Glu274 and p.Glu379 (E235 and E340) [9] and there are four glycosylation sites at p.Asn58, p.Asn98, p.Asn185 and p.Asn309 (N19, N59, N146 and N270) [29,30].

While glucocerebrosidase is expressed in all tissues, the levels of GBA mRNA are variable in cell lines. High, moderate, low and negligible levels have been reported in epithelial, fibroblast, macrophage and B-cell lines, respectively [26,31–33]. In addition, mRNA levels have not always correlated well with the level of enzyme activity detected; i.e., high gene expression has been associated with high or low enzyme activity in different cell types and the converse has also been observed [32,34]. Discrepancies also have been detected in the ability of certain cell lines to efficiently translate GBA mRNA or process the resultant translation product into an active enzyme upon ectopic expression [35]. These data suggest the presence of a coordinated control system for the expression of glucocerebrosidase at the level of transcription, translation and posttranslation modification and transport.

The 5'genomic region of GBA contains consensus sequences for two TATA boxes and two CAAT-like boxes [4,26,32]. A 650-bp *Sac*I fragment encompassing this region is necessary and sufficient to direct expression of GBA and reporter constructs in cultured cell lines [4,32,36]. Oligonucleotides corresponding to pre-dicted transcription factor DNA-recognition sites OCT, AP-1, PEA3 and CBP have been shown to be bound by protein complexes in fibroblast and B-cell lines [36]. When compared to the corresponding 5' sequence of psGBA, this region of the functional gene is 8- to 10-fold more potent in driving reporter gene expression [4,26].

IDENTIFICATION OF MUTANT ALLELES

The first two mutations described in GBA, c.1448T>C (L444P) and c.1226A>G (N370S), were identified in the late 1980s [6,37]. These two alleles continue to be among the most prevalent throughout much of the world. Initially, groups screened patient cohorts for the presence or absence of these two base changes using Southern blotting, and then PCR. Subsequently, restriction digests and sequencing of genomic and cDNAs demonstrated that patients harbor a wide range of different mutations, including missense, nonsense, frameshift and splice-site mutations, as well as insertions, deletions and recombinant alleles. The identification of mutant alleles in GBA is laborious because primers must be designed to amplify the functional gene, but not the pseudogene. For over a decade, most laboratories used PCR-based screening to identify a panel of three to seven known mutations. Among type 1 patients of Ashkenazi Jewish ancestry, this was fairly efficient and about 90% of mutant alleles could be detected when screening for five to six mutations [13,14,38]. In non-Jewish cohorts, however, screening strategies failed to identify a significant portion of patient alleles, especially among patients with neuronopathic forms of the disorder. With the advent of automated sequencing, new mutations in GBA have been found frequently, and, to date, more than 200 have been identified (Table 2.1).

One difficulty in assessing the novelty of an identified sequence variant has been the lack of consistent nomenclature in the Gaucher literature. In this chapter, we have adopted the recommendations of the Human Genome Variation Society (HGVS, http://www.hgvs.org). In Table 2.1, we have attempted to identify alleles by the name assigned at the time of first report, cDNA and genomic nucleotide change, and the resultant protein alteration. Following the recommendation that all variants should be described at the most basic level, mutations discussed in the text are referred to by cDNA nomenclature, with the traditional name of the allele included in parentheses to aid in recognition. Two prominent changes have been made to the genomic ("g.") and protein ("p.") numbering system. The genomic nucleotide number is based on the GenBank reference sequence J03059.1; i.e., it is purely arbitrary with "g.1" being the first nucleotide in the database reference file. In addition, the current suggestion is to identify amino acid residues according to the primary translation product, not a processed mature protein. Consequently, the amino acid numbering begins with the first methionine of the longer leader sequence of glucocerebrosidase, which increases the traditional numbering system by 39 residues.

Another factor confounding the difficulties in genotyping in Gaucher disease is the presence of recombinant alleles [15–19,39]. Such alleles have been designated as "complex" alleles [17], "pseudopattern" or "psi" [16], "rec" for "recombinant" [15], "fusion" alleles [18] and "chimeric" alleles [40]. The high degree of sequence identity and the close physical proximity of GBA and its pseudogene contribute to recombination between the two loci. It has been shown that the sites of these recombination events are variable, ranging from intron 2 to exon 11 [39,41,42].

Table 2.1 Mutations in GBA

Allele Name	cDNA [a]	gDNA [a]	Exon	Protein [a,b]	Disease Types [c]	Refererence
	Substitutions					
W(-4)X	c.108G>A	g.1934G>A	2	p.Trp36X	1	[80]
V15L	c.160G>T	g.2538G>T	3	p.Val54Leu	(1)	[63]
V15M	c.160G>A	g.2538G>A	3	p.Val54Met	1	[91]
C16S	c.164G>C	g.2542G>C	3	p.Cys55Ser	2, 2/3 int.	[92]
D24N	c.187G>A	g.2565G>A	3	p.Asp63Asn	1	[68]
F37V	c.226T>G	g.2604T>G	3	p.Phe76Val	(1)	[93]
E41K	c.238G>A	g.2616G>A	3	p.Glu80Lys	2	[94]
T43I	c.245C>T	g.2623C>T	3	p.Thr82Ile	1	[95]
G46E	c.254G>A	g.2632G>A	3	p.Gly85Glu	1	[63]
R47X	c.256C>T	g.2634C>T	3	p.Arg86X	1	[91]
R48W	c.259C>T	g.2637C>T	3	p.Arg87Trp	1	[95]
K74X	c.337A>T	g.2838A>T	4	p.Lys113X	2	[96]
K79N	c.354G>C	g.2855G>C	4	p.Lys118Asn	1, 3	[97]
A90T	c.385G>A	g.2886G>A	4	p.Ala129Thr	1	[14]
L105R	c.431T>G	g.2932T>G	4	p.Leu144Arg	1	[98]
S107L	c.437C>T	g.2938C>T	4	p.Ser146Leu	2, 2/3 int., 3	[99]
F109V	c.442T>G	g.2943T>G	4	p.Phe148Val	1	[100]
G113E	c.455G>A	g.3921G>A	5	p.Gly152Glu	1	[80]
N117D	c.466A>G	g.3932A>G	5	p.Asn156Asp	1	[14]
I119T	c.473T>C	g.3939T>C	5	p.Ile158Thr	1	[64]
I119S	c.473T>G	g.3939T>G	5	p.Ile158Ser	1	[68]
R120W	c.475C>T	g.3941C>T	5	p.Arg159Trp	1, 2	[93,97,101]
R120Q	c.476G>A	g.3942G>A	5	p.Arg159Gln	1, 2	[25]
P122S	c.481C>T	g.3947C>T	5	p.Pro161Ser	3	[89]
P122L	c.482C>T	g.3948C>T	5	p.Pro161Leu	2	[102]
M123T	c.485T>C	g.3951T>C	5	p.Met162Thr	1	[91]
R131C	c.508C>T	g.3974C>T	5	p.Arg170Cys	1, 2	[94]

Table 2.1 Mutations in GBA (Continued)

Allele Name	cDNA [a]	gDNA [a]	Exon	Protein [a,b]	Disease Types [c]	Reference
R131L	c.509G>T	g.3975G>T	5	p.Arg170Leu	1, 2	[103]
T134P	c.517A>C	g.3983A>C	5	p.Thr173Pro	1	[80]
T134I	c.518C>T	g.3984C>T	5	p.Thr173Ile	1	[14]
Y135X	c.522T>G	g.3988T>G	5	p.Tyr174X	1	[14]
A136E	c.524C>A	g.3990C>A	5	p.Ala175Glu	NA	[104]
K157Q	c.586A>C	g.4052A>C	5	p.Lys196Gln	1, 2	[19,45]
P159T	c.592C>A	g.4268C>A	6	p.Pro198Thr	1	[105]
P159L	c.593C>T	g.4269C>T	6	p.Pro198Leu	1	[99]
I161S	c.599T>G	g.4275T>G	6	p.Ile200Ser	1	[106]
H162P	c.602A>C	g.4278A>C	6	p.His201Pro	1	[104]
R163X	c.604C>T	g.4280C>T	6	p.Arg202X	1, 2	[99]
Q169X	c.622C>T	g.4298C>T	6	p.Gln208X	1	[80]
R170C	c.625C>T	g.4301C>T	6	p.Arg209Cys	1	[14]
R170P	c.626G>C	g.4302G>C	6	p.Arg209Pro	1	[64]
S173X	c.635C>G	g.4311C>G	6	p.Ser212X	1	[95]
A176D	c.644C>A	g.4320C>A	6	p.Ala215Asp	1, 3	[46]
P178S	c.649C>T	g.4325C>T	6	p.Pro217Ser	2	[107]
W179X	c.653G>A	g.4329G>A	6	p.Trp218X	1	[96]
P182T	c.661C>A	g.4337C>A	6	p.Pro221Thr	1	[46]
P182L	c.662C>T	g.4338C>T	6	p.Pro221Leu	1	[50,68]
W184R	c.667T>C	g.4343T>C	6	p.Trp223Arg	1, 2	[14,100,108]
N188S	c.680A>G	g.4356A>G	6	p.Asn227Ser	1, 3	[63]
N188K	c.681T>G	g.4357T>G	6	p.Asn227Lys	1, 2	[64]
G189V	c.683G>T	g.4359G>T	6	p.Gly228Val	(1)	[109]
A190T	c.685G>A	g.4361G>A	6	p.Ala229Thr	3	[14]
A190E	c.686C>A	g.4362C>A	6	p.Ala229Glu	2	[42]
V191G	c.689T>G	g.4365T>G	6	p.Val230Gly	1	[110]
V191E	c.689T>A	g.4365T>A	6	p.Val230Glu	1	[68]
G195W	c.700G>T	g.4376G>T	6	p.Gly234Trp	1, 3	[40]

	cDNA	Genomic	Protein	Exon		Ref.
G195E	c.701G>A	g.4377G>A	p.Gly234Glu	6	1, 2	[96]
S196P	c.703T>C	g.4379T>C	p.Ser235Pro	6	1, 2	[42,51]
K198E	c.709A>G	g.4385A>G	p.Lys237Glu	6	2/3 int.	[104]
G202R	c.721G>A	g.4397G>A	p.Gly241Arg	6	1, 2, 2/3 int., 3	[46]
G202E	c.722G>A	g.4398G>A	p.Gly241Glu	6	1	[51]
Y205C	c.731A>G	g.4407A>G	p.Tyr244Cys	6	3	[111]
Y212H	c.751T>C	g.4427T>C	p.Tyr251His	6	1	[89]
F213I	c.754T>A	g.4430T>A	p.Phe252Ile	6	1, 2, 2/3 int., 3	[112]
F213C	c.755T>G	g.4431T>G	p.Phe252Cys	6	NA	[113]
F216Y	c.764T>A	g.4995T>A	p.Phe255Tyr	7	1, 3	[114]
T231R	c.809C>G	g.5040C>G	p.Thr270Arg	7	2	[105]
E233X	c.814G>T	g.5045G>T	p.Glu272X	7	2	[115]
S237P	c.826T>C	g.5057T>C	p.Ser276Pro	7	1	[64]
G243V	c.845G>T	g.5076G>T	p.Gly282Val	7	NA	[116]
F251L	c.870C>A	g.5101C>A	p.Phe290Leu	7	2	[104]
H255Q	c.882T>G	g.5113T>G	p.His294Gln	7	2	[42]
R257X	c.886C>T	g.5117C>T	p.Arg296X	7	1	[80]
R257Q	c.887G>A	g.5118G>A	p.Arg296Gln	7	1, 2, 2/3 int.	[46]
F259L	c.894C>A	g.5125C>A	p.Phe298Leu	7	2	[42]
G265D	c.911G>A	g.5142G>A	p.Gly304Asp	7	1	[106]
P266L	c.914C>T	g.5145C>T	p.Pro305Leu	7	2	[91]
P266R	c.914C>G	g.5145C>G	p.Pro305Arg	7	1	[117]
S271N	c.929G>A	g.5160G>A	p.Ser310Asn	7	1	[96]
L279P	c.953T>C	g.5184T>C	p.Leu318Pro	7	NA	[116]
R285C	c.970C>T	g.5201C>T	p.Arg324Cys	7	1, 3	[46]
R285H	c.971G>A	g.5202G>A	p.Arg324His	7	1, 2	[42]
P289L	c.983C>T	g.5214C>T	p.Pro328Leu	7	1	[118]
K303I	c.1025A>T	g.6127A>T	p.Lys342Ile	8	1	[64]
Y304C	c.1028A>G	g.6130A>G	p.Tyr343Cys	8	2, 2/3 int.	[42]
Y304X	c.1029T>G	g.6131T>G	p.Tyr343X	8	3	[14]
A309V	c.1043C>T	g.6145C>T	p.Ala348Val	8	1	[19]
H311R	c.1049A>G	g.6151A>G	p.His350Arg	8	2	[119]
W312C	c.1053G>T	g.6155G>T	p.Trp351Cys	8	1	[19]

Table 2.1 Mutations in GBA (Continued)

Allele Name	cDNA [a]	gDNA [a]	Exon	Protein [a,b]	Disease Types [c]	Reference
Y313H	c.1054T>C	g.6156T>C	8	p.Tyr352His	1	[120]
D315H	c.1060G>C	g.6162G>C	8	p.Asp354His	1	[117]
A318D	c.1070C>A	g.6172C>A	8	p.Ala357Asp	1	[117]
T323I	c.1085C>T	g.6187C>T	8	p.Thr362Ile	1	[118]
L324P	c.1088T>C	g.6190T>C	8	p.Leu363Pro	1	[64]
G325R	c.1090G>A	g.6192G>A	8	p.Gly364Arg	2, 3	[15]
G325W	c.1090G>T	g.6192G>T	8	p.Gly364Trp	1	[47]
E326K	c.1093G>A	g.6195G>A	8	p.Glu365Lys	A, 1	[46]
L336P	c.1124T>C	g.6226T>C	8	p.Leu375Pro	1	[91]
A341T	c.1138G>A	g.6240G>A	8	p.Ala380Thr	1, 3	[99]
C342G	c.1141T>G	g.6243T>G	8	p.Cys381Gly	2, 3	[15]
C342R	c.1141T>C	g.6243T>C	8	p.Cys381Arg	1	[98]
Q350X	c.1165C>T	g.6267C>T	8	p.Gln389X	1	[104]
V352L	c.1171G>C	g.6273G>C	8	p.Val391Leu	1	[96]
R353G	c.1174C>G	g.6276C>G	8	p.Arg392Gly	3	[121]
G355D	c.1181G>A	g.6283G>A	8	p.Gly394Asp	2	[113,122]
S356F	c.1184C>T	g.6286C>T	8	p.Ser395Phe	1	[113]
R359X	c.1192C>T	g.6294C>T	8	p.Arg398X	1, 2	[123]
R359Q	c.1193G>A	g.6295G>A	8	p.Arg398Gln	1	[124]
Y363C	c.1205A>G	g.6307A>G	8	p.Tyr402Cys	2	[102]
S364R	c.1207A>C	g.6309A>C	8	p.Ser403Arg	2	[91]
S364N	c.1208G>A	g.6310G>A	8	p.Ser403Asn	1	[47]
S364T	c.1208G>C	g.6310G>C	8	p.Ser403Thr	1	[19]
S366G	c.1213A>G	g.6315A>G	8	p.Ser405Gly	1	[109]
S366N	c.1214G>A	g.6316G>A	8	p.Ser405Asn	1	[99]
S366T	c.1214G>C	g.6316G>C	8	p.Ser405Thr	NA	[125]
T369M	c.1223C>T	g.6325C>T	8	p.Thr408Met	1	[97]
N370S	c.1226A>G	g.6728A>G	9	p.Asn409Ser	1	[37]
L371V	c.1228C>G	g.6730C>G	9	p.Leu410Val	1	[126]

V375L	c.1240G>T	9	g.6742G>T	p.Val414Leu	1	[127]
G377S	c.1246G>A	9	g.6748G>A	p.Gly416Ser	1, 3	[128]
W378G	c.1249T>G	9	g.6751T>G	p.Trp417Gly	1	[46]
D380N	c.1255G>A	9	g.6757G>A	p.Asp419Asn	1	[46]
D380H	c.1255G>C	9	g.6757G>C	p.Asp419His	NA	[129]
D380A	c.1256A>C	9	g.6758A>C	p.Asp419Ala	2	[130]
W381X	c.1259G>A	9	g.6761G>A	p.Trp420X	1	[131]
N382K	c.1263C>A	9	g.6765C>A	p.Asn421Lys	2	[102]
L383R	c.1265T>G	9	g.6767T>G	p.Leu422Arg	2	[102]
L385P	c.1271T>C	9	g.6773T>C	p.Leu424Pro	2	[102]
P387L	c.1277C>T	9	g.6779C>T	p.Pro426Leu	1	[132]
G389E	c.1283G>A	9	g.6785G>A	p.Gly428Glu	2	[80]
P391L	c.1289C>T	9	g.6791C>T	p.Pro430Leu	1	[80]
N392I	c.1292A>T	9	g.6794A>T	p.Asn431Ile	2	[80]
W393R	c.1294T>A	9	g.6796T>A	p.Trp432Arg	1	[99]
V394L	c.1297G>T	9	g.6799G>T	p.Val433Leu	1, 2/3 int., 3	[133]
R395P	c.1301G>C	9	g.6803G>C	p.Arg434Pro	1	[100]
N396T	c.1304A>C	9	g.6806A>C	p.Asn435Thr	1	[134]
V398L	c.1309G>C	9	g.6811G>C	p.Val437Leu	3	[135]
V398F	c.1309G>T	9	g.6811G>T	p.Val437Phe	2	[119]
D399N	c.1312G>A	9	g.6814G>A	p.Asp438Asn	1, 2, 3	[123]
D399Y	c.1312G>T	9	g.6814G>T	p.Asp438Tyr	1	[14]
P401L	c.1319C>T	9	g.6821C>T	p.Pro440Leu	1	[136]
I402F	c.1321A>T	9	g.6823A>T	p.Ile441Phe	3	[104]
I402T	c.1322T>C	9	g.6824T>C	p.Ile441Thr	1	[127]
D409H	c.1342G>C	9	g.6844G>C	p.Asp448His	1, 2, 3	[15,133,137]
D409V	c.1343A>T	9	g.6845A>T	p.Asp448Val	3	[133]
F411I	c.1348T>A	9	g.6850T>A	p.Phe450Ile	(1)	[106]
Y412H	c.1351T>C	9	g.6853T>C	p.Tyr451His	1	[80]
K413Q	c.1354A>C	9	g.6856A>C	p.Lys452Gln	1	[109]
Q414X	c.1357C>T	9	g.6859C>T	p.Gln453X	1	[97]
P415R	c.1361C>G	9	g.6863C>G	p.Pro454Arg	2	[33]
M416V	c.1363A>G	9	g.6865A>G	p.Met455Val	2	[102]

Table 2.1　Mutations in GBA (Continued)

Allele Name	cDNA[a]	gDNA[a]	Exon	Protein[a,b]	Disease Types[c]	Reference
F417V	c.1366T>G	g.6868T>G	9	p.Phe456Val	1	[138]
Y418C	c.1370A>G	g.6872A>G	9	p.Tyr457Cys	1	[139]
K425E	c.1390A>G	g.7261A>G	10	p.Lys464Glu	3	[124]
R433G	c.1414A>G	g.7285A>G	10	p.Arg472Gly	1	[109]
L444P	c.1448T>C	g.7319T>C	10	p.Leu483Pro	1, 2, 3	[6]
L444R	c.1448T>G	g.7319T>G	10	p.Leu483Arg	2	[140]
A446P	c.1453G>C	g.7324G>C	10	p.Ala485Pro	A, 1	[64]
H451R	c.1469A>G	g.7340A>G	10	p.His490Arg	1	[105]
V460M	c.1495G>A	g.7366G>A	10	p.Val499Met	1	[141]
N462K	c.1503C>G	g.7374C>G	10	p.Asn501Lys	2	[66]
R463C	c.1504C>T	g.7375C>T	10	p.Arg502Cys	1, 3	[16]
D474Y	c.1537G>T	g.7502G>T	11	p.Asp513Tyr	2	[142]
G478S	c.1549G>A	g.7514G>A	11	p.Gly517Ser	1	[89]
T491I	c.1589C>T	g.7554C>T	11	p.Thr530Ile	3	[135]
R496C	c.1603C>T	g.7568C>T	11	p.Arg535Cys	1	[124]
R496H	c.1604G>A	g.7569G>A	11	p.Arg535His	1	[89]
Insertions						
84GG	c.84dupG	g.1910dupG	2	p.Leu29AlafsX18	1, 2, 3	[143]
122CC	c.121dupC	g.2499dupC	3	p.Arg41ProfsX6	1	[144]
c.153−154insTACAGC	c.148_153dupTACAGC	g.25262531dupTACAGC	3	p.Tyr51_Ser52dup	1	[14]
D127X	c.394dupA	g.2895dupA	4	p.Ilc132AsnfsX14	1	[105]
500insT	c.500dupT	g.3966dupT	5	p.Ser168LeufsX12	1	[80]
c.841−842insTGA	c.839_841dupTGA	g.5070_5072dupTGA	7	p.Leu280_Ser281insMet	2	[122,145]
1098insA	c.1098dupA	g.6200dupA	8	p.His367ThrfsX69	1	[120]
c.1122−1123insTG	c.1121_1122dupTG	g.6223_6224dupTG	8	p.Leu375CysfsX20	1	[14]
c.1326insT	c.1326dupT	g.6828dupT	9	p.Val443CysfsX26	1	[51]
c.1515_1516insAGTGA GGGCAAT	c.1515_1516insAGTGA GGGCAAT	g.7480_7481insAGTGA GGGCAAT	11	p.Lys505_Asp506ins SerGluGlyAsn	2	[146]

Deletions

Mutation	cDNA	gDNA	Exon	Protein	n	Reference
72delC	c.73delC	g.1898delC	2	p.Leu25SerfsX66	1	[89]
203C del	c.203delC	g.2581delC	3	p.Pro68ArgfsX23	1	[46]
c.222_224delTAC	c.222_224delTAC	g.2600_2602delTAC	3	p.Thr75del	1	[14]
c.330delA	c.330delA	g.2831delA	4	p.Glu111AsnfsX7	2	[42]
c.533delC	c.533delC	g.3999delC	5	p.Pro178LeufsX22	2	[147]
595-596delCT	c.595_596delCT	g.4271_4272delCT	6	p.Leu199AspfsX62	NA	[129]
V214X	c.741delC	g.4417delC	6	p.Trp248GlyfsX6	1	[105]
898delG	c.898delG	g.5129delG	7	p.Ala300ProfsX4	NA	[129]
914C del	c.914delC	g.5145delC	7	p.Pro305LeufsX31	1	[95]
c.953delT	c.953delT	g.5184delT	7	p.Leu318ProfsX18	1	[91]
g5255delT	c.1029delT	g.6131delT	8	p.Tyr343X	1	[148]
L354X	c.1148delG	g.6250delG	8	p.Gly383AlafsX11	3	[105]
c.1214delG,C	c.1214_1215delGC	g.6316_6317delGC	8	p.Ser405AsnfsX30	1	[91]
1450del2	c.1451_1452delAC	g.7322_7323delAC	10	p.Asp484GlyfsX21	1	[96]
1447-1466 del 20, ins TG	c.1447_1466delinsTG	g.7318_7337delinsTG	10	p.Leu483TrpfsX49	2	[140]
c.1510delT,C,T	c.1510_1512delTCT	g.7475_7477delTCT	11	p.Ser504del	1	[91]
total	c.-229_*563del	g.1230_8118del	1-11		1	[149]

Splice Junction

Mutation	cDNA	gDNA	Protein	n	Reference
IVS2+1G>A	c.115+1G>A	g.1942G>A	Not determined	1, 2, 3	[150,151]
IVS2+1G>T	c.115+1G>T	g.1942G>T	Not determined	1	[64]
IVS5+1G>T	c.588+1G>T	g.4055G>T	Not determined	1	[80]
g.4252C>G	c.589-13C>G	g.4252C>G	p.Ile197LeufsX4	1	[152]
g.4426A>G	c.750A>G	g.4426A>G	p.Tyr251_Lys254del	1	[152]
IVS8(-11delC)(-14T>A)	c.1225-10delC; c.1225-14T>A	g.6717delC; g.6713T>A	p.Asn409SerfsX20	1	[153]

Table 2.1 Mutations in GBA (Continued)

Allele Name	cDNA[a]	gDNA[a]	Exon	Protein[a,b]	Disease Types[c]	Reference
IVS10−1G>A, R463Q	c.1505G>A	g.7376G>A		p.Arg502GlnfsX2	1, 2, 3	[154]
IVS10+2T>A	c.1505+2T>A	g.7378T>A		Not determined	3	[104]
IVS10(+2)	c.1505+2T>G	g.7378T>G		Not determined	1, 2	[95]

[a] All mutations are described as recommended at http://www.hgvs.org/mutnomen (modified February 27, 2005) [155,156]. GBA cDNA nucleotides ("c.") are numbered with the adenine of the first ATG translation initiation codon as nucleotide +1 (GenBank reference sequence NM_000157.1). Genomic nucleotide positions ("g.") start with nucleotide +1 based on the numbering used in the database reference files for GBA (GenBank J03059.1) and psGBA (GenBank J03060.1). Amino acid designations ("p.") are based on the primary GBA translation product, including the 39-residue signal peptide.

[b] Protein sequences have been deduced from primary sequence and not experimentally confirmed; e.g., many nonsense- and frameshift-containing mRNAs are not translated into protein but are degraded by nonsense-mediated decay [157].

[c] Gaucher disease types: 1, type 1, nonneuronopathic; (1), type 1 diagnosed in early childhood, development of subsequent neurological signs not excluded; 2, type 2, acute neuronopathic; 2/3 int., type 2-type 3 intermediate presentation [57]; 3, type 3, chronic neuronopathic; A, asymptomatic; NA, not available.

Table 2.2 Complex and Recombinant Alleles of GBA

Allele Name	Fig. 2	cDNA [a]	gDNA [a]	First X-over [d]	Second X-over [e]	Exons Affected	Protein [a,b]	Disease Types [c]	Reference
Complex Alleles									
c.(-203)A>G + IVS4-2a>g	—	c.-203A>G; c.-455-2A>G	g.1256A>G; g.3919A>G	—	—		Not determined	1	[91]
G202R + M361I	—	c.721G>A; c.1200G>A	g.4397G>A; g.6302G>A	—	—	6, 8	p.Gly241Arg; p.Met400Ile	2	[105]
H255Q + D409H	—	c.882T>G; c.1342G>C	g.5113T>G; g.6844G>C	—	—	7, 9	p.His294Gln; p.Asp448His	1, 2/3 int.	[158]
N370S + S448P	—	c.1226A>G; c.1579T>C	g.6728A>G; g.7144T>C	—	—	9, 11	p.Asn409Ser; p.Ser527Pro	2	[159]
c.1379G>A; c.1469A>G	—	c.1379G>A; c.1469A>G	g.6881G>A; g.7340A>G	—	—	9, 10	p.Gly460Asp; p.His490Arg	1	[68]
g.7319T>C + g.7741T>C	—	c.1448T>C; c.*165T>C	g.7319T>C; g.7741T>C	—	—	10	p.Leu483Pro	1	[49]
Complex Alleles with c.1093G>A(E326K) or c.1223C>T(T369M)									
c.203-204insC + E326K	—	c.203dupC; c.1093G>A	g.2581dupC; g.6195G>A	—	—	3, 8	p.Thr69AspfsX12	1	[14]
R120W + T369M	—	c.475C>T; c.1223C>T	g.3941C>T; g.6325C>T	—	—	5, 8	p.Arg159Trp; p.Thr408Met	1	[44]
D140H + E326K	—	c.535G>C; c.1093G>A	g.4001G>C; g.6195G>A	—	—	5, 8	p.Asp179His; p.Glu365Lys	1	[45]
N188S + E326K	—	c.680A>G; c.1093G>A	g.4356A>G; g.6195G>A	—	—	6, 8	p.Asn227Ser; p.Glu365Lys	2/3 int.	[47]
E326K + G377S	—	c.1093G>A; c.1246G>A	g.6195G>A; g.6748G>A	—	—	8, 9	p.Glu365Lys; p.Gly416Ser	3	[104]

Table 2.2 Complex and Recombinant Alleles of GBA (Continued)

Allele Name	Fig. 2	cDNA [a]	gDNA [a]	First X-over [d]	Second X-over [e]	Exons Affected	Protein [a,b]	Disease types [c]	Reference
E326K + L444P	—	c.1093G>A; c.1448T>C	g.6195G>A; g.7319T>G	—	—	8, 10	p.Glu365Lys; p.Leu483Pro	2	[115]
T369M + D409H	—	c.1223C>T; c.1342G>C	g.6325C>T; g.6844G>C	—	—	8, 9	p.Thr408Met; p.Asp448His	1	[51]
Recombinant Alleles Beginning Before Intron 8									
Rec (int-2)	Fig. 2A	c.119_121delins GT and other pseudogene sequence	g.2497_J03060.1: g.1336del	Intron 2	—	3–11	p.Ala40GlyfsX23	1	[41]
Rec A, Rec 4b	Fig. 2B	c.320C>T; c.329_333del CAGAA and other pseudogene sequence	g.2767_J03060.1: g.1601del	Intron 3	—	4–11	p.Thr107Ile; p.Gln112ValfsX32	2	[39,42,160]
Complex C, Rec 5a	Fig. 2C	c.475C>T; c.667T>C;ᶠ c.681T>G; c.689T>G; c.703T>C; c.721G>A; c.754T>A	g.3941_4430con J03060.1:g.2129_2618	Intron 4– Exon 5	Exon 6– Intron 6	5–6	p.Arg159Trp; p.Trp223Arg; p.Asn227Lys; p.Val230Gly; p.Ser235Pro; p.Gly241Arg; p.Phe252Ile	1	[19]

Rec 5b g.4179_5042con J03060.1:2367_2911	Fig. 2C	c.475C>T; c.680_681delins GG;† c.689T>G; c.703T>C; c.721G>A; c.754T>A; c.812C>G; c.921C>T; c.928A>G; c.1026A>G; c.1038C>T; c.1090G>A; c.1103G>A; c.1149C>T; c.1263_1317del and other pseudogene sequence	g.3817_J03060.1: g.2005del	Intron 4	—	5–11	p.Arg159Trp; p.Asn227Arg; p.Val230Gly; p.Ser235Pro; p.Gly241Arg; p.Phe252Ile; p.Ala271Gly; p.Ser310Gly; p.Gly364Arg; p.Arg368His; p.Leu422ProfsX4	3	[39]
g.4179_5042con J03060.1:2367_2910	Fig. 2C	c.680_681delins GG;† c.689T>G; c.703T>C; c.721G>A; c.754T>A		Intron 5	Exon 7	6–7	p.Asn227Arg; p.Val230Gly; p.Ser235Pro; p.Gly241Arg; p.Phe252Ile	1	[68]

Table 2.2　Complex and Recombinant Alleles of GBA (Continued)

Allele Name	Fig. 2	cDNA [a]	gDNA [a]	First X-over [d]	Second X-over [e]	Exons Affected	Protein [a,b]	Disease types [c]	Reference
Rec I	Fig. 2D	c.812C>G; c.921C>T; c.928A>G; c.1026A>G; c.1038C>T; c.1090G>A; c.1103G>A; c.1149C>T; c.1263_1317del and other pseudogene sequence	g.4641_J03060.1: g.2828	Intron 6	—	7–11	p.Ala271Gly; p.Ser310Gly; p.Gly364Arg; p.Arg368His; p.Leu422ProfsX4	1	[161]
Recombinant Alleles Beginning with c.1263_1317del (RecΔ55)									
1263-1317 del, 1255del55, RecΔ55	Fig. 2E	c.1263_1317del	g.6764_6820con J03060.1:g.4356_4357[h]	Intron 8– Exon 9	Exon 9– Intron 9	9	p.Leu422ProfsX4	1,2	[89,130]
Rec[1263del55;1342G>C]	Fig. 2E	c.1263_1317del; c.1342G>C	g.6763_6844con J03060.1:g.4357_4381	Intron 8– Exon 9	Exon 9– Intron 9	9	p.Leu422ProfsX4	1	[40]
c1263del+RecTL, rec(g4889-6506), Rec B, Rec 3a or b	Fig. 2E	c.1263_1317del; c.1342G>C; c.1448T>C; c.1483G>C; c.1497G>C	g.6763_7368con J03060.1:g.4357_4905 OR g.6763_J03060.1: g.4356del	Intron 8– Exon 9 Intron 8– Exon 9	Exon 10 –3'UTR —	9–11	p.Leu422ProfsX4	1, 2, 3	[39,42,51,66]

Recombinant Alleles Beginning with c.1342G>C (D409H)

Allele	Fig.	Mutations		Genomic		Exons	Protein	Figs	Refs
RecTL, Complex B, Rec C, Rec 2a or b	Fig. 2E	c.1342G>C; c.1448T>C; c.1483G>C; c.1497G>C	Exon 9 Exon 9	g.6844_7368con J03060.1:g.4381_4905 OR g.6844_J03060.1:g.4380del	Exon 10 -3'UTR —	9–10	p.Asp448His; p.Leu483Pro; p.Ala495Pro	1, 2, 3	[15,17,39,42]
Complex B + g.7668G>A + g.7678T>C	Fig. 2E	c.1342G>C; c.1448T>C; c.1483G>C; c.1497G>C; c.*92G>A; c.*102T>C	Exon 9	g.6844_7678con J03060.1:g4381_5216	3'UTR	9–11	p.Asp448His; p.Leu483Pro; p.Ala495Pro	3	[49]

Recombinant Alleles Beginning with c.1448T>C (L444P)

Allele	Fig.	Mutations		Genomic		Exons	Protein	Figs	Refs
RecNciI, pseudo pattern, Complex A, Rec F, Rec 1a or b	Fig. 2F	c.1448T>C; c.1483G>C; c.1497G>C	Intron 9– Exon 10	g.7319_7368con J03060.1:g.4856_4905 OR g.7319_J03060.1:g.4855del	Exon 10 -3'UTR Intron 9– Exon 10	10–11	p.Leu483Pro; p.Ala495Pro	1, 2, 3	[15–17, 39,42]
Rec D	Fig. 2F	c.1448T>C; c.1483G>C; c.1497G>C; c.*92G>A	Intron 9	g.7159_J03060.1:g.4695del	—	10–11	p.Leu483Pro; p.Ala495Pro	1, 3	[42]
Rec E	Fig. 2F	c.1448T>C; c.1483G>C; c.1497G>C; c.*92G>A	Intron 9	g.7192_J03060.1:g.4728del	—	10–11	p.Leu483Pro; p.Ala495Pro	1, 2	[42,94]

Table 2.2 Complex and Recombinant Alleles of GBA (Continued)

Allele Name	Fig. 2	cDNA [a]	gDNA [a]	First X-over [d]	Second X-over [e]	Exons Affected	Protein [a,b]	Disease types [c]	Reference
AZRecTL	Fig. 2F	c.1448T>C; c.1483G>C; c.1497G>C; c.*92G>A	g.7319_J03060.1: g.4855del	Intron 9–Exon 10	—	10–11	p.Leu483Pro; p.Ala495Pro	1	[18]
RecA456P	Fig. 2F	c.1448T>C; c.1483G>C;	g.7319_7354con J03060.1:g.4856_4891	Intron 9–Exon 10	Exon 10	10	p.Leu483Pro; p.Ala495Pro	1	[66]
L444P + V460V[g]	Fig. 2F	c.1448T>C; c.1497G>C	g.7319T>G; g.7368G>C	—	—	10	p.Leu483Pro	1	[51]
N370S + L444P + A456P	Fig. 2F	c.1226A>G; c.1448T>C; c.1483G>C	g.6728A>G; g.7319_7354con J03060.1:g.4856_4891	—	—	9, 10	p.Asn409Ser; p.Leu483Pro; p.Ala495Pro	1	[162]
Complex A + g.7668G>A + g.2113-2114delAG	Fig. 2F	c.115+172_173 delAG; c.1448T>C; c.1483G>C; c.1497G>C; c.*92G>A	g.2113_2114delAG; g.7319_7668con J03060.1:g.4856_5205	Intron 9–Exon 10	3'UTR	10–11	p.Leu483Pro; p.Ala495Pro	1	[49]
Recombinant Alleles Beginning in Exon 11 3'UTR									
L444P + Rec G, L444P + Rec 6b	Fig. 2G	c.1448T>C; c.*92G>A	g.7319T>G; g.7667_J03060.1: g.5205del	Intron 10–3'UTR	—	10	p.Leu483Pro	2	[39,42]

a, b, c See Table 2.1.

d Site of first crossover event.

e Site of second crossover. (—) Indicates a nonexistent or undescribed crossover.

f Pseudogene variant for this position [52].

g Possible recombination with pseudogene variant J03060.1:g.4891C>G, resulting in absence of the c.1483G>C (A456P) mutation [52].

h For this gene conversion allele, genomic DNA numbering is based on nucleotides bracketing 55 bp deleted in the pseudogene.

Moreover, the mechanisms resulting in recombinant alleles also vary. Direct sequencing and Southern blot analyses have demonstrated that these include reciprocal and nonreciprocal recombination [39] (Figure 2.2).

Many groups identify one recombinant allele, traditionally referred to as Rec*Nci*I [17], by screening for the pseudogene-derived alterations c.1448T>C (L444P) and c.1483G>C (A456P). This approach is not adequate, however, to accurately describe the many different possible recombinant alleles that include these point mutations. A precise description of a recombinant allele can only be achieved by a combination of direct sequencing and Southern blot analysis. When pseudogene-derived mutations such as c.1342G>C (D409H) or c.1448T>C are identified, the entire gene should be sequenced, and the known exonic and intronic mismatches between GBA and psGBA used to define the length of incorporated pseudogene sequence. In addition, a Southern blot is necessary to establish whether the rearrangement is the result of gene conversion or due to a duplication or deletion [39]. Reciprocal recombination results in altered banding patterns on Southern blots, reflecting deleted or duplicated segments, while nonreciprocal gene conversions usually introduce pseudogene sequence into GBA without altering the intergenic restriction sites used for Southern analysis (Figure 2.3). Table 2.2 summarizes the many recombinant alleles described to date.

Once again, the traditional nomenclature for recombinant alleles is confusing. When first identified, many recombinant alleles were described only by the missense mutations they carried, and the type and extent of recombination with psGBA was not determined. For example, the allele classically described as Rec*Nci*I, which includes substitutions c.1448T>C, c.1483G>C and c.1497G>C, can result from either reciprocal or nonreciprocal recombination, and may include pseudogene tracts upstream or downstream of the missense mutations reported (Figure 2.2F). When not specified in the literature, we have defined these recombinant alleles by both nonreciprocal and reciprocal mechanisms that would account for the point mutations observed, using the shortest stretch of genomic sequence that would result in the reported changes. Gene conversions have been designated by "con" after an indication of the first and last nucleotides of GBA affected by the conversion, followed by the new nucleotides derived from psGBA (GenBank J03060.1). Nucleotide numbering for reciprocal recombination events includes the breakpoints for GBA and psGBA followed by "del" to indicate the loss of intervening sequences. Recombinant alleles for which the breakpoints have been mapped are included as separate entries in the table.

In addition, complex alleles that do not appear to be derived from the pseudogene have been identified. These alleles may represent multiple substitution events but also could originate from single disease-causing substitutions in *cis* with polymorphisms. Two specific alterations, c.1093G>C (E326K) and c.1223C>T (T369M), have been found in patients primarily in *cis* with other identified mutations [43,44]. Early reports of these alterations as single disease-causing mutations predate more complete mutation identification techniques [45,46] and, in some cases, have been corrected in subsequent publications after identification of a second mutation on the same allele [47,48]. Zhao et al. [49] reported a family in which a c.1093G>C allele was found in two unaffected individuals who carried either c.721G>A (G202R) or

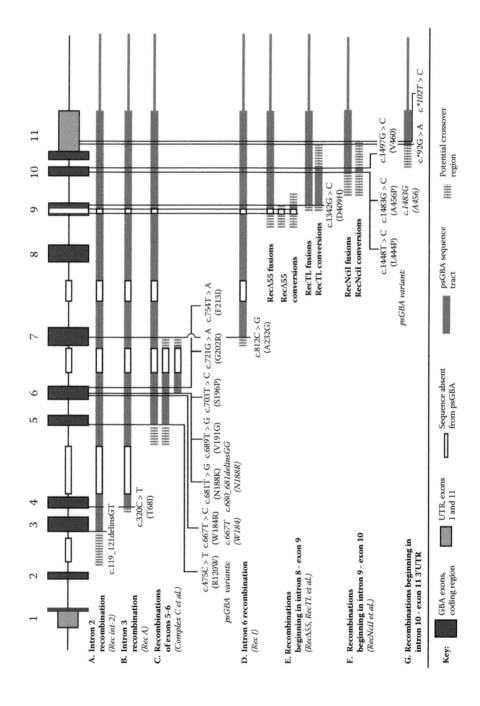

Figure 2.2 Recombinant alleles observed in GBA. The exonic and intronic structure of GBA is shown. Recombinant alleles documented in Table 2.2 are represented by groups A–G, with pseudogene-derived mismatches and amino acid changes associated with the allele indicated below. Typical nonreciprocal (gene conversion) and reciprocal (gene fusion or duplication) recombinants are shown. The exact locations of the breakpoints have not been mapped for many recombinant alleles. Regions of sequence identity between GBA and psGBA that flank the observed mismatches, in which recombination could occur, are depicted as barred lines. Pseudogene sequence variations that can alter the combination of amino acid changes seen in a recombinant allele are shown in italics.

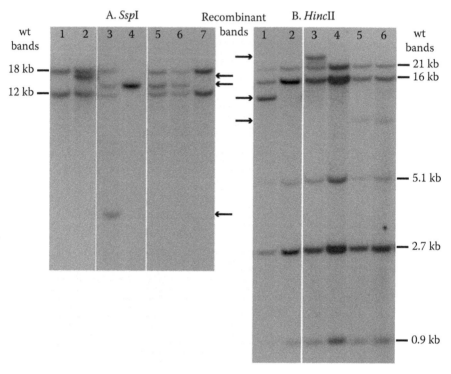

Figure 2.3 Composite Southern blots illustrating different recombinant alleles. Genomic DNAs digested with either *Ssp*I (A) or *Hinc*II (B) and hybridized to a GBA cDNA probe reveal recombinant alleles that result from fusions or duplications. Recombinant alleles that result from gene conversion may show a normal banding pattern on Southern if the conversion does not disrupt the restriction site. Lanes A1 and B4 are from control individuals, demonstrating the expected size bands, while lanes A7 and B2 are from patients with gene conversion alleles. The remaining lanes show a variety of recombinant alleles observed in our patient cohort. (From Tayebi, N. et al., *Am. J. Hum. Genet.*, 72, 519, 2003. With permission.)

c.1448T>C (L444P) on their second allele. Neither of the two affected individuals in this family had the c.1093G>C alteration. While both c.1093G>C and c.1223C>T reportedly have an impact on enzyme activity [50,51], each has been found in approximately 1% of control individuals [43,44].

Complex alleles also may correspond to unrecognized recombinant alleles that were not detected due to pseudogene variation or ascertainment method [52]. For example, the complex allele described by Latham et al. [19] contained seven mutations in exons 5 and 6 of GBA, six of which were present in the allelic pseudogene. The single nucleotide substitution not attributed to the pseudogene, c.667T>C, was flanked by psGBA sequences on either side. In addition, while c.681T>G was detected, the adjacent c.680A>G substitution that would be expected from a recombination with the pseudogene was absent. Subsequently, polymorphic variations in the psGBA sequence have been documented [52] and this complex allele can be explained by a single gene conversion event with a psGBA allele containing polymorphic sequence J03060.1:g.2531T>C; g.2544G>A (Figure 2.2C). Similarly, a complex allele reported to harbor c.1448T>C and c.1497G>C, but not c.1483G>C [51], may have resulted from recombination with another psGBA variant (J03060.1:g.4891C>G) (Figure 2.2F).

As might be expected, of the many mutations described, the great majority are private mutations that are rarely encountered. Among the more common single mutations are c.115+1G>A (IVS2+1), c.84dupG (84GG), c.475C>T (R120W), c.680A>G (N188S), c.721G>A (G202R), c.754T>A (F213I), c.887G>A (R257Q), c.1090G>A (G325R), c.1226A>G (N370S), c.1246G>A (G377S), c.1342G>C (D409H), c.1448T>C (L444P), c.1504C>T (R463C), c.1604G>A (R496H) and c.1263_1317del. Among Ashkenazi Jewish patients with type 1 Gaucher disease, screening for c.1226A>G, c.1448T>C, c.84dupG, c.115+1G>A, c.1504C>T and c.1604G>A accounts for over 90% of mutant alleles [13,14,38]. It has been reported that c.1226A>G makes up approximately 70% of the mutant alleles in this population, and c.84dupG about 10% [8]. Among non-Ashkenazi Jewish patients, however, screening for a handful of mutations, even among patients with type 1 disease, fails to identify 25–50% of the mutant alleles [7,14].

The distribution of mutations varies among patients of different ethnicities. Approximating allelic distributions is difficult because many diagnostic laboratories do not distinguish between the point mutation c.1448T>C (L444P) and recombinant alleles that include this mutation. Among patients with type 2 Gaucher disease, the mutation profile is quite different, with recombinant and rare alleles being particularly prevalent [42]. In patients with type 3, those primarily manifesting visceral manifestations often carry c.1448T>C and/or c.1504C>T (R463C), while c.680A>G (N188S), c.1246G>A (G377S) and c.1297G>T (V394L) are more commonly seen in patients with myoclonic epilepsy, often together with a null or recombinant allele [14,53]. Mutation c.1342G>C (D409H) is associated with a rare phenotype that includes calcification or fibrosis of the cardiac valves, corneal opacities, hydrocephalus and dysmorphic features [54–56].

Despite these generalizations, there are major difficulties in predicting patient phenotype from genotype. Many of the mutations described above are found in patients with more than one type of Gaucher disease (Table 2.1) [11–14]. The primary exception is mutation c.1226A>G (N370S), which is encountered only in patients with type 1 disease. Of note, the observed frequency of c.1226A>G homozygotes is considerably less than expected when calculated from the allele frequency in the Ashkenazi Jewish population, suggesting that the majority of these individuals do not reach medical attention [13].

Genotype-phenotype studies are also complicated by the ever-increasing spectrum of clinical phenotypes observed in Gaucher disease. The division into the three classic types can be cumbersome, with some patients defying categorization [57]. In fact, the associated phenotypes can be seen as a continuum (Figure 2.4), with the major distinction being the presence and degree of neurological involvement [58]. To further confound the issue, clinically similar patients have many different genotypes, even in unique subgroups of patients.

Another difficulty with genotype-phenotype correlation in this disorder is that individuals sharing the same genotype can have different disease manifestations, clinical courses and responses to therapy [58]. These differences have been observed

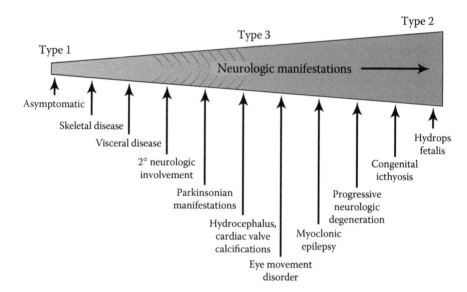

Figure 2.4   Gaucher disease as a phenotypic continuum. Gaucher disease presents with a range of phenotypes; some patients have clinical evidence of brain involvement, while others do not. There is a "grey zone," indicated by the curved lines, in which it is not clear whether the neurological involvement is a direct result of the enzyme deficiency. (Reproduced from Sidransky, E., *Mol. Gen. Metab.*, 83, 6, 2004. With permission.)

even in siblings and twins [59,60]. Patients with genotype [c.1226A>G]+ [c.1226A>G] (N370S/N370S) have phenotypes varying from asymptomatic adults to children with significant organomegaly, growth delay or bone disease. A recent series of 35 patients with genotype [c.1448T>C]+[c.1448T>C] (L444P/L444P), each confirmed not to have a recombinant allele, also demonstrated significant phenotypic variability, ranging from death in early childhood to autism to successful college students [61]. The age at diagnosis for each was under 2 years and all exhibited slowed horizontal eye movements. Some trends related to ethnic background were observed, with more cognitive deficits noted among subjects of Hispanic or African American ancestry, suggesting variable expressivity. The residual enzyme activity ranged from 1–30% and did not correlate with the severity of symptoms. Studies of both discordant sib pairs and discordant patients sharing the same genotype ultimately will help to identify genetic modifiers influencing phenotype in patients with Gaucher disease.

One approach to making genotypic predictions is to focus on phenotypes encountered in individuals with homozygous genotypes. Approximately 30 different homozygous genotypes have been reported in the literature (Table 2.3). There is clearly a dosage effect with some alleles, such as c.354G>C (K79N), c.680A>G (N188S) or c.1246G>A (G377S), where homozygosity for the mutation results in type 1 disease, while compound heterozygosity with a null allele leads to a type 3 phenotype [53,62–64]. Likewise, individuals homozygous for c.754T>A (F213I) or c.1448T>C (L444P) usually develop chronic neuronopathic Gaucher disease [14,65], but either mutation together with a null allele is more likely to have a type 2 phenotype [42]. Mutation c.1504C>T (R463C) is particularly confusing, as homozygotes have been described with type 1 disease [66], yet when inherited with a null allele, either type 1 or type 3 ensues [14].

While expression of specific mutant alleles has been attempted by different groups, the results have been variable and difficult to compare [47,50,67–70]. Even the amount of residual enzymatic activity does not appear to be associated with the patient phenotype. Thus, while the long awaited crystal structure of glucocerebrosidase can be useful in making some predictions regarding functional domains of the enzyme [9], the identification of genetic or environmental modifiers still appears to be critical.

GBA POLYMORPHISMS

While proving invaluable for genomic diversity studies, the characterization of GBA polymorphisms has not shed light on disease severity. Early characterization of the glucocerebrosidase gene revealed a restriction fragment length polymorphism upon digestion with *Pvu*II [71]. Alleles that generated a 1.1 kb fragment were designated Pv1.1⁺ and those that lacked this fragment were designated Pv1.1⁻. In time, this single polymorphism, localized to a g.4813G>A substitution in intron 6 [72], was expanded to include eleven additional polymorphisms, eight in intronic sequences and three in the 5'-untranslated region. Each of these sequence variations are in linkage disequilibrium and constitute two major haplotypes [73]. Pv1.1⁻ and Pv1.1⁺ occur with frequencies of roughly 70% and 30% in the Caucasian population, respectively, and with inverted frequencies in Asian and African populations [71,74–77]. The common c.1226A>G (N370S) mutation has been found in linkage disequilibrium with the Pv1.1⁻ haplotype [72,78–82], while c.84dupG is identified with Pv1.1⁺ [72,73,82]. In contrast, the panethnic c.1448T>C (L444P) mutation has been observed with both of these common GBA haplotypes [46,80,83,84].

An additional GBA polymorphism has been isolated from Portuguese Gaucher type 1 patients, a g.5470G>A transition superimposed on the Pv1.1⁻ haplotype; however, this polymorphism has been limited to the Portuguese and Spanish populations [79,85]. Two novel nucleotide repeats, a CT repeat (5GC3.2) and a tetranucleotide AAAT repeat (ITG6.2) have been defined 3.2 kb upstream and 9.8 kb downstream of the glucocerebrosidase gene, respectively [86]. Variation in the GBA pseudogene includes seventeen single nucleotide substitutions, a three-nucleotide deletion, and a polyadenine tract [52,87]. As mentioned in the discussion of recombinant alleles, awareness of these polymorphisms in psGBA can aid in the recognition of psGBA to GBA gene conversion events [52].

Examination of a highly polymorphic trinucleotide repeat in the gene for pyruvate kinase (PKLR), which is closely linked to the glucocerebrosidase gene [75], has revealed that 96% of alleles with the c.1226A>G (N370S) mutation also carry one allelic form of the PKLR polymorphism, which is present in only 6.7% in the control population [82]. A second PKLR polymorphism was characterized in 195 Gaucher patients with genotype [c.1226A>G]+[c.1226A>G] and all were found to

Table 2.3 Reported Clinical Correlates of GBA Mutations Found in the Homozygous State

Allele Name	cDNA [a]	Disease Types in Homozygotes [c]	Comments	Reference
C16S	c.164G>C	2	Hydrops	[163]
K79N	c.354G>C	1	Compound heterozygosity with g.6844_7678conJ03060.1:g4381_5215 results in type 3	[62,97]
P122S	c.481C>T	3	Index case identified as type 1 at age 3; re-classified as type 3 at age 12	[89,90]
R131C	c.508C>T	2		[105]
R131L	c.509G>T	2	Collodion baby phenotype	[103]
P159L	c.593C>T	1		[99]
R170C	c.625C>T	1	Compound heterozygosity with c.1448T>C results in type 1	[14,105]
R170P	c.626G>C	1	Compound heterozygosity with c.115+1G>A results in type 1	[14,64,105]
N188S	c.680A>G	1	Compound heterozygosity with c.1448T>C results in type 1; with g.7319_7368conJ03060.1:g.4856_4905(RecNciI), type 3	[14,63]
S196P	c.703T>C	2	Neonatal Gaucher disease	[42]
K198E	c.709A>G	2/3 int.	Index case died at age 2, myoclonic epilepsy	[104]
G202R	c.721G>A	2	Compound heterozygosity with c.1448T>C results in type 2 or 3	[105]
F213I	c.754T>A	3	Compound heterozygosity with c.1448T>C results in type 3; with g.7319_7368conJ03060.1:g.4856_4905(RecNciI), type 2	[65,111]
P266L	c.914C>T	2		[91]
H311R	c.1049A>G	2	Hydrops fetalis	[119]
R353G	c.1174C>G	3		[105]
S366G	c.1213A>G	1		[109]
N370S	c.1226A>G	1	Compound heterozygosity with all other mutations results in type 1	[37]
L371V	c.1228C>G	1		[126]
V375L	c.1240G>T	1		[127]
G377S	c.1246G>A	1	Compound heterozygosity with c.1448T>C results in type 1; with c.1029delT, type 3 with myoclonic epilepsy	[40,53,65,134,148]
N396T	c.1304A>C	1		[164]
I402T	c.1322T>C	1		[127]

D409H	c.1342G>C	3	Unique phenotype. Compound heterozygosity with c.1448T>C results in type 1; with g.7319_7368conJ03060.1:g.4856_4905(RecNcil), unique type 3 or type 2	[14,56,80]
K413Q	c.1354A>C	(1)	Index case age 8 when reported as type 1	[109]
L444P	c.1448T>C	1, 2, 3 usually 3	Compound heterozygosity with many recombinant alleles results in type 2	[42,61,165,166]
A446P	c.1453G>C	A, 1	Two homozygous siblings; one (age 31) presented as type 1 with severe skeletal involvement, the other (age 29), asymptomatic	[64]
R463C	c.1504C>T	1	Compound heterozygosity with c.1448T>C and with g.7319_7368conJ03060.1:g.4856_4905(RecNcil) results in either type 1 or type 3	[14,66]
R496C	c.1603C>T	1		[65]
c.533delC	c.533delC	2		[147]
Many recombinant alleles		2	Homozygosity for several recombinant alleles results in perinatal lethal type 2	[42]

a, c See Table 2.1.

share homozygosity for the c.1705C PKLR haplotype [81]. In addition, a c.628T>C transition in the metaxin gene has been associated with c.1226A>G [88]. Haplotype analysis of the GBA region in 17 worldwide populations has revealed linkage disequilibrium that is higher than in other genomic regions of the same size [76]; it has been speculated that this linkage disequilibrium may extend over a broad region in the long arm of chromosome 1.

GENETIC COUNSELING

Despite the large number of GBA mutations identified, there are clearly great limitations in our ability to make prognostic predictions from genotypic data. Table 2.1 includes the different mutations reported in patients with type 1, 2 and 3 Gaucher disease. This approach has limited utility, however, since the combination of mutations on both alleles is important in defining the phenotype and the vast majority of the identified alterations have not been found in the homozygous form. Determining the "severity" of a given mutation, especially those found in only a few individuals, is dependent both on the mutation on the second allele and on an accurate assessment of patient phenotype. For example, patients who present in infancy or childhood without neurological signs have been reported to be type 1 in the original description of the causative alleles. In some cases, however, the development of neurological signs later in life has been reported, resulting in a reclassification of the patients as type 3 [89,90]. Caution must be exercised in prediction of disease severity or prognosis for those patients harboring rare alleles or alleles associated with more than one phenotype. Nevertheless, Table 2.3 summarizes some of the generalizations drawn from patients with homozygous genotypes. Even among homozygotes sharing the same mutations, it is quite clear that the point mutations, while important, are not the entire picture. Other factors, including complex alleles, alternate substrates, contiguous genes, environmental or infectious exposures and a multitude of potential genetic modifiers, are likely to contribute to the diverse phenotypes encountered in patients with Gaucher disease [58]. The identification of these modifiers and the mechanisms of their effects on phenotypic expression in Gaucher disease remains a major challenge, but will significantly improve our understanding of genotype/phenotype relationships and have important implications for other Mendelian disorders.

REFERENCES

1. Barneveld, R.A. et al., Assignment of the gene coding for human beta-glucocerebrosidase to the region q21-q31 of chromosome 1 using monoclonal antibodies, *Hum. Genet.*, 64, 227, 1983.
2. Ginns, E.I. et al., Isolation of cDNA clones for human beta-glucocerebrosidase using the lambda gt11 expression system, *Biochem. Biophys. Res. Commun.*, 123, 574, 1984.
3. Sorge, J. et al., Molecular cloning and nucleotide sequence of human glucocerebrosidase cDNA, *Proc. Natl. Acad. Sci. U.S.A.*, 82, 7289, 1985.

4. Horowitz, M. et al., The human glucocerebrosidase gene and pseudogene: structure and evolution, *Genomics*, 4, 87, 1989.

5. Winfield, S.L. et al., Identification of three additional genes contiguous to the gluco-cerebrosidase locus on chromosome 1q21: implications for Gaucher disease, *Genome Res.*, 7, 1020, 1997.

6. Tsuji, S. et al., A mutation in the human glucocerebrosidase gene in neuronopathic Gaucher's disease, *N. Engl. J. Med.*, 316, 570, 1987.

7. Beutler, E. and Gelbart, T., Gaucher disease mutations in non-Jewish patients, *Br. J. Haematol.*, 85, 401, 1993.

8. Grabowski, G.A. and Horowitz, M., Gaucher's disease: molecular, genetic and enzy-mological aspects, *Baillieres Clin. Haematol.*, 10, 635, 1997.

9. Dvir, H. et al., X-ray structure of human acid-beta-glucosidase, the defective enzyme in Gaucher disease, *EMBO Rep.*, 4, 704, 2003.

10. Zimran, A. et al., Prediction of severity of Gaucher's disease by identification of mutations at DNA level, *Lancet*, 2, 349, 1989.

11. Beutler, E., Gaucher disease: new molecular approaches to diagnosis and treatment, *Science*, 256, 794, 1992.

12. Sidransky, E. et al., DNA mutational analysis of type 1 and type 3 Gaucher patients: how well do mutations predict phenotype? *Hum. Mutat.*, 3, 25, 1994.

13. Grabowski, G.A., Gaucher disease: gene frequencies and genotype/phenotype corre-lations, *Genet. Test*, 1, 5, 1997.

14. Koprivica, V. et al., Analysis and classification of 304 mutant alleles in patients with type 1 and type 3 Gaucher disease, *Am. J. Hum. Genet.*, 66, 1777, 2000.

15. Eyal, N., Wilder, S., and Horowitz, M., Prevalent and rare mutations among Gaucher patients, *Gene*, 96, 277, 1990.

16. Hong, C.M. et al., Sequence of two alleles responsible for Gaucher disease, *DNA Cell Biol.*, 9, 233, 1990.

17. Latham, T. et al., Complex alleles of the acid beta-glucosidase gene in Gaucher disease, *Am. J. Hum. Genet.*, 47, 79, 1990.

18. Zimran, A. et al., A glucocerebrosidase fusion gene in Gaucher disease. Implications for the molecular anatomy, pathogenesis, and diagnosis of this disorder, *J. Clin. Invest.*, 85, 219, 1990.

19. Latham, T.E. et al., Heterogeneity of mutations in the acid beta-glucosidase gene of Gaucher disease patients, *DNA Cell Biol.*, 10, 15, 1991.

20. Long, G.L. et al., Structure and organization of the human metaxin gene (MTX) and pseudogene, *Genomics*, 33, 177, 1996.

21. Adolph, K.W. et al., Structure and organization of the human thrombospondin 3 gene (THBS3), *Genomics*, 27, 329, 1995.

22. Armstrong, L.C. et al., Metaxin is a component of a preprotein import complex in the outer membrane of the mammalian mitochondrion, *J. Biol. Chem.*, 272, 6510, 1997.

23. Thomas, J.W. et al., Comparative analyses of multi-species sequences from targeted genomic regions, *Nature*, 424, 788, 2003.

24. LaMarca, M.E. et al., Multi-species comparative sequence analysis of the glucocer-ebrosidase gene locus, *Am. J. Hum. Genet.*, 74, A1579, 2004.

25. Graves, P.N. et al., Gaucher disease type 1: cloning and characterization of a cDNA encoding acid beta-glucosidase from an Ashkenazi Jewish patient, *DNA*, 7, 521, 1988.

26. Reiner, O., Wigderson, M., and Horowitz, M., Structural analysis of the human glucocerebrosidase genes, *DNA*, 7, 107, 1988.

27. Sorge, J.A. et al., The human glucocerebrosidase gene has two functional ATG initiator codons, *Am. J. Hum. Genet.*, 41, 1016, 1987.
28. Pasmanik-Chor, M. et al., Overexpression of human glucocerebrosidase containing different-sized leaders, *Biochem. J.*, 317 (Pt 1), 81, 1996.
29. Martin, B.M., Sidransky, E., and Ginns, E.I., Gaucher's disease: advances and challenges, *Adv. Pediatr.*, 36, 277, 1989.
30. Berg-Fussman, A. et al., Human acid beta-glucosidase. N-glycosylation site occupancy and the effect of glycosylation on enzymatic activity, *J. Biol. Chem.*, 268, 14861, 1993.
31. Reiner, O. et al., Efficient in vitro and in vivo expression of human glucocerebrosidase cDNA, *DNA,* 6, 101, 1987.
32. Reiner, O. and Horowitz, M., Differential expression of the human glucocerebrosidase-coding gene, *Gene,* 73, 469, 1988.
33. Wigderson, M. et al., Characterization of mutations in Gaucher patients by cDNA cloning, *Am. J. Hum. Genet.*, 44, 365, 1989.
34. Doll, R.F. and Smith, F.I., Regulation of expression of the gene encoding human acid beta-glucosidase in different cell types, *Gene,* 127, 255, 1993.
35. Xu, Y.H., Wenstrup, R., and Grabowski, G.A., Effect of cellular type on expression of acid beta-glucosidase: implications for gene therapy in Gaucher disease, *Gene Ther.*, 2, 647, 1995.
36. Moran, D., Galperin, E., and Horowitz, M., Identification of factors regulating the expression of the human glucocerebrosidase gene, *Gene,* 194, 201, 1997.
37. Tsuji, S. et al., Genetic heterogeneity in type 1 Gaucher disease: multiple genotypes in Ashkenazic and non-Ashkenazic individuals, *Proc. Natl. Acad. Sci. U.S.A.*, 85, 2349, 1988.
38. Beutler, E. et al., Gaucher disease: gene frequencies in the Ashkenazi Jewish population, *Am. J. Hum. Genet.*, 52, 85, 1993.
39. Tayebi, N. et al., Reciprocal and nonreciprocal recombination at the glucocerebrosidase gene region: implications for complexity in Gaucher disease, *Am. J. Hum. Genet.*, 72, 519, 2003.
40. Sarria, A.J. et al., Detection of three rare (G377S, T134P and 1451delAC), and two novel mutations (G195W and Rec[1263del55;1342G>C]) in Spanish Gaucher disease patients. Mutation in brief no. 251. Online, *Hum. Mutat.*, 14, 88, 1999.
41. Cormand, B. et al., A new gene-pseudogene fusion allele due to a recombination in intron 2 of the glucocerebrosidase gene causes Gaucher disease, *Blood Cells Mol. Dis.*, 26, 409, 2000.
42. Stone, D.L. et al., Glucocerebrosidase gene mutations in patients with type 2 Gaucher disease, *Hum. Mutat.*, 15, 181, 2000.
43. Park, J.K. et al., The E326K mutation and Gaucher disease: mutation or polymorphism? *Clin. Genet.*, 61, 32, 2002.
44. Walker, J.M. et al., Glucocerebrosidase mutation T369M appears to be another polymorphism, *Clin. Genet.*, 63, 237, 2003.
45. Eyal, N. et al., Three unique base pair changes in a family with Gaucher disease, *Hum. Genet.*, 87, 328, 1991.
46. Beutler, E., Demina, A., and Gelbart, T., Glucocerebrosidase mutations in Gaucher disease, *Mol. Med.*, 1, 82, 1994.
47. Torralba, M.A. et al., Identification and characterization of a novel mutation c.1090G>T (G325W) and nine common mutant alleles leading to Gaucher disease in Spanish patients, *Blood Cells Mol. Dis.*, 27, 489, 2001.

48. Torralba, M.A. et al., Erratum: Identification and characterization of a novel mutation c.1090G>T (G325W) and nine common mutant alleles leading to Gaucher disease in Spanish patients (vol 27(2), 489-95, 2001), *Blood Cells Mol. Dis.*, 27, 713, 2001.

49. Zhao, H. et al., Gaucher's disease: identification of novel mutant alleles and genotype-phenotype relationships, *Clin. Genet.*, 64, 57, 2003.

50. Montfort, M. et al., Functional analysis of 13 GBA mutant alleles identified in Gaucher disease patients: Pathogenic changes and "modifier" polymorphisms, *Hum. Mutat.*, 23, 567, 2004.

51. Hodanova, K. et al., Analysis of the beta-glucocerebrosidase gene in Czech and Slovak Gaucher patients: mutation profile and description of six novel mutant alleles, *Blood Cells Mol. Dis.*, 25, 287, 1999.

52. Martinez-Arias, R. et al., Glucocerebrosidase pseudogene variation and Gaucher disease: recognizing pseudogene tracts in GBA alleles, *Hum. Mutat.*, 17, 191, 2001.

53. Park, J.K. et al., Myoclonic epilepsy in Gaucher disease: genotype-phenotype insights from a rare patient subgroup, *Pediatr. Res.*, 53, 387, 2003.

54. Uyama, E. et al., Hydrocephalus, corneal opacities, deafness, valvular heart disease, deformed toes and leptomeningeal fibrous thickening in adult siblings: a new syndrome associated with β-glucocerebrosidase deficiency and a mosaic population of storage cells, *Acta Neurol. Scand.*, 86, 407, 1992.

55. Abrahamov, A. et al., Gaucher's disease variant characterised by progressive calcification of heart valves and unique genotype, *Lancet*, 346, 1000, 1995.

56. Chabas, A. et al., Unusual expression of Gaucher's disease: cardiovascular calcifications in three sibs homozygous for the D409H mutation, *J. Med. Genet.*, 32, 740, 1995.

57. Goker-Alpan, O. et al., Phenotypic continuum in neuronopathic Gaucher disease: an intermediate phenotype between type 2 and type 3, *J. Pediatr.*, 143, 273, 2003.

58. Sidransky, E., Gaucher disease: complexity in a "simple" disorder, *Mol. Genet. Metabol.*, 83, 6, 2004.

59. Lachmann, R.H. et al., Twin pairs showing discordance of phenotype in adult Gaucher's disease, *Qjm*, 97, 199, 2004.

60. Amato, D. et al., Gaucher disease: variability in phenotype among siblings, *J. Inherit. Metab. Dis.*, 27, 659, 2004.

61. Goker-Alpan, O. et al., Divergent phenotypes in Gaucher disease implicate the role of modifiers, *J. Med. Genet.*, 42, e37, 2005.

62. Zhao, H. et al., Gaucher disease: *in vivo* evidence for allele dose leading to neuronopathic and nonneuronopathic phenotypes, *Am. J. Med. Genet.*, 116A, 52, 2003.

63. Kim, J.W. et al., Gaucher disease: identification of three new mutations in the Korean and Chinese (Taiwanese) populations, *Hum. Mutat.*, 7, 214, 1996.

64. Germain, D.P. et al., Exhaustive screening of the acid beta-glucosidase gene, by fluorescence-assisted mismatch analysis using universal primers: mutation profile and genotype/phenotype correlations in Gaucher disease, *Am. J. Hum. Genet.*, 63, 415, 1998.

65. Ida, H. et al., Mutation screening of 17 Japanese patients with neuropathic Gaucher disease, *Hum. Genet.*, 98, 167, 1996.

66. Hatton, C.E. et al., Mutation analysis in 46 British and Irish patients with Gaucher's disease, *Arch. Dis. Child.*, 77, 17, 1997.

67. Pasmanik-Chor, M. et al., Expression of mutated glucocerebrosidase alleles in human cells, *Hum. Mol. Genet.*, 6, 887, 1997.

68. Miocic, S. et al., Identification and functional characterization of five novel mutant alleles in 58 Italian patients with Gaucher disease type 1, *Hum. Mutat.*, 25, 100, 2005.

69. Grace, M.E. et al., Analysis of human acid beta-glucosidase by site-directed mutagenesis and heterologous expression, *J. Biol. Chem.*, 269, 2283, 1994.
70. Hodanova, K. et al., Transient expression of wild-type and mutant glucocerebrosidases in hybrid vaccinia expression system, *Eur. J. Hum. Genet.*, 11, 369, 2003.
71. Sorge, J. et al., Heterogeneity in type I Gaucher disease demonstrated by restriction mapping of the gene, *Proc. Natl. Acad. Sci. U.S.A.*, 82, 5442, 1985.
72. Zimran, A., Gelbart, T., and Beutler, E., Linkage of the PvuII polymorphism with the common Jewish mutation for Gaucher disease, *Am. J. Hum. Genet.*, 46, 902, 1990.
73. Beutler, E., West, C., and Gelbart, T., Polymorphisms in the human glucocerebrosidase gene, *Genomics*, 12, 795, 1992.
74. Masuno, M. et al., Restriction fragment length polymorphism analysis in healthy Japanese individuals and Japanese families with Gaucher disease, *Acta Paediatr. Jpn.*, 31, 158, 1989.
75. Glenn, D., Gelbart, T., and Beutler, E., Tight linkage of pyruvate kinase (PKLR) and glucocerebrosidase (GBA) genes, *Hum. Genet.*, 93, 635, 1994.
76. Mateu, E. et al., PKLR⁻ GBA region shows almost complete linkage disequilibrium over 70 kb in a set of worldwide populations, *Hum. Genet.*, 110, 532, 2002.
77. Cormand, B. et al., Genetic fine localization of the beta-glucocerebrosidase (GBA) and prosaposin (PSAP) genes: implications for Gaucher disease, *Hum. Genet.*, 100, 75, 1997.
78. Lacerda, L. et al., The N370S mutation in the glucocerebrosidase gene of Portuguese type 1 Gaucher patients: linkage to the PvuII polymorphism, *J. Inherit. Metab. Dis.*, 17, 85, 1994.
79. Amaral, O. et al., Distinct haplotype in non-Ashkenazi Gaucher patients with N370S mutation, *Blood Cells Mol. Dis.*, 23, 415, 1997.
80. Cormand, B. et al., Molecular analysis and clinical findings in the Spanish Gaucher disease population: putative haplotype of the N370S ancestral chromosome, *Hum. Mutat.*, 11, 295, 1998.
81. Demina, A., Boas, E., and Beutler, E., Structure and linkage relationships of the region containing the human L-type pyruvate kinase (PKLR) and glucocerebrosidase (GBA) genes, *Hematopathol. Mol. Hematol.*, 11, 63, 1998.
82. Rockah, R. et al., Linkage disequilibrium of common Gaucher disease mutations with a polymorphic site in the pyruvate kinase (PKLR) gene, *Am. J. Med. Genet.*, 78, 233, 1998.
83. Tuteja, R. et al., 1448C mutation linked to the Pv1.1⁻ genotype in Italian patients with Gaucher disease, *Hum. Mol. Genet.*, 2, 781, 1993.
84. Iwasawa, K., Ida, H., and Eto, Y., Differences in origin of the 1448C mutation in patients with Gaucher disease, *Acta Paediatr. Jpn.*, 39, 451, 1997.
85. Rodriguez-Mari, A. et al., New insights into the origin of the Gaucher disease-causing mutation N370S: extended haplotype analysis using the 5GC3.2, 5470 G/A, and ITG6.2 polymorphisms, *Blood Cells Mol. Dis.*, 27, 950, 2001.
86. Lau, E.K. et al., Two novel polymorphic sequences in the glucocerebrosidase gene region enhance mutational screening and founder effect studies of patients with Gaucher disease, *Hum. Genet.*, 104, 293, 1999.
87. Martinez-Arias, R. et al., Sequence variability of a human pseudogene, *Genome Res.*, 11, 1071, 2001.
88. LaMarca, M.E. et al., A novel alteration in metaxin 1, F202L, is associated with N370S in Gaucher disease, *J. Hum. Genet.*, 49, 220, 2004.
89. Beutler, E., Gelbart, T., and West, C., Identification of six new Gaucher disease mutations, *Genomics*, 15, 203, 1993.

90. Sinclair, G., Choy, F.Y., and Ferreira, P., Heterologous expression and characterization of a rare Gaucher disease mutation (c.481C > T) from a Canadian aboriginal population using archival tissue samples, *Mol. Genet. Metab.*, 74, 345, 2001.

91. Alfonso, P. et al., Mutation prevalence among 51 unrelated Spanish patients with Gaucher disease: identification of 11 novel mutations, *Blood Cells Mol. Dis.*, 27, 882, 2001.

92. Bodamer, O.A. et al., Variant Gaucher disease characterized by dysmorphic features, absence of cardiovascular involvement, laryngospasm, and compound heterozygosity for a novel mutation (D409H/C16S), *Am. J. Med. Genet.*, 109, 328, 2002.

93. Choy, F.Y., Humphries, M.L., and Shi, H., Identification of two novel and four uncommon missense mutations among Chinese Gaucher disease patients, *Am. J. Med. Genet.*, 71, 172, 1997.

94. Sinclair, G., Choy, F.Y., and Humphries, L., A novel complex allele and two new point mutations in type 2 (acute neuronopathic) Gaucher disease, *Blood Cells Mol. Dis.*, 24, 420, 1998.

95. Beutler, E. et al., Five new Gaucher disease mutations, *Blood Cells Mol. Dis.*, 21, 20, 1995.

96. Grace, M.E., Desnick, R.J., and Pastores, G.M., Identification and expression of acid beta-glucosidase mutations causing severe type 1 and neurologic type 2 Gaucher disease in non-Jewish patients, *J. Clin. Invest.*, 99, 2530, 1997.

97. Beutler, E. et al., Gaucher disease: four families with previously undescribed mutations, *Proc. Assoc. Am. Physicians*, 108, 179, 1996.

98. Choy, F.Y. et al., Identification of two novel mutations, L105R and C342R, in Type I Gaucher disease, *Clin. Genet.*, 61, 229, 2002.

99. Demina, A. and Beutler, E., Six new Gaucher disease mutations, *Acta Haematol.*, 99, 80, 1998.

100. Amaral, O. et al., Gaucher disease: expression and characterization of mild and severe acid beta-glucosidase mutations in Portuguese type 1 patients, *Eur. J. Hum. Genet.*, 8, 95, 2000.

101. Chabas, A. et al., Neuronopathic and non-neuronopathic presentation of Gaucher disease in patients with the third most common mutation (D409H) in Spain, *J. Inherit. Metab. Dis.*, 19, 798, 1996.

102. Tang, N.L. et al., Novel mutations in type 2 Gaucher disease in Chinese and their functional characterization by heterologous expression, *Hum. Mutat.*, 26, 59, 2005.

103. Stone, D.L. et al., Type 2 Gaucher disease: the collodion baby phenotype revisited, *Arch. Dis. Child. Fetal. Neonatal. Ed.*, 82, F163, 2000.

104. Orvisky, E. et al., The identification of eight novel glucocerebrosidase (GBA) mutations in patients with Gaucher disease, *Hum. Mutat.*, 19, 458, 2002.

105. Filocamo, M. et al., Analysis of the glucocerebrosidase gene and mutation profile in 144 Italian gaucher patients, *Hum. Mutat.*, 20, 234, 2002.

106. Cormand, B. et al., Mutation analysis of Gaucher disease patients from Argentina: high prevalence of the RecNciI mutation, *Am. J. Med. Genet.*, 80, 343, 1998.

107. Choy, F.Y. and Wei, C., Identification of a new mutation (P178S) in an African-American patient with type 2 Gaucher disease, *Hum. Mutat.*, 5, 345, 1995.

108. Choy, F.Y. et al., Novel point mutation (W184R) in neonatal type 2 Gaucher disease, *Pediatr. Dev. Pathol.*, 3, 180, 2000.

109. Ida, H. et al., Mutation prevalence among 47 unrelated Japanese patients with Gaucher disease: identification of four novel mutations, *J. Inherit. Metab. Dis.*, 20, 67, 1997.

110. Choy, F.Y., Humphries, M.L., and Ben-Yoseph, Y., A novel mutation (V191G) in a German-British type 1 Gaucher disease patient. Mutations in brief no. 131. Online, *Hum. Mutat.*, 11, 411, 1998.

111. Choy, F.Y., Wong, K., and Shi, H.P., Glucocerebrosidase mutations among Chinese neuronopathic and non-neuronopathic Gaucher disease patients, *Am. J. Med. Genet.*, 84, 484, 1999.

112. Kawame, H. and Eto, Y., A new glucocerebrosidase-gene missense mutation responsible for neuronopathic Gaucher disease in Japanese patients, *Am. J. Hum. Genet.*, 49, 1378, 1991.

113. Cooper, A., Church, H.J., and Wraith, J.E., Gaucher disease, *Hum. Genet.*, 108, 82, 2001.

114. Beutler, E. and Gelbart, T., Gaucher disease associated with a unique KpnI restriction site: identification of the amino-acid substitution, *Ann. Hum. Genet.*, 54 (Pt 2), 149, 1990.

115. Grace, M.E. et al., Non-pseudogene-derived complex acid beta-glucosidase mutations causing mild type 1 and severe type 2 gaucher disease, *J. Clin. Invest.*, 103, 817, 1999.

116. Eblan, M. and Sidransky, E., Unpublished data.

117. Walley, A.J., Ellis, I., and Harris, A., Three unrelated Gaucher's disease patients with three novel point mutations in the glucocerebrosidase gene (P266R, D315H and A318D), *Br. J. Haematol.*, 91, 330, 1995.

118. He, G.S., Grace, M.E., and Grabowski, G.A., Gaucher disease: four rare alleles encoding F213I, P289L, T323I, and R463C in type 1 variants, *Hum. Mutat.*, 1, 423, 1992.

119. Stone, D.L. et al., Is the perinatal lethal form of Gaucher disease more common than classic type 2 Gaucher disease? *Eur. J. Hum. Genet.*, 7, 505, 1999.

120. Cormand, B. et al., Two novel (1098insA and Y313H) and one rare (R359Q) mutations detected in exon 8 of the beta-glucocerebrosidase gene in Gaucher's disease patients, *Hum. Mutat.*, 7, 272, 1996.

121. Parenti, G. et al., A novel mutation of the beta-glucocerebrosidase gene associated with neurologic manifestations in three sibs, *Clin. Genet.*, 53, 281, 1998.

122. Tsai, F.J. et al., Mutation analysis of type II Gaucher disease in five Taiwanese children: identification of two novel mutations, *Acta Paediatr. Taiwan*, 42, 231, 2001.

123. Beutler, E. and Gelbart, T., Two new Gaucher disease mutations, *Hum. Genet.*, 93, 209, 1994.

124. Kawame, H. et al., Rapid identification of mutations in the glucocerebrosidase gene of Gaucher disease patients by analysis of single-strand conformation polymorphisms, *Hum. Genet.*, 90, 294, 1992.

125. Church, H.J., Cooper, A., and Wraith, J.E., Gaucher disease, *Hum. Genet.*, 108, 83, 2001.

126. Shamseddine, A. et al., Novel mutation, L371V, causing multigenerational Gaucher disease in a Lebanese family, *Am. J. Med. Genet.* A, 125, 257, 2004.

127. Cormand, B. et al., Two new mild homozygous mutations in Gaucher disease patients: clinical signs and biochemical analyses, *Am. J. Med. Genet.*, 70, 437, 1997.

128. Laubscher, K.H. et al., Use of denaturing gradient gel electrophoresis to identify mutant sequences in the beta-glucocerebrosidase gene, *Hum. Mutat.*, 3, 411, 1994.

129. Cabrera-Salazar, M.A. et al., Mutation analysis of 25 Columbian Gaucher disease patients using dHPLC: report of three novel mutations, *Am. J. Hum. Genet.*, 74, A1819, 2004.

130. Walley, A.J. and Harris, A., A novel point mutation (D380A) and a rare deletion (1255del55) in the glucocerebrosidase gene causing Gaucher's disease, *Hum. Mol. Genet.*, 2, 1737, 1993.

131. Ponce, E. and Grabowski, G.A., W381Ter: Novel mutation in the coding sequence of the acid beta-glucosidase gene of a Type 1 Gaucher patient who developed neutralizing antibodies, *Am. J. Hum. Genet.*, 67, 287, 2000.

132. Morar, B. and Lane, A., The molecular characterization of Gaucher disease in South Africa, *Clin. Genet.*, 50, 78, 1996.

133. Theophilus, B.D. et al., Comparison of RNase A, a chemical cleavage and GC-clamped denaturing gradient gel electrophoresis for the detection of mutations in exon 9 of the human acid beta-glucosidase gene, *Nucleic Acids Res.*, 17, 7707, 1989.

134. Amaral, O. et al., Type 1 Gaucher disease: identification of N396T and prevalence of glucocerebrosidase mutations in the Portuguese, *Hum. Mutat.*, 8, 280, 1996.

135. Seeman, P.J. et al., Two new missense mutations in a non-Jewish Caucasian family with type 3 Gaucher disease, *Neurology*, 46, 1102, 1996.

136. Wasserstein, M.P. et al., Type 1 Gaucher disease presenting with extensive mandibular lytic lesions: identification and expression of a novel acid beta-glucosidase mutation, *Am. J. Med. Genet.*, 84, 334, 1999.

137. Ohashi, T. and Eto, Y., Molecular analysis of Japanese Gaucher disease, *J. Inherit. Metab. Dis.*, 12, 355, 1989.

138. Choy, F.Y. et al., A new missense mutation in glucocerebrosidase exon 9 of a non-Jewish Caucasian type 1 Gaucher disease patient, *Hum. Mol. Genet.*, 3, 821, 1994.

139. Tuteja, R. et al., Y418C: a novel mutation in exon 9 of the glucocerebrosidase gene of a patient with Gaucher disease creates a new Bgl I site, *Hum. Genet.*, 94, 314, 1994.

140. Uchiyama, A. et al., New Gaucher disease mutations in exon 10: a novel L444R mutation produces a new NciI site the same as L444P, *Hum. Mol. Genet.*, 3, 1183, 1994.

141. Cooper, A. and Wraith, J.E., Gene symbol: GBA, disease: Gaucher disease 1, *Hum. Genet.*, 108, 82, 2001.

142. Choy, F.Y., Humphries, M.L., and Ben-Yoseph, Y., Gaucher type 2 disease: identification of a novel transversion mutation in a French-Irish patient, *Am. J. Med. Genet.*, 78, 92, 1998.

143. Beutler, E. et al., Identification of the second common Jewish Gaucher disease mutation makes possible population-based screening for the heterozygous state, *Proc. Natl. Acad. Sci. U.S.A.*, 88, 10544, 1991.

144. Choy, F.Y., Humphries, M.L., and Ferreira, P., Novel insertion mutation in a non-Jewish Caucasian type 1 Gaucher disease patient, *Am. J. Med. Genet.*, 68, 211, 1997.

145. Wu, J.Y. et al., Identification of a novel three-nucleotide insertion mutation (c.841-842insTGA) in the acid beta-glucosidase gene of a Taiwan Chinese patient with type II Gaucher disease, *Hum. Mutat.*, 17, 238, 2001.

146. Felderhoff-Mueser, U., et al., Intrauterine onset of acute neuropathic type 2 Gaucher disease: identification of a novel insertion sequence, *Am. J. Med. Genet.*, 128A, 138, 2004.

147. Tayebi, N. et al., Prenatal lethality of a homozygous null mutation in the human glucocerebrosidase gene, *Am. J. Med. Genet.*, 73, 41, 1997.

148. Germain, D.P., Kaneski, C.R., and Brady, R.O., Mutation analysis of the acid beta-glucosidase gene in a patient with type 3 Gaucher disease and neutralizing antibody to alglucerase, *Mutat. Res.*, 483, 89, 2001.

149. Beutler, E. and Gelbart, T., Erroneous assignment of Gaucher disease genotype as a consequence of a complete gene deletion, *Hum. Mutat.*, 4, 212, 1994.

150. He, G.S. and Grabowski, G.A., Gaucher disease: A G+1----A+1 IVS2 splice donor site mutation causing exon 2 skipping in the acid beta-glucosidase mRNA, *Am. J. Hum. Genet.*, 51, 810, 1992.

151. Beutler, E. et al., Mutations in Jewish patients with Gaucher disease, *Blood*, 79, 1662, 1992.

152. Dominissini, S. et al., Charaterization of two novel GBA mutations causing Gaucher disease that lead to aberrant RNA species by functional splicing assays, *Hum. Mutat.*, 27, 119, 2006.

153. Romano, M. et al., Functional characterization of the novel mutation IVS 8 (-11delC) (-14T>A) in the intron 8 of the glucocerebrosidase gene of two Italian siblings with Gaucher disease type I, *Blood Cells Mol. Dis.*, 26, 171, 2000.

154. Ohshima, T. et al., A novel splicing abnormality in a Japanese patient with Gaucher's disease, *Hum. Mol. Genet.*, 2, 1497, 1993.

155. den Dunnen, J.T. and Antonarakis, S.E., Mutation nomenclature extensions and suggestions to describe complex mutations: a discussion, *Hum. Mutat.*, 15, 7, 2000.

156. Antonarakis, S.E., Recommendations for a nomenclature system for human gene mutations. Nomenclature Working Group, *Hum. Mutat.*, 11, 1, 1998.

157. Maquat, L.E., When cells stop making sense: effects of nonsense codons on RNA metabolism in vertebrate cells, *RNA*, 1, 453, 1995.

158. Filocamo, M. et al., Homozygosity for a non-pseudogene complex glucocerebrosidase allele as cause of an atypical neuronopathic form of Gaucher disease, *Am. J. Med. Genet.* A, 134, 95, 2005.

159. Filocamo, M. et al., Somatic mosaicism in a patient with Gaucher disease type 2: implication for genetic counseling and therapeutic decision-making, *Blood Cells Mol. Dis.*, 26, 611, 2000.

160. Reissner, K. et al., Type 2 Gaucher disease with hydrops fetalis in an Ashkenazi Jewish family resulting from a novel recombinant allele and a rare splice junction mutation in the glucocerebrosidase locus, *Mol. Genet. Metab.*, 63, 281, 1998.

161. Filocamo, M. et al., Identification of a novel recombinant allele in three unrelated Italian Gaucher patients: implications for prognosis and genetic counseling, *Blood Cells Mol. Dis.*, 26, 307, 2000.

162. Beutler, E. et al., Three Gaucher-disease-producing mutations in a patient with Gaucher disease: mechanism and diagnostic implications, *Acta Haematol.*, 104, 103, 2000.

163. Church, H.J. et al., Homozygous loss of a cysteine residue in the glucocerebrosidase gene results in Gaucher's disease with a hydropic phenotype, *Eur. J. Hum. Genet.*, 12, 975, 2004.

164. Amaral, O. et al., Homozygosity for two mild glucocerebrosidase mutations of probable Iberian origin, *Clin. Genet.*, 56, 100, 1999.

165. Masuno, M. et al., Non-existence of a tight association between a 444leucine to proline mutation and phenotypes of Gaucher disease: high frequency of a NciI polymorphism in the non-neuronopathic form, *Hum. Genet.*, 84, 203, 1990.

166. Cormand, B. et al., Gaucher disease in Spanish patients: analysis of eight mutations, *Hum. Mutat.*, 5, 303, 1995.

CHAPTER 3

Cell Biology and Biochemistry of Acid β-Glucosidase: The Gaucher Disease Enzyme

Gregory A. Grabowski, Andrzej Kazimierczuk, and Benjamin Liou

CONTENTS

INTRODUCTION

The sphingolipidoses have provided the opportunity to examine derangements of glycosphingolipid metabolism at the molecular and cell biologic levels. Of these disorders, the variants of Gaucher disease are the most frequent in the Western World and have served as prototypes for enzyme, substrate depletion, molecular chaperone, and other direct treatment strategies. The phenotypes of Gaucher disease (as detailed elsewhere in this volume) result from the defective functions of acid β-glucosidase (glucocerebrosidase, GCase) and the accumulation of its major natural substrates, glucosyl ceramide, and glucosylsphingosine, the deacylated analogue of glucosyl ceramide. The basic cell biology and enzymologies of wild-type and mutant GCases have been evaluated to understand the cellular effects of the Gaucher disease allelic variants and their expression at the phenotypic level. This review evaluates the current state of understanding of GCase variant structure/function and their cell biologic properties. There remain great deficits in our understanding of this simple glycosidase and the resultant altered gene products leading to Gaucher disease.

CELL BIOLOGY OF GCASE

Expression of GCase mRNA in Various Tissues

GCase is expressed at various levels in all nucleated mammalian cells. The highest levels of activity are in ectodermal-derived tissues and are lowest in those from the mesoderm (for review see[1]). *In situ* hybridization with antisense GCase riboprobes showed differential expression in mouse CNS and visceral tissues.[2] Limited immunostaining for GCase protein suggests a similar, but not identical, distribution in the human CNS.[3] In mouse embryonic CNS, generalized low level expression was present. Shortly before birth, more discrete localization of high level signals appear in neurons, particularly those of the cerebellar Purkinje cell layer, neurons and glial cells of the cerebral cortex, and large hippocampal pyramidal cells.[2] In adult visceral tissues, the highest mRNA signals were in the outer mature epidermal layers of the skin, epithelial cells of the stomach, villus epithelial cells of the small intestine, osteoblasts and chondrocytes of developing bone, and epithelial cells of the kidney tubules, pancreatic ducts and epididymis.[2] Similar detailed studies are not available across the human developmental spectrum.

In human CNS, a rostral to caudal gradient of glucosyl ceramide accumulation was found in the neuronopathic variants of Gaucher disease.[4] Such a pattern was not coincident with the GCase RNA expression in CNS of mice[2] or of immunocrossreactivity in humans.[3] Using a monoclonal antibody to human GCase, a gradient of decreasing signal was found in the hippocampal pyramidal cells. The highest levels were in the C4 region and lower levels in C1. Another discrete stepdown in expression was found between C1 and C2 regions. These observations indicate that pathologic involvement might be determined by factors influencing the amount or types of substrate(s) (see below) presented to lysosomes during development, relative to the level of residual enzyme in particular cells. However, the examination of GCase mRNA and protein during development and across regions is incomplete. Such correlations with the activity of GC synthase and the regional/cellular levels of GC are not available, but essential to understanding the pathogenesis of GD.

GCase Translation and Lysosomal Targeting

A MW~60,000 naked peptide is translated from the GBA mRNA. An 80 kDa cytoplasmic protein that bound to coding sequences of the GCase mRNA was identified and shown to inhibit *in vitro* and *ex vivo* translation of this mRNA.[5–7] Characterization of this gene showed identity to ILF3/NF90. This protein was termed TCP/ILF3. Its translational control functions are modulated by PKC isoforms and inhibition of TCP/ILF3 phosphorylation leads to enhanced translation of GCase in cultured fibroblasts and HepG2 cells.[7] This protein functions by preventing engagement of GCase mRNA with polysomes.[5] The exact role of this protein in normal physiologic function of GCase is unknown, but the translation of GCase mRNAs is inhibited significantly in overexpressed systems with obvious implications for efficient enzyme production and gene therapy.

The GCase mRNA encodes the mature protein monomer and two leader sequences (39 and 20 amino acids) each following a functional ATG initiator.[8,9] Co-translational glycosylation occurs, and four of the five potential N-glycosylation consensus sequences acquire occupancy (Asn 19, 59, 146, and 270[10]). The glycosylation sequence beginning with Asn 462 is not occupied. Occupancy of Asn 19 is essential for GCase to develop a catalytically active conformer.[11] Deglycosylation of preformed GCase leads to only small losses of activity, but the de-decorated enzyme is very susceptible to proteolysis.[12] The oligosaccharide structure of the natural placental enzyme has complex oligosaccharide modification of branched core mannosyl structures.[13] The natural placental or CHO cell produced GCase have been modified by exoglycosidase treatment to expose α-mannosyl residues for binding to the macrophage mannose receptor.[14–16] Characterization of these trimmed structures showed fucosylation of the core of Asn 146 with oligomannose structure (Man6) predominately on Asn 19.[10]

Metabolic labeling studies of cultured fibroblasts showed that the oligosaccharides of wild-type GCase are remodeled to complex and high mannose side chains as the protein is transported from the *cis* to *trans* Golgi and into the lysosome.[17,18] The type and extent of oligosaccharide modifications vary with the tissue source and have not been characterized in detail. In these cells, the transit time from the ER to the lysosome is ~3 hours and the enzyme has a t_f ~ 60 hours in the lysosome.[17] The half-lives of GCase in other tissues are not known. The effect of oligosaccharide variation on such GCase function or cellular properties appears minor, but requires more extensive investigation.

The presence of adequate concentrations of saposin C, the natural protein activator of GCase is essential to the stability of GCase to proteolysis in the lysosome.[19] Importantly, saposin C and GCase must reside in the same compartment since a physical interaction appears needed for this protective effect. Such a segregation may account for the partial deficiency of GCase in Niemann-Pick C disease.[20]

Unlike freely soluble lysosomal proteins, GCase becomes membrane associated[21] and does not contain mannose-6-phosphate for targeting to the lysosome. The majority of newly synthesized human GCase remains nonmembrane associated (80%) for ~4 hours. However, by 24 hours the retained intracellular enzyme is ~90% membrane associated, in lysosomes, in glycosylated or unglycosylated forms [21];

GCase lysosomal targeting is oligosaccharide independent. Preliminary studies in these laboratories implicate, as yet to be characterized peptide sequences in the COOH-terminal third of GCase, as essential to lysosomal localization.[22] At wild-type expression levels, GCase is not secreted from cells in significant amounts, but when overexpressed, GCase is secreted in large amounts (60–80%).[7,21,23] Tunicamycin treatment leads to synthesis of an unglycosylated GCase and a complete block of its secretion.[21] These results suggest a positive signal for secretion of GCase out of cells may be oligosaccharide mediated.

Studies of GCase variants in cultured cells from GD patients are available in detail only for the enzymes expressed from the N370S and L444P alleles.[17,24–26] The N370S GCase is present at protein levels, has transit times, and has a half-life equivalent to those of the wild-type enzyme. In comparison, the L444P enzyme only has wild-type M_r and initial oligosaccharide modification.[17] The half-life of the L444P enzyme is <4 hours.[17,25] Conflicting data exist for the G202R variant enzyme. One study found Golgi retention of abundant amounts of this variant in homozygotes, while an unstable enzyme was detected in heteroallelic variants (G202R/L444P).[27,28] Thus, the nature of the heteroallele or other factors (i.e., culture conditions) may influence the properties of the mutant enzymes. Such details are not available for most of the ~200 point mutant GCases in cells from GD patients, but mistrafficking has been observed for several variants.[29,30]

Half-Life, Survival and Fate of Injected GCase

The fates and distributions of exogenously supplied GCase are of considerable interest to enzyme and gene therapy. The available data are from studies in wild-type mice. α-Mannosyl terminated placental or CHO GCase was used to evaluate uptake and survival in tissues.[31–33] Uptake was oligosaccharide mediated. This is in contrast to endogenously synthesized enzyme that does not require oligosaccharides for lysosomal targeting.[21] The tissue half-lives for activity and protein were biphasic following bolus intravenous injections with half-lives ~1.5 and 12–14 hours. However, the molecular rate constant for the 4-MUG substrate was about 50–60% of the enzyme before injection.[32] This probably is due to pH inactivation and/or aggregation in the plasma and during transit from the cell surface to the lysosome.[32] The α-mannosyl terminated GCase is preferentially taken up by the macrophage mannose receptor into Kupffer cells,[31,32] but since these cells are only ~15% of hepatic cells, the majority of this enzyme is taken up by other receptors into hepatocytes and sinusoidal lining cells.[31] Detailed studies have not been completed in GD mouse models.

Characterization of Acid β-Glucosidase

Wild-type human GCase is encoded by a single structural locus (GBA) at chromosome 1q21.[34,35] Chemically determined or cDNA predicted amino acid sequences from various tissues or cultured cell libraries, except for cloning artifacts, are identical (see Chapter 2). The GCase amino acid sequence is highly conserved over many species. Among mammalian species, GCase variation in amino acid sequence

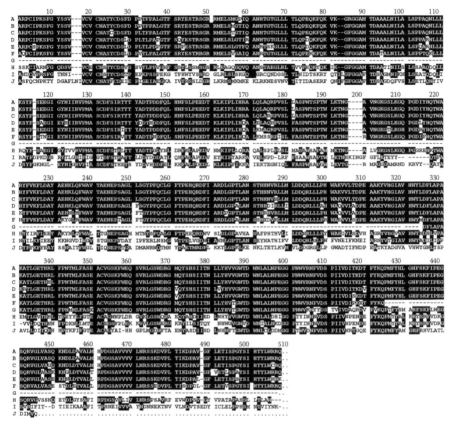

Figure 3.1 Amino acid sequence alignments of GCases from several species. The blackened regions indicate amino acid identity. Amino acid similarity and nonhomology are indicated as black letters and white background. The amino acid identities between these species are as follows: A) Human (100%), B) Chimpanzee (99.7%), C) Pongo (97.7%), D) Dog (91.1%), E) Pig (89.2%), F) Mouse (85%), G) Rat (~40%, H) Fugu (45.5%), I) Honey bee (19.7%) and J) *C. elegans* (13.2%). The canine amino acid sequence was predicted from our determined sequence of the canine cDNA.

is less than 15% (see Figure 3.1). This increases to 50% and >70% from Fugu to *C. elegans* and honey bee. The presence and placement of the five *N*-glycosylation consensus sequences in several mammalian species are strictly preserved, as are the amino acids surrounding the active site residues, E235 and E340.[36,37]

The native state of GCase in cells and tissues has not been defined. The detergent and/or organic solvent extractions needed for solubilization and removal of endogenous lipids disrupts noncovalent interactions. The reported molecular weights of the extracted enzyme have varied from about 60,000 to 450,000.[38–41] Also, molecular weight analyses under nondenaturing conditions have been hampered by the anomalous migration of GCase in polyacrylamide or agarose gels. The native molecular weight in tissues, estimated by *in situ* radioinactivation of the enzymatic activity,[38,39] were consistent with either a monomer or dimer. Consequently, the nature of the GCase in cells remains unresolved, even though the physical state of the enzyme in

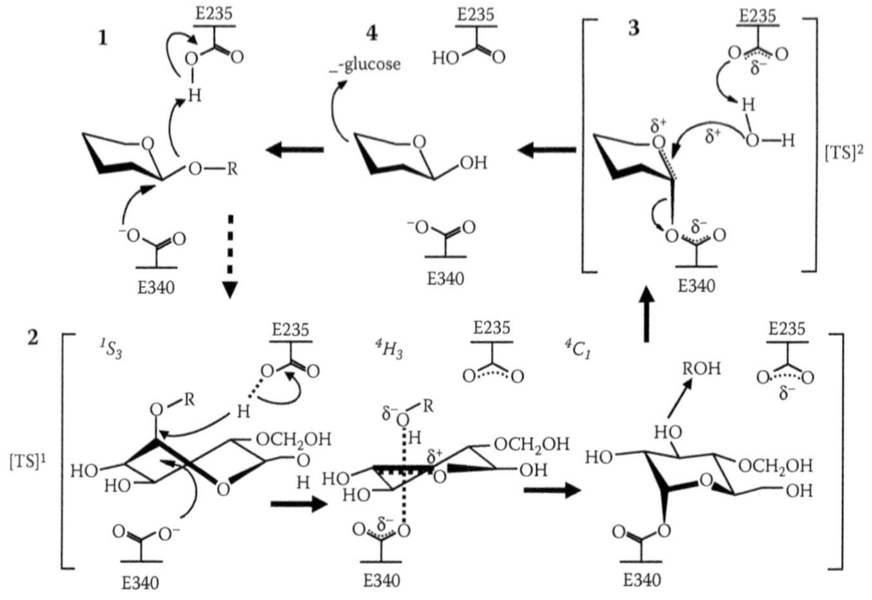

Figure 3.2 Schematic of the reaction mechanism for GCase. E235 and E340 are the acid/base and nucleophile in catalysis, respectively (See Figure 3.2). The transition state, $[TS]^1$, is modeled after recent crystallographic trapping of intermediates for retaining glycosidases.[54] $[TS]^1$ includes passage through conformations of the pyranose ring, including $^1S_3$, $^4H_3$, and $^4C_1$. Passage through $[TS]^2$ with retention of configuration at the anomeric carbon is facilitated by regeneration of the acid at E235 and the nucleophile at E340. The δ refers to partial charges (+ or -) generated during the transition states. The R group for the natural substrate of GCase, glucosylceramide, is ceramide.

the cell has significant implications for understanding the enzyme's function. The enzyme is monomeric during *in vitro* studies with detergents. Purified GCase has variable reported specific activities ($\sim5 \infty 10^5$ to $5 \infty 10^6$ nmol/h/mg protein) primarily due to differences in assay conditions. Using active site titration to quantify the amount of active enzyme, the catalytic rate constants, k_{cat}, were estimated for several glucosylceramides and 4-alkyl-umbelliferyl-glucosides as 2350 ± 1054/min.[42,43]

GCase is a member of the retaining glycosidases whose catalytic cycle proceeds by a two-step, double displacement mechanism (Figure 3.2).[44-46] During the first step (glucosylation), a nucleophilic residue (E340) attacks the O-glycosidic anomeric linkage at C1 with protonic assistance from E235. Water is added to this complex, and the ceramide is released; the enzyme-glucosyl complex is formed. Site-directed mutagenesis of E340 to E340G destroys the nucleophilic capacity of this residue and diminishes catalytic activity by at least 10^5-fold.[47] Following covalent attachment of glucose, the covalent bond is attacked by H_2O in a base assisted cleavage resulting from OH^- generated by H^+ donation to E235. Glucose is released with retention of configuration. The acid of the acid/base, E235 is thus regenerated as is the nucleophile, E340.

Hydrophobic cluster analyses of families of glycosidases suggested that E235 is the acid/base involved in catalysis by assisting cleavage of the O-glycosidic bond.[37,47–49] The loss of enzyme activity is not a test of participation in active site function since all of the mutations that have been found in Gaucher disease patients lead to significant loss and, in some cases, absolute loss of catalytic activity. For all the mutant enzymes studied to date, the catalytic rate constants have been 7 to > 100-fold decreased.[42,43,50,51] Because of their neutral natures, e.g., V, L, I or G, many of these mutations cannot participate in the cleavage during the catalytic cycle. Thus, functional studies were needed to support E235 as the acid/base in catalysis. Site-directed mutagenesis and kinetic analyses delineated the relationship of the pK_a of the leaving group and activity supported identification of this residue's function.[37] The E235G GCase had nearly zero activity toward the 4-methylumbilliferyl- and p-nitrophenyl-β-glucoside substrates. Glycine cannot participate actively in the catalytic cycle. Detailed biochemical,[37] phylogenetic,[44] and structural[52,53] support for E235 as the acid/base are available.

Recent x-ray crystallographic studies have provided details of the transition states for several glycosidases.[54] For retaining β-glucosidases, the transition states have strong oxocarbenium ionlike character. In addition, passage of the pyranose ring proceeds through $^1S_3$, to $^4H_3$ to $^4C_1$ conformations (Figure 3.2).[54] These suggest interesting new possibilities for potent molecular chaperones with therapeutic potential.

Detergent/Bile Salts and Negatively Charged Phospholipid Modifiers

In the absence of detergents or negatively charged lipids, delipidated human, bovine, and murine GCases have little hydrolytic activity toward 4MU-Glc or glycosylceramide substrates (for review see[55]). Many negatively charged detergents and bile acids activate GCase and have activation constants (K_a) near their respective CMCs (for review see[55]). Such agents will continue to have roles in the enzymology of GCases from normal and Gaucher disease tissues since numerous assay procedures for the diagnosis of this disease and its carrier state are based on their use.[1] However, they are not of physiologic importance and their *in vitro* use obscures the effects of many components for *in vivo* GCase activity (i.e., saposin C).

Because of the *in vivo* membrane association of GCase, investigators reasoned that specific membrane-derived lipids might be required for enzymatic activity. Ho et al.,[56] suggested that the negatively charged phospholipids might be the natural "activators" of this enzyme. Dale et al.[57] and Berent and Radin[58] demonstrated that GCase activity toward 4MU-Glc was enhanced by negatively charged phospholipids in the following series: phosphatidylserine > phosphatidylinositol > phosphatidic acid, whereas phosphatidyl-ethanolamine or -choline had no effect on enzymatic activity. These results have been verified by numerous investigators in human, bovine, and murine sources.[55] Using artificial membrane systems, optimal activity required the presence of negatively charged phospholipids to which GCase directly binds. Sarmientos et al.[59] indicated that maximal activity was obtained with about 19 mol% of phosphatidylserine in liposomes, a content of the same order of magnitude as found in lysosomal membranes. Characterization of the lipids associated with GCase from different tissues could provide an interesting insight into the enzyme's *in vivo* functional hydrophobic environment. Although remaining a useful reagent, phosphatidylserine is not present in significant amounts on the inner lysosomal membrane. However, mono- and di-phosphorylated phosphatidyl-inos-itol variants and lyso-bis-phosphatidic acid are present in the lysosomal-endosomal complex and do activate GCase at least as well as phosphatidylserine.[22] The nature of the physiologic lipid activator on the inner lysosomal membrane remains uni-dentified.

Mechanism of Saposin C Activation of GCase

Saposin C is the natural activator of GCase since its selective absence leads to GC storage and a Gaucher-like disease.[60,61] The mechanism of saposin C activation of GCase was evaluated using fluorescence spectroscopy and circular dichroism.[62] Addition of saposin C, which contains no tryptophan, into mixtures of GCase-PS liposomes was accompanied by significant changes in the enzyme's tryptophan emission spectrum. The quantum yield and blue-shift in emission spectrum increased as the concentration of saposin C increased. In the presence of negatively charged phospholipids, detectable CD conformational changes in GCase, and/or saposin C result from the interactions between these two proteins.[62,63]

Saposin C's mechanism of GCase activation includes direct interaction with the enzyme, binding to enzyme surfaces that become exposed following phosphati-dylserine-GCase association, and destabilization of such membranes at acidic pH that promote GCase binding.[58,64] The lipid membrane binding of saposin C is essen-tial. Two nonexclusive consequences result: 1) the binding causes conformational changes that lead to reorientation of the GCase activation domain for saposin C thereby promoting its association with GCase. 2) The lipid membrane interaction induces the conformational change of saposin to expose hydrophobic regions of amphipathic helices (H2 and H3) in saposin C. The former is needed for specificity of the interaction and the latter may be required for membrane binding.

The region including amino acid residues 48 to 63 (H-3) is needed for GCase activation and binding.[65] FRET analyses indicate that this domain is on the surface of the lipid membrane. The nonhelical regions between H-3 and -4 or H-1 and -2 may be important for lipid-induced conformational changes in saposin C. Impor-tantly, saposin C induces reorganization of lipid bilayers, perhaps promoting the GCase binding to membranes. Vaccaro, et al.[66] have suggested that enhanced enzyme activity requires larger liposomes with lesser curvature. Sandhoff and colleagues have proposed a scooting model for GCase/saposin C cleavage of glucosylceramide that require smaller, more highly curved surfaces.[67] Direct fluorescence and FRET data support the former model. These models are not mutually exclusive since the fusion of small liposomes induced by saposin C may provide a mechanism for GCase transfer between vesicles undergoing fusion and fission.[62]

Substrate Specificity

The natural substrates for GCase are a mixture of *N*-acyl-sphingosyl-1-*O*-β-D-glucosides, glucosylceramides, with varying fatty acid acyl and sphingosyl moieties depending upon the tissue source. GCase is specific for D-glucose since L-glucosyl derivatives are not hydrolyzed.[68] Glucosylsphingosine, the deacylated analog of glucosylceramide, is a minor substrate and is present in all tissues from severely affected Gaucher disease patients.[69-71] The sphingosyl moiety of these substrates varies between 18 and 22 carbons in length. The C_{18}-sphingosine is in the greatest abundance.[69-71] Glucosylsphingosine and its N-alkylated derivatives have a positive charge on the C2 nitrogen of sphingosine and are hydrolyzed at 25- to 100-fold lesser rates than glucosylceramides.[72,73] Maximal hydrolytic rates of glucosylceramide were obtained with fatty acid acyl chains of 8 to 18 carbon atoms. These rates were about 10-fold greater than those found with the formoyl derivative. The length of the fatty acid acyl chain has little effect on the apparent K_M values.[74] Sarmientos et al.[59] showed that the D-glucosyl-D-*erythro*-ceramides had somewhat lower $K_M$app values and higher V_{max} values than the corresponding D-glucosyl-L-*threo* derivatives. The effects of variations in the sphingosyl chain length on GCase activity are not known in detail either *in vitro* or *in vivo*. Such variation in glucosylceramide in various tissues could lead in part to differential effects on organ involvement if GCase variants have altered substrate specificities.

The pH optimum for hydrolysis has varied somewhat with the composition of the assay mixtures, i.e., micellar, phospholipid dispersions, or liposomal. The V_{max} profiles for the glucosylceramide or 4MU-Glc substrates conformed to a diprotic model with a monoprotonated active form. The ionizable groups had pK_a values of about 4.7 and 6.7 with a V_{max} optimum at pH~5.75. Murine, bovine, and human GCases are similar. In the presence of negatively charged phospholipids and saposin C, more acidic optima were obtained (see[55,75] for review).

Inhibitor Studies

With the increasing interest in molecular chaperonins for therapy, understanding the interactions of inhibitors at the active site is important to moving this field forward. Extensive work by Grabowski and co-workers[55,62] and others[59,76] is summarized below. A kinetic model of the active site of GCase would include binding sites for the glycon head group, the sphingosyl moiety, and the fatty acid acyl chain of glycosyl ceramide. Interaction at the glycon binding site requires a specific bonding with each hydroxyl group on glucose, while the sphingosyl and fatty acid acyl binding sites would accommodate hydrophobic chains up to about 18 carbon atoms in length. These latter sites are proposed to consist of multiple subsites that accommodate single methylene groups on the alkyl chains of substrates and inhibitors. This model is similar to those proposed for the binding of globotriaosylceramide to saposin B[77,78] and phospholipids to phospholipid transfer protein.[79] Although additional supporting structural data are required for these models, they may serve as prototypes for binding sites on lipid transfer proteins and partial or complete active sites of glycosphingolipid hydrolases. Co-crystallization of inhibitors with varying chain lengths with GCase are needed to more fully determine the structure of the substrate binding site(s) and the conformation of the active site (see below).

Compared with N-alkyl-deoxynojirimycins, the corresponding glucosylamines were 20- to 100-fold more potent inhibitors with overall inhibitory constants of

~100–150 pM.[80] Thus, the movement of the nitrogen from the 5 position in the pyranose ring (deoxynojirimycins) to the anomeric carbon (β-glucosylamine) altered the nature of the inhibition, i.e., the alkyl-β-glucosylamines were slow tight-binding inhibitors whereas the N-alkyl-deoxynojirimycins were rapidly reversible.[80] Analysis of these progress curves indicated that these extremely potent inhibitors induced an isomerization of the enzyme after the formation of the initial collision complex. The development of conformation specific alkyl-glycon transition statelike analogues should allow for more accurate descriptions of the kinetic interactions at the active site.

Molecular Enzymology of GCase

Alleles encoding the N370S mutant enzyme are quite common in nonneurono-pathic Gaucher disease (type 1) patients. The presence of such alleles in affected patients appears to be protective against the development of nervous system disease.[81,82] The N370S variant has characteristic kinetic abnormalities that include a diminished binding of the transition state analogues.[42,43,51,83–85] Because of this, N370S has been the target of attempts to "activate" the residual enzyme using molecular chaperonins.[86]

We undertook detailed mechanistic analyses of the role of N370 in the catalytic cycle.[37] By placing multiple substitutions at the N370 residue, selected kinetic abnormalities were shown to be associated uniquely with the N370S substitution. These results imply that, the local conformational effects prescribe specific alignments in the active site that are modified by the assay environment, and indicate a substantial effect on the catalytic cycle. Examination of the active site glucosylation and deglucosylation steps suggest that both components are abnormal in the catalytic cycle.[37] These functional effects might not be expected from the proposed location of N370 outside of the active site pocket obtained from the initial crystal structure.[52,53]

For all mutant enzymes found in Gaucher disease patients, the $K_m$app values have been normal or mildly changed.[42,43,51,87] For Michaelis-Menten kinetics, the $K_m$app approximates K_s, the thermodynamic binding constant. However, the discrepancy between selected $K_i$app (2- to 5-fold increase) for potent competitive inhibitors, and the $K_m$app for substrates indicates that for the N370S mutant GCase, $K_m$app is a kinetic constant, rather that a thermodynamic constant. This implies that the $K_i$app values obtained with enzymes containing specific substituted residues at 370 more likely reflects active site interaction than do the $K_m$app values. These above studies provide guidance for interpretation of crystallization to evaluate effects of the N370S substitution on GCase function.

Studies presented here and additional studies with mutant enzymes in combination with refined crystal structures of active GCases are providing greater insight into the potential for modeling GCases and their evolution to more active functional enzymes for therapy. Of importance is the lack of significant polymorphic variation of human GCases. The rare allele encoding E326K is the only variant described to date that is associated with a normal phenotype when in *trans* to another disease allele.[28] All other mutations, albeit found in affected individuals, lead to highly defective enzymes (see Chapter 2 for a mutation list). Studies shown here with variously substituted residues at 370 also indicate that it will be difficult to obtain GCase variants that retain or have enhanced function, and that only selected mutants with globally abnormal folding and/or normal catalytic mechanisms would be amenable to chaperone approaches to correction. The mutations in Gaucher disease are spread throughout the 3-D space occupied by GCase and indeed involve all the domains predicted from the reported[52] and our[89] crystal structures of GCase. These results suggest rather global changes in the enzyme function or proteolytic stability will be obtained by most natural or designed mutations.

COMPARATIVE STUDIES OF VARIANT GCASES

A comparison of GCases from seven mammalian species indicates little variation in the sequences at the amino acid level (Figure 3.1). Indeed, less than 15% variation is found between human to chimpanzee, pongo, dog, pig, mouse, and rat GCases. The greatest difference is a stretch of four amino acids between human and mouse. In addition, none of the known 200+ Gaucher disease mutations are duplicated in any of these mammalian species. Extremely strong conservation is present for the amino acid sequence, location and sequence of the N-glycosylation consensus sequences, and the areas surrounding the acid/base *(E235)* and the nucleophile *(E340)* in catalysis. At more distant evolutionary relationships, comparison of human GCase with that from Fugu, the last common ancestor of the vertebrates, shows ~45% amino acid identity with preservation of acid/base and nucleophile sequences. Significant stretches of amino acids differ between human and Fugu. However, when placed on the crystal structure,[52,53] Fugu and other species show preservation of remarkable degrees of homology (Figure 3.1). This is also true at more distant evolutionary distances as shown by comparing human GCase and those from honey bee and *C. elegans* (Figure 3.3). These enzymes show less than 30% amino acid identity, but with some minor gaps there remains substantial domain preservation. The central area of conservation is similar to that between human and *C. elegans* and includes domain 2 that contains the catalytic center. Clear distinctions between those sequences that are conserved and those that are nonconserved indicate that about half of GCase is heavily conserved specifically around the active site structure. The placement of the first four cysteines and the sequences surrounding these cysteines are essentially identical. In comparison, the cysteines at 126 and 248 are preserved in all mammalian species but only 126 is present in Fugu. Neither of these two cysteines is preserved in most distantly related honey bee and *C. elegans*. In contrast, cysteine 342, near the catalytic nucleophile, E340, is strictly conserved across all species from human to worm. When compared to the crystal structure of β-xylanase from *Erwinia chrysanthemi*, human GC shows a highly similar structure with many areas of direct overlap in 3-D space, even though the linear sequences are >80% nonidentical. The active site nucleophile and acid/base from human GCase directly superimpose over E253 and E165 of *E. chrysanthemi*, respectively. These findings indicate a highly conserved 3-D structure at the active site over great evolutionary distances (500 M years).

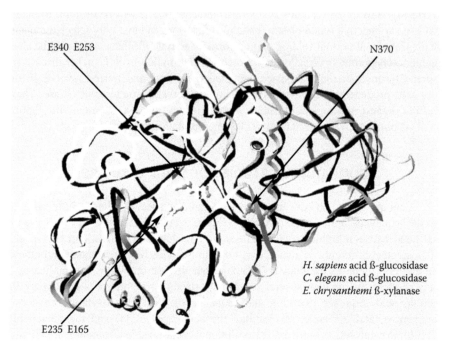

E340 E253

N370

H. sapiens acid ß-glucosidase
C. elegans acid ß-glucosidase
E. chrysanthemi ß-xylanase

E235 E165

Figure 3.3   Superimposition of the crystal structures of human GCase[52,53] and *Erwinia chrysanthemi* β-xylanase.[88] The GCase structure is represented in black and gray, and
that for β-xylanase is in white. The gray represents homologous primary amino
acid segments between GCases from human and *C. elegans*. The black depicts
amino acid sequences that are not homologous between GCases of these two
species. E235 and E340 are the acid/base and nucleophile in catalysis for human
GCase; these are conserved between human, *C. elegans* and *Erwinia chrysanthemi*. E165 and E253 are the respective counterparts in *Erwinia chrysanthemi*
β-xylanase (white). Interestingly, large sections of domain 2 are homologous
between human GCase and *E. chrysanthemi* β-xylanase but not the GCases from
human and *C. elegans*. This later finding is also true for the GCase from Fugu.
Overall, in the regions surrounding the active site, the structure is remarkably
preserved.

To evaluate the importance of the cysteine residues in human GCase, each of
the cysteines was individually mutated to the isosteric serine and expressed in the
baculovirus system. The Cys to Ser changes at positions 4, 16, 19, and 26 individually
led to catalytically inactive proteins that were relatively stable to proteolysis. Mutations at Cys 126 or 248 to Ser produced enzymes with significant catalytic activity;
C126S was a normally active enzyme. However, C342G was a catalytically inactive
enzyme, suggesting this residue's function related to the nearby catalytic nucleophile.
This cysteine is preserved throughout phylogeny from human to *C. elegans*. Two
particular amino acid changes between human and Fugu GCases also are of interest:
1) an Asp for Asn at 370. This change also occurs in *C. elegans*, mosquito, and
drosophila. When N370D was created in the human GCase, a catalytically active
enzyme was produced that under specific reaction conditions led to an increased
activity enzyme.[37] 2) At human GCase amino acid 90, a Gaucher disease mutation,

an A90T, is mimicked in Fugu GCase, suggesting that this mutation may not cause major abnormalities in the enzyme, although this remains to be tested.

An extensive study of over 40 mutations found in patients with Gaucher disease was undertaken to understand the nature of the catalytic and/or stability defects resulting from these mutations.[22] In sum, three types of abnormalities were found when the human GCases bearing these mutations were expressed in the baculovirus system. Group 1 contained N370S, V394L, and I161N that produced abnormal active site function with decreased binding of specific transition state analogs and altered activation (enhanced) by phosphatidylserine. Group 2 enzymes had about 10–20% of the intrinsic catalytic activity relative to wild-type, but were likely to be unstable to proteolysis in human cells. Group 3 mutations led to the production of stable GCase proteins that were catalytically inactive. These were at residues not involved in the catalytic site, i.e., M123V, K198E, M416V, and R463P.[89]

The importance of understanding the effects of Gaucher disease mutations on catalytic function of the enzyme relates specifically to the growing field of molecular chaperones for the preservation and indeed the enhancement of residual catalytic activity of some enzymes. Such studies of mutant enzymes may provide insight into whether or not specific chaperones would need to be developed for very specific mutations or whether generalized chaperones would be available for many of the mutations. Continuing characterization of the kinetic and *ex vivo* stability of various Gaucher disease mutations will assume greater importance if more general molecular chaperonins are to be developed for enhancement of existing residual activity in live cells of affected individuals. Importantly, N-alkyl-amino- or -imino sugars (β-glucosylamines, 5-deoxy-5-imino-glucose [nojirimycins]) have some properties of transition state analogues and have had small effects (1.5- to 1.7-fold) as chaperonins on the N370S enzyme[86] (this has been reproduced in this laboratory), no effect on the L444P enzyme, and very large enhancements of the mouse V394L enzyme in cultured skin fibroblasts.[22] Such studies suggest that the overall effect of the chaperonin might be greatest on enzymes that retain much of the wild-type conformation and stability.

CONCLUSIONS

For a simple, lysosomal hydrolase involved in a lysosomal storage disease, GCase has provided a wealth of biological data important to the molecular biology of glucosphingolipids and to understanding the mechanisms of reactions of lipases and their cofactors. The identification of mutations in patients with Gaucher disease has provided the impetus to study mutant GCases, which in, conjunction with modern biochemical techniques, provide heretofore unavailable insight into this enzyme's activity in the normal and mutant state. Continued elucidation of this enzyme's role, the need and functions of its cofactor, saposin C, and of the molecular enzymology involved with its catalytic activity should provide a rich and fertile resource of general reaction mechanisms for these enzyme groups.

REFERENCES

1. Beutler, E. and Grabowski, G.A., Gaucher disease, in *The Metabolic and Molecular Bases of Inherited Diseases*, 8th ed., Scriver, C.R., Beaudet, A.L., Sly, W.S., and Valle, D. Eds., McGraw-Hill, New York, 2001, 3635.
2. Ponce, E. et al., Temporal and spatial expression of murine acid β-glucosidase mRNA, *Mol. Genet. Metab.*, 74, 426, 2001.
3. Wong, K. et al., Neuropathology provides clues to the pathophysiology of Gaucher disease, *Mol. Genet. Metab.*, 82, 192, 2004.
4. Kaye, E.M. et al., Type 2 and type 3 Gaucher disease: a morphological and biochemical study, *Ann. Neurol.*, 20, 223, 1986.
5. Xu, Y.H., Busald, C., and Grabowski, G.A., Reconstitution of TCP80/NF90 translation inhibition activity in insect cells, *Mol. Genet. Metab.*, 70, 106, 2000.
6. Xu, Y.H. and Grabowski, G.A., Molecular cloning and characterization of a translational inhibitory protein that binds to coding sequences of human acid β-glucosidase and other mRNAs, *Mol. Genet. Metab.*, 68, 441, 1999.
7. Xu, Y.H. and Grabowski, G.A., Translational inefficiency of acid β-glucosidase mRNA in transgenic mammalian cells, *Mol. Genet. Metab.*, 64, 87, 1998.
8. Sorge, J.A. et al., The human glucocerebrosidase gene has two functional ATG initiator codons, *Am. J. Hum. Genet.*, 41, 1016, 1987.
9. Sorge, J. et al., Molecular cloning and nucleotide sequence of human glucocerebrosidase cDNA, *Proc. Natl. Acad. Sci. U.S.A.*, 82, 7289, 1985.
10. Edmunds, T., Glucocerebrosidase: Ceredase and Cerezyme, in *Directory of Therapeutic Enzymes*, Walsh, M. Ed., Taylor & Francis, Boca Raton, FL, 2005.
11. Berg-Fussman, A. et al., Human β-glucosidase: N-glycosylation site occupancy and the effect of glycosylation on enzymatic activity, *J. Biol. Chem.*, 268, 14861, 1993.
12. Grace, M.E. and Grabowski, G.A., Human acid β-glucosidase: Glycosylation is required for catalytic activity, *Biochem. Biophys. Res. Commun.*, 168, 771, 1990.
13. Takasaki, S. et al., Structure of the N-asparagine-linked oligosaccharide units of human placental β-glucocerebrosidase, *J. Biol. Chem.*, 259, 10112, 1984.
14. Fiete, D., Beranek, M.C. and Baenziger, J.U., The macrophage/endothelial cell mannose receptor cDNA encodes a protein that binds oligosaccharides terminating with SO4-4-GalNAcb1,4GlcNAcb or Man at independent sites, *Proc. Natl. Acad. Sci. U.S.A.*, 94, 11256, 1997.
15. Pontow, S.E., Kery, V., and Stahl, P.D., Mannose receptor, *Int. Rev. Cytol.*, 137, 221, 1997.
16. Lee, S.J. et al., Mannose receptor-mediated regulation of serum glycoprotein homeostasis, *Science*, 295, 1898, 2002.
17. Bergmann, J.E. and Grabowski, G.A., Posttranslational processing of human lysosomal acid β-glucosidase: a continuum of defects in Gaucher disease type 1 and type 2 fibroblasts, *Am. J. Hum. Genet.*, 44, 741, 1989.
18. Erickson, A.H., Ginns, E.I., and Barranger, J.A., Biosynthesis of the lysosomal enzyme glucocerebrosidase, *J. Biol. Chem.*, 260, 14319, 1985.
19. Sun, Y., Qi, X., and Grabowski, G.A., Saposin C is required for normal resistance of acid β-glucosidase to proteolytic degradation, *J. Biol. Chem.*, 278, 31918, 2003.
20. Salvioli, R. et al., Glucosylceramidase mass and subcellular localization are modulated by cholesterol in Niemann-Pick disease type C, *J. Biol. Chem.*, 279, 17674, 2004.
21. Leonova, T. and Grabowski, G.A., Fate and sorting of acid β-glucosidase in transgenic mammalian cells, *Mol. Genet. Metab.*, 70, 281, 2000.

22. Grabowski, G.A., Unpublished observation, 2005.
23. Liu, C. et al., Long-term expression and secretion of human glucocerebrosidase by primary murine and human myoblasts and differentiated myotubes, *J. Mol. Med.*, 76, 773, 1998.
24. Beutler, E., Kuhl, W., and Sorge, J., Cross-reacting material in Gaucher disease fibroblasts, *Proc. Natl. Acad. Sci. U.S.A.*, 81, 6506, 1984.
25. Beutler, E., Kuhl, W., and Sorge, J., Glucocerebrosidase "processing" and gene expression in various forms of Gaucher disease, *Am. J. Hum. Genet.*, 37, 1062, 1985.
26. Martin, B.M. et al., Glycosylation and processing of high levels of active human glucocerebrosidase in invertebrate cells using a baculovirus expression vector, *DNA* 7, 99, 1988.
27. Zimmer, K.P. et al., Intracellular transport of acid β-glucosidase and lysosome-associated membrane proteins is affected in Gaucher's disease (G202R mutation), *J. Path.*, 188, 407, 1999.
28. Zhao, H. et al., Gaucher's disease: identification of novel mutant alleles and genotype-phenotype relationships, *Clin. Genet.*, 64, 57, 2003.
29. Schmitz, M. et al., Impaired trafficking of mutants of lysosomal glucocerebrosidase in Gaucher's disease, *Int. J. Biochem. Cell. Biol.*, 37, 2310, 2005.
30. Ron, I. and Horowitz, M., ER retention and degradation as the molecular basis underlying Gaucher disease heterogeneity, *Hum. Mol. Genet.*, 14, 2387, 2005.
31. Bijsterbosch, M.K. et al., Quantitative analysis of the targeting of mannose-terminal glucocerebrosidase. Predominant uptake by liver endothelial cells, *Eur. J. Biochem.*, 237, 344, 1996.
32. Xu, Y.H. et al., Turnover and distribution of intravenously administered mannose-terminated human acid β-glucosidase in murine and human tissues, *Pediatr. Res.*, 39, 313, 1996.
33. Murray, G.J. and Jin, F-S., Immunoelectron microscopic localization of mannose-terminal glucocerebrosidase in lysosomes of rat liver Kupffer cells, *J. Histochem. Cytochem.*, 43, 149, 1995.
34. Horowitz, M. et al., The human glucocerebrosidase gene and pseudogene: structure and evolution, *Genomics*, 4, 87, 1989.
35. Ginns, E.I. et al., Gene mapping and leader polypeptide sequence of human glucocerebrosidase: implications for Gaucher disease, *Proc. Natl. Acad. Sci. U.S.A.*, 82, 7101, 1985.
36. Miao, S. et al., Identification of Glu340 as the active site nucleophile in human glucocerebrosidase by use of electrospray tandem mass spectrometry, *J. Biol. Chem.* 269, 10975, 1994.
37. Liou, B., Mechanistic studies of normal and mutant acid β-glucosidases, M.S., University of Cincinnati, 1996.
38. Maret, A. et al., Modifications of the molecular weight of membrane-bound nonspecific β-glucosidase in type 1 Gaucher disease determined *in situ* by the radiation inactivation method, *Biochim. Biophys. Acta.*, 799, 91, 1984.
39. Maret, A. et al., Modification of subunit interaction in membrane-bound acid β-glucosidase from Gaucher disease, *FEBS Lett.*, 160, 93, 1983.
40. Salvayre, R. et al., Comparative properties of the three groups of splenic β-glucosidase molecular forms, *Prog. Clin. Biol. Res.*, 95, 443, 1982.
41. Maret, A. et al., Properties of the molecular forms of β-glucosidase and β-glucocerebrosidase from normal human and Gaucher disease spleen. *Eur. J. Biochem.*, 115, 455, 1981 (in French, author's translation).

42. Grace, M.E. et al., Analyses of catalytic activity and inhibitor binding of human acid β-glucosidase by site-directed mutagenesis. Identification of residues critical to catalysis and evidence for causality of two Ashkenazi Jewish Gaucher disease type 1 mutations, *J. Biol. Chem.* 265, 6827, 1990.

43. Grace, M.E. et al., Analysis of human acid β-glucosidase by site-directed mutagenesis and heterologous expression, *J. Biol. Chem.*, 269, 2283, 1994.

44. Paal, K., Ito, M., and Withers, S.G., Paenibacillus sp. TS12 glucosylceramidase: kinetic studies of a novel sub-family of family 3 glycosidases and identification of the catalytic residues, *Biochem. J.*, 378, 141, 2004.

45. Kempton, J.B. and Withers, S.G., Mechanism of Argobacterium β-glucosidase: kinetic studies, *Biochemistry*, 31, 9961, 1992.

46. Withers, S.G. et al., Mechanistic consequences of mutation of the active site nucleophile Glu358 in Argobacterium β-glucosidase, *Biochemistry*, 31, 9979, 1992.

47. Grace, M.E. et al., Non-pseudogene-derived complex acid β-glucosidase mutations causing mild type 1 and severe type 2 gaucher disease, *J. Clin. Invest.*, 103, 817, 1999.

48. Amaral, O. et al., Gaucher disease: expression and characterization of mild and severe acid β-glucosidase mutations in Portuguese type 1 patients, *Eur. J. Hum. Genet.*, 8, 95, 2000.

49. Henrissat, B. et al., Conserved catalytic machinery and the prediction of a common fold for several families of glycosyl hydrolases, *Proc. Natl. Acad. Sci. U.S.A.*, 92, 7090, 1995.

50. Ohashi, T. et al., Characterization of human glucocerebrosidase from different mutant alleles, *J. Biol. Chem.*, 266, 3661, 1991.

51. Grace, M.E. et al., Gaucher disease: heterologous expression of two alleles associated with neuronopathic phenotypes, *Am. J. Hum. Genet.*, 49, 646, 1991.

52. Dvir, H. et al., X-ray structure of human acid-β-glucosidase, the defective enzyme in Gaucher disease, *EMBO*, Rep 4, 704, 2003.

53. Premkumar, L. et al., X-ray structure of human acid β-glucosidase covalently bound to conduritol-B-epoxide. Implications for Gaucher disease, *J. Biol. Chem.*, 280, 23815, 2005.

54. Davies, G.J., Ducros, V.M-A., and Zechel, D.I., Mapping the conformational itinerary of b-glycosidase by x-ray crystallography, *Biochem. Soc. Trans.*, 31, 523, 2003.

55. Grabowski, G.A., Gatt, S., and Horowitz, M., Acid β-glucosidase: enzymology and molecular biology of Gaucher disease, *Crit. Rev. Biochem. Mol. Biol.*, 25, 385, 1990.

56. Ho, M.W. et al., Glucocerebrosidase: reconstitution of activity from macromolecular components, *Biochem. J.*, 131, 173, 1973.

57. Dale, G.L. et al., Large scale purification of glucocerebrosidase from human placentas, *Birth Defects*, 16, 33, 1980.

58. Berent, S.L. and Radin, N.S., Mechanism of activation of glucocerebrosidase by co-β-glucosidase (glucosidase activator protein), *Biochim. Biophys. Acta.*, 664, 572, 1981.

59. Sarmientos, F., Schwarzmann, G., and Sandhoff, K., Specificity of human glucosylceramide β-glucosidase towards synthetic glucosphingolipids inserted into liposomes. Kinetic studies in a detergent-free assay system, *Eur. J. Biochem.*, 160, 527, 1986.

60. Christomanou, H. et al., Activator protein deficient Gaucher's disease. A second patient with the newly identified lipid storage disorder, *Klin Wochenschr*, 67, 999, 1989.

61. Christomanou, H., Aignesberger, A., and Linke, R.P., Immunochemical characterization of two activator proteins stimulating enzymic sphingomyelin degradation *in vitro*. Absence of one of them in a human Gaucher disease variant, *Biol. Chem. Hoppe. Seyler.*, 367, 879, 1986.

62. Qi, X. and Grabowski, G.A., Acid β-glucosidase: intrinsic fluorescence and conformational changes induced by phospholipids and saposin C, *Biochemistry*, 37, 11544, 1998.

63. Qi, X. and Grabowski, G.A., Molecular and cell biology of acid β-glucosidase and prosaposin, *Prog. Nucl. Acid. Res. Mol. Biol.*, 66, 203, 2001.

64. Vaccaro, A.M. et al., pH-Dependent conformational properties of saposins and their interations with phospholipid membranes, *J. Biol. Chem.*, 270, 30576, 1995.

65. Qi, X. et al., Functional organization of saposin C: definition of the neurotrophic and acid β-glucosidase activation regions, *J. Biol. Chem.*, 271, 6874, 1996.

66. Vaccaro, A.M. et al., Saposin C induces pH-dependent destabilization and fusion of phosphatidylserine-containing vesicles, *FEBS Lett.* 349, 181, 1994.

67. Wilkening, G., Linke, T., and Sandhoff, K., Lysosomal degradation on vesicular membrane surfaces. Enhanced glucosylceramide degradation by lysosomal anionic lipids and activators, *J. Biol. Chem.*, 273, 30271, 1998.

68. Sarmientos, F., Schwarzmann, G., and Sandhoff, K., Direct evidence by carbon-13 NMR spectroscopy for the erythro configuration of the sphingoid moiety in Gaucher cerebroside and other natural sphingolipids, *Eur. J. Biochem.*, 146, 59, 1985.

69. Nilsson, O. and Svennerholm, L., Accumulation of glucosylceramide and glucosyl-sphingosine (psychosine) in cerebrum and cerebellum in infantile and juvenile Gaucher disease, *J. Neurochem.*, 39, 709, 1982.

70. Nilsson, O. and Svennerholm, L., Characterization and quantitative determination of gangliosides and neutral glycosphingolipids in human liver, *J. Lipid Res.*, 23, 327, 1982.

71. Nilsson, O. et al., The occurrence of psychosine and other glycolipids in spleen and liver from the three major types of Gaucher's disease, *Biochim. Biophys. Acta.*, 712, 453, 1982.

72. Vaccaro, A.M., Kobayashi, T., and Suzuki, K., Comparison of synthetic and natural glucosylceramides as substrate for glucosylceramidase assay, *Clin. Chim. Acta.*, 118, 1, 1982.

73. Vaccaro, A.M., Muscillo, M., and Suzuki, K., Characterization of human glucosyl-sphingosine glucosyl hydrolase and comparison with glucosylceramidase, *Eur. J. Biochem.*, 146, 315, 1985.

74. Osiecki-Newman, K. et al., Human acid β-glucosidase: use of inhibitors, alternative substrates and amphiphiles to investigate the properties of the normal and Gaucher disease active sites, *Biochim. Biophys. Acta.*, 915, 87, 1987.

75. Grabowski, G.A., Gaucher disease: gene frequencies and genotype/phenotype correlations, *Genet. Test.*, 1, 5, 1997.

76. Sarmientos, F., Schwarzmann, G., and Sandhoff, K., Specificity of human glucosyl-ceramide β-glucosidase towards synthetic glucosylsphingolipids inserted into liposomes. Kinetic studies in a detergent-free assay system, *Eur. J. Biochem.*, 160, 527, 1986.

77. Fluharty, C.B. et al., Comparative lipid binding study on the cerebroside sulfate activator (Saposin B), *J. Neurosci. Res.*, 63, 82, 2001.

78. Ahn, V.E. et al., Crystal structure of saposin B reveals a dimeric shell for lipid binding, *Proc. Natl. Acad. Sci. U.S.A.*, 100, 38, 2003.

79. Van Paridon, P.A., Visser, A.J., and Wirtz, K.W., Binding of phospholipids to the phosphatidylinositol transfer protein from bovine brain as studied by steady-state and time-resolved fluorescence spectroscopy, *Biochim. Biophys. Acta.*, 898, 172, 1987.
80. Greenberg, P. et al., Human acid β-glucosidase: use of sphingosyl and N-alkyl-glucosylamine inhibitors to investigate the properties of the active site, *Biochim. Biophys. Acta.*, 1039, 12, 1990.
81. Theophilus, B. et al., Gaucher disease: molecular heterogeneity and phenotype-genotype correlations, *Am. J. Hum. Genet.*, 45, 212, 1989.
82. Latham, T. et al., Complex alleles of the acid β-glucosidase gene in Gaucher disease, *Am. J. Hum. Genet.*, 47, 79, 1990.
83. Gonzales, M.L. et al., Activation of human spleen glucocerebrosidases by monoacylglycol sulfates and diacylglycerol sulfates, *Arch. Biochem. Biophys.*, 262, 345, 1988.
84. Basu, A. et al., Activators of spleen glucocerebrosidase from controls and patients with various forms of Gaucher's disease, *J. Biol. Chem.*, 259, 1714, 1984.
85. Glew, R.H. et al., Enzymic differentiation of neurologic and nonneurologic forms of Gaucher's disease, *J. Neuropathol. Exp. Neurol.*, 41, 630, 1982.
86. Sawkar, A.R. et al., Chemical chaperones increase the cellular activity of N370S β-glucosidase: a therapeutic strategy for Gaucher disease, *Proc. Natl. Acad. Sci. U.S.A.*, 99, 15428, 2002.
87. Osiecki-Newman, K. et al., Human acid β-glucosidase: inhibition studies using glucose analogues and pH variation to characterize the normal and Gaucher disease glycon binding sites, *Enzyme*, 40, 173, 1988.
88. Bedarkar, S. et al., Crystallization and preliminary x-ray diffraction analysis of the catalytic domain of Cex, an exo-β-1,4-glucanase and β-1, 4-xylanase from the bacterium Cellulomonas fimi, *J. Mol. Biol.*, 228, 693, 1992.
89. Liou, B. et al., Analyses of variant acid β-glucosidases: effects of Gaucher disease mutaitons, *J. Biol. Chem.*, 281, 4242, 2006.

CHAPTER 4

Saposin C and Other Sphingolipid Activator Proteins

Silvia Locatelli Hoops, Thomas Kolter, and Konrad Sandhoff

CONTENTS

INTRODUCTION

Sphingolipid Activator Proteins are essential for lysosomal glycosphingolipid degradation. Glycosphingolipids (GSLs) are composed of a hydrophobic membrane anchor, to which an oligosaccharide head group of one to about ten sugar residues can be attached. Together with other membrane compounds, lipids and proteins, GSLs are continuously degraded in the acidic compartments of the cells, the endosomes and the lysosomes. There, the different macromolecules are cleaved into their building blocks, which are able to leave the lysosome.

According to a recent model of endocytosis and membrane digestion[1] (Figure 4.1b), the degradation of plasma membrane-derived lipids occurs on the surface of intra-endosomal and intra-lysosomal vesicles. Many observations support this model.[2,3]

During the process of endocytosis, the lipid composition of the involved membranes changes drastically. Internal lysosomal membranes are characterized by low cholesterol levels and high levels of an unusual lysosomal lipid, BMP (bis [monoacylglycerol] phosphate).[2] Other anionic lipids like phosphatidylinositol and dolichol

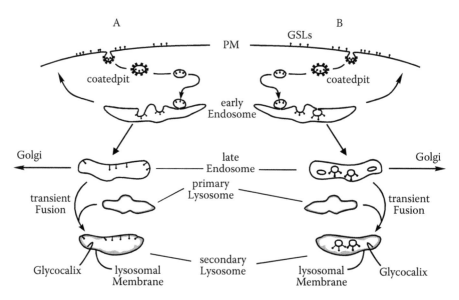

Figure 4.1 Models of endocytosis and lysosomal digestion of sphingolipids and glycosphingolipids (GSL) in the plasma membrane. A) Conventional model: the degradation of the GSLs of plasma membrane occurs selectively within the lysosomal membrane. B) Alternative model: invagination of those parts of the endosomal membranes enriched in components derived from the plasma membrane occurs. The so-formed intra-endosomal vesicles can become intra-lysosomal vesicles after successive processes of membrane fission and fusion. Thus, the glycoconjugates to be degraded face the lysosol on the outer leaflet of intra-lysosomal vesicles. In contrast to the lysosomal perimeter membrane, which is covered by a thick glycocalix, there is no glycocalix impairing the action of the hydrolases. PM: plasma membrane. B: GSL. (From Sandhoff, K. and Kolter, T., *Trends Cell. Biol.*, 6, 98, 1996. With permission.)

phosphate, albeit in smaller amounts than BMP, are also found within the lysosomal compartment.

Glycosphingolipid degradation starts with the sequential cleavage of monosaccharide residues from the nonreducing end of the oligosaccharide part of the glycosphingolipids by the action of acid exohydrolases. These enzymes are soluble in the lysosol (the lumen of the lysosome), whereas their lipid substrates are embedded in membrane structures. In the case of GSLs with short oligosaccharid chains of four or less sugars,[4] the lipids to be degraded are no longer accessible to their respective enzymes for steric reasons. The degradation of these GSLs requires not only the presence of hydrolytic enzymes, but of additional factors: Sphingolipid Activator Proteins (SAPs) and anionic lipids. SAPs mediate the interaction between the water-soluble enzyme and the membrane-bound lipid substrate. Until now, five SAPs are known: the GM2 activator protein (GM2AP) and the four saposins or Saps A, B, C, and D, which derive from a common protein precursor called pSap, Sap precursor or prosaposin. *In vivo*, the enzymatic hydrolysis of most membrane-bound sphingolipids is also stimulated by the anionic lysosomal phospholipid BMP and by other negatively charged lysosomal lipids.[4-8] Apart from the composition of the vesicles, additional requirements for degradation of glycolipids are given by vesicle size and the lateral pressure.

HISTORY OF SPHINGOLIPID ACTIVATOR PROTEINS

The discovery of the five SAPs known to date extended over three decades.[9] The first SAP, Sap-B, was discovered by Mehl and Jatzkewitz in 1964.[10] Sap-C was first isolated from the spleen of a patient affected by Gaucher disease in 1971.[11] It stimulates the activity of glucosylceramide-β-glucosidase toward natural and synthetic substrates of the enzyme.[6,12,13]

The existence of an essential cofactor for the degradation of the ganglioside GM2 by β-hexosaminidase A was postulated in 1977.[14] This was done on the basis that the brain of a human patient of amaurotic idiocy stored two gangliosides, GM2 and GA2, despite the fact that the enzymes required for the degradation of these lipids were present. Enzymes isolated from the patient were able to catalyse the cleavage of the storage material in the presence of a surfactant. The "endogenous detergent," absent in the tissues of the patient, (GM2-AP) could then be identified[15] and purified.[16] The elucidation of this AB variant of GM2 gangliosidosis demonstrated the physiological relevance of Sphingolipid Activator Proteins and established the basis for the analysis of further activator protein deficiencies.

In 1986, Fujibayashi and Wenger[17,18] showed that both Sap-B and Sap-C originate from a large precursor protein. By cloning and protein sequencing,[19,20] the mRNA of a precursor protein was identified that coded for both Sap-B and Sap-C, and in addition for a newly discovered one, Sap-D. Sap-A was the last activator protein discovered, which also originated from this common precursor protein prosaposin or pSap.[21-23] In this chapter we will focus on Sap-C, whose deficiency leads to an abnormal juvenile form of Gaucher disease,[24,25] and on its precursor.

SAPOSINS AND THEIR PRECURSOR

General Aspects of Saposins

The saposins A–D[9] are four homologous glycoproteins of 8–11 kDa, with iso-electric points of about 4.3. They consist of approximately 80 amino acids, occur in dimeric forms, and are not enzymatically active. With their six highly conserved cystein residues, all saposins form three disulfide bridges,[26] which make the activators heat-stable and protease-resistant. They also contain one conserved N-glycosylation site,[27] with the exception of Sap-A, which has an additional one.

The saposins belong to a family of proteins with different functions but similar properties, such as lipid-binding and membrane-perturbing properties.[28] These so-called saposin-like proteins (SAPLIP) are characterized by a common structural motif. Other members of this group are NK-Lysin,[29] acid sphingomyelinase, the pore forming peptide of *Entamoeba histolytica*,[30,31] the small subunit of human acyloxy-acylhydrolase, and the three Sap-like domains of pulmonary surfactant protein B precursor.[28]

Biosynthesis of Saposins: The Sap Precursor

Saposins are formed within the acidic compartment of the cell after proteolytical processing of a glycoprotein of 73 kDa called prosaposin or pSap (Figure 4.2).[19,22–32,33]

During the biosynthesis of prosaposin, the 53 kDa polypeptide chain is cotranslationally glycosylated at all five potential glycosylation sites to yield a 65 kDa molecule.[32] After modification of the sugar moieties in the Golgi apparatus, the complete precursor of 73 kDa is formed and can be intracellularly targeted to the lysosomes via the mannose-6-P-receptor pathway or via sortilin.[34,35] Alternatively, it can be first secreted to the extracellular space and subsequently endocytosed by the same or by neighboring cells. At least three receptors could be involved in the re-uptake process[36]: the mannose-6-phosphate receptor (M-6-P), the mannose receptor, and especially the low-density-lipoprotein-receptor-related-protein (LRP).

Proteolytic processing of the precursor protein to the mature saposins starts in the early endosomes. Although the four Saps share a high degree of homology and show similar properties, their specificities and modes of action are quiet different.

Saposin C

Biochemical Properties

Human Sap-C is a glycoprotein with an isoelectric point of pH 4.3 to 4.4.[37] It consists of a polypeptide chain of 80 amino acids. This corresponds to a molecular weight of 8.9 kDa.[38] One complex-type carbohydrate chain is N-linked on Asn 22.[38,39] The molecular mass depends on the degree of glycosylation and can reach 12 kDa, six cysteine residues[40] form three intramolecular disulfide bridges, the positions of

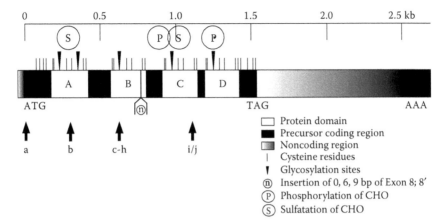

Figure 4.2   Structure of the prosaposin cDNA.[9] Prosaposin is encoded by a gene on chromosome 10. The cDNA of prosaposin codes for a sequence of 524, 526, or 527 amino acids, including a signal peptide of 16 amino acids and four domains corresponding to the mature Saps. The existence of three isoformes of the protein is a consequence of alternative splicing, which produce pSap mRNAs with all nine bases of exon 8, mRNA with only the last six bases, or mRNAs lacking them. The positions of the cysteine residues are marked by vertical bars and the positions of the N-glycosylation sites by arrow heads. CHO = carbohydrates. a,[92,93] b,[58] c,[100] d,[62,101] e,[102] f,[103,104] g,[105] h,[94] i,[25] and j[96] are the position of the known mutations leading to diseases. (From Fürst, W. and Sandhoff, K., *Biochim. Biophys. Acta*, 1126, 1, 1992. With permission.)

which have been identified by mass spectrometry.[41] The first cysteine is linked to the last one, the second to the second-last, and the third to the fourth, without intersection of the bridges.

Function

Saposin C is essential for the *in vivo* degradation of glucosylceramide by glucosylceramide-β-glucosidase in the lysosomal compartment (Figure 4.3).[11] The binding of saposin C to glucosylceramide-β-glucosidase not only stimulates the enzyme, but simultaneously protects it from proteolytic degradation inside the cell.[42] In addition, an allosteric stimulation of glucosylceramide-β-glucosidase toward lipid- and water-soluble synthetic substrates *in vitro* has also been described.[11] Sap-C also plays a role in the *in vivo* degradation of ceramide catalysed by acid ceramidase.[43]

In vitro, saposin C has been demonstrated to stimulate several reactions, e.g., the conversion of lactosylceramide to glucosylceramide by GM1-β-galactosidase.[44] Since saposin B can also mediate this reaction, the isolated deficiency of one of these Saps does not cause an accumulation of the respective glycosphingolipid. Galactosylceramide-β-galactosidase[45,46] and acid sphingomyelinase[24,45,46] are also stimulated by saposin C *in vitro*.

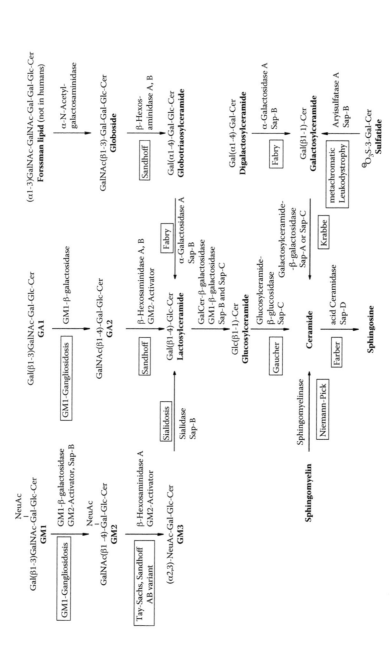

Figure 4.3 Degradation of selected sphingolipids in the lysosomes of the cells. The eponyms of individual inherited diseases (in boxes) are given. Activator proteins required for the respective degradation step *in vivo* are indicated. Variant AB, AB variant of GM2 gangliosidosis (deficiency of GM2-activator protein); SAP, sphingolipid-activator protein. (From Kolter, T. and Sandhoff, K., *Brain Pathol.*, 8, 79, 1998. With permission.)

Structure and Proposed Mechanism of Action

Sap-C adopts the saposin-fold common to the other members of the family. Its solution structure[47] consists of five α-helices, which form the half of a sphere by packing the first helix against the remaining four. The hydrophobic residues of the protein are contained within the core, whereas all charged amino acids, predominantly negative ones, are exposed to the solvent. In the presence of SDS, the protein adopts an open conformation with an exposed hydrophobic pocket. Its structure is very similar to a monomer in the saposin B homodimer structure.[48]

Saposin C participates in the degradation of glucosylceramide. Three components are needed in this process: the enzyme glucosylceramide-β-glucosidase, acidic phospholipids, and saposin C. The enzymatic reaction takes place at least in part on the membrane surface of intralysosomal membranes (Figure 4.4).

Acidic phospholipids, such as phosphatidic acid, phosphatidylserine, or phosphatidylinositol, efficiently activate glucosylceramide-β-glucosidase toward glucosylceramide or synthetic substrates even in the absence of Sap-C. They induce a conformational change of the enzyme, leading to the formation of enzyme aggregates.[6,7,49,50]

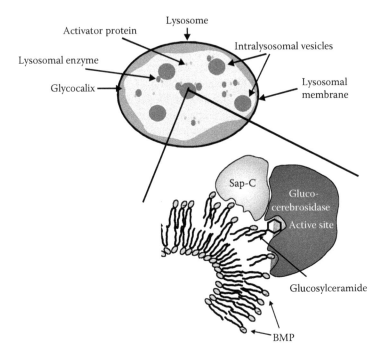

Figure 4.4 **(See color insert following page 304.)** Model for the degradation of membrane-bound glucosylceramide by glucosylceramide-β-glucosidase (glucocerebrosidase) and Sap-C. (From Wilkening, G., Linke, T., and Sandhoff, K., *J. Biol. Chem.*, 273, 30271, 1998. With permission.)

The binding of Sap-C to membranes is a reversible process that is triggered by a decrease of the pH. Under acidic conditions no conformational changes occur, but surface negative charges of Sap-C are partially neutralized,[47] allowing the interaction with the membrane. Even if the presence of acidic phospholipids has a positive effect on binding, saposin C can also associate with neutral phospholipid-containing vesicles.[51]

After binding, at least two processes take place: a direct interaction of saposin C with the enzyme,[6,11,13] and a reorganization of the membrane.

Acidic phospholipids and Sap-C seem to bind at different sites of glucosylceramide-β-glucosidase. After activation of the enzyme by these lipids, the binding of saposin C causes additional conformational changes in glucosylceramide-β-glucosidase, leading to an optimal hydrolytical activity. Sap-C was reported to form 1:1 complexes with glucosylceramide-β-glucosidase in the presence of phosphatidylserine.[6] Glucosylceramide-β-glucosidase activation sites and binding sites on Sap-C have been identified.[52] In this case, a Sap-C: glucosylceramide-β-glucosidase complex with a stoichiometry of 4:2 has been postulated.

Two types of membrane restructuring are induced by saposin C: formation of patchlike structural domains, and membrane destabilization.[53] The former process is accompanied by a thickness increase of the membrane and is independent of acidic phospholipids. The latter process causes a reduction of membrane thickness and depends on the presence and concentration of acidic phospholipids.[4,54] Destabilization of the membrane caused by Sap-C facilitates the association of glucosylceramide-β-glucosidase with the substrate. Saposin C has also been found to have fusogenic activity. The activator induces fusion of BMP-containing phospholipid vesicles and in this way might be involved in the regulation of multivesicular bodies formation in cells.[55] The fusogenic domain is located at the amino-terminal half of saposin C.[56] Interactions between glucosylceramide-β-glucosidase, Sap-C, and lysosomal phospholipids such as BMP seem to be further modulated by the cholesterol content of the membrane.[57]

Saposin A, B, and D

Sap-A is essential for the degradation of galactosylceramide by galactosylceramide-β-galactosidase, as demonstrated in a 6-month-old infant girl suffering from Krabbe disease.[58] In this case the deletion of a conserved valine of the Sap-A protein has been proposed to be the cause of abnormal galactosylceramide metabolism and abnormal myelination in the cerebral white matter. In addition, mice carrying a different mutation in the Sap-A domain of the Sap precursor accumulate galactosylceramide and exhibit a phenotype resembling a late-onset form of Krabbe disease.[59] *In vitro*, Sap-A also stimulates glucosylceramide-β-glucosidase by a similar mechanism, but to a lower extent than Sap-C,[60,61] probably in a synergistic manner.[13]

Sap-B has been shown to stimulate the hydrolysis of about 20 different naturally occurring or chemically modified glycolipids by 6 different enzymes of human, plant, or bacterial origin. Its action in the degradation process of sulfatide by arylsulfatase A has to be stressed, since the inherited defect of Sap-B leads to the storage of this lipid. The Sap-B deficiency manifests itself as an atypical form of metachromatic leukodystrophy, with late infantile or juvenile onset.[62] In addition, Sap-B is required for the degradation of globotriaosylceramide and digalactosylceramide in vivo,[63] as well as of ganglioside GM3 and lactosylceramide.[64] The mode of action of Sap-B can be compared to that of a physiological detergent. The large hydrophobic cavity enclosed by the shell-like homodimeric structure of Sap-B[65] would permit the interaction with the membrane-anchored substrate and its presentation to the water-soluble enzyme.

In vivo, Sap-D stimulates the lysosomal degradation of ceramide by acid ceramidase. This has been first demonstrated in prosaposin-deficient human fibroblasts where Sap-D reduces the amount of accumulated ceramide.[43] This was confirmed in saposin D deficient mice, which have been generated recently.[66] In this case, ceramides, particularly those containing hydroxy fatty acids, accumulate in the kidney and the brain of the animals. Until now, no human disease is known that is caused by the isolated deficiency of saposin D.

ADDITIONAL FUNCTIONS

Additional Functions of Prosaposin

Prosaposin and the mature Saps are differentially distributed in organs and tissues.[67] In addition, the Sap precursor has been found in several body fluids.[68,69] This suggests that prosaposin might have functions on its own.

Neurotrophic/Neuroprotective Properties — Prosaposin has been shown to stimulate neurite outgrowth and choline acetyltransferase activity in murine and human neuroblastoma cells[70] and to prevent cell death of neurons.[40,71] Several in vivo studies support these observations. For example, human prosaposin promoted regeneration of transsected sciatic nerves.[72] The region responsible for the neurotrophic/neuroprotective function of prosaposin has been localized in the amino-terminal region of the Sap-C domain.[73] A variety of prosaptides, synthetic peptides corresponding to the putative neurotrophic region, have been constructed. However, the investigation of the effect of one prosaptide in the validated rabbit spinal cord ischemia model showed that the prosaptide exacerbates isquemia-induced behavioral deficits.[74] Therefore, these properties of pSap have to be further investigated.

Myelinotrophic Properties — In mice and humans lacking prosaposin, myelin formation is deficient. In agreement with this observation, prosaposin, Sap-C, and prosaptides were able to increase sulfatide production in oligodendrocytes and in primary and transformed Schwann cells. Prosaposin might promote myelin lipid synthesis and prolong cell survival in these cell types.[75]

Myotrophic Properties — Prosaposin has been immunolocalized to skeletal, cardiac, and smooth muscle. Its amount apparently depends on the degree of innervation. A myoprotective activity of prosaposin has been suggested.[76] Nevertheless, in order to prove this hypothesis, so far only experiments using prosaptides have been performed.

Function in the Male Reproductive System — The rat sulfated glycoprotein I (SGP-1) is believed to be the equivalent of human prosaposin. SGP-1 is secreted as an uncleaved protein into the seminiferous tubule by Sertoli cells. Here, prosaposin might support sperm maturation by degradation of lipids in residual bodies and/or in the modification of membrane lipids during this process. In agreement with these observations, mice lacking prosaposin showed reduced spermiogensis and an under-developed male reproductive system.[77] Additional studies in human, boar, and bull sperm showed increased fertility after treatment with pSap.[78]

Function in the Skin — As a precursor of saposins, prosaposin is essential for forming the water permeability barrier of the skin (see below).

Role of Saposin C in the Skin

Saposin C contributes to the formation of the water permeability barrier of the skin, by stimulating the formation of complex ceramides from complex glucosylce-ramides by glucosylceramide-β-glucosidase. In addition, together with other Saps, it probably participates in the rearrangement of the extracellular stacks of lipid layers during the formation of the stratum corneum.[79]

The epidermis acts as a barrier that prevents transepidermal water loss of terres-trial animals. In particular, the integrity of the stratum corneum (SC), the outermost layer of the epidermis, is essential for maintaining the permeability barrier function of the skin. The SC consists of flattened cells called corneocytes embedded in an extracellular matrix of lipids and proteins. The lipid composition of this multilamel-lar matrix is given by long chain ceramides, free fatty acids and cholesterol. The ceramides can be generated either by *de novo* biosynthesis or by degradation of their precursors, e.g., glucosylceramide.[80] For this process, glucosylceramide-β-glucosi-dase and saposin C are essential requirements,[81,82] both of which are upregulated in the neonatal epidermis of mice.[83,84] Deficiency of this degrading system leads to a modified lipid composition in the SC-interstices so that they are not hydrophobic enough to form a functioning permeability barrier. In a neonatally type of Gaucher disease, the patients develop a defective skin phenotype (Collodian Baby) due to excessive epidermal water loss.[82,85] This demonstrates the importance of glucosyl-ceramide-β-glucosidase in the epidermis. The participation of saposins could be demonstrated analyzing the mouse model of prosaposin deficiency.[86] These mice showed ichthyotic skin phenotype with red and wrinkled skin (similar to the skin of the glucosylceramide-β-glucosidase deficient Gaucher mice), morphological alter-ations of the SC and of the interstice below it, increased amounts of glucosylcera-mides and protein-bound ω-hydroxyglucosylceramides, and a simultaneous decrease of ceramides, free and protein-bound hydroxy-fatty acids and free and protein-bound ω-hydroxyceramides. Accumulation of these lipids was also found in glucosylcera-mide-β-glucosidase deficient mice.[81]

In agreement with the known lipid transfer properties of Saps, they also seem to participate in the arrangement of the extruded lipids into smooth, linear lipid arrays, in the intercellular spaces at the stratum corneum-stratum granulosum interface.

Role of Saposin C in the Immune System

Saposin C plays a role in lipid antigen presentation by CD1b immune receptor molecules. It is now established that peptide as well as lipid antigens can be recognized by T lymphocytes. Whereas peptides are recognized in the context of major histocompatibility complex (MHC) molecules, lipids are presented to T cells by CD1 molecules.[87] The five different human CD1 proteins (CD1a-CD1e) are characterized by a unique hydrophobic cavity that binds lipid antigens in both the secretory and endosomal compartments. Posttranslationally they are probably loaded with nonantigenic phosphatidylinositol in the endoplasmic reticulum, which can be exchanged by other lipids in the acidic compartments of the cell. In the lysosomes, lipid antigens have first to be removed from inner membranes into which they are embedded and then transported to CD1 molecules through a hydrophilic environment. Therefore, the participation of lipid transfer proteins is an essential requirement for this process. There is evidence that sphingolipid activator proteins are required for this loading process. Human CD1b apparently requires especially Sap-C to present different types of antigens.[88] A model has been proposed in which Sap-C disturbs intralysosomal membranes and extracts lipids for loading on CD1b molecules, which itself can bind to Sap-C.[88]

In addition to CD1b, antigen presentation by human[89] and mouse CD1d[90] has been studied. In these cases other sphingolipid activator proteins are involved in the loading of the immune receptor. The state of our knowledge in this regard is incomplete. Further research is required to reach a better understanding of the mechanisms involved in this process.

SAPOSIN DEFICIENCIES

Deficiency of Prosaposin

The Sap-precursor deficiency is a fatal infantile storage disorder caused by the absence of all four saposins. The clinical picture of the first known case of prosaposin deficiency was similar to those of patients suffering from Gaucher disease, type 2. The disease is characterized by severe neurological symptoms, accompanied by hepatosplenomegaly. In all human patients, but also in the Sap-precursor knockout mouse,[9,86] there were defects at multiple sites in the sphingolipid catabolic pathway. The accumulated sphingolipids include ceramide, glucosylceramide, lactosylceramide, ganglioside GM3, galactosylceramide, sulfatides, digalactosylceramide, and globotriaosylceramide, together with an accumulation of intralysosomal membrane.[91] In prosaposin deficient fibroblasts, it could be demonstrated that treatment with human Sap-precursor completely reversed the storage.[3]

Until now, two different mutations in five human patients have been reported. These mutations lead to a complete deficiency of the whole Sap-precursor protein and consequently of all four Saps. A patient who died at the age of 16 weeks and his fetal brother,[91,92] as well as an infant with death at 17 weeks,[93] carried homoallelic mutation of the start codon. In the remaining patients, who belong to two unrelated families of the same district of eastern Slovakia, homoallelic deletion within the Sap-B domain led to a frame-shift and a premature stop codon.[94]

pSap deficiency can be diagnosed by immunochemical methods, by demonstration of a metabolic defect, or by performing processing studies using antisera against more than one Sap protein.[9]

Saposin C Deficiency

Sap-C deficiency (Gaucher activator deficiency) is a Gaucher-like disease with nearly normal glucosylceramide-β-glucosidase activity. Until now, only two individuals have been found to suffer from this kind of Gaucher disease.[24,95] In both cases, Sap-C deficiency was due to point mutations (see Figure 4.2) in the Sap-C domain of one of the pSap alleles, leading to substitution of the same cysteine residue by phenylalanine[25] and glycine.[96] Due to the loss of a disulfide bridge variant Sap-C molecules are probably more labile to proteolysis than the wild type proteins. The variant Sap precursor, however, was apparently normally synthesized and processed; therefore the other Saps were unaffected. Apparently in both patients no prosaposin mRNA was transcribed from the other allele. The resulting absence of Sap-C causes accumulation of glucosylceramide mainly in brain,[97] liver, and spleen. The level of glucosylceramide-β-glucosidase appeared to be normal.[95]

The clinical findings in Sap-C deficiency are similar to those in Gaucher disease, type 3, which is a juvenile visceromegalic and neuronopathic disease. The additional particular symptoms of the two Sap-C deficient patients known are discussed below.

The first patient described[24] started walking late and was never able to stand up from a sitting position. At age 4, she displayed abnormal speech and gait. Ocular pigmentation and skin pigmentation were irregular. Her muscles were weak and atrophic. She showed spasticity and seizures before death at 14 years of age. The second patient[95,97] developed normally until age 8 years when epileptic fits started. During the next years, the neurologic signs increased with myoclonic jerks, frequent generalized seizures, slurred speech, ocular apraxia and loss of motor and intellectual abilities. The boy died with bronchopneumonia in an epileptic status at age 15.5.

The diagnosis of Sap-C deficiency may be verified by an immunochemical analysis of extracts of tissue or cultured skin fibroblasts from the patient[24,95] or by molecular biology studies.[22]

Due to the pioneering work of R.O. Brady, a therapy of the adult form of Gaucher disease (type 1) caused by glucosylceramide-β-glucosidase deficiency is available. For the development of effective therapeutic approaches also for the treatment of activator protein deficiencies and other sphingolipidoses, a complete understanding of the lysosomal sphingolipid degradation is essential.

REFERENCES

1. Fürst, W. and Sandhoff, K., Activator proteins and topology of lysosomal sphingolipid catabolism, *Biochim. Biophys. Acta,* 1126, 1, 1992.

2. Möbius, W. et al., Recycling compartments and the internal vesicles of multivesicular bodies harbor most of the cholesterol found in the endocytic pathway, *Traffic*, 4, 222, 2003.
3. Burkhardt, J.K. et al., Accumulation of sphingolipids in SAP-precursor (prosaposin)-deficient fibroblasts occurs as intralysosomal membrane structures and can be completely reversed by treatment with human SAP-precursor, *Eur. J. Cell Biol.*, 73, 10, 1997.
4. Wilkening, G. et al., Degradation of membrane-bound ganglioside GM1. Stimulation by bis(monoacylglycero)phosphate and the activator proteins SAP-B and GM2-AP, *J. Biol. Chem.*, 275, 35814, 2000.
5. Werth, N. et al., Degradation of membrane-bound ganglioside GM2 by beta-hexosaminidase A. Stimulation by GM2 activator protein and lysosomal lipids, *J. Biol. Chem.*, 276, 12685, 2001.
6. Berent, S.L. and Radin, N.S., Mechanism of activation of glucocerebrosidase by co-beta-glucosidase (glucosidase activator protein), *Biochim. Biophys. Acta*, 664, 572, 1981.
7. Sarmientos, F., Schwarzmann, G., and Sandhoff, K., Specificity of human glucosylceramide beta-glucosidase towards synthetic glucosylsphingolipids inserted into liposomes. Kinetic studies in a detergent-free assay system, *Eur. J. Biochem.*, 160, 527, 1986.
8. Salvioli, R. et al., Further studies on the reconstitution of glucosylceramidase activity by Sap C and anionic phospholipids, *FEBS Lett.*, 472, 17, 2000.
9. Sandhoff K. et al., Sphingolipid activator proteins, in *The Metabolic and Molecular Bases of Inherited Disease,* Scriver, C.R., Beaudet, A.L., Sly, W.S., and Valle, D., Eds., McGraw-Hill, New York, 2004, chapter 134.
10. Mehl, E. and Jatzkewitz, H., Eine Cerebrosidsulfatase aus Schweineniere, *Hoppe Seylers Z. Physiol. Chem.*, 339, 260, 1964.
11. Ho, M.W. and O'Brien, J.S., Gaucher's disease: deficiency of 'acid'-glucosidase and reconstitution of enzyme activity *in vitro*, *Proc. Natl. Acad. Sci. U.S.A.*, 68, 2810, 1971.
12. Prence, E. et al., Further studies on the activation of glucocerebrosidase by a heat-stable factor from Gaucher spleen, *Arch. Biochem. Biophys.*, 236, 98, 1985.
13. Fabbro, D. and Grabowski, G.A., Human acid beta-glucosidase. Use of inhibitory and activating monoclonal antibodies to investigate the enzyme's catalytic mechanism and saposin A and C binding sites, *J. Biol. Chem.*, 266, 15021, 1991.
14. Sandhoff, K., The biochemistry of sphingolipid storage diseases, *Angew. Chem. Int. Ed. Engl.*, 16, 273, 1977.
15. Conzelmann, E. and Sandhoff, K., AB variant of infantile GM2 gangliosidosis: deficiency of a factor necessary for stimulation of hexosaminidase A-catalyzed degradation of ganglioside GM2 and glycolipid GA2, *Proc. Natl. Acad. Sci. U.S.A.*, 75, 3979, 1978.
16. Conzelmann, E. and Sandhoff, K., Purification and characterization of an activator protein for the degradation of glycolipids GM2 and GA2 by hexosaminidase A, *Hoppe Seylers Z. Physiol. Chem.*, 360, 1837, 1979.
17. Fujibayashi, S. and Wenger, D.A., Synthesis and processing of sphingolipid activator protein-2 (SAP-2) in cultured human fibroblasts, *J. Biol. Chem.*, 261, 15339, 1986.
18. Fujibayashi, S. and Wenger, D.A., Biosynthesis of the sulfatide/GM1 activator protein (SAP-1) in control and mutant cultured skin fibroblasts, *Biochim. Biophys. Acta*, 875, 554, 1986.

19. Fürst, W., Machleidt, W., and Sandhoff, K., The precursor of sulfatide activator protein is processed to three different proteins, *Biol. Chem. Hoppe Seyler,* 369, 317, 1988.

20. Furst, W. et al., The complete amino-acid sequences of human ganglioside GM2 activator protein and cerebroside sulfate activator protein, *Eur. J. Biochem.,* 192, 709, 1990.

21. Morimoto, S. et al., Saposin A: second cerebrosidase activator protein, *Proc. Natl. Acad. Sci. U.S.A.,* 86, 3389, 1989.

22. O'Brien, J.S. et al., Coding of two sphingolipid activator proteins (SAP-1 and SAP-2) by same genetic locus, *Science,* 241, 1098, 1988.

23. Nakano, T., Sandhoff, K., Stumper, J., Christomanou, H., and Suzuki, K., Structure of full-length cDNA coding for sulfatide activator, a Co-beta-glucosidase and two other homologous proteins: two alternate forms of the sulfatide activator, *J. Biochem. (Tokyo),* 105, 152 (1989).

24. Christomanou, H., Aignesberger, A., and Linke, R.P., Immunochemical characterization of two activator proteins stimulating enzymic sphingomyelin degradation *in vitro.* Absence of one of them in a human Gaucher disease variant, *Biol. Chem. Hoppe Seyler,* 367, 879, 1986.

25. Schnabel, D., Schröder, M., and Sandhoff, K., Mutation in the sphingolipid activator protein 2 in a patient with a variant of Gaucher disease, *FEBS Lett.,* 284, 57, 1991.

26. Vaccaro, A.M. et al., Structural analysis of saposin C and B. Complete localization of disulfide bridges, *J. Biol. Chem.,* 270, 9953, 1995.

27. Kishimoto, Y., Hiraiwa, M., and O'Brien, J.S., Saposins: structure, function, distribution, and molecular genetics, *J. Lipid Res.,* 33, 1255, 1992.

28. Munford, R.S., Sheppard, P.O., and O'Hara, P.J., Saposin-like proteins (SAPLIP) carry out diverse functions on a common backbone structure, *J. Lipid Res.,* 36, 1653, 1995.

29. Liepinsh, E. et al., Saposin fold revealed by the NMR structure of NK-lysin, *Nat. Struct. Biol.,* 4, 793, 1997.

30. Hecht, O. et al., Solution structure of the pore-forming protein of Entamoeba histolytica, *J. Biol. Chem.,* 279, 17834, 2004.

31. Leippe, M. et al., Ancient weapons: the three-dimensional structure of amoebapore A, *Trends Parasitol.,* 21, 5, 2005.

32. Nakano, T. et al., Structure of full-length cDNA coding for sulfatide activator, a Co-beta-glucosidase and two other homologous proteins: two alternate forms of the sulfatide activator, *J. Biochem. (Tokyo),* 105, 152, 1989.

33. Hiraiwa, M. et al., Isolation, characterization, and proteolysis of human prosaposin, the precursor of saposins (sphingolipid activator proteins), *Arch. Biochem. Biophys.,* 304, 110, 1993.

34. Lefrancois, S. et al., The lysosomal trafficking of sphingolipid activator proteins (SAPs) is mediated by sortilin, *Embo J.,* 22, 6430, 2003.

35. Lefrancois, S., Canuel, M., Zeng, J., and Morales, C.R., Inactivation of sortilin (a novel lysosomal sorting receptor) by dominant negative competition and RNA interference, *Biol. Proced. Online,* 7, 17, 2005.

36. Vielhaber, G., Hurwitz, R., and Sandhoff, K., Biosynthesis, processing, and targeting of sphingolipid activator protein (SAP) precursor in cultured human fibroblasts. Mannose 6-phosphate receptor-independent endocytosis of SAP precursor, *J. Biol. Chem.,* 271, 32438, 1996.

37. Berent, B.L. and Radin, N.S., beta-Glucosidase activator protein from bovine spleen ("coglucosidase"), *Arch. Biochem. Biophys.,* 208, 248, 1981.

38. Kleinschmidt, T., Christomanou, H., and Braunitzer, G., Complete amino-acid sequence and carbohydrate content of the naturally occurring glucosylceramide activator protein (A1 activator) absent from a new human Gaucher disease variant, *Biol. Chem. Hoppe Seyler,* 368, 1571, 1987.
39. Sano, A. and Radin, N.S., The carbohydrate moiety of the activator protein for glucosylceramide beta-glucosidase, *Biochem. Biophys. Res. Commun.,* 154, 1197, 1988.
40. Kleinschmidt, T., Christomanou, H., and Braunitzer, G., Complete amino-acid sequence and carbohydrate content of the naturally occurring glucosylceramide activator protein (A1 activator) absent from a new human Gaucher disease variant, *Biol. Chem. Hoppe Seyler,* 368, 1571, 1987.
41. Vaccaro, A.M. et al., Structural analysis of saposin C and B. Complete localization of disulfide bridges, *J. Biol. Chem.,* 270, 9953, 1995.
42. Sun, Y., Qi, X., and Grabowski, G.A., Saposin C is required for normal resistance of acid beta-glucosidase to proteolytic degradation, *J. Biol. Chem.,* 278, 31918, 2003.
43. Klein, A. et al., Sphingolipid activator protein D (sap-D) stimulates the lysosomal degradation of ceramide *in vivo, Biochem. Biophys. Res. Commun.,* 200, 1440, 1994.
44. Zschoche, A. et al., Hydrolysis of lactosylceramide by human galactosylceramidase and GM1-beta-galactosidase in a detergent-free system and its stimulation by sphingolipid Eur J activator proteins, sap-B and sap-C. Activator proteins stimulate lactosylceramide hydrolysis, *Eur. J. Biochem.,* 222, 83, 1994.
45. Wenger, D.A., Sattler, M., and Roth, S., A protein activator of galactosylceramide beta-galactosidase, *Biochim. Biophys. Acta,* 712, 639, 1982.
46. Poulos, A. et al., Studies on the activation of sphingomyelinase activity in Niemann-Pick type A, B, and C fibroblasts: enzymological differentiation of types A and B, *Pediatr. Res.,* 18, 1088, 1984.
47. de Alba, E., Weiler, S., and Tjandra, N., Solution structure of human saposin C: pH-dependent interaction with phospholipid vesicles, *Biochemistry,* 42, 14729, 2003.
48. Hawkins, C.A., Alba, E., and Tjandra, N., Solution structure of human saposin C in a detergent environment, *J. Mol. Biol.,* 346, 1381, 2005.
49. Ho, M.W. and Rigby, M., Glucocerebrosidase: stoichiometry of association between effector and catalytic proteins, *Biochim. Biophys. Acta,* 397, 267, 1975.
50. Ho, M.W. and Light, N.D., Glucocerebrosidase: reconstitution from macromolecular components depends on acidic phospholipids, *Biochem. J.,* 136, 821, 1973.
51. Vaccaro, A.M. et al., pH-dependent conformational properties of saposins and their interactions with phospholipid membranes, *J. Biol. Chem.,* 270, 30576, 1995.
52. Weiler, S. et al., Identification of the binding and activating sites of the sphingolipid activator protein, saposin C, with glucocerebrosidase, *Protein Sci.,* 4, 756, 1995.
53. You, H.X., Qi, X., and Yu, L., Direct AFM observation of saposin C-induced membrane domains in lipid bilayers: from simple to complex lipid mixtures, *Chem. Phys. Lipids,* 132, 15, 2004.
54. Wilkening, G., Linke, T., and Sandhoff, K., Lysosomal degradation on vesicular membrane surfaces. Enhanced glucosylceramide degradation by lysosomal anionic lipids and activators, *J. Biol. Chem.,* 273, 30271, 1998.
55. Chu, Z., Witte, D.P., and Qi, X., Saposin C-LBPA interaction in late-endosomes/lysosomes, *Exp. Cell Res.,* 303, 300, 2005.
56. Qi, X. and Chu, Z., Fusogenic domain and lysines in saposin C, *Arch. Biochem. Biophys.,* 424, 210, 2004.

57. Salvioli, R. et al., Glucosylceramidase mass and subcellular localization are modulated by cholesterol in Niemann-Pick disease type C, *J. Biol. Chem.*, 279, 17674,
2004.

58. Spiegel, R. et al., A mutation in the saposin A coding region of the prosaposin gene
in an infant presenting as Krabbe disease: first report of saposin A deficiency in
humans, *Mol. Genet. Metab.*, 84, 160, 2005.

59. Matsuda, J. et al., A mutation in the saposin A domain of the sphingolipid activator
protein (prosaposin) gene results in a late-onset, chronic form of globoid cell leukodystrophy in the mouse, *Hum. Mol. Genet.*, 10, 1191, 2001.

60. Morimoto, S. et al., Saposin A: second cerebrosidase activator protein, *Proc. Natl.
Acad. Sci. U.S.A.*, 86, 3389, 1989.

61. Morimoto, S. et al., Interaction of saposins, acidic lipids, and glucosylceramidase, *J.
Biol. Chem.*, 265, 1933, 1990.

62. Kretz, K.A. et al., Characterization of a mutation in a family with saposin B deficiency: a glycosylation site defect, *Proc. Natl. Acad. Sci. U.S A.*, 87, 2541, 1990.

63. Li, S.C. et al., Activator protein required for the enzymatic hydrolysis of cerebroside
sulfate. Deficiency in urine of patients affected with cerebroside sulfatase activator
deficiency and identity with activators for the enzymatic hydrolysis of GM1 ganglioside and globotriaosylceramide, *J. Biol. Chem.*, 260, 1867, 1985.

64. Conzelmann, E., Lee-Vaupel, M., and Sandhoff, K., The physiological roles of activator proteins for lysosomal glycolipid degradation, in *Lipid storage disorders,* Salvayre, R., Douste-Blazy, L., and Gatt, S., Eds., Plenum Publishing, New York, 1988,
p. 323.

65. Ahn, V.E. et al., Crystal structure of saposin B reveals a dimeric shell for lipid binding,
Proc. Natl. Acad. Sci. U.S.A., 100, 38, 2003.

66. Matsuda, J. et al., Mutation in saposin D domain of sphingolipid activator protein
gene causes urinary system defects and cerebellar Purkinje cell degeneration with
accumulation of hydroxy fatty acid-containing ceramide in mouse, *Hum. Mol. Genet.*,
13, 2709, 2004.

67. Sano, A. et al., Sphingolipid hydrolase activator proteins and their precursors, *Biochem. Biophys. Res. Commun.*, 165, 1191, 1989.

68. Kondoh, K. et al., Isolation and characterization of prosaposin from human milk,
Biochem. Biophys. Res. Commun., 181, 286, 1991.

69. Hineno, T. et al., Secretion of sphingolipid hydrolase activator precursor, prosaposin,
Biochem. Biophys. Res. Commun., 176, 668, 1991.

70. O'Brien, J.S. et al., Identification of the neurotrophic factor sequence of prosaposin,
Faseb J., 9, 681, 1995.

71. Tsuboi, K., Hiraiwa, M., and O'Brien, J.S., Prosaposin prevents programmed cell
death of rat cerebellar granule neurons in culture, *Brain Res. Dev. Brain Res.*, 110,
249, 1998.

72. Kotani, Y. et al., Prosaposin facilitates sciatic nerve regeneration *in vivo*, *J. Neurochem.*, 66, 2019, 1996.

73. Kotani, Y. et al., A hydrophilic peptide comprising 18 amino acid residues of the
prosaposin sequence has neurotrophic activity *in vitro* and *in vivo*, *J. Neurochem.*,
66, 2197, 1996.

74. Lapchak, P.A. et al., Prosaptide exacerbates ischemia-induced behavioral deficits *in
vivo*; an effect that does not involve mitogen-activated protein kinase activation,
Neuroscience, 101, 811, 2000.

75. Hiraiwa, M. et al., Cell death prevention, mitogen-activated protein kinase stimulation, and increased sulfatide concentrations in Schwann cells and oligodendrocytes by prosaposin and prosaptides, *Proc. Natl. Acad. Sci. U.S.A.,* 94, 4778, 1997.

76. Rende, M. et al., Prosaposin is immunolocalized to muscle and prosaptides promote myoblast fusion and attenuate loss of muscle mass after nerve injury, *Muscle Nerve,* 24, 799, 2001.

77. Morales, C.R., and Badran, H., Prosaposin ablation inactivates the MAPK and Akt signaling pathways and interferes with the development of the prostate gland, *Asian J. Androl.,* 5, 57, 2003.

78. Amann, R.P., Seidel, G.E., Jr., and Brink, Z.A., Exposure of thawed frozen bull sperm to a synthetic peptide before artificial insemination increases fertility, *J. Androl.,* 20, 42, 1999.

79. Schuette, C.G. et al., Sphingolipid activator proteins: proteins with complex functions in lipid degradation and skin biogenesis, *Glycobiology,* 11, 81R, 2001.

80. Holleran, W.M. et al., Processing of epidermal glucosylceramides is required for optimal mammalian cutaneous permeability barrier function, *J. Clin. Invest.,* 91, 1656, 1993.

81. Doering, T., Proia, R.L., and Sandhoff, K., Accumulation of protein-bound epidermal glucosylceramides in beta-glucocerebrosidase deficient type 2 Gaucher mice, *FEBS Lett.,* 447, 167 (1999).

82. Doering, T., Holleran, W.M., Potratz, A., Vielhaber, G., Elias, P.M., Suzuki, K., and Sandhoff, K., Sphingolipid activator proteins are required for epidermal permeability barrier formation, *J. Biol. Chem.,* 274, 11038, 1999.

83. Gallala, H., Macheleidt, O., Doering, T., Schreiner, V., and Sandhoff, K., Nitric oxide regulates synthesis of gene products involved in keratinocyte differentiation and ceramide metabolism, *Eur. J. Cell Biol.,* 83, 667, 2004.

84. Doering, T., Brade, H., and Sandhoff, K., Sphingolipid metabolism during epidermal barrier development in mice, *J. Lipid Res.,* 43, 1727, 2002.

85. Sidransky, E., Sherer, D.M., and Ginns, E.I., Gaucher disease in the neonate: a distinct Gaucher phenotype is analogous to a mouse model created by targeted disruption of the glucocerebrosidase gene, *Pediatr. Res.,* 32, 494, 1992.

86. Fujita, N. et al., Targeted disruption of the mouse sphingolipid activator protein gene: a complex phenotype, including severe leukodystrophy and wide-spread storage of multiple sphingolipids, *Hum. Mol. Genet.,* 5, 711, 1996.

87. Kronenberg, M., Presenting fats with SAPs, *Nat. Immunol.,* 5, 126, 2004.

88. Winau, F. et al., Saposin C is required for lipid presentation by human CD1b, *Nat. Immunol.,* 5, 169, 2004.

89. Kang, S.J. and Cresswell, P., Saposins facilitate CD1d-restricted presentation of an exogenous lipid antigen to T cells, *Nat. Immunol.,* 5, 175, 2004.

90. Zhou, D. et al., Editing of CD1d-bound lipid antigens by endosomal lipid transfer proteins, *Science,* 303, 523, 2004.

91. Bradova, V. et al., Prosaposin deficiency: further characterization of the sphingolipid activator protein-deficient sibs. Multiple glycolipid elevations (including lactosylceramidosis), partial enzyme deficiencies and ultrastructure of the skin in this generalized sphingolipid storage disease, *Hum. Genet.,* 92, 143, 1993.

92. Schnabel, D. et al., Simultaneous deficiency of sphingolipid activator proteins 1 and 2 is caused by a mutation in the initiation codon of their common gene, *J. Biol. Chem.,* 267, 3312, 1992.

93. Elleder, M. et al., Prosaposin deficiency — a rarely diagnosed, rapidly progressing, neonatal neurovisceral lipid storage disease. Report of a further patient, *Neuropediatrics*, 36, 171, 2005.

94. Hulkova, H. et al., A novel mutation in the coding region of the prosaposin gene leads to a complete deficiency of prosaposin and saposins, and is associated with a complex sphingolipidosis dominated by lactosylceramide accumulation, *Hum. Mol. Genet.*, 10, 927, 2001.

95. Christomanou, H. et al., Activator protein deficient Gaucher's disease. A second patient with the newly identified lipid storage disorder, *Klin Wochenschr*, 67, 999, 1989.

96. Rafi, M.A. et al., Mutational analysis in a patient with a variant form of Gaucher disease caused by SAP-2 deficiency, *Somat. Cell. Mol. Genet.*, 19, 1, 1993.

97. Pampols, T., Pineda, M., Giros, M.L., Ferrer, I., Cusi, V., Chabas, A., Sanmarti, F.X., Vanier, M.T., and Christomanou, H., Neuronopathic juvenile glucosylceramidosis due to sap-C deficiency: clinical course, neuropathology and brain lipid composition in this Gaucher disease variant, *Acta Neuropathol. (Berl.)*, 97, 91, 1999.

98. Sandhoff, K. and Kolter, T. Topology of glycosphingolipid degradation, *Trends Cell. Biol.*, 6, 98, 1996.

99. Kolter, T. and Sandhoff, K., Recent advances in the biochemistry of sphingolipidoses, *Brain Pathol.*, 8, 79, 1998.

100. Henseler, M. et al., Analysis of a splice site mutation in the SAP precursor gene of a patient with metachromatic leukodystrophy, *Am. J. Hum. Genet.*, 58, 65, 1996.

101. Rafi, M.A. et al., Detection of a point mutation in sphingolipid activator protein-1 mRNA in patients with a variant form of metachromatic leukodystrophy, *Biochem. Biophys. Res. Commun.*, 166, 1017, 1990.

102. Holtschmidt, H. et al., Sulfatide activator protein. Alternative splicing that generates three mRNAs and a newly found mutation responsible for a clinical disease, *J. Biol. Chem.*, 266, 7556, 1991.

103. Zhang, X-L. et al., Insertion in the mRNA of a metachromatic leukodystrophy patient with sphingolipid activator protein-1 deficiency, *Proc. Natl. Acad. Sci. U.S.A.*, 87, 1426, 1990.

104. Zhang, X.L. et al., The mechanism for a 33-nucleotide insertion in messenger RNA causing sphingolipid activator protein (SAP-1)-deficient metachromatic leukodystrophy, *Hum. Genet.*, 87, 211, 1991.

105. Wrobe, D. et al., A non-glycosylated and functionally deficient mutant (N215H) of the sphingolipid activator protein B (SAP-B) in a novel case of metachromatic leukodystrophy (MLD), *J. Inherit. Metab. Dis.*, 23, 63, 2000.

CHAPTER 5

The X-Ray Structure of Human Acid-β-Glucosidase: Implications for Second-Generation Enzyme Replacement Therapy

Lakshmanane Premkumar, Israel Silman, Joel L. Sussman, and Anthony H. Futerman

CONTENTS

INTRODUCTION

Acid β-glucosidase (EC 3.2.1.45, D-glucosyl-N-acylsphingosine glucohydrolase, glucocerebrosidase, GlcCerase), the enzyme defective in Gaucher disease, is a peripheral lysosomal membrane protein that hydrolyzes the β-glucosyl linkage of glucosylceramide (GlcCer). GlcCerase requires the coordinate action of saposin C and negatively charged lipids for maximal activity.[1,2] Enzyme replacement therapy (ERT) with Cerezyme®, a recombinant human GlcCerase,[3] is the main treatment for Type 1 Gaucher disease. Although attempts at structural prediction had been made earlier,[4,5] the lack of an experimental 3-D structure of GlcCerase hampered attempts to establish its catalytic mechanism and to analyze the structural relationships between the mutations, levels of residual enzyme activity, and disease severity. The recent determination in our laboratories of the x-ray structures of GlcCerase at 2.0 Å resolution[6] and, subsequently, of a conjugate with an irreversible inhibitor,[7] conduritol B epoxide (CBE), has paved the way for detailed structural analysis of GlcCerase. In this chapter we will review the two structures, and discuss the insights that they provide that may help in designing second-generation GlcCerases for ERT.

THE THREE-DIMENSIONAL STRUCTURE OF GlcCerase

GlcCerase comprises three noncontiguous domains (Figure 5.1A) that could not be predicted from the primary amino acid sequence. It has a characteristic $(\beta/\alpha)_8$ (TIM) barrel containing the catalytic residues, designated as domain III, and two closely associated β-sheets, resembling an immunoglobulin (Ig) architecture, designated as domain II. In addition, it possesses an unusual domain containing one major β-stranded antiparallel β-sheet that is flanked by a perpendicular N-terminal β-strand and loop, designated as domain I. This domain is formed by the two β-strands from the N-terminus and the two antiparallel β-strands from an insertion between β-strand 8 and α-helix 8 of the TIM barrel (Figure 5.1A). Inter-domain interactions between domains I and II, which are connected by a long loop, and between domains II and III, which are connected by a hinge, do not seem to be significant in the crystal structure, whereas domain I tightly interacts with catalytic domain III (Figure 5.1A).

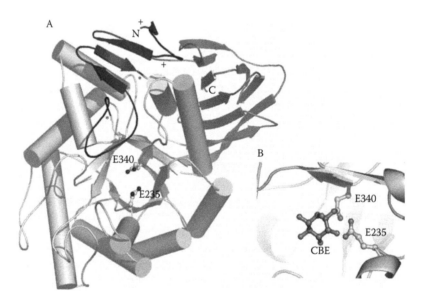

Figure 5.1 Crystal structure of GlcCerase. (A) Domains are shaded in decreasing black color (domain I, black; domain II, gray; and domain III, silver). The boundaries of domain I are indicated by +; residues 384–414 are indicated by *. The catalytic residues, E235 and E340, as well as the N- and C-termini, are also indicated. (B) Binding of CBE to the catalytic nucleophile, E340. The acid/base catalyst, E235, is also shown.

Site-directed mutagenesis and homology modeling of GlcCerase[4,5] had suggested that E235 is the acid/base catalyst, and tandem mass spectrometry had identified E340 as the nucleophile.[8] These two residues (Figure 5.1A) are located near the C-termini of strands 4 and 7 in domain III, with an average distance between their carboxyl oxygens of 5.2Å, consistent with retention of the anomeric carbon upon cleavage, rather than inversion.[9] Moreover, in crystals of Cerezyme® into which CBE had been soaked, the distance between C1 of the cyclohexitol and E340$O\varepsilon2$ is 1.43Å, confirming E340 as the active-site nucleophile (Figure 5.1B). Moreover, the epoxide oxygen of CBE, oriented similarly to the cyclohexitol ring, is within hydrogen-bonding distance of E235O, consistent with the role of E235 as the acid/base catalyst.[10] In addition, substrate docking shows that only the glucose moiety and the adjacent glycoside bond of GlcCer fit within the active-site pocket, implying that the two GlcCer hydrocarbon chains either remain embedded in the lipid bilayer during catalysis or, alternatively, interact with saposin C.

Analysis of a surface plot of GlcCerase revealed additional interesting features. First of all, there is an annulus of hydrophobic residues surrounding the entrance to the active site.[6] Secondly, there is a patch of positive electrostatic potential running around the interface of domain II and domain III (unpublished data). Either or both of these surface features might be involved in the interaction of GlcCerase with the lyosomal membrane or with saposin C.

A LID AT THE ENTRANCE TO THE ACTIVE SITE OF GlcCerase

To determine whether substrate or inhibitor binding could induce conformational change(s) in GlcCerase, and to try to gain insight into possible roles of the noncatalytic domains, we compared the structures of native GlcCerase and of the CBE/GlcCerase complex. Structural comparison showed that GlcCerase does not undergo a global structural change upon binding CBE, suggesting that the GlcCerase structure can be used as a starting point for designing structure-based drugs aimed at restoring the activity of defective GlcCerase. However, a significant difference between native GlcCerase and CBE/GlcCerase was observed with respect to the conformation of two loops, S345-E349 (loop 1) and V394-D399 (loop 2) (Figure 5.2). Interestingly, both loops are located on the surface of GlcCerase, at the entrance to the active site, and display two alternative conformations in the native GlcCerase structure (Figure 5.2A). Specifically, a conformational difference is seen in the positions of N396 and F397 in loop 2, and a more modest difference is seen in the conformations of K346 and E349 in loop 1 (Figure 5.2B and Figure 5.2C). In contrast, the CBE/GlcCerase structure adopts only one of these two conformations. In the conformation adopted in the CBE/GlcCerase structure, N396 and F397 are positioned such that access to the active site is not restricted (Figure 5.2E). However, in the alternative conformation, which is displayed only in the native GlcCerase structure, the side-chains of N396 and F397 swing over and block the entrance to the active site (Figure 5.2D), suggesting that this loop serves as a lid regulating access to the active site. Thus, these two loops allow GlcCerase to exist in either an open or closed conformation, depending on the orientation of the two loops.

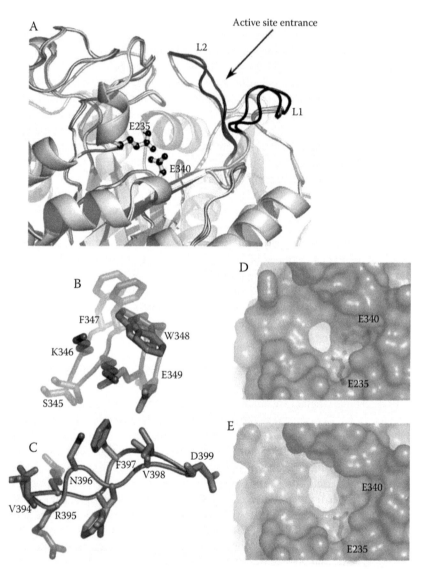

Figure 5.2 Entrance to the active site of GlcCerase is regulated by two surface loops. (A) The alternative conformations of the two loops, L1 (S345-E348) and L2 (V394-D399) are indicated. (B,C) Conformational changes in individual residues in loop 1 (B) and loop 2 (C). (D,E) Surface of GlcCerase illustrating the closed conformation (D), in which the surface lid restricts access to the active site, and the open conformation (E).

Interestingly, a lid has also been detected in a number of glycosyltransferases[11] in which one or two flexible loops near the substrate binding site undergo a marked conformational change from an open to a closed conformation upon binding the donor substrate. The flexibility of these loops is crucial for the catalytic activity of these enzymes.[11]

MUTATIONS IN GlcCerase LEADING TO GAUCHER DISEASE

Most mutations in GlcCerase appear to either partially or entirely decrease catalytic activity[12] or to reduce GlcCerase stability[13] or both. Prior to elucidation of the 3-D structure of GlcCerase, no clear correlation was apparent between the location of particular mutants within the sequence and the severity of clinical symptoms. Even now that the 3-D structure is available, no clear relationship is immediately discernible between the spatial location of most of the ~200 known GlcCerase mutations and disease severity, with one or two exceptions. A number of mutations (e.g., H311R, A341T, and C342G) located near the active site, result in severe disease, as might be predicted, but the majority of the mutations are spread throughout all three domains (Figure 5.3).

No 3-D structures are yet available for any mutant GlcCerase. Conjecture as to the mechanism by which catalytic activity might be compromised by a given mutation is thus limited to structural predictions and *in silico* mutational analysis. We have attempted such analysis for all the mutations (Table 5.1), focusing particularly on the six most common single amino-acid substitutions (Figure 5.4). One of the

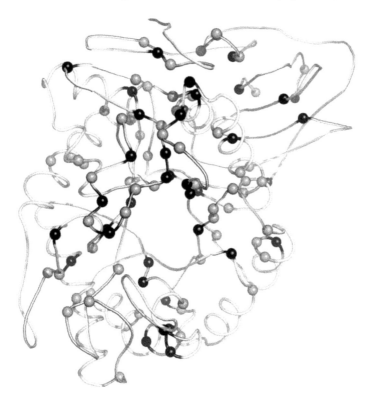

Figure 5.3   Distribution of mutations caused by single amino-acid substitutions in the 3-D structure of GlcCerase. Mutations causing severe (type II and III) and mild (type I) Gaucher disease are shown as black and gray spheres, respectively.

Table 5.1 Possible Structural Features in GlcCerase Mutations Leading to Gaucher Disease. This table excludes the six most common mutations, which are presented in more detail in Figure 5.4

Mutation	Phenotype[1]	Location of Mutation (domain)	Distance (Å) from E340 and E235[2]	Molecular Effect[3]
F37V	Mild	II	>15	HY decrease
T43I	Mild	II	>15	HB lost
G46E	Mild	II	>15	OP and BS
R48W	Mild	II	>15	BP and HB lost
K79N	Mild, Severe	III	10	HB lost and SB lost
S107L	Severe	III	>15	HB lost and OP
G113E	Mild	III	>15	BC, OP and BS
N117D	Mild	III	>15	HB lost
I119T	Mild	III	10	HY decrease
R120Q	Mild, Severe	III	4	HB lost and SB lost
R120W	Mild, Severe	III	4	BP, OP, HB lost and SB lost
P122S	Severe	III	10	HY decrease
R131L	Mild, Severe	III	>15	SB lost
T134I	Mild	III	15	HB lost and OP
T134P	Mild	III	15	HB lost and BS
K157Q	Mild, Severe	III	15	SB lost
I161S	Mild	III	>15	HY decrease
R170C	Mild	III	>15	HB lost
R170P	Mild	III	>15	HB lost and BS
A176D	Mild, Severe	III	10	BC, OP and HY decrease
P178S	Severe	III	6	HY decrease
P182T	Mild	III	15	OP
W184R	Mild, Severe	III	>15	HY decrease
N188K	Mild, Severe	III	>15	HB lost
V191G	Mild	III	>15	HY decrease
G195E	Mild, Severe	III	10	BC, OP and BS
G195W	Mild, Severe	III	10	OP and BS
S196P	Mild, Severe	III	>15	HB lost and BS
G202R	Mild, Severe	III	>15	ESR, OP and BS
G202E	Mild	III	>15	OP and BS
Y205C	Severe	III	>15	HY decrease
Y212H	Mild	III	15	HB lost and HY decrease
F213I	Mild, Severe	III	15	HY decrease
F216V	Mild, Severe	III	15	HY decrease
S237P	Mild	III	4	HB lost
R257Q	Mild, Severe	III	15	OP and SB lost
F259L	Severe	III	15	HY decrease
G265D	Mild	III	>15	BC and OP
S271N	Mild	III	>15	BP and HB lost
R285H	Mild, Severe	III	10	OP and HY decrease
R285C	Mild, Severe	III	10	HB lost and HY decrease
P289L	Mild	III	>15	OP and BS
Y304C	Severe	III	>15	HY decrease
W312C	Mild	III	4	HY decrease
Y313H	Mild	III	4	HB lost and HY decrease
D315H	Mild	III	>15	OP, HB lost and SB lost
A318D	Mild	III	>15	BC, OP and HY decrease
T323I	Mild	III	15	HB lost and OP

Table 5.1   Possible Structural Features in GlcCerase Mutations Leading to Gaucher Disease. This table excludes the six most common mutations, which are presented in more detail in Figure 5.4. (Continued)

Mutation	Phenotype[1]	Location of Mutation (domain)	Distance (Å) from E340 and E235[2]	Molecular Effect[3]
L324P	Mild	III	15	HB lost and HY decrease
G325R	Severe	III	>15	ESR
R353G	Severe	III	>15	HB lost and SB lost
R359Q	Mild	III	15	BP, HB lost and SB lost
S364T	Mild	III	15	BP and HB lost
G377S	Mild, Severe	III	15	OP
W378G	Mild	III	10	HY decrease
D380A	Severe	III	6	HB lost
D380N	Mild	III	6	OP
G389E	Severe	I	15	BC, OP and BS
N392I	Severe	I	10	HB lost and OP
W393R	Mild	I	15	HY decrease
D399Y	Mild	I	10	BP, OP, HB lost and SB lost
D399N	Mild, Severe	I	10	BP, HB lost and SB lost
P401L	Mild	I	15	OP
I402T	Mild	I	15	HY decrease
F411I	Mild	I	>15	HY decrease
Y412H	Mild	I	>15	HY decrease
K413Q	Mild	I	>15	SB lost
P415R	Severe	III	>15	BC, ESR and OP
F417V	Mild	III	15	OP and HY decrease
Y418C	Mild	III	>15	HB lost and HY decrease
K425E	Severe	III	>15	HB lost
R433G	Mild	II	>15	HY decrease
A446	Mild	II	>15	BS
D409H	Mild, Severe	II	>15	HB lost
A456P	Mild, Severe	II	>15	HB lost and BS
N462K	Severe	II	>15	BC, HB lost and OP
D474Y	Severe	II	>15	HB lost, OP and SB lost
G478S	Mild	II	>15	OP, BS

[1] In some cases, phenotype assignment as mild (type 1) or severe (types 2 and 3) is based on a few, or sometimes only one, individuals.
[2] The average distance from the catalytic nucleophile E340, and the acid/base catalyst E235.
[3] SB, salt bridge(s); HB, hydrogen bond(s); HY, hydrophobic interaction; PO, polar interaction; ESR, electrostatic repulsion; OP, overpacking (i.e., introduction of a larger side chain, making unavoidable short atomic contacts of <2.5 Å in the interior of a protein or <2.0 Å on the surface); BS, backbone strain; BC, buried charge; BP, buried polar residue.

most frequent mutations is N370S, which accounts for 70% of mutant alleles in Ashkenazi Jews and 25% in non-Jewish patients. N370S predisposes to Type 1 disease and precludes neurological involvement, suggesting that it causes relatively minor changes in GlcCerase structure and hence in catalytic activity. Consistent with this is the assignment of N370 to the longest α-helix (helix 7) in GlcCerase, which is located at the interface of domains II and III, but too far from the active site to participate directly in catalysis. *In silico* mutational analysis of N370S suggests that this mutation is likely to decrease the hydrophobic and polar interactions at this site

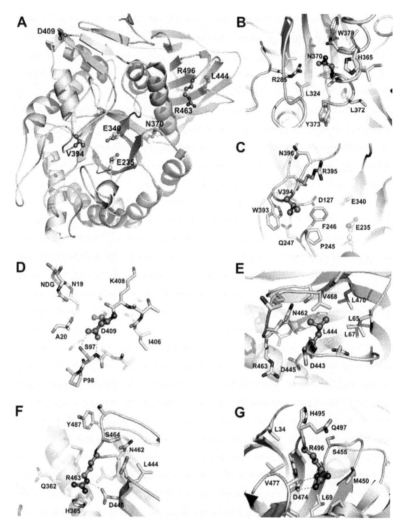

Figure 5.4   The six most common mutations and possible structural alterations. (A) Locations of the residues on the 3-D structure are shown in ball-and-stick format. A more detailed view of each individual mutation is shown in the following panels: (B) N370, which is located on α-helix 7 at the interface of domains II and III, and upon mutation to S, is likely to moderately decrease hydrophobic and polar interactions; (C) V394, which is located on a loop that acts as an active site lid, and upon mutation to L is likely to enhance hydrophobic interactions of the closed conformation, resulting in restricted access to the active site; (D) D409, which is located on noncatalytic domain I, and upon mutation to V will result in loss of hydrogen bonds; (E) L444, which is located at the hydrophobic core of domain II, and upon mutation to P, will cause loss of a hydrogen bond, resulting in decreased hydrophobic interactions and introduction of backbone strain, which might affect the stability and folding of the enzyme; (F) R463, located at the surface of the hydrophobic core of domain II, which upon mutation to C will result in the loss of a salt bridge; (G) R496, located at the hydrophobic core of domain II, which upon mutation to H will result in loss of a salt bridge as well as in decreased hydrophobic interactions. Hydrogen bonds are indicated by dashed lines.

(Figure 5.4B). Another mutation, V394L (Figure 5.4C), located at the active site lid (see above), is likely to destabilize the open conformation or stabilize the closed conformation, and thereby limit substrate access to the active site. Thus, hydrophobic interactions are enhanced between L394 and W393 and F246, resulting in stabilization of the closed conformation; conversely, this same mutation destabilizes the open conformation since the larger side chain cannot interact with the aromatic side chains of W393 and F246.[14] Mutation L444P (Figure 5.4E), which always predisposes to severe disease (i.e., type 2 or 3),[1] is located on domain II, distant from the active site (Figure 5.4A). Earlier studies suggested that L444P produces unstable protein,[13] suggesting that the correct folding of domain II is required for optimal protein stability. Mutations D409V (Figure 5.4D), R463C (Figure 5.4F), and R496H (Figure 5.4G), most likely also reduce the stability of the enzyme, rather than directly affecting catalysis, as all three are distant (>25 Å) from the active site (Figure 5.4).

We have observed a number of mutational "hot spots" (i.e., clusters of mutations in the same area of the protein). One such hot spot is located on the 18-residue α-helix 7 (residues 356–373). Mutations of five amino acid residues on this helix, viz. R359Q,[15] S364T,[16] S366G,[17] T369M,[18] and N370S[1] result in Gaucher disease. Another mutational hot spot is loop 2 of the active-site lid (loop 2), with V394L,[1] R395P,[19] N396T,[20] V398L/F,[21,22] and D399N[23] all resulting in Gaucher disease. In silico analysis of the loop mutations is consistent with either destabilization of the open conformation or stabilization of the closed conformation, thus limiting substrate access to the active site. For instance, R395P destabilizes the open conformation due to loss of a stabilizing salt bridge with E388; N396T results in additional hydrogen bonding between T396 and the carbonyl oxygens of residues E388 and G389 in the closed conformation, but in reduced H-bonding with D127 in the open conformation.

As mentioned earlier, structural and biochemical studies aimed at elucidating the structural basis for the biological action(s) of the mutations causing Gaucher disease have been hampered by the lack of an efficient heterologous expression system for GlcCerase and its mutants. Consequently, we have used a web-based database that predicts the functional effects of nonsynonymous single nucleotide polymorphisms (SNPs), based on structure and sequence analysis to analyze the possible molecular effects of the mutations in the GlcCerase gene, which result in Gaucher disease.[24] Table 5.1 lists the possible molecular changes that may occur in GlcCerase for single nucleotide substitutions, based on high confidence predictions from the SNPs3D database (http://www.snps3d.org).

CONCLUSIONS AND FUTURE PERSPECTIVES

How does the availability of the GlcCerase structure assist development of improved ERT for Gaucher disease?[25] Firstly, the tools of structure-based drug design should permit rational design of activators to enhance activity of both mutant enzyme(s), such as the common N370S mutation, and of enzyme administered in ERT, so as to enhance its stability and/or activity. Since about two thirds of patients homozygous for N370S are asymptomatic, and thus presumably have sufficient residual enzyme activity to prevent development of symptoms, only a small degree of enhancement of activity would be required to yield levels of catalytic activity adequate to overcome the effects of N370S and, possibly, of other mutations. Design of activators might require structural comparison of native GlcCerase with mutant enzymes or with enzyme co-crystallized with small molecules, similar to that achieved with CBE.

The second approach, and the more demanding, is the engineering of more stable or more active GlcCerase. Although few studies have been published that system-atically examine the fate of GlcCerase after infusion (the main study was performed with Ceredase®[26] — a first-generation, placental GlcCerase), the enzyme is cleared from the blood within a few minutes, and has a half-life in the bone marrow of ~14 hours.[1] Engineering a more stable enzyme, or an enzyme with higher catalytic activity, would reduce the number of infusions required and likely also the cost of treatment. Site-directed mutagenesis and directed evolution are the main tools that one might utilize to achieve this objective. Characterization of the noncatalytic domains, I and II, and the catalytic domain TIM barrel, domain III, could help with this approach. Thus, it would be worthwhile to systematically manipulate domains I or II, by their removal or replacement. This could help to resolve such issues as the roles of domain I and II, or the role of the sequence between β-strand 8 and α-helix 8 of domain III that forms a part of domain I (384–414). Furthermore, pro-duction of an enzyme consisting only of the catalytic domain would resolve the roles of the noncatalytic domains as potential regulatory domains. Substituting amino acids near or around the active site could dramatically affect enzyme activity, and mutations of amino acids, or deletion/modification of structural determinants in the regulatory domains such as insertion of suitably positioned disulfide bridges, might influence stability. In this regard, one important open question concerns identification of the molecular mechanism by which saposin C enhances GlcCerase activity. Resolving this issue will require co-crystallization of saposin C with GlcCerase. Moreover, the availability of the GlcCerase structure permits a rational approach to the production of small molecules for use as pharmacological chaperones (see Chapter 19). In summary, the 3-D structure of Cerezyme should pave the way for an exciting new era in ERT, involving the use of small molecules or modified enzyme to increase ERT efficacy.

REFERENCES

1. Beutler, E and Grabowski, GA. Gaucher Disease, in *The Metabolic and Molecular Bases of Inherited Disease*, Scriver, CR, Sly, WS, Childs, B, Beaudet, AL, Valle, D, Kinzler, KW, and Vogelstein, B, Eds., McGraw-Hill, New York, Vol. II, 2001, pp. 3635–3668.
2. Grabowski, GA, Gatt, S, and Horowitz, M. Acid β-Glucosidase: Enzymology and molecular biology of Gaucher disease. *Critical Rev. Biochem. Mol. Biol.*, 25, 385–414, 1990.

3. Grabowski, GA, Barton, NW, Pastores, G, Dambrosia, JM, Banerjee, TK, McKee, MA, Parker, C, Schiffmann, R, Hill, SC, and Brady, RO. Enzyme therapy in type 1 Gaucher disease: comparative efficacy of mannose-terminated glucocerebrosidase from natural and recombinant sources. *Ann. Intern. Med.*, 122, 33–39, 1995.

4. Fabrega, S, Durand, P, Codogno, P, Bauvy, C, Delomenie, C, Henrissat, B, Martin, BM, McKinney, C, Ginns, EI, Mornon, JP, and Lehn, P. Human glucocerebrosidase: heterologous expression of active site mutants in murine null cells. *Glycobiology*, 10, 1217–1224, 2000.

5. Fabrega, S, Durand, P, Mornon, JP, and Lehn, P. The active site of human glucocerebrosidase: structural predictions and experimental validations. *J. Soc. Biol.* (France), 196, 151–160, 2002.

6. Dvir, H, Harel, M, McCarthy, AA, Toker, L, Silman, I, Futerman, AH, and Sussman, JL. X-ray structure of human acid-β-glucosidase, the defective enzyme in Gaucher disease. *EMBO Rep.*, 4, 704–709, 2003.

7. Premkumar, L, Sawkar, AR, Boldin-Adamsky, S, Toker, L, Silman, I, Kelly, JW, Futerman, AH, and Sussman, JL. X-ray structure of human acid-β-glucosidase covalently bound to conduritol-B-epoxide. Implications for Gaucher disease. *J. Biol. Chem.*, 280, 23815–23819, 2005.

8. Miao, S, McCarter, JD, Grace, ME, Grabowski, GA, Aebersold, R., and Withers, SG. Identification of Glu340 as the active-site nucleophile in human glucocerebrosidase by use of electrospray tandem mass spectrometry. *J. Biol. Chem.*, 269, 10975–10978, 1994.

9. Davies, G and Henrissat, B. Structures and mechanisms of glycosyl hydrolases. *Structure*, 3, 853–859, 1995.

10. Henrissat, B, Callebaut, I, Fabrega, S, Lehn, P, Mornon, JP, and Davies, G. Conserved catalytic machinery and the prediction of a common fold for several families of glycosyl hydrolases. *Proc. Natl. Acad. Sci. U.S.A.*, 92, 7090–7094, 1995.

11. Qasba, PK, Ramakrishnan, B, and Boeggeman, E. Substrate-induced conformational changes in glycosyltransferases. *Trends Biochem. Sci.*, 30, 53–62, 2005.

12. Meivar-Levy, I, Horowitz, M, and Futerman, AH. Analysis of glucocerebrosidase activity using N-(1-[14C]Hexanoyl)-D-erythro-glucosylsphingosine demonstrates a correlation between levels of residual enzyme activity and the type of Gaucher disease. *Biochem. J.*, 303, 377–382, 1994.

13. Grace, ME, Newman, KM, Scheinker, V, Berg-Fussman, A, and Grabowski, GA. Analysis of human acid β-glucosidase by site-directed mutagenesis and heterologous expression. *J. Biol. Chem.*, 269, 2283–2291, 1994.

14. Ron, I, Dagan, A, Gatt, S, Pasmanik-Chor, M, and Horowitz, M. Use of fluorescent substrates for characterization of Gaucher disease mutations. *Blood Cells Mol. Dis.*, 35, 57–65, 2005.

15. Kawame, H, Hasegawa, Y, Eto, Y, and Maekawa, K. Rapid identification of mutations in the glucocerebrosidase gene of Gaucher disease patients by analysis of single-strand conformation polymorphisms. *Hum. Genet.*, 90, 294–296, 1992.

16. Latham, TE, Theophilus, BD, Grabowski, GA, and Smith, FI. Heterogeneity of mutations in the acid beta-glucosidase gene of Gaucher disease patients. *DNA Cell. Biol.*, 10, 15–21, 1991.

17. Ida, H, Rennert, OM, Kawame, H, Maekawa, K, and Eto, Y. Mutation prevalence among 47 unrelated Japanese patients with Gaucher disease: identification of four novel mutations. *J. Inherit. Metab. Dis.*, 20, 67–73, 1997.

18. Beutler, E, Gelbart, T, Balicki, D, Demina, A, Adusumalli, J, Elsas, L, 2nd, Grinzaid, KA, Gitzelmann, R, Superti-Furga, A, Kattamis, C, and Liou, BB. Gaucher disease: four families with previously undescribed mutations. *Proc. Assoc. Am. Physicians*, 108, 179–184, 1996.

19. Amaral, O, Marcao, A, Sa Miranda, M, Desnick, RJ, and Grace, ME. Gaucher disease: expression and characterization of mild and severe acid β-glucosidase mutations in Portuguese type 1 patients. *Eur. J. Hum. Genet.*, 8, 95–102, 2000.

20. Amaral, O, Pinto, E, Fortuna, M, Lacerda, L, and Sa Miranda, MC. Type 1 Gaucher disease: identification of N396T and prevalence of glucocerebrosidase mutations in the Portuguese. *Hum. Mutat.*, 8, 280–281, 1996.

21. Seeman, PJ, Finckh, U, Hoppner, J, Lakner, V, Liebisch, I, Grau, G, and Rolfs, A. Two new missense mutations in a non-Jewish Caucasian family with type 3 Gaucher disease. *Neurology*, 46, 1102–1107, 1996.

22. Stone, DL, van Diggelen, OP, de Klerk, JB, Gaillard, JL, Niermeijer, MF, Willemsen, R, Tayebi, N, and Sidransky, E. Is the perinatal lethal form of Gaucher disease more common than classic type 2 Gaucher disease? *Eur. J. Hum. Genet.*, 7, 505–509, 1999.

23. Beutler, E and Gelbart, T. Two new Gaucher disease mutations. *Hum. Genet.*, 93, 209–210, 1994.

24. Yue, P, Li, Z, and Moult, J. Loss of protein structure stability as a major causative factor in monogenic disease. *J. Mol. Biol.*, 353, 459–473, 2005.

25. Futerman, AH, Sussman, JL, Horowitz, M, Silman, I, and Zimran, A. New directions in the treatment of Gaucher disease. *Trends Pharmacol. Sci.*, 25, 147–151, 2004.

26. Mistry, PK, Wraight, EP, and Cox, TM. Therapeutic delivery of proteins to macrophages: implications for treatment of Gaucher's disease. *Lancet*, 348, 1555–1559, 1996.

CHAPTER 6

Cellular Pathology in Gaucher Disease

Anthony H. Futerman

CONTENTS

In Gaucher disease, ~200 different mutations have been described in the gene encoding lysosomal glucocerebrosidase (GlcCerase) (see Chapter 2). Irrespective of the mutation, glucosylceramide (GlcCer) is degraded much more slowly in cells from Gaucher patients than in normal cells and as a consequence, accumulates intracellularly, primarily in cells of mononuclear phagocyte origin. These GlcCer-laden macrophages, known as "Gaucher cells," are the classical hallmark of the disease. Since GlcCer is an important constituent of biological membranes and is a key intermediate in the biosynthetic and degradative pathways of complex glycosphingolipids (see Chapter 7), its accumulation in Gaucher disease is likely to have severe pathological consequences. However, and somewhat surprisingly, the pathological mechanism leading from GlcCer accumulation to Gaucher disease has not been established; this is true for most lysosomal storage diseases (LSDs) in which the biochemical and cellular pathways leading from substrate accumulation to disease have not been delineated [1].

I will now review two aspects of the cellular pathology of Gaucher disease. The first concerns type 1 Gaucher disease, which is essentially a macrophage disorder, lacking primary central nervous system involvement. The second concerns the neuronal forms of the disease, type 2, the acute neuronopathic form, characterized by neurological impairment in addition to visceral symptoms, and type 3, which is also characterized by neurological involvement, but neurological symptoms generally appear later in life than in type 2 [2]. Clinical descriptions of each of these forms of Gaucher disease are discussed elsewhere in this book. In this chapter, after a brief description of levels of GlcCer accumulation in different tissues in Gaucher disease patients, I will focus on recent advances in determining the pathological conse- quences of GlcCer accumulation in macrophages in type 1 disease, and in neurons in types 2 and 3 Gaucher disease.

GLUCOSYLCERAMIDE ACCUMULATION

Residual levels of GlcCerase activity in Gaucher disease patients have been estimated at between 5–25% of normal activity, depending on the substrate used and the conditions of the reaction; in rare cases, Gaucher disease can be caused by mutations in the saposin C domain of the prosaposin gene [3], which encodes the saposin C activator protein that is required for optimal GlcCerase activity. Most of the known GlcCerase mutations partially or entirely decrease catalytic activity, or are believed to reduce GlcCerase stability [4]. As a consequence, GlcCer accumulates in essentially every tissue where its levels have been measured. For instance, GlcCer accumulates to levels of ~30–40 mmol/kg tissue in spleen obtained from all three types of Gaucher disease patients, and glucosylsphingosine (GlcSph), the deacylated form of GlcCer, which is not usually detectable in normal tissues, accumulates to lower but significant levels of ~0.1–0.2 mmol/kg [5]. GlcSph is found at higher levels in the brains of type 2 and 3 Gaucher disease patients [6], suggesting a potential pathological role for this lipid in the brain [7]. The fatty acid composition of GlcCer differs between the brain and peripheral tissues, with a prevalence of stearic acid in the central nervous system, and palmitic acid in GlcCer of peripheral tissues, imply- ing a different metabolic or cellular origin of GlcCer in these different tissues [8]. GlcCer levels are also elevated in the plasma of patients with Gaucher disease [9,10].

Although GlcCer levels are elevated they are not high enough to account for changes in tissue mass and/or tissue pathology. Thus, whereas the size of the spleen increases up to 25-fold in Gaucher disease patients, GlcCer accounts for <2% of the additional tissue mass [11], implying that other biochemical pathways must be activated in Gaucher disease and contribute to changes in tissue mass and the development of pathology, as discussed below.

CELLULAR PATHOLOGY: MACROPHAGES

The cellular pathology of Gaucher disease begins in lysosomes, membrane- bound organelles that consist of a limiting, external membrane and intra-lysosomal vesicles. Endogenous and exogenous macromolecules, including GlcCer, are deliv- ered to lysosomes by processes such as endocytosis, pinocytosis, phagocytosis and autophagocytosis [12] and the lysosomal proteins themselves, at least the soluble hydrolases, are targeted to lysosomes mainly via the mannose-6-phosphate receptor

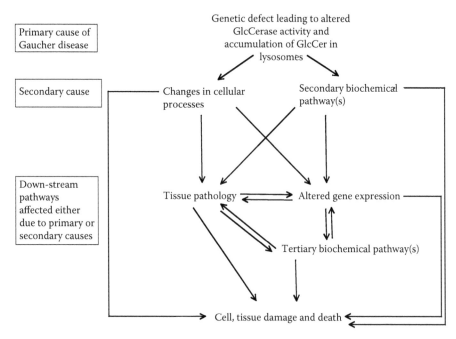

Figure 6.1 Possible pathways of Gaucher disease pathology. Gaucher disease is character-ized by the intra-lysosomal accumulation of GlcCer, which is the primary cause of disease, but the extensive range of disease symptoms indicates that many secondary biochemical and cellular pathways must also be activated. Thus, GlcCer accumulation affects a secondary biochemical or cellular pathway, which then subsequently causes tissue pathology, altered gene expression, and activation of tertiary biochemical pathways, as yet, unidentified.

[13]. The mechanism by which GlcCerase is targeted from its site of synthesis in the endoplasmic reticulum (ER) to lysosomes is not known [14].

As stated earlier, little is known about how GlcCer accumulation in lysosomes leads to cellular pathology (Figure 6.1). One vital, but as yet unanswered question, is whether GlcCer mediates all of its pathological effects from within the lysosome, or whether some GlcCer can escape the lysosome and thereby interact with bio-chemical and cellular pathways located in other organelles; this issue is discussed in more detail below. However, subsequent to its accumulation, GlcCer causes many cellular responses, particularly in Gaucher cells, macrophages that actively phago-cytose other cells, especially senescent blood cells from the circulation [15–17]. The macrophage origin of Gaucher cells has been demonstrated in many studies, includ-ing demonstration of pre-Gaucher monocytes and monocytoid cells with character-istic cytoplasmic inclusions [18], the detection of surface macrophage markers [19,20], and intense phagocytic activity [17]. Gaucher cells are about 20 to 100 μm in diameter and have small, usually eccentrically placed nuclei and cytoplasm with characteristic crinkles or striations. Moreover, all cells of the mononuclear phagocyte system, and especially tissue macrophages of the liver (Kupffer cells), bone (osteo-clasts), the central nervous system (microglia, cerebrospinal fluid macrophages), lungs (alveolar macrophages), spleen, lymph nodes, bone marrow, gastro-intestinal and genito-urinary tracts, pleura, peritoneum, and others, can be affected in Gaucher disease [21]. Gaucher-like cells are well described in various haematological malignancies unrelated to Gaucher disease, including Hodgkin's disease, non-Hodgkin's lymphoma, multiple myeloma (MM) and chronic myelogenous leukemia (CML) [21].

Since macrophages are the main cell type affected in Gaucher disease, some effort has been invested to determine how and why macrophage biology is altered. It is now apparent that pathology is caused not just by the burden of GlcCer storage, but by macrophage activation. Thus, levels of interleukin-1β (IL-1β), interleukin-1 receptor antagonist, interleukin-6 (IL-6), tumor necrosis factor-α (TNFα), and soluble interleukin-2 receptor (sIL-2R), are elevated in serum of Gaucher patients [22], as are CD14 and M-CSF [23]. These changes could potentially explain some of the pathological features since IL-1β, TNFα, IL-6 and Il-10 may contribute to osteopenia, IL-1β, TNFα and IL-6 may contribute to activation of coagulation and hypermetabolism, and IL-6 and IL-10 to gammopathies [24] and multiple myeloma [22]. Changes in levels of other macrophage-derived markers have also been reported in the plasma of Gaucher disease patients. However, on macrophages themselves, expression of pro-inflammatory mediators is not always apparent [19], although markers characteristic of alternatively activated macrophages are found. Finally, chitotriosidase, a human chitinase produced by activated macrophages, is markedly elevated and commonly used to examine Gaucher disease severity and improvement upon treatment [25,26]. Other hematological manifestations unconnected to macrophages, such as decreased levels of coagulation factors [27,28] and decreased platelet aggregation [29], have also been reported.

To conclude, the main unresolved mechanistic question is whether altered macrophage function is responsible for all of the pathological manifestations in all tissues where pathology is observed, or whether secondary biochemical changes caused directly by GlcCer accumulation in the specific tissues also play a role in pathological development. For instance, in the central nervous system, there is evidence of infiltrating macrophages [30], but neurons themselves are also known to be defective, at least with respect to calcium homeostasis, as discussed below.

CELLULAR PATHOLOGY: NEURONS

Over the past few years, considerable information has emerged implying that at least one major cellular pathway, namely intracellular Ca^{2+}-homeostasis, is altered in neurons in a number of LSDs, including Gaucher disease. That Ca^{2+} might be involved is perhaps not surprising since Ca^{2+} plays an important role in regulating a great variety of neuronal processes. Moreover, altered Ca^{2+}-homeostasis is a hallmark of brain pathophysiology in several diseases, such as epilepsy [31], stroke [32], and Alzheimer's disease [33].

The ER is the major intracellular Ca^{2+} store in neurons [34]. Ca^{2+} is released from the ER to the cytosol via two types of Ca^{2+}-release channels that reside in the ER membrane, the Ca^{2+}-gated Ca^{2+}-release channel, the ryanodine receptor (RyaR)

[35], and the inositol 1,4,5-trisphosphate-gated Ca^{2+}-release channel, the IP_3-receptor (IP_3R) [36]. Ca^{2+} is pumped from the cytosol into the ER lumen via the sarco/endo-plasmic reticulum Ca^{2+}-ATPase (SERCA) [37].

In 1999, we demonstrated altered Ca^{2+}-homeostasis [38] in neurons in a neu-ronopathic model of Gaucher disease [39]. Upon incubation of neurons with con-duritol B-epoxide (CBE), a GlcCerase inhibitor [40], enhanced Ca^{2+}-release was detected from intracellular stores in response to caffeine, an agonist of the RyaR, which resulted in increased sensitivity to neurotoxic agents, especially glutamate. The increased sensitivity to neurotoxic agents could be reversed either by inhibition of sphingolipid synthesis [38] or by exogenously added GlcCerase [41]. Moreover, preincubation with antagonistic concentrations of ryanodine completely blocked the toxic effects of a number of neurotoxins, demonstrating that Ca^{2+}-release from the ER via the RyaR was responsible for neuronal death [41].

We next analyzed the effect of GlcCer on Ca^{2+}-release and uptake in microsomes isolated from rat brain. When added to microsomes, GlcCer had no effect by itself on either Ca^{2+}-release or uptake [42]. However, it significantly and specifically augmented Ca^{2+}-release induced by RyaR agonists. In addition, an increase in the frequency of "Ca^{2+} sparks" (quanta of Ca^{2+}-release caused either by spontaneous or Ca^{2+}-induced channel opening) was also observed in microsomes incubated with GlcCer; this observation is of some importance since Ca^{2+} sparks were suggested to be of key importance in Ca^{2+}-signaling in the brain [43]. Finally, GlcCer was the only glycosphingolipid of those tested that enhanced agonist-induced Ca^{2+}-release via the RyaR, suggesting a highly specific and regulated mode of action of GlcCer on the RyaR.

Importantly, the amount of exogenously added GlcCer required to sensitize the RyaR was similar to levels detected in brain microsomes obtained post mortem from Gaucher disease patients [42], and we therefore decided to examine whether agonist-induced Ca^{2+}-release is also altered in human Gaucher tissue. For these analyses, we obtained brain tissue (the temporal lobe) post mortem from three type 1 Gaucher patients, five type 2, and three type 3 patients, and compared them with 3 to 5 control brains. A linear correlation was observed between microsomal GlcCer levels and the extent of agonist-induced Ca^{2+}-release via the RyaR, and some correlation was observed with the severity of the neurological symptoms. Thus, 43% of the Ca^{2+} in microsomes was released from type 2 brains, which contained 27 nmol GlcCer/mg of protein, compared to 27% Ca^{2+}-release in type 3 brains, which contained 11 nmol GlcCer/mg of protein, 28% Ca^{2+}-release in type 1 brains, which contained 9 nmol GlcCer/mg of protein, and 18% Ca^{2+}-release in control brains, which contained 2 nmol GlcCer/mg of protein [44]. These findings are consistent with our suggestion that defective Ca^{2+}-homeostasis may be one of the mechanisms responsible for neuropathophysiology in type 2 Gaucher disease. Moreover, Ca^{2+}-induced Ca^{2+}-release (CICR) was consistently detected in type 2 brain microsomes, with CICR particularly evident in samples that displayed high levels of agonist-induced Ca^{2+}-release. Thus, a cumulative effect of enhanced agonist-induced Ca^{2+}-release and CICR from internal stores would result in a significant elevation of cytosolic Ca^{2+} in neurons that have accumulated GlcCer. Interestingly, skeletal muscle is apparently not involved in Gaucher disease pathogenesis, and cardiac pathology is only occasionally observed [45], even though all of these tissues contain high levels of the RyaR, albeit different isoforms to that in the brain, suggesting that GlcCer interacts with a brain-specific RyaR isoform. Of the three RyaR isoforms, RyaR2 is the most widely expressed in the brain, although high concentrations of RyaR1 and RyaR3 are found in glutaminergic brain areas such as the cerebellar Purkinje cell layer and the hippocampus [46].

GlcSph also stimulated Ca^{2+}-release from microsomes, but by a completely different mechanism to GlcCer [42,47]. GlcSph directly induced Ca^{2+}-release in an agonist-independent manner, suggesting that GlcSph is itself an agonist of a Ca^{2+}-release channel. In addition, various other lysoglycosphingolipids and lysosphingolipids induced Ca^{2+}-release, but all in a G-protein coupled receptor (GPCR)-independent manner. This latter point is particularly important since previous studies had demonstrated that the second messengers, sphingosine-1-phosphate (S1P) and sphingosylphosphorylcholine (SPC), act as extra-cellular ligands for GPCRs [48], and mobilize Ca^{2+} from intracellular stores via phospholipase C (PLC) activation and inositol 1,4,5-trisphosphate generation [49]. It was also reported that sphingosine, sphingosine-1-phosphate (S1P), galactosylsphingosine [50] and SPC [51,52] mobilize intracellular Ca^{2+} in a GPCR- and PLC-independent manner. However, sphingosine [53,54] and SPC [55,56] were shown to inhibit Ca^{2+}-release from intracellular stores in other studies. Our data demonstrated that galactosylsphingosine- and lactosylsphingosine-induced Ca^{2+}-release could be completely blocked by ryanodine, demonstrating that these lysoglycosphingolipids are agonists of the RyaR. In contrast, GlcSph mobilized Ca^{2+} from microsomes by a RyaR-independent mechanism that was mediated at least in part by the IP_3R. Whether the effects of the lysoglycosphingolipids on intracellular Ca^{2+}-mobilization are relevant for understanding the molecular basis of neuronal dysfunction in the glycosphingolipid storage disorders is a matter of debate. The concentrations of lysoglycosphingolipids required to elicit Ca^{2+}-release [42,47] were significantly higher than those found in tissues from GSL storage disease patients (see, for instance [6,57–59]). We therefore concluded that despite the specificity of lysoglycosphingolipid-induced Ca^{2+}-release with respect to ER Ca^{2+}-channels and pumps [42,47], these effects are unlikely to account for the pathophysiology of neuronal dysfunction in glycosphingolipid storage diseases, including Gaucher disease.

In summary, we suggest that defective Ca^{2+}-homeostasis caused by GlcCer accumulation in neurons may be responsible, at least in part, for the neuropathophysiology observed in the acute neuronopathic form of Gaucher disease. If this proposal is borne out by further experimental evidence, then the acute neuronopathic form of Gaucher disease would join a growing list of neurodegenerative diseases in which Ca^{2+}-homeostasis is central to disease mechanism [60]. Understandably, we have not been able to test whether changes in levels of cytosolic Ca^{2+} occur in living patients, but the accumulative effect of both enhanced agonist-induced Ca^{2+}-release and CICR should presumably result in a significant elevation of cytosolic Ca^{2+}. This would be particularly pronounced in glutaminergic brain areas, since glutamate-induced cell death in cultured neurons that accumulate GlcCer can be inhibited by blocking Ca^{2+}-mobilization via the RyaR [41]. Moreover, data obtained using the same brain specimens as used in our study demonstrated neurodegeneration and gliosis in hippocampal layers CA2 to CA4 [61] (see also Chapter 13), but not in CA1, areas which both have significant glutaminergic input. This regional difference correlated remarkably with the high levels of CICR-susceptible neurons enriched in RyaRs in CA2-4 and low levels in CA1 [62]. In addition, our data imply that some neurological symptoms in Gaucher disease could be alleviated by the use of Ca^{2+} blockers, such as dantrolene, similar to that attempted in epilepsy [63] and other neurodegenerative diseases in which Ca^{2+}-mobilization is altered [64].

Interestingly, Ca^{2+}-homeostasis is also altered in at least two other LSDs involving sphingolipid accumulation. Thus, altered brain Ca^{2+}-homeostasis is observed in a Sandhoff disease mouse model, which accumulates ganglioside GM2, although the mechanism is different to that observed in Gaucher disease. In brain microsomes obtained from these mice, the rate of Ca^{2+}-uptake into the ER, via SERCA, was dramatically reduced, but no difference in the rate of Ca^{2+}-release via the RyaR was observed [65]. Likewise, Ca^{2+}-homeostasis is also altered in the cerebellum in a mouse model of Niemann-Pick A disease, the ASM-/- mouse, due to decreased SERCA protein expression [66]; levels of the IP_3R are also decreased. Altered Ca^{2+}-homeostasis has been demonstrated in a recent study on the GM1 gangliosidosis mouse, which resulted in activation of the unfolded protein response [67]. It therefore appears that altered Ca^{2+}-homeostasis may be a common pathological theme in neuronal forms of sphingolipid storage diseases, including neuronal forms of Gaucher disease.

Altered Ca^{2+}-homeostasis is unlikely to be the only pathway affected in neuronal forms of Gaucher disease. We have also demonstrated that another major cellular pathway, namely phospholipid metabolism, is altered in Gaucher disease. A significant elevation of phosphatidylcholine (PC) synthesis was observed in brain tissue and neurons from a mouse model of Gaucher disease, in a chemically induced model of neuronopathic Gaucher disease [68], and also in a chemically induced model of Gaucher macrophages [69]. In the case of neurons, we suggested that GlcCer accumulation leads to enhanced PC synthesis, which subsequently results in increased axonal growth rates. In contrast, reduced levels of phospholipid synthesis were observed in the Sandhoff mouse model [70], which correlated with decreased rates of axonal growth [71]. Our data therefore suggested a direct correlation between the type of accumulating glycosphingolipid, its effect on phospholipid synthesis, and rates of axonal growth, and implied that altered phospholipid metabolism might also play a role in disease pathogenesis.

CONCLUSION AND FUTURE PROSPECTS

In this chapter, I have discussed what is known about the pathological mechanisms leading from GlcCer accumulation in macrophages and in neurons to disease development. Although the information available at present is not very extensive, it is to be hoped that the renewed interest in the roles that sphingolipids and glycosphingolipids play in normal physiology will stimulate research into their pathological roles upon their accumulation in sphingolipid storage diseases, and, in particular, in Gaucher disease.

REFERENCES

1. Futerman, AH and van Meer, G. The cell biology of lysosomal storage disease. *Nature Rev. Mol. Cell. Biol.*, 5, 554–565, 2004.
2. Beutler, E and Grabowski, GA. Gaucher Disease, in *The Metabolic and Molecular Bases of Inherited Disease.* Scriver, CR, Sly, WS, Childs, B, Beaudet, AL, Valle, D, Kinzler, KW, and Vogelstein, B, Eds., McGraw-Hill, New York, Vol. II, 2001, pp. 3635–3668.
3. Horowitz, M and Zimran, A. Mutations causing Gaucher disease. *Hum. Mutat.*, 3, 1–11, 1994.
4. Grace, ME, Newman, KM, Scheinker, V, Berg-Fussman, A, and Grabowski, GA. Analysis of human acid β-glucosidase by site-directed mutagenesis and heterologous expression. *J. Biol. Chem.*, 269, 2283–2291, 1994.
5. Nilsson, O, Mansson, JE, Hakansson, G, and Svennerholm, L. The occurrence of psychosine and other glycolipids in spleen and liver from the three major types of Gaucher's disease. *Biochim. Biophysica Acta*, 712, 453–463, 1982.
6. Orvisky, E, Park, JK, LaMarca, ME, Ginns, EI, Martin, BM, Tayebi, N, and Sidransky, E. Glucosylsphingosine accumulation in tissues from patients with Gaucher disease: correlation with phenotype and genotype. *Mol. Genet. Metab.*, 76, 262–270, 2002.
7. Suzuki, K. Twenty-five years of the 'psychosine hypothesis': a personal perspective of its history and present status. *Neurochem. Res.*, 23, 251–259, 1998.
8. Gornati, R, Berra, B, Montorfano, G, Martini, C, Ciana, G, Ferrari, P, Romano, M, and Bembi, B. Glycolipid analysis of different tissues and cerebrospinal fluid in type II Gaucher disease. *J. Inher. Metabol. Dis.*, 25, 47–55, 2002.
9. Nilsson, O, Hakansson, G, Dreborg, S, Groth, CG, and Svennerholm, L. Increased cerebroside concentration in plasma and erythrocytes in Gaucher disease: significant differences between type I and type III. *Clin. Genet.*, 22, 274–279, 1982.
10. Gornati, R, Bembi, B, Tong, X, Boscolo, R, and Berra, B. Total glycolipid and glucosylceramide content in serum and urine of patients with Gaucher's disease type 3 before and after enzyme replacement therapy. *Clinica Chimica Acta*, 271, 151–161, 1998.
11. Cox, TM. Gaucher disease: understanding the molecular pathogenesis of sphingolipidoses. *J. Inher. Metabol. Dis.*, 24, Suppl. 2:106–121, 2001.
12. Sabatini, DD and Adesnik, MB, Eds., *The Biogenesis of Membranes and Organelles*, McGraw-Hill, New York, 2001.
13. Aerts, JM, Hollak, C, Boot, R, and Groener, A. Biochemistry of glycosphingolipid storage disorders: implications for therapeutic intervention. *Phil. Trans. Royal Soc. London*, 358, 905–914, 2003.
14. Rijnboutt, S, Aerts, HM, Geuze, HJ, Tager, JM, and Strous, GJ. Mannose 6-phosphate-independent membrane association of cathepsin D, glucocerebrosidase, and sphingolipid-activating protein in HepG2 cells. *J. Biol. Chem.*, 266, 4862–4868, 1991.
15. Bitton, A, Etzell, J, Grenert, JP, and Wang, E. Erythrophagocytosis in Gaucher cells. *Arch. Pathol. Lab. Med.*, 128, 1191–1192, 2004.
16. Naito, M, Takahashi, K, and Hojo, H. An ultrastructural and experimental study on the development of tubular structures in the lysosomes of Gaucher cells. *Lab. Invest.*, 58, 590–598, 1988.
17. Pennelli, N, Scaravilli, F, and Zacchello, F. The morphogenesis of Gaucher cells investigated by electron microscopy. *Blood*, 34, 331–347, 1969.

18. Parkin, J and Brunning, R. Pathology of the Gaucher cell, in *Gaucher Disease: a Century of Delineation and Research*, Desnick, R, Gatt, S, and Grabowski, G, Eds., Alan R. Liss, New York, 1982, p. 151.

19. Boven, LA, van Meurs, M, Boot, RG, Mehta, A, Boon, L, Aerts, JM, and Laman, JD. Gaucher cells demonstrate a distinct macrophage phenotype and resemble alternatively activated macrophages. *Am. J. Clin. Pathol.*, 122, 359–369, 2004.

20. Florena, AM, Franco, V, and Campesi, G. Immunophenotypical comparison of Gaucher's and pseudo-Gaucher cells. *Pathol. Int.*, 46, 155–160, 1996.

21. Zimran, A, Ed., *Gaucher's Disease*. Balliere Tindall, 1997.

22. Barak, V, Acker, M, Nisman, B, Kalickman, I, Abrahamov, A, Zimran, A, and Yatziv, S. Cytokines in Gaucher's disease. *Eur. Cytokine Network*, 10, 205–210, 1999.

23. Hollak, CE, Evers, L, Aerts, JM, and van Oers, MH. Elevated levels of M-CSF, sCD14 and IL8 in type 1 Gaucher disease. *Blood Cells, Mol. Dis.*, 23, 201–212, 1997.

24. Brautbar, A, Elstein, D, Pines, G, Abrahamov, A, and Zimran, A. Effect of enzyme replacement therapy on gammopathies in Gaucher disease. *Blood Cells, Mol. Dis.*, 32, 214–217, 2004.

25. Hollak, CE, van Weely, S, van Oers, MH, and Aerts, JM. Marked elevation of plasma chitotriosidase activity. A novel hallmark of Gaucher disease. *J. Clin. Invest.*, 93, 1288–1292, 1994.

26. Renkema, GH, Boot, RG, Strijland, A, Donker-Koopman, WE, van den Berg, M, Muijsers, AO, and Aerts, JM. Synthesis, sorting, and processing into distinct isoforms of human macrophage chitotriosidase. *Eur. J. Biochem.*, 244, 279–285, 1997.

27. Barone, R, Giuffrida, G, Musso, R, Carpinteri, G, and Fiumara, A. Haemostatic abnormalities and lupus anticoagulant activity in patients with Gaucher disease type I. *J. Inher. Metab. Dis.*, 23, 387–390, 2000.

28. Hollak, CE, Levi, M, Berends, F, Aerts, JM, and van Oers, MH. Coagulation abnormalities in type 1 Gaucher disease are due to low-grade activation and can be partly restored by enzyme supplementation therapy. *Brit. J. Haematol.*, 96, 470–476, 1997.

29. Gillis, S, Hyam, E, Abrahamov, A, Elstein, D, and Zimran, A. Platelet function abnormalities in Gaucher disease patients. *Am. J. Hematol.*, 61, 103–106, 1999.

30. Wong, K, Sidransky, E, Verma, A, Mixon, T, Sandberg, GD, Wakefield, LK, Morrison, A, Lwin, A, Colegial, C, Allman, JM, and Schiffmann, R. Neuropathology provides clues to the pathophysiology of Gaucher disease. *Mol. Gen. Metabol.*, 82, 192–207, 2004.

31. Stefani, A, Spadoni, F, and Bernardi, G. Voltage-activated calcium channels: targets of antiepileptic drug therapy? *Epilepsia*, 38, 959–965, 1997.

32. Small, DL, Morley, P, and Buchan, AM. Biology of ischemic cerebral cell death. *Prog. Cardiovasc. Dis.*, 42, 185–207, 1999.

33. Mattson, MP and Chan, SL. Neuronal and glial calcium signaling in Alzheimer's disease. *Cell. Calcium*, 34, 385–397, 2003.

34. Henkart, M. Identification and function of intracellular calcium stores in neurons. *Fed. Proc.*, 39, 2776–2777, 1980.

35. Fill, M and Copello, JA. Ryanodine receptor calcium release channels. *Physiol. Rev.*, 82, 893–922, 2002.

36. Mikoshiba, K. The InsP3 receptor and intracellular $Ca2+$ signaling. *Curr. Opin. Neurobiol.*, 7, 339–345, 1997.

37. Misquitta, CM, Mack, DP, and Grover, AK. Sarco/endoplasmic reticulum $Ca2+$ (SERCA)-pumps: link to heart beats and calcium waves. *Cell. Calcium*, 25, 277–290, 1999.

38. Korkotian, E, Schwarz, A, Pelled, D, Schwarzmann, G, Segal, M, and Futerman, AH. Elevation of intracellular glucosylceramide levels results in an increase in endoplasmic reticulum density and in functional calcium stores in cultured neurons. *J. Biol. Chem.*, 274, 21673–21678, 1999.

39. Schwarz, A, Rapaport, E, Hirschberg, K, and Futerman, AH. A regulatory role for sphingolipids in neuronal growth. Inhibition of sphingolipid synthesis and degradation have opposite effects on axonal branching. *J. Biol. Chem.*, 270, 10990–10998, 1995.

40. Legler, G and Bieberich, E. Active site directed inhibition of a cytosolic beta-glucosidase from calf liver by bromoconduritol B epoxide and bromoconduritol F. *Arch Biochem. Biophys.*, 260, 437–442, 1988.

41. Pelled, D, Shogomori, H, and Futerman, AH. The increased sensitivity of neurons with elevated glucocerebroside to neurotoxic agents can be reversed by imiglucerase. *J. Inherit. Metab. Dis.*, 23, 175–184, 2000.

42. Lloyd-Evans, E, Pelled, D, Riebeling, C, Bodennec, J, de-Morgan, A, Waller, H, Schiffmann, R, and Futerman, AH. Glucosylceramide and glucosylsphingosine modulate calcium mobilization from brain microsomes via different mechanisms. *J. Biol. Chem.*, 278, 23594–23599, 2003.

43. Melamed-Book, N, Kachalsky, SG, Kaiserman, I, and Rahamimoff, R. Neuronal calcium sparks and intracellular calcium "noise." *Proc. Natl. Acad. Sci, U.S.A.*, 96, 15217–15221, 1999.

44. Pelled, D, Trajkovic-Bodennec, S, Lloyd-Evans, E, Sidransky, E, Schiffmann, R, and Futerman, AH. Enhanced calcium release in the acute neuronopathic form of Gaucher disease. *Neurobiol. Dis.*, 18, 83–88, 2005.

45. Smith, RL, Hutchins, GM, Sack, GH, Jr., and Ridolfi, RL. Unusual cardiac, renal and pulmonary involvement in Gaucher's disease. Intersitial glucocerebroside accumulation, pulmonary hypertension and fatal bone marrow embolization. *Am. J. Med.*, 65, 352–360, 1978.

46. Mori, F, Fukaya, M, Abe, H, Wakabayashi, K, and Watanabe, M. Developmental changes in expression of the three ryanodine receptor mRNAs in the mouse brain. *Neurosci. Lett.*, 285, 57–60, 2000.

47. Lloyd-Evans, E, Pelled, D, Riebeling, C, and Futerman, AH. Lyso-glycosphingolipids mobilize calcium from brain microsomes via multiple mechanisms. *Biochem. J.*, 375, 561–565, 2003.

48. Fukushima, N, Ishii, I, Contos, JJ, Weiner, JA, and Chun, J. Lysophospholipid receptors. *Annu. Rev. Pharmacol. Toxicol.*, 41, 507–534, 2001.

49. Ghosh, TK, Bian, J, and Gill, DL. Intracellular calcium release mediated by sphingosine derivatives generated in cells. *Science*, 248, 1653–1656, 1990.

50. Liu, R, Farach-Carson, MC, and Karin, NJ. Effects of sphingosine derivatives on MC3T3-E1 pre-osteoblasts: psychosine elicits release of calcium from intracellualr stores. *Biochem. Biophys. Res. Commun.*, 214, 676–684, 1995.

51. Calcerrada, MC, Miguel, BG, Catalan, RE, and Martinez, AM. Sphingosylphosphorylcholine increases calcium concentration in isolated brain nuclei. *Neurosci. Res.*, 33, 229–232, 1999.

52. Dettbarn, C, Betto, R, Salviati, G, Sabbadini, R, and Palade, P. Involvement of ryanodine receptors in sphingosylphosphorylcholine-induced calcium release from brain microsomes. *Brain Res.*, 669, 79–85, 1995.

53. Sabbadini, RA, Betto, R, Teresi, A, Fachechi-Cassano, G, and Salviati, G. The effects of sphingosine on sarcoplasmic reticulum membrane calcium release. *J. Biol. Chem.*, 267, 15475–15484, 1992.

54. Sharma, C, Smith, T, Li, S, Schroepfer, GJ, Jr., and Needleman, DH. Inhibition of Ca2+ release channel (ryanodine receptor) activity by sphingolipid bases: mechanism of action. *Chem. Phys. Lipids*, 104, 1–11, 2000.

55. Uehara, A, Yasukochi, M, Imanaga, I, and Berlin, JR. Effect of sphingosylphosphorylcholine on the single channel gating properties of the cardiac ryanodine receptor. *FEBS Lett.*, 460, 467–471, 1999.

56. Yasukochi, M, Uehara, A, Kobayashi, S, and Berlin, JR. Ca2+ and voltage dependence of cardiac ryanodine receptor channel block by sphingosylphosphorylcholine. *Pflugers Arch*, 445, 665–673, 2003.

57. Bodennec, J, Bodennec-Trajkovic, S, and Futerman, AH. Simultaneous quantification of lyso-neutral glycosphingolipids and neutral glycosphingolipids by N-acetylation with [3H]acetic anhydride. *J. Lipid. Res.*, 44, 1413–1419, 2003.

58. Nilsson, O, Mansson, JE, Hakansson, G, and Svennerholm, L. The occurrence of psychosine and other glycolipids in spleen and liver from the three major types of Gaucher's disease. *Biochim. Biophys. Acta*, 712, 453–463, 1982.

59. Rodriguez-Lafrasse, C and Vanier, MT. Sphingosylphosphorylcholine in Niemann-Pick disease brain: accumulation in type A but not in type B. *Neurochem. Res.*, 24, 199–205, 1999.

60. Berridge, MJ, Bootman, MD, and Lipp, P. Calcium — a life and death signal. *Nature*, 395, 645–648, 1998.

61. Sandberg, GD, Wong, K, Morrison, AL, Colegial, CH, Mena, H, and Schiffmann, R. Neuropathologic clues to laminar neuronal cell loss in neuronopathic Gaucher disease. *J. Neuropathol. Exp. Neurol.*, 57, 60, 1998.

62. Verma, A, Hirsch, DJ, and Snyder, SH. Calcium pools mobilized by calcium or inositol 1,4,5-trisphosphate are differentially localized in rat heart and brain. *Mol. Biol. Cell.*, 3, 621–631, 1992.

63. Pal, S, Sun, D, Limbrick, D, Rafiq, A, and DeLorenzo, RJ. Epileptogenesis induces long-term alterations in intracellular calcium release and sequestration mechanisms in the hippocampal neuronal culture model of epilepsy. *Cell. Calcium*, 30, 285–296, 2001.

64. Mattson, MP, LaFerla, FM, Chan, SL, Leissring, MA, Shepel, PN, and Geiger, JD. Calcium signaling in the ER: its role in neuronal plasticity and neurodegenerative disorders. *Trends Neurosci.*, 23, 222–229, 2000.

65. Pelled, D, Lloyd-Evans, E, Riebeling, C, Jeyakumar, M, Platt, FM, and Futerman, AH. Inhibition of calcium uptake via the sarco/endoplasmic reticulum Ca2+-ATPase in a mouse model of Sandhoff disease and prevention by treatment with N-butyldeoxynojirimycin. *J. Biol. Chem.*, 278, 29496–29501, 2003.

66. Ginzburg, L and Futerman, AH. Defective calcium homeostasis in the cerebellum in a mouse model of Niemann-Pick A disease. *J. Neurochem.*, 2005.

67. Tessitore, A, del, PMM, Sano, R, Ma, Y, Mann, L, Ingrassia, A, Laywell, ED, Steindler, DA, Hendershot, LM, and d'Azzo, A. GM1-ganglioside-mediated activation of the unfolded protein response causes neuronal death in a neurodegenerative gangliosidosis. *Mol. Cell.*, 15, 753–766, 2004.

68. Bodennec, J, Pelled, D, Riebeling, C, Trajkovic, S, and Futerman, AH. Phosphatidylcholine synthesis is elevated in neuronal models of Gaucher disease due to direct activation of CTP:phosphocholine cytidylyltransferase by glucosylceramide. *Faseb J.*, 16, 1814–1816, 2002.

69. Trajkovic-Bodennec, S, Bodennec, J, and Futerman, AH. Phosphatidylcholine metabolism is altered in a monocyte-derived macrophage model of Gaucher disease but not in lymphocytes. *Blood Cells Mol. Dis.*, 33, 77–82, 2004.

70. Buccoliero, R, Bodennec, J, Van Echten-Deckert, G, Sandhoff, K, and Futerman, AH. Phospholipid synthesis is decreased in neuronal tissue in a mouse model of Sandhoff disease. *J. Neurochem.*, 90, 80–88, 2004.
71. Pelled, D, Riebeling, C, van Echten-Deckert, G, Sandhoff, K, and Futerman, AH. Reduced rates of axonal and dendritic growth in embryonic hippocampal neurones cultured from a mouse model of Sandhoff disease. *Neuropathol. Appl. Neurobiol.*, 29, 341–349, 2003.

CHAPTER 7

The Biochemistry and Cellular Biology of Sphingolipids and Glucosylceramide

James A. Shayman

CONTENTS

INTRODUCTION

Glucosylceramide (GlcCer) is a member of a very heterogeneous group of lipids termed glycosphingolipids. These lipids are distinct from other cellular lipids due to the presence of a ceramide backbone in addition to a carbohydrate head group. GlcCer is the base cerebroside for most of the more than 300 mammalian glycosphingolipids characterized to date. Many of these glycosphingolipids, including GlcCer, have been implicated in a variety of cellular functions. These functions include, but are not limited to, serving as cellular receptors mediating cell-cell, cell-matrix, cell-toxin, and cell-pathogen interactions and serving as mediators of the regulation of cell signaling and membrane protein sorting. Glycosphingolipids also appear to be important components of membrane "rafts" that along with cholesterol are important platforms for cell signaling and recognition functions.

The synthesis and metabolism of GlcCer is tightly regulated at the cellular level based not only on its metabolic pathways, but also because of the separation of these pathways and trafficking of GlcCer on a subcellular level. *A priori*, any factors that regulate the substrates for GlcCer formation (ceramide and UDP-glucose) or metabolism, including degradation or further glycosylation for more complex glycosphingolipid synthesis, may impact on cellular GlcCer content and function. Therefore, an attempt to understand the role GlcCer in the pathogenesis of Gaucher disease, and how GlcCer levels may be targeted by different therapeutic strategies, should begin with an understanding of the biochemistry and cellular biology of sphingolipids in general.

LONG CHAIN BASE AND CERAMIDE SYNTHESIS

Sphingolipids are defined as being any and all compounds that contain a long chain sphingoid base. Long chain bases containing 18 and 20 carbons are the most commonly found.[1] Of these, a C4-C5 double bond in the *trans* D-*erythro* conformation predominates followed by bases lacking the double bond.

Sphingolipid synthesis begins with the formation of the sphingoid amine. The first step is acknowledged to be the condensation of serine and palmitoyl-CoA to form (2S)-3-ketosphinganine.[2] The anabolic enzyme, serine palmitoyltransferase, is a pyridoxal 5'-phosphate-dependent enzyme that demonstrates distinct specificities for various acyl-CoAs. Palmitoyl-CoA is preferred over stearoyl-CoA by a ratio of 5:1.[3] The relative abundance of 18 and 20 carbon long chain bases in mammalian cells is likely the result of this difference in substrate specificity. The 3-keto group is subsequently reduced to the 3-hydroxyl by 3-ketosphinganine reductase to form (2S-3R)-sphinganine.[4] This reaction is NADPH dependent. The gene encoding the enzyme responsible for this reduction has recently been identified as the follicular lymphoma variant translocation-1 (FVT-1) gene.[5] Ceramide is then formed by the acylation of sphinganine to produce ceramide. In mammalian cells the fatty acids present in ceramide vary, but they are typically long. The most common fatty acids are palmitic (C16:0) and stearic (C18:0). α-Hydroxylated fatty acids are often present as well. The gene and products that likely represent cellular sphinganine:N-acyl transferase (dihydroceramide synthase) activities have recently been identified. The yeast proteins, Lag1p and Lac1p were identified as being required for the synthesis of very long chain ceramides.[6] A mammalian homolog of these proteins, UOG1, was subsequently observed to regulate the synthesis of stearoyl-containing sphingolipids.[7] In search for genes that might encode the Lag1p motif, two additional candidate genes were then identified, named TRH1 and TRH4. TRH4 (also known as LASS5) expression increases the synthesis of sphinganine-containing ceramides.[8]

Ceramides containing sphinganine as their long chain base are the precursors for other cell ceramides. Sphinganine-containing ceramides are referred to as dihydroceramides. However, in most cellular sphingolipids that contain a ceramide base, sphingenine (sphingosine) is the long chain base. The Δ4-*trans*-unsaturation of sphinganine most likely results from the activity of two recently identified, endo-plasmic reticulum associated desaturases, DES1 and DES2.[9]

CERAMIDE METABOLISM

Ceramide is the precursor for GlcCer, the primary sphingolipid that accumulates in Gaucher disease. In addition, ceramide is a central intermediate in the formation of several other biologically important sphingolipids (Figure 7.1). Ceramide has important biological activities in its own right. Most prominently, ceramide has been studied as a mediator of apoptosis[10] and of cell cycle arrest.[11] A number of potential targets for ceramide mediated cell death have been identified. These have been extensively reviewed by others.[12,13]

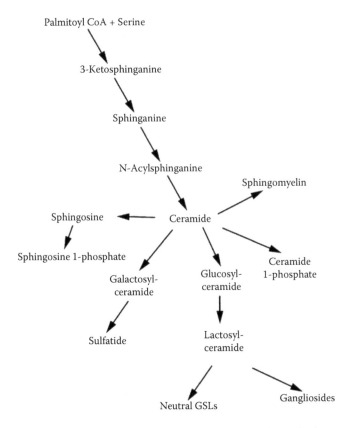

Figure 7.1   Primary pathways for sphingolipid synthesis. Ceramide synthesis proceeds from long chain base synthesis. Ceramide is a key intermediate in the formation of glycosphingolipids, sphingomyelin, sphingosine 1-phosphate, and ceramide 1-phosphate.

Ceramide may be deacylated by a ceramidase to form free sphingosine. The cleavage of ceramide occurs at the amide bond resulting in the formation of sphingosine and free fatty acid. Ceramidases have historically been characterized as acidic, neutral, and alkaline based on their pH optima. An acidic ceramidase was first described in 1963 by Gatt.[14] Subsequently, this activity was discovered to be absent in patients with Farber disease.[15] The enzyme exists in its mature form as a ca. 50 kDa polypeptide consisting of an α subunit of 13 kDa and β subunit of 40 kDa.[16,17] The full-length cDNA encodes a 55 kDa precursor that is proteolytically processed into the two subunits. The mature enzyme primarily exists within the lysosome, but activity can be measured extracellularly as well.

Neutral ceramidases have more recently been purified, cloned, and characterized in several species, including bacteria, Drosophila, zebrafish, mouse, rat, and human.[18–23] A primary structure that is distinct from the acid ceramidase supports the view that these enzymes arose from different ancestral genes. However, among the types of neutral ceramidases, differences between species have been noted as to whether the enzyme is soluble or membrane associated. When isolated as a soluble protein, the mouse enzyme is 94 kDa; when isolated as a membrane-associated protein, the enzyme is 112 kDa. The latter mouse ceramidase is a type II integral membrane protein in which there is an N-terminal hydrophobic sequence. An adjacent mucin box that is O-glycosylated serves to anchor the enzyme to the membrane.[24]

Ceramidase activities are important in regulating the levels of ceramide within cells. Free sphingosine is not formed de novo, rather it can only be formed by the hydrolysis of ceramide. Thus ceramidases also function to regulate the levels of sphingosine and its metabolite sphingosine 1-phosphate. The latter is formed by the phosphorylation of sphingosine by sphingosine kinase.[25] Significant work has been focused on the biological roles of sphingosine 1-phosphate mediated through the S1P (Edg) family of G protein coupled receptors.[26] Thus a key role for ceramidases in general and for neutral ceramidases in particular may be the regulation of levels of ceramide, sphingosine, and sphingosine 1-phosphate within nonlysosomal compartments. In support of this view are the observations that neutral ceramidases can be activated by both cytokines and growth factors.

Ceramide is also metabolized to sphingomyelin, a major phospholipid within cells. Sphingomyelin synthesis occurs via a phosphatidylcholine:ceramide cholinephosphotransferase that transfers the phosphorylcholine group from phosphatidylcholine to the 1-hydroxyl of ceramide.[27] Two products are formed, sphingomyelin and diacylglycerol. Using a functional cloning strategy in yeast, two mammalian sphingomyelin synthases (SMSs) were recently identified. SMS1 is localized to the Golgi; SMS2 resides in the plasma membrane.[28]

Other pathways for ceramide metabolism have been identified that may produce additional bioactive products. A ceramide kinase has recently been cloned.[29] A transacylation pathway resulting in the formation of 1-O-acylceramide has also been described.[30] This pathway appears to be mediated by an acidic lysosomal phospholipase A2.[31] The biological significance of 1-O-acylceramide is unknown.

GLUCOSYLCERAMIDE SYNTHESIS

The glycosylation of ceramide leads to a series of greater than 300 mammalian glycosphingolipids. The transfer of a glucose residue from UDP-glucose is the first glycosylation step for the vast majority of glycosphingolipids. In mammals, the transfer of a galactose residue from UDP-galactose is restricted to a limited number of specialized tissues. UDP-Gal:ceramide galactosyl transferase is an ER-associated enzyme. Following the transport of galactosylceramide to the Golgi, sulfation may occur at the 3 position of galactose to form sulfatide.[32] Glycosphingolipids containing galactosylceramide as the base cerebroside are primarily restricted to the myelin sheath surrounding axons. In addition, these glycolipids are found in renal tubular epithelia and in the intestine.

GlcCer, by contrast, is ubiquitous. UDP-glucose:ceramide glucosyltransferase (GlcCer synthase) catalyzes the formation of GlcCer. This enzyme has been the target of small molecule inhibitors designed to block the synthesis of GlcCer and other GlcCer-based glycosphingolipids. Thus a more comprehensive review of its structure and properties is warranted.

GlcCer synthase was originally cloned by Ichikawa et al. and was mapped to chromosome 9q31 in humans.[33] The enzyme has been subsequently cloned from many organisms, including mouse,[34] rat,[35] drosophila,[36] and *C. elegans*.[37] Functional analyses have also led to the identification of GlcCer synthase genes in plants (*Gossypium arboreum* [cotton]) and various fungi (*Magnaporthe grisea, Candida albicans*, and *Pichia pastoris*).[38] GlcCer synthase is a type III protein but bears limited structural homology to other proteins. The protein contains a type II glycosyltransferase domain. This domain is observed in a diverse family of enzymes, transferring sugar from UDP-glucose, UDP-N-acetyl-galactosamine, GDP-mannose or CDP-abequose, to a range of substrates, including cellulose, dolichol phosphate, and teichoic acids. The mouse gene, termed *Ugcg*, is 32 kilobases in length and is comprised of 9 exons and 8 introns.

Structural and topological studies of GlcCer synthase have revealed that the enzyme is localized to the cytosolic side of the Golgi.[39] Both the recombinant enzyme and enzyme isolated from liver Golgi membranes migrate as a 38 kDa protein on SDS-polyacrylamide electrophoresis. When Golgi membranes are treated with crosslinking reagents, however, an antiGlcCer synthase antibody recognizes a 50 kDa polypeptide, consistent with the oligmerization of the enzyme with a small, ca. 15 kDa protein. This characterization and function of this accessory protein were not reported at the time this chapter was written.

A limited amount of work has been performed on the characterization of the active site of GlcCer synthase. The amino acid-specific reagent, NEM, is reported to inactivate the enzyme. Replacement of C207 with alanine leads to a reduction in activity consistent with C207 being the target of NEM. Similarly, diethylpyrocarbonate inactivates the enzyme. This inactivation is blocked by an H193A conversion or by preincubation with excess UDP-glucose. These observations suggest that H193 is near the UDP-glucose binding site. When aligned with other glycosyltransferases, a conserved active site motif consisting of D1, D2, D3, (Q/R)XXRW found in GTF2 β-glycosyltransferases.[40] These amino acids correspond to D-92, D144, D-236,

R-272, R-275, and W-276 in the rat synthase. The substitution of each residue with alanine results in the absence or near total loss of synthase activity. Mutagenesis of H193 of the rat synthase also abolishes the inhibitory effect of PDMP.[35]

HIGHER ORDER GLYCOSPHINGOLIPIDS

GlcCer may be transported unchanged to the plasma membrane or may be further metabolized. The catabolism of GlcCer to ceramide by β-glucocerebrosidase is the subject of several chapters in this volume and will not be further discussed. More complex glycosphingolipids are produced by the sequential addition of single sugars by the use of activated nucleotide precursors. In mammalian tissues the first sugar added is galactose. The product of this reaction is lactosylceramide (Galβ1-4GlcCer). The enzyme that catalyzes this reaction is lactosylceramide synthase. This enzyme is also found in the Golgi. However, it is localized to the Golgi lumen. Thus GlcCer formed on the cytosolic side of the Golgi is required to flip into the luminal side to allow for more complex glycosphingolipids to be synthesized.

Several enzymes are involved in the synthesis of the oligosaccharide backbone. By convention, a series of root names has been assigned to the various glycosphingolipids based on the type and sequence of sugars (Table 7.1). Additional sugars may be added to the periphery of the oligosaccharide backbone. These include sialic acids and fucose groups. When sialic acids are added, the resultant glycosphingolipids are termed gangliosides. Svennerholm categorized these glycolipids based on the number of sialic acids present and their migration pattern on chromatography (Table 7.2).[41] These gangliosides may be alternatively categorized as containing no (O series), one (a series), two (b series), or three (c series) sialic acids. These sialic acids may be further modified by O-acetylation, N-glycosylation, and lactonization.[42]

The biological diversity and potential importance of glycosphingolipids in general and gangliosides in particular are supported by a wide range of patterns of expression between species and cell types. The regulation of ganglioside synthesis has been the subject of considerable study and depends in part on the localization of particular glycosyltransferases within the Golgi and the observation that the stepwise glycosylation of precursor glycolipids is accomplished by a finite number of enzymes of limited specificity.[43] Thus the sialyltransferases that form GM3 from

Table 7.1   Root Names and Structures for Complex Glycosphingolipids

Root	Symbol	Structure
Ganglio	Gg	Galβ1-3GalNAcβ1-4Galβ1-4Glcβ1-Cer
Lacto	Lc	Galβ1-3GlcNAcβ1-4Galβ1-4Glcβ1-Cer
Neolacto	nLc	Galβ1-4GlcNAcβ1-4Galβ1-4Glcβ1-Cer
Globo	Gb	GalNAcβ1-3Galα1-4Galβ1-4Glcβ1-Cer
Isoglobo	iGb	GalNAcβ1-3Galα1-3Galβ1-4Glcβ1-Cer
Mollu	Mu	GlcNAcβ1-2Manα1-3Manβ1-4Glcβ1-Cer
Arthro	At	GalNAcβ1-4GlcNAcβ1-3Manβ1-4Glcβ1-Cer

Table 7.2   The Svennerholm System for Ganglioside Nomenclature

Abbreviation	Structure
GM4	NeuAcα2-3Galβ1-Cer
GM3	NeuAcα2-3Galβ1-4Glcβ1-Cer
GM2	GalNAcβ1-4(NeuAcα2-3)Galβ1-4Glcβ1-Cer
GM1a	Galβ1-GalNAcβ1-4(NeuAcα2-3)Galβ1-4Glcβ1-Cer
GM1b	NeuAcα2-3Galβ1-GalNAcβ1-4Galβ1-4Glcβ1-Cer
GD3	(NeuAcα2-8NeuAcα2-3)Galβ1-4Glcβ1-Cer
GD2	GalNAcβ1-4(NeuAcα2-8NeuAcα2-3)Galβ1-4Glcβ1-Cer
GD1a	NeuAcα2-3Galβ1-3GalNAcβ1-4(NeuAcα2-3)Galβ1-4Glcβ1-Cer
GD1b	Galβ1-3GalNAcβ1-4(NeuAcα2-8NeuAcα2-3)Galβ1-4Glcβ1-Cer
GT1a	NeuAcα2-8NeuAcα2-3 Galβ1-3GalNAcβ1-4(NeuAcα2-3)Galβ1-4Glcβ1-Cer
GT1b	NeuAcα2-3Galβ1-3GalNAcβ1-4 (NeuAcα2-8NeuAcα2-3)Galβ1-4Glcβ1-Cer
GT1c	Galβ1-3GalNAcβ1-4(NeuAcα2-8NeuAcNeuAcα2-8NeuAcα2-3)Galβ1-4Glcβ1-Cer

lactosylceramide (SAT I), GD3 from GM3 (SAT II), and GT3 from GD3 (SAT III) are highly specific for their substrates. By contrast, the stepwise glycosylation of the lactosylceramide, GM3, GD3, and GT3, precursors for the O- a-, b-, and c- series gangliosides, respectively, is performed by a limited number of glycosyltransferases (Figure 7.2).

These anabolic glycosyltransferases are membrane bound and react with their glycolipid substrates by diffusion within the lipid bilayer. By contrast, the glycosidases that catabolized glycolipids are soluble enzymes that act on the membrane bound substrates. These enzymes often require activator proteins, termed saposins, for activity.[44]

SITES OF CELLULAR METABOLISM AND SORTING

Long chain base and ceramide synthesis occurs in the endoplasmic reticulum where the early enzymes including serine palmitoyltransferase, 3-ketosphinganine reductase, and dihydroceramide synthase are localized.[45] Ceramide appears to spontaneously cross the ER membrane and can be converted to galactosylceramide when galactosylceramide synthase is present. Ceramide more often is transported to the Golgi via a vesicular route to the early Golgi where it is inserted into the membrane facing the cytosol.[46,47] Transport is dependent on a 68 kDa protein termed CERT. Ceramide is then converted to GlcCer by GlcCer synthase localized on the cytosolic side of the Golgi membrane.[48]

GlcCer then moves to the luminal side of the *cis*-Golgi by an uncharacterized flippase. The subsequent glycosylations with the additions of the core and peripheral carbohydrates occurs within the Golgi stacks. The current view is that a gradient distribution a glycosyltransferases exists with early glycosylations occurring in the *cis*- and *medial*-Golgi and later glycosylations occurring in the *trans*-Golgi network.[49] Two mutually compatible hypotheses exist for explaining the subcellular sites of sphingolipid glycosylation. The first hypothesis states that the transmembrane domain of each glycosyltransferase precludes the inclusion of the enzymes in the

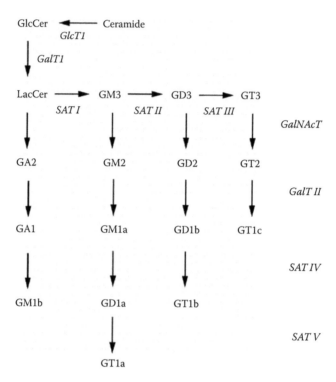

Figure 7.2 Combinatorial gangliosides synthesis. The pathways for the major gangliosides listed in Table 7.2 are shown. The catalytic enzymes are italicized. The abbreviations used are as follows: GlcT1, glucosylceramide synthase; GalT1, lactosylceramide synthase; GalNAcT, N-acetylgalactosamine transferase; SAT I-V, sialyltransferase 1 through 5 respectively. (Adapted from Kolter, T., Proia, R.L., and Sandhoff, K., *J. Biol. Chem.*, 277, 25859, 2002. With permission.)

vesicular trafficking of nascent glycosphingolipids from one Golgi stack to another. The second hypothesis states that sequential glycosylation occurs as part of multicomponent complexes.

Some evidence exists that glycosylation and deglycosylation events may occur within the plasma membrane.[50] For example, a sialidase has been identified in the plasma membrane of many cells that desialates gangliosides.[51,52] One such sialidase may be anchored by a glycerophosphatidylinositide, placing it within a sphingolipid-enriched membrane raft.[53]

Glycosphingolipid recycling has been well documented by use of radiolabeled or fluorescently tagged lipids. This may occur through two initial routes. Glycosphingolipids, particularly gangliosides, may circulate extracellularly and be reincorporated through an endocytic process. Alternatively, plasma membrane-associated glycolipids may also endocytose. Studies on exogenously added gangliosides are consistent with the incorporation of glycosphingolipids into a sorting endosome. From here glycolipids may traffic through one of at least three distinct pathways. They may directly recycle in an unaltered form to the plasma membrane. They may sort to the Golgi apparatus where further glycosylation may occur. Alternatively, they may sort to late endosomes and lysosomes to undergo degradation.[49]

PHYSICAL PROPERTIES AND MEMBRANE RAFTS

Investigators have long noted the existence of significant biophysical differences between sphingolipids and glycerolipids. More recently, these differences have been studied in an attempt to understand how sphingolipids may function uniquely within biological membranes. A major difference between sphingolipids and glycerolipids is the high degree of saturation of both the fatty acid in amide linkage and the long chain base. Because the 4-5 double bond in sphingosine is in the *trans* configuration, the saturation of most ceramides is preserved. This property results in greater packing of sphingolipids compared to glycerolipids within the membrane. This packing is further enhanced by the number of chemical groups that can function as hydrogen bond donors and acceptors. This includes the region between the ceramide backbone and the polar phosphorylcholine or carbohydrate head groups. In addition, the sugar head groups can interact with one another. Finally, cholesterol preferentially interacts with sphingolipids over glycerolipids through van der Waals interactions.[54]

The combination of increased hydrogen bonding and denser packing can be measured as an increase in melting temperature. Melting temperature represents the temperature above which a bilayer switches from a gel or solid-ordered state to a fluid or liquid crystalline state. Sphingolipids have a higher melting temperature than glycerolipids. On this basis, it is believed that sphingolipids, present at much lower concentrations within the membrane, are able to self-aggregate in the presence of glycerolipids. Support for this view has been derived from studies on lipid phase separations in which a gel phase or solid-ordered phase, lipid-ordered phase, and lipid-disordered phase can be measured. In such artificial systems, the distribution of proteins containing glycosylphosphatidylinositol anchors can be demonstrated to be affected by the presence of gangliosides.[55–57]

Based on the biophysical differences between glycerolipids and sphingolipids, investigators have postulated that these phase separations exist within biological membranes. In support of this view are several papers describing the enrichment of sphingolipids in membrane preparations extracted with cold detergents such as Triton X-100.[58,59] Microscopic evidence has shown that GM1 and GM3 are clustered on erythrocyte membranes, and other optical techniques have been used to study the real-time movement of probes into and out of domains within live cells. These methods provide support for the existence of microdomains or "lipid rafts" within cells. However, the size, shape, and dynamics of these rafts remain to be fully elucidated.

CELLULAR BIOLOGY OF SPHINGOLIPIDS

Much has been written about the functions of sphingolipids over the last 20 years. A comprehensive treatment of the role of sphingolipids in cellular biology and the pathogenesis of disease is well beyond the scope of this brief review. Much of this work has focused on the role of sphingolipids in cell signaling processes. Collectively one may divide these studies into three general groups. These groups include sphingolipids as intracellular second messengers, sphingolipids as cell differentiation markers and receptors, and sphingolipids as modulators of signal transduction.

Ceramide has been identified as a potential signal for apoptosis,[10] cell growth arrest,[11] and stress responses[60] for many years. Traditionally ceramide has been viewed as a second messenger akin to diglyceride and protein kinase C. Unlike diglyceride, however, the identification of a single class of cellular targets for ceramide action has been difficult. Cellular targets that have been identified as sites for ceramide include but are not limited to PP2A,[61] PP1,[62] c-Raf,[63] PKCζ,[64] and KSR1.[65]

Sphingosine-1-phosphate has been identified as a bioactive lipid that functions as an extracellular ligand and an intracellular second messenger.[66] S1P binds to one of five G protein coupled Edg/S1P receptors. The biological responses are mediated in part through the mobilization of intracellular calcium. These responses are profound and vary from cell proliferation, immune cell trafficking, and vascular permeability.[67] Ceramide-1-phosphate appears to act intracellularly. Ceramide 1-phosphate has been associated with a number of biological activities. These activities include mitogenesis, membrane fusion (including that associated with phagolysosome formation),[68] calcium signaling,[69] and the regulation of phospholipase A2.[70]

Glycosphingolipids have long been studied by cell biologists. Glycosphingolipids have been characterized as blood group antigens, including those comprising the ABH, P, Lewis, and Ii systems.[71] Glycolipids are also markers of cellular development and differentiation, and they are associated with various diseases, including cancer.[72] However, glycosphingolipids are not simply differentiation and disease associated markers, they are critical for several cell recognition processes. These include cell-cell recognition, cell-matrix interactions, and cell-pathogen interactions. Cell-pathogen interactions are particularly interesting and important. Glycosphingolipids are well established as the cell entry points for viruses, bacteria, and bacterial toxins. These include HIV[73] and rotavirus,[74] *H. pylori*[75] and *N. gonorrheae*[76] and Shiga toxin[77] and Cholera toxin, respectively.[78] Blocking these interactions has been proposed as a therapeutic strategy for treating cancer and a variety of infectious diseases.

Glycosphingolipids have been implicated as mediators of receptor-based signaling events. The appreciation that the physical properties of sphingolipids may result in their self-association into microdomains or rafts has made earlier observations on the modulation of tyrosine kinase[79] and phospholipase C[80] by gangliosides and other GlcCer-based lipids more comprehensible. The activities of a variety of growth factors including EGF, PDGF, and insulin have been reported to be markedly altered by ganglioside GM3 in a structure-dependent fashion. Indeed the isolation of glycosphingolipid-enriched rafts is associated with the recovery of a variety of signaling proteins, including src kinases, phospholipase C, and tyrosine kinases. Thus sphingolipid enriched domains have been suggested to be "hot spots" for cell signaling events.

CONCLUSIONS AND UNANSWERED QUESTIONS

A great deal of progress has been made in understanding the biochemistry and cellular biology of sphingolipids over the last 20 years. During this time a marked evolution in the view of the role and importance of sphingolipids has occurred. Sphingolipids, including GlcCer, are not simply unusual molecules that accumulate within the lysosomes of unfortunate individuals that inherit defective hydrolytic enzymes, but they play important roles in normal physiology and in the pathophysiology of more common diseases such as cancer, diabetes, and atherosclerosis.

Growing insight into the molecular biology of the genes that encode the enzymes that regulate sphingolipid metabolism and of the structural biology of these enzymes has provided new opportunities to better understand the pathogenesis of Gaucher disease and of other sphingolipidosis. This work has led to the development of better models of these disorders and, most importantly, has resulted in the early development of therapeutics that offer new strategies for the treatment of these diseases.

Whilst these efforts have been noteworthy, fundamental questions for each sphingolipidosis remain unanswered. Specifically, how is it that the lysosomal accumulation of a particular sphingolipid results in such dramatic and differing phenotypes for each of these diseases? The low β-glucocerebrosidase activity and GlcCer accumulation in Type I Gaucher disease, for example, causes massive hepatomegaly and splenomegaly. Yet the accumulation of GlcCer is primarily restricted to the macrophage. By contrast, Fabry disease is marked by the accumulation of globotriaosylceramide in the kidneys, vasculature, and skin of affected patients. These individuals suffer from profound cardiovascular complications. Similar observations can be made for each of the sphingolipids storage diseases.

A fundamental and unanswered question is whether these sphingolipids cause these associated pathologies exclusively through their accumulation within the lysosome or whether excess sphingolipid is present in other cellular compartments such as lipid rafts of the plasma membrane. Conceivably, alterations in raft content may change the functional properties of signaling complexes or the receptor properties of glycosphingolipids. Despite years of work and thousands of publications on the sphingolipidoses, this question does not appear to have been systematically addressed.

In the setting of impaired β-glucocerebrosidase activity there are many ways in which one might conceive of glycosphingolipids accumulating outside of the lysosome. First, under conditions of impaired degradation, GlcCer and other glycosphingolipids might traffic less rapidly to the lysosome and be retained in other cellular compartments. Indeed some studies have suggested that glycosphingolipid may be found in the endoplasmic reticulum and at the mitochondrial membrane. Alternatively, glycosphingolipids that have accumulated in the lysosome may "leak out" and sort to other cell organelles.

One might conjecture as to the consequences of aberrant glycosphingolipid accumulation in various organelles. An increase in ganglioside GD3 at the mitochondrial membrane has been associated with TNF-α induced apoptosis.[81] GlcCer and glucosyl-sphingosine have been reported to increase the release of calcium from ryanodine-sensitive stores in the endoplasmic reticulum.[82] Finally, as reviewed above, neutral glycosphingolipids and gangliosides might directly alter critical cell specific signaling events at the level of the plasma membrane.

An equally important question concerns the biochemical and physiological consequences of inhibiting key steps in sphingolipid metabolism. The development of potent and specific inhibitors of GlcCer synthase has provided a tool for evaluating the consequences of ceramide accumulation and of inhibition of GlcCer based glycosphingolipids. Knockout mice have been similarly used to evaluate both the functional roles of glycosphingolipids and to provide proof of principle for treating glycosphingoliposes through inhibition of synthetic pathways. The interpretation of these "proof of concept" studies, however, is limited by a poor understanding of the compartments within which sphingolipids substrates are accumulating or within which sphingolipids are depleted. Until better tools are developed for evaluating changes in sphingolipid mass and distribution within these compartments, our understanding of the pathophysiology of these diseases and ability to rationally develop new therapeutics will be constrained.

REFERENCES

1. Karlsson, K.A., On the chemistry and occurrence of sphingolipid long-chain bases, *Chem. Phys. Lipids,* 5, 6, 1970.
2. Brady, R.O. and Koval, G.J., The enzymatic synthesis of sphingosine, *J. Biol. Chem.,* 233, 26, 1958.
3. Williams, R.D., Wang, E., and Merrill, A.H., Jr., Enzymology of long-chain base synthesis by liver: characterization of serine palmitoyltransferase in rat liver microsomes, *Arch. Biochem. Biophys.,* 228, 282, 1984.
4. Ternes, P., Franke, S., Zahringer, U., Sperling, P., and Heinz, E., Identification and characterization of a sphingolipid delta 4-desaturase family, *J. Biol. Chem.,* 277, 25512, 2002.
5. Kihara, A. and Igarashi, Y., FVT-1 is a mammalian 3-ketodihydrosphingosine reductase with an active site that faces the cytosolic side of the endoplasmic reticulum membrane, *J. Biol. Chem.,* 279, 49243, 2004.
6. Schorling, S. et al., Lag1p and Lac1p are essential for the Acyl-CoA-dependent ceramide synthase reaction in Saccharomyces cerevisae, *Mol. Biol. Cell.,*12, 3417, 2001.
7. Venkataraman, K. et al., Upstream of growth and differentiation factor 1 (uog1), a mammalian homolog of the yeast longevity assurance gene 1 (LAG1), regulates N-stearoyl-sphinganine (C18-(dihydro)ceramide) synthesis in a fumonisin B1-independent manner in mammalian cells, *J. Biol. Chem.,* 277, 35642, 2002.
8. Riebeling, C. et al., Two mammalian longevity assurance gene (LAG1) family members, trh1 and trh4, regulate dihydroceramide synthesis using different fatty acyl-CoA donors, *J. Biol. Chem.,* 278, 43452, 2003.
9. Omae, F. et al., Identification of an essential sequence for dihydroceramide C-4 hydroxylase activity of mouse DES2, *FEBS Lett.,* 576, 63, 2004.
10. Obeid, L.M. et al., Programmed cell death induced by ceramide, *Science,* 259, 1769, 1993.

11. Rani, C.S. et al., Cell cycle arrest induced by an inhibitor of glucosylceramide synthase. Correlation with cyclin-dependent kinases, *J. Biol. Chem.*, 270, 2859, 1995.

12. Pettus, B.J., Chalfant, C.E., and Hannun, Y.A., Ceramide in apoptosis: an overview and current perspectives, *Biochim. Biophys. Acta.*, 1585, 114, 2002.

13. Levade, T. et al., Ceramide in apoptosis: a revisited role, *Neurochem. Res.*, 27, 601, 2002.

14. Gatt, S., Enzymic Hydrolysis and Synthesis of Ceramides, *J. Biol. Chem.*, 238, 3131, 1963.

15. Dulaney, J. et al., The biochemical defect in Farber's disease, *Adv. Exp. Med. Biol.*, 68, 403, 1976.

16. Schuchman, E.H. et al., Human acid sphingomyelinase. Isolation, nucleotide sequence and expression of the full-length and alternatively spliced cDNAs, *J. Biol. Chem.*, 266, 8531, 1991.

17. Ferlinz, K. et al., Human acid ceramidase: processing, glycosylation, and lysosomal targeting, *J. Biol. Chem.*, 276, 35352, 2001.

18. Tani, M. et al., Purification and characterization of a neutral ceramidase from mouse liver. A single protein catalyzes the reversible reaction in which ceramide is both hydrolyzed and synthesized, *J. Biol. Chem.*, 275, 3462, 2000.

19. Tani, M. et al., Molecular cloning of the full-length cDNA encoding mouse neutral ceramidase. A novel but highly conserved gene family of neutral/alkaline ceramidases, *J. Biol. Chem.*, 275, 11229, 2000.

20. Okino, N. et al., Molecular cloning, sequencing, and expression of the gene encoding alkaline ceramidase from Pseudomonas aeruginosa. Cloning of a ceramidase homologue from Mycobacterium tuberculosis, *J. Biol. Chem.*, 274, 36616, 1999.

21. Yoshimura, Y. et al., Molecular cloning and functional analysis of zebrafish neutral ceramidase, *J. Biol. Chem.*, 279, 44012, 2004.

22. Yoshimura, Y. et al., Molecular cloning and characterization of a secretory neutral ceramidase of Drosophila melanogaster, *J. Biochem.* (Tokyo), 132, 229, 2002.

23. Mitsutake, S. et al., Purification, characterization, molecular cloning, and subcellular distribution of neutral ceramidase of rat kidney, *J. Biol. Chem.*, 276, 26249, 2001.

24. Tani, M., Iida, H., and Ito, M., O-glycosylation of mucin-like domain retains the neutral ceramidase on the plasma membranes as a type II integral membrane protein, *J. Biol. Chem.*, 278, 10523, 2003.

25. Liu, H. et al., Sphingosine kinases: a novel family of lipid kinases, *Prog. Nucleic Acid Res. Mol. Biol.*, 71, 493, 2002.

26. Pyne, S. and Pyne, N.J., Sphingosine 1-phosphate signalling and termination at lipid phosphate receptors, *Biochim. Biophys. Acta.*, 1582, 121, 2002.

27. Ullman, M.D. and Radin, N.S., The enzymatic formation of sphingomyelin from ceramide and lecithin in mouse liver, *J. Biol. Chem.*, 249, 1506, 1974.

28. Huitema, K. et al., Identification of a family of animal sphingomyelin synthases, *Embo J.*, 23, 33, 2004.

29. Sugiura, M. et al., Ceramide kinase, a novel lipid kinase. Molecular cloning and functional characterization, *J. Biol. Chem.*, 277, 23294, 2002.

30. Abe, A., Shayman, J.A., and Radin, N.S., A novel enzyme that catalyzes the esterification of N-acetylsphingosine. Metabolism of C2-ceramides, *J. Biol. Chem.*, 271, 14383, 1996.

31. Hiraoka, M., Abe, A., and Shayman, J.A., Cloning and characterization of a lysosomal phospholipase A2, 1-O-acylceramide synthase, *J. Biol. Chem.*, 277, 10090, 2002.

32. Tadano-Aritomi, K. et al., Kidney lipids in galactosylceramide synthase-deficient mice. Absence of galactosylsulfatide and compensatory increase in more polar sulfoglycolipids, *J. Lipid Res.*, 41, 1237, 2000.

33. Ichikawa, S. et al., Expression cloning of a cDNA for human ceramide glucosyltransferase that catalyzes the first glycosylation step of glycosphingolipid synthesis, *Proc. Natl. Acad. Sci. U.S.A.*, 93, 4638, 1996.

34. Ichikawa, S., Ozawa, K., and Hirabayashi, Y., Molecular cloning and expression of mouse ceramide glucosyltransferase, *Biochem. Mol. Biol. Int.*, 44, 1193, 1998.

35. Wu, K. et al., Histidine-193 of rat glucosylceramide synthase resides in a UDP-glucose- and inhibitor (D-threo-1-phenyl-2-decanoylamino-3-morpholinopropan-1-ol)-binding region: a biochemical and mutational study, *Biochem. J.*, 341, 395, 1999.

36. Kohyama-Koganeya, A. et al., Drosophila glucosylceramide synthase: a negative regulator of cell death mediated by proapoptotic factors, *J. Biol. Chem.*, 279, 35995, 2004.

37. Leipelt, M. et al., Characterization of UDP-glucose:ceramide glucosyltransferases from different organisms, *Biochem. Soc. Trans.*, 28, 751, 2000.

38. Leipelt, M. et al., Glucosylceramide synthases, a gene family responsible for the biosynthesis of glucosphingolipids in animals, plants, and fungi, *J. Biol. Chem.*, 276, 33621, 2001.

39. Marks, D.L. et al., Oligomerization and topology of the Golgi membrane protein glucosylceramide synthase, *J. Biol. Chem.*, 274, 451, 1999.

40. Marks, D.L. et al., Identification of active site residues in glucosylceramide synthase. A nucleotide-binding catalytic motif conserved with processive beta-glycosyltransferases, *J. Biol. Chem.*, 276, 26492, 2001.

41. Chester, M.A., IUPAC-IUB Joint Commission on Biochemical Nomenclature (JCBN). Nomenclature of glycolipids — recommendations 1997, *Eur. J. Biochem.*, 257, 293, 1998.

42. Traving, C. and Schauer, R., Structure, function and metabolism of sialic acids, *Cell. Mol. Life Sci.*, 54, 1330, 1998.

43. Kolter, T., Proia, R.L., and Sandhoff, K., Combinatorial ganglioside biosynthesis, *J. Biol. Chem.*, 277, 25859, 2002.

44. Wilkening, G., Linke, T., and Sandhoff, K., Lysosomal degradation on vesicular membrane surfaces. Enhanced glucosylceramide degradation by lysosomal anionic lipids and activators, *J. Biol. Chem.*, 273, 30271, 1998.

45. Mandon, E.C. et al., Subcellular localization and membrane topology of serine palmitoyltransferase, 3-dehydrosphinganine reductase, and sphinganine N-acyltransferase in mouse liver, *J. Biol. Chem.*, 267, 11144, 1992.

46. Hanada, K. et al., Molecular machinery for non-vesicular trafficking of ceramide, *Nature*, 426, 803, 2003.

47. Kumagai, K. et al., CERT mediates intermembrane transfer of various molecular species of ceramides, *J. Biol. Chem.*, 280, 6488, 2005.

48. Coste, H., Martel, M.B., and Got, R., Topology of glucosylceramide synthesis in Golgi membranes from porcine submaxillary glands, *Biochim. Biophys. Acta.*, 858, 6, 1986.

49. Tettamanti, G., Ganglioside/glycosphingolipid turnover: new concepts, *Glycoconj J.*, 20, 301, 2004.

50. Goi, G. et al., Membrane anchoring and surface distribution of glycohydrolases of human erythrocyte membranes, *FEBS Lett.*, 473, 89, 2000.

51. Schengrund, C.L., Jensen, D.S., and Rosenberg, A., Localization of sialidase in the plasma membrane of rat liver cells, *J. Biol. Chem.*, 247, 2742, 1972.

52. Schengrund, C.L. and Rosenberg, A., Intracellular location and properties of bovine brain sialidase, *J. Biol. Chem.*, 245, 6196, 1970.
53. Chiarini, A. et al., Human erythrocyte sialidase is linked to the plasma membrane by a glycosylphosphatidylinositol anchor and partly located on the outer surface, *Glycoconj J.*, 10, 64, 1993.
54. Boggs, J.M., Lipid intermolecular hydrogen bonding: influence on structural organization and membrane function, *Biochim. Biophys. Acta.*, 906, 353, 1987.
55. Silvius, J.R., Cholesterol modulation of lipid intermixing in phospholipid and glycosphingolipid mixtures. Evaluation using fluorescent lipid probes and brominated lipid quenchers, *Biochemistry*, 31, 3398, 1992.
56. Silvius, J.R., del Giudice, D., and Lafleur, M., Cholesterol at different bilayer concentrations can promote or antagonize lateral segregation of phospholipids of differing acyl chain length, *Biochemistry*, 35, 15198, 1996.
57. Brown, D.A. and London, E., Structure of detergent-resistant membrane domains: does phase separation occur in biological membranes? *Biochem. Biophys. Res. Commun.*, 240, 1, 1997.
58. Brown, D.A. and Rose, J.K., Sorting of GPI-anchored proteins to glycolipid-enriched membrane subdomains during transport to the apical cell surface, *Cell*, 68, 533, 1992.
59. Simons, K. and Ikonen, E., Functional rafts in cell membranes, *Nature*, 387, 569, 1997.
60. Chang, Y., Abe, A., and Shayman, J.A., Ceramide formation during heat shock: a potential mediator of alpha B-crystallin transcription, *Proc. Natl. Acad. Sci. U.S.A.*, 92, 12275, 1995.
61. Dobrowsky, R.T. et al., Ceramide activates heterotrimeric protein phosphatase 2A, *J. Biol. Chem.*, 268, 15523, 1993.
62. Chalfant, C.E. et al., Long chain ceramides activate protein phosphatase-1 and protein phosphatase-2A. Activation is stereospecific and regulated by phosphatidic acid, *J. Biol. Chem.*, 274, 20313, 1999.
63. Huwiler, A. et al., Ceramide-binding and activation defines protein kinase c-Raf as a ceramide-activated protein kinase, *Proc. Natl. Acad. Sci. U.S.A.*, 93, 6959, 1996.
64. Muller, G. et al., PKC zeta is a molecular switch in signal transduction of TNF-alpha, bifunctionally regulated by ceramide and arachidonic acid, *Embo J.*, 14, 1961, 1995.
65. Zhang, Y. et al., Kinase suppressor of Ras is ceramide-activated protein kinase, *Cell*, 89, 63, 1997.
66. Watterson, K. et al., Pleiotropic actions of sphingosine-1-phosphate, *Prog. Lipid Res.*, 42, 344, 2003.
67. Hla, T., Physiological and pathological actions of sphingosine 1-phosphate, *Semin. Cell. Dev. Biol.*, 15, 513, 2004.
68. Hinkovska-Galcheva, V.T. et al.,The formation of ceramide-1-phosphate during neutrophil phagocytosis and its role in liposome fusion, *J. Biol. Chem.*, 273, 33203, 1998.
69. Tornquist, K. et al., Ceramide 1-phosphate enhances calcium entry through voltage-operated calcium channels by a protein kinase C-dependent mechanism in GH4C1 rat pituitary cells, *Biochem. J.*, 380 (Pt 3), 661, 2004.
70. Pettus, B.J. et al., Ceramide 1-phosphate is a direct activator of cytosolic phospholipase A2, *J. Biol. Chem.*, 279, 11320, 2004.
71. Hakomori, S., Antigen structure and genetic basis of histo-blood groups A, B and O: their changes associated with human cancer, *Biochim. Biophys. Acta.*, 1473, 247, 1999.
72. Hakomori, S., Glycosylation defining cancer malignancy: new wine in an old bottle, *Proc. Natl. Acad. Sci. U.S.A.*, 99, 10231, 2002.

73. McAlarney, T. et al., Characteristics of HIV-1 gp120 glycoprotein binding to gly-colipids, *J. Neurosci. Res.*, 37, 453, 1994.

74. Delorme, C. et al., Glycosphingolipid binding specificities of rotavirus: identification of a sialic acid-binding epitope, *J. Virol.*, 75, 2276, 2001.

75. Lingwood, C.A., Glycolipid receptors for verotoxin and Helicobacter pylori: role in pathology, *Biochim. Biophys. Acta.,* 1455, 375, 1999.

76. Hugosson, S. et al., Glycosphingolipid binding specificities of Neisseria meningitidis and Haemophilus influenzae: detection, isolation, and characterization of a binding-active glycosphingolipid from human oropharyngeal epithelium, *J. Biochem.* (Tokyo), 124, 1138, 1998.

77. Lingwood, C.A., Verotoxins and their glycolipid receptors, *Adv. Lipid Res.*, 25, 189, 1993.

78. Spiegel, S., Fluorescent derivatives of ganglioside GM1 function as receptors for cholera toxin, *Biochemistry*, 24, 5947, 1985.

79. Bremer, E.G. et al., Ganglioside-mediated modulation of cell growth, growth factor binding, and receptor phosphorylation, *J. Biol. Chem.,* 259, 6818, 1984.

80. Shu, L. and Shayman, J.A., Src kinase mediates the regulation of phospholipase C-gamma activity by glycosphingolipids, *J. Biol. Chem.,* 278, 31419, 2003.

81. Garcia-Ruiz, C. et al., Trafficking of ganglioside GD3 to mitochondria by tumor necrosis factor-alpha, *J. Biol. Chem.,* 277, 36443, 2002.

82. Ginzburg, L., Kacher, Y., and Futerman, A.H., The pathogenesis of glycosphingolipid storage disorders, *Semin. Cell. Dev. Biol.*, 15, 417, 2004.

CHAPTER 8

The Development of Enzyme Replacement Therapy for Lysosomal Diseases: Gaucher Disease and Beyond

Edward H. Schuchman and Silvia Muro

CONTENTS

HISTORICAL OVERVIEW OF ENZYME REPLACEMENT THERAPY FOR LYSOSOMAL STORAGE DISEASES

The first identification of lysosomes occurred in the late 1950s and early 1960s with the pioneering cell fractionation work of de Duve and co-workers [1]. Lysosomes have been found in almost all mammalian cell types and are composed of a self-limiting lipid membrane enclosing several dozen hydrolytic enzymes required to degrade complex molecules [for review, see Reference 2]. Embedded in the lipid membrane are also found transport proteins, receptors, and ion pumps required to maintain the acidic environment of the organelle. In addition, several small "activator" molecules have been found within lysosomes that influence the activities of specific hydrolytic enzymes.

Shortly after the identification of lysosomes, several of the enzymes residing within this organelle were identified and purified, and by the late 1960s more than two dozen such "lysosomal" enzymes had been reported. This work on the purification and characterization of lysosomal enzymes remained an intense focus of research throughout the late 1960s and 1970s. In the early 1970s, the seminal work of Sly, Neufeld, and Kornfeld led to the identification of a specific trafficking system that governed the transport of most lysosomal proteins to the organelle via a specific modification (phosphorylation) of mannose residues on their N-linked oligosaccharide chains [for reviews, see References 3,4]. This modification subsequently led to "uptake" by mannose-6-phosphate (M6P) receptors present within lysosomal membranes. In addition, during this time it became known that most cell types released small amounts of lysosomal enzymes and that these "secreted" forms could be internalized via M6P receptors present on the cell surface. This led to the concept of enzymatic "cross-correction," i.e., the idea that small amounts of enzyme released from normal cells could be internalized by diseased cells and correct the metabolic defect in patients with lysosomal storage disorders [e.g., 5–8].

The first human disease associated with the deficiency of a specific lysosomal enzyme was Pompe disease, due to the deficient activity of alpha glucosidase activity [9]. Since then, more than 40 such "lysosomal storage disorders" have been identified [10]. All lysosomal diseases are characterized by the accumulation of specific complex molecules within cells and tissues. Since lysosomes are ubiquitous organelles, most lysosomal storage diseases affect multiple cell types and organ systems. However, their clinical presentation can vary considerably depending on the specific metabolic pathway affected and the type of mutations inherited by individual patients. About half of all lysosomal diseases have central nervous system involvement.

The diagnosis of most lysosomal diseases can be carried out by measuring the specific activities of individual lysosomal enzymes in cultured cells from affected patients. Such early diagnostic studies revealed that the amount of "residual" enzymatic activity in patients with severe forms of lysosomal diseases was only moderately reduced when compared to individuals with mild forms of the same disease. However, in both types of patients the amount of residual activity was usually about 1% or less when compared to normal individuals. This observation led investigators to conclude that very small amounts of functional enzymatic activity could have significant clinical effects if present in the proper target sites of pathology, an observation that has been confirmed by recent treatment studies in animal models and human patients (see below).

ENZYME REPLACEMENT THERAPY FOR GAUCHER DISEASE: LESSONS LEARNED

Shortly after the demonstration of enzymatic "cross-correction" by Neufeld and colleagues, the concept of treating lysosomal diseases by direct enzyme replacement and/or the transplantation of enzyme-producing cells evolved. Early studies of bone marrow transplantation revealed that engrafting such cells into patients with specific lysosomal enzyme deficiencies could provide a source of circulating enzymes that could be taken up by target cells of pathology and lead to metabolic correction [for reviews, see 11,12]. The success of these procedures relied on the degree of engraftment and the accessibility of the target cells to the circulating enzymes. However, these transplantation studies also revealed that this approach relied on the availability of suitable donor individuals and, even if such individuals were available, that transplantation was often associated with significant morbidity and/or mortality. Although such hematopoietic cell transplantation procedures have improved considerably since these initial attempts, particularly with the availability of cord blood stem cells, they are still limited by these realities.

Attempts at enzyme replacement therapy were first undertaken in the early 1970s [e.g., 13,14; see 15 for an early review]. Although these early attempts were severely hampered by the limited availability of the purified, normal enzymes, they did provide "proof of principle" for the approach, and showed that intravenously administered enzymes could be taken up by target cells of pathology and lead to reductions of pathological substrates. In addition, these early studies revealed the critical importance of the N-linked oligosaccharide chains on lysosomal stability and cell-type specific uptake, and the need to improve enzyme targeting to specific cell types in individual diseases (e.g., macrophages in Gaucher disease, see below). These early enzyme replacement studies also revealed that circulating lysosomal enzymes could not effectively cross the blood brain barrier, and that new approaches would be needed for the neurodegenerative, lysosomal disease forms.

Throughout the 1970s and 1980s, one of the lysosomal diseases that attracted many of these early therapeutic endeavors was Type 1 (nonneurological) Gaucher disease, due to the deficient activity of the enzyme beta glucosidase [14]. This enzyme is normally required to hydrolyze the glucose residue from the glycosphingolipid, glucosylceramide, and in patients with Gaucher disease this lipid accumulates in the cells, tissues, and fluids. The primary cellular site of pathology in Type 1 Gaucher disease is the macrophage, and in affected individuals the bone marrow and reticuloendothelial organs become infiltrated with lipid laden "foam" cells also known as "Gaucher" cells. Type 1 Gaucher disease is one of the most common lysosomal diseases, with an estimated carrier frequency within the Ashkenazi Jewish community of about 1:25, making it a prime focus for early therapeutic endeavors.

The seminal work of Brady and colleagues in the 1960s and 1970s had identified the enzymatic defect in Gaucher disease patients and developed the first purification methods for human beta glucosidase [14,16,17]. Using this enzyme they also investigated its uptake by Gaucher disease macrophages and found that it was independent of the M6P trafficking system and poorly internalized. This led to a method of modifying the N-linked oligosaccharide chains to remove complex sugars (e.g., sialic acid), thus exposing terminal mannose residues. This "high mannose" form of the enzyme could be recognized by mannose receptors abundant on macrophage membranes, was internalized readily by Gaucher disease macrophages, and resulted in rapid substrate depletion.

Using the tools available in the "premolecular" era of the 1970s and 1980s, only small amounts of "mannose-terminated" beta glucosidase could be prepared from human tissues. However, the early cell culture and short-term clinical studies using this enzyme were encouraging, and more extensive clinical trials in Gaucher disease patients were undertaken using a commercial "high mannose" beta glucosidase from human placentae, named Ceredase [18,19]. This enzyme was quickly shown to be extremely effective at reducing the liver and spleen volumes in Type 1 Gaucher disease patients, resulting in improved growth and activity. In addition, enzyme therapy markedly improved the hematological parameters in the treated patients and led to substantial improvements in bone density and other clinical findings.

There are currently over 3000 Type 1 Gaucher disease patients receiving enzyme replacement therapy worldwide (now using a recombinant form of beta glucosidase called Cerezyme). In general, it is considered an extremely safe and effective form of therapy for these patients, many of whom suffered from severe, life-threatening complications before its availability. Numerous lessons have been learned from the experiences with Gaucher disease. First, and foremost, the experience with Gaucher disease has validated earlier animal and clinical studies, and demonstrated the efficacy of enzyme replacement therapy for lysosomal diseases. Although the cost of this therapy is high and requires life-long, intravenous infusions, the clinical improvements are remarkable and the treatment is "cost-saving" in the sense that it reduces the number of subsequent hospitalizations and the requirement for long-term care of many Gaucher disease patients. In addition, the success of Gaucher disease enzyme therapy placed lysosomal storage diseases "on the map" with biotechnology companies and opened the door for the development of similar approaches with several other diseases (see below). Prior to Gaucher disease, drug development for such "orphan" diseases was generally considered unprofitable. Presently, however, there are at least four different companies and numerous academic laboratories partnering to develop a variety of therapeutic approaches for lysosomal diseases, a remarkable shift in a short period of time.

From a scientific viewpoint, we have also learned that while enzyme replacement therapy can result in marked clinical improvements in most treated patients, it is not a "cure" for these individuals and there is significant "patient-specific" variability in the response. In addition, as noted above, based on the uptake properties of the infused enzyme, there are also significant organ-specific variations in the response, an observation that has been confirmed by recent clinical trials for several other lysosomal diseases (see below). Thus, while this therapeutic approach has been shown to be safe and effective, there is room for improvement by modifying the infused enzymes to enhance stability, activity, and/or cell type-specific uptake, and by altering/enhancing the enzyme delivery procedures to deliver therapeutic activity levels to target tissues that are not currently accessible (e.g., central nervous system). These represent areas of active, current research.

THE FUTURE: NEW TECHNOLOGIES FOR ENZYME TARGETING AND DELIVERY

By the mid-1980s, the specific enzyme deficiencies were known for most lysosomal diseases and purification procedures had been developed for many of these proteins. However, the availability of these enzymes relied on access to human tissues, and even when such tissues were available the quantity of pure material obtained was very small. Thus, despite the fact that the first enzyme replacement studies were carried out in human patients in the early 1970s, follow-up studies could not be pursued because of the lack of enzymes.

Molecular genetic technology markedly changed this reality and had a major impact on the development of enzyme therapy for lysosomal diseases throughout the mid-1980s and 1990s. The full-length sequences encoding the first lysosomal enzyme (beta glucuronidase), deficient in the lysosomal storage disorder known as "Sly" disease, or mucopolysaccharidoses (MPS) type VII, was cloned in the early 1980s [20]. Over the ensuing 10 years, the cDNAs and genes encoding most other lysosomal enzymes were obtained and expression systems were developed to produce many of the recombinant proteins. Furthermore, numerous animal models became available during this time, providing excellent experimental systems in which to evaluate new therapies [for review, see 21]. Thus, by the mid-1990s there was a convolution of these technologies and reagents, providing lysosomal disease researchers with the ability to vigorously pursue the development of therapeutic approaches.

Much of this renewed research effort was devoted to the development of enzyme replacement therapy. To date, enzyme replacement therapy is available for four lysosomal diseases, Type 1 Gaucher disease (beta glucosidase deficiency), Fabry disease (alpha galactosidase deficiency), and Mucopolysacchridoses type I (Hurler/Scheie; alpha iduronidase deficiency), and VI (Maroteaux-Lamy; N-acetyl-galactosamine-4-sulfatase deficiency), and is under development for Mucopolysaccharidosis Type II (Hunter; iduronide sulfatase deficiency) Pompe disease (alpha glucosidase deficiency), and Niemann-Pick disease type B (acid sphingomyelinase deficiency).

Despite the remarkable success of enzyme replacement therapy, however, researchers and clinicians have recognized the need for improvement. Major goals include more effective delivery to specific cellular sites of pathology, including the brain, improved enzyme production, and less invasive and cost-effective delivery methods. Several new and exciting technologies are under investigation to address these issues, and the remainder of this review will be devoted to this topic.

Enzyme Fusion Proteins

Although several systems are currently available for the large-scale production of recombinant enzymes (e.g., Chinese hamster ovary cells, Sf9 insect cells, transgenic animals, etc.), improper or sub-optimal glycosylation of these recombinant proteins can sometimes result in poor delivery via cell receptors (e.g., mannose or M6P receptors), (Figure 8.1A), resulting in inefficient cell targeting and substrate clearance [22–25]. One approach to cope with this obstacle is the generation of peptide-targeted fusion enzymes (Figure 8.1B). Several bi-functional, chimeric proteins have recently been designed that contain the catalytically active lysosomal enzyme (the enzymatic moiety) fused to a ligand-derived peptide (the targeting moiety) that can be recognized by cell surface receptors in a glycosylation-independent manner [26–28]. These strategies will be briefly reviewed below.

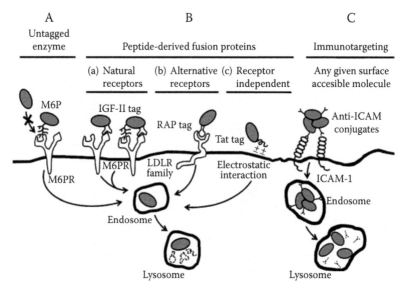

Figure 8.1   Strategies of enzyme replacement therapy for lysosomal disorders. (A) Intracellular delivery of recombinant lysosomal enzymes depends on glycosylation-mediated pathways, e.g., recognition of mannose-6-phosphate (M6P) residues on the recombinant enzymes by M6P receptors (M6PR) located in the cell surface, triggering subsequent clathrin-dependent uptake and lysosomal trafficking. (B) Sub-optimal delivery can be improved by recombinant peptide-derived fusion enzymes, which may be designed to target: (a) the natural receptors (e.g., M6PR), which can be recognized by insulin-like growth factor (IGF-II); (b) alternative receptors (e.g., LDLR family), as those targeted by receptor associated protein (RAP); or (c) fusion proteins designed to simply bind to negative charged plasma membrane (e.g., through positively charged Tat peptide). (C) Immunotargeting is an attractive alternative to improve lysosomal enzyme delivery, which may be attained by coupling recombinant enzymes to antibodies to any given surface molecule capable of undergo endocytic uptake (e.g., Intercellular adhesion molecule 1 or ICAM-1). This strategy offers the possibility to generate multimeric conjugates, thus further increasing targeting and the intracellular delivery capacity of this system.

Fusion Proteins Targeted to the M6P Receptor

LeBowitz and colleagues [27] synthesized a fusion protein containing recombinant beta-glucuronidase linked at its carboxyl terminus to a peptide derived from human insulin-like growth factor II (IGF-II). This strategy is based on the fact that IGF-II can bind to the same cation-independent M6P receptor that is involved in lysosomal enzyme internalization via M6P-containing oligosaccharide chains [29]. Thus, chimeric beta-glucuronidase-IGF-II may be able to efficiently target M6P receptors independent of its sugar moieties (Figure 8.1B, panel a).

Improved interaction of beta-glucuronidase-IGF-II with M6P receptors was demonstrated in skin fibroblasts isolated from MPS VII patients, and also the chimeric enzyme was more efficiently delivered to target organs in a MPS VII mouse model, highlighting the relevance of this strategy. Interestingly, IGF-II recognizes a domain on the M6P receptor distinct from the one to which M6P binds [29]; thus, both the M6P residues present on the enzyme and IGF-II can contribute to M6P receptor recognition. Indeed, binding of the beta-glucuronidase-IGF-II chimera to M6P receptors was partially inhibited by excess IGF-II or M6P [27]. Importantly, treatment of this fusion protein with endoglycosidase F1 to remove sugar moieties on the enzyme rendered binding entirely independent of the enzyme's M6P residues, but still permitted uptake. Thus, bypassing glycosylation-mediated targeting of recombinant enzymes and/or enhancing the avidity of glycosylated enzymes to bind natural receptors (e.g., the M6P receptor) are promising approaches to improve the distribution and uptake of administered recombinant enzymes.

Fusion Proteins Targeted to Alternative Receptors

Strategies utilizing chimeric lysosomal enzymes may also target alternative receptors to achieve cell delivery (Figure 8.1B, panel b). An example of such an approach was recently reported by Prince and colleagues [28], who synthesized recombinant alpha-L-iduronidase or alpha-glucosidase (deficient in Hurler/Scheie syndromes or MPS I, and Pompe disease, respectively) as fusion proteins tagged with a peptide derived from receptor associated protein (RAP). This endoplasmic reticulum chaperon protein binds to several LDL receptor family members and prevents their association to physiological ligands [30]. Thus, when injected intravenously, RAP has the ability to mediate uptake of chimeric fusion proteins into tissues via recognition of such receptors, which, similar to natural receptors, leads to internalization of the chimera by clathrin-dependent mechanisms and delivery to lysosomes [31].

Consistent with these intrinsic properties, chimeric RAP fused to the amino terminus of alpha-L-iduronidase or alpha-glucosidase effectively delivered these recombinant enzymes into several cell types, including patient-isolated skin fibroblasts, rat glioma cells, and murine myoblasts [28]. Uptake of the fusion constructs greatly exceeded that of the untagged enzymes and could be mediated by several receptors, including LDLR-related protein 1 or LRP1, apoER2, VLDLR, and megalin. Remarkably, enzyme delivery was effective despite the apparent lack of M6P and high mannose residues in the RAP-alpha-glucosidase chimera. Additionally, in the case of RAP-alpha-L-iduronidase, which contained mannose and M6P residues, binding was prevented by excess, free RAP, but not by competitors of the M6P receptor. Thus, RAP provides glycosylation-independent targeting and the possibility of delivering recombinant enzymes to a variety of cell surface receptors. Some of these, such as megalin, may further facilitate accessibility to intractable target organs (e.g., the brain) by mediating transcytosis pathways.

Receptor-Independent Fusion Proteins

Improved distribution of recombinant lysosomal enzymes also may be obtained by fusion to protein transduction domains (PTDs) such as those derived from the HIV transcription factor, Tat [32,33]. In contrast to the peptide-based systems described above, which require specific recognition by cell surface receptors, basic charges in Tat mediate electrostatic interactions with membrane phospholipids and polysaccharides, which then facilitates its uptake into the cells (Figure 8.1B, panel c). Although the mechanism of Tat internalization remains elusive, several pieces of evidence indicate that this may occur as a result of various routes, including direct membrane penetration and internalization concomitant to endocytosis. Theoretically, recombinant enzymes could be delivered by the later pathway to any given cell within the organism.

Lysosomal enzyme delivery using Tat peptides is effective when determined in both cultured cells and animal models, as shown by Xia and colleagues [26]. In this study the investigators generated a chimeric form of beta-glucuronidase, fused at its carboxyl terminus to Tat, and demonstrated internalization by A549 and NIH3T3 cells. Uptake was only partially competed by M6P, likely due to the presence of such residues in the recombinant enzyme moiety of the fusion protein. Furthermore, beta-glucuronidase-Tat, expressed from adenoviral vectors after administration into mice by several routes, was broadly distributed among nontransduced cells, indicating that the chimera was secreted from transduced cells and effectively participated in "cross-correction." Although the usefulness of Tat fusion proteins for enzyme replacement therapy has not yet been tested in animal models, it is tempting to speculate that the lack of specificity of Tat-derived peptides for any one cell type may in fact improve biodistribution of the administered enzymes.

Thus, the use of lysosomal enzyme fusion proteins represents a novel approach to enhance receptor-mediated uptake of these enzymes and/or to alter the natural receptor targeting and improve delivery to cellular sites of pathology. Moreover, by bypassing the natural glycosylation-mediated targeting system of lysosomal enzymes, recombinant enzymes may be produced in systems (e.g., yeast or bacteria) other than those currently used. Benefits of peptide-tagged enzyme chimeras also may be extended to gene therapy strategies to improve biodistribution and uptake of enzymes secreted by the transduced cells. Enhanced delivery of recombinant lysosomal enzymes to target sites of disease by these means may permit reduction of the therapeutic dose and subsequent dose-dependent adverse effects, altogether enhancing the efficacy of treatment for lysosomal disorders. However, in all of these strategies the success of this approach will rely on the stability and activity of the chimeric protein, which is likely to vary greatly among different lysosomal enzymes and in different target tissues.

Unconventional Approaches for Lysosomal Enzyme Replacement Therapy

"Noninternalizable" Molecules as Pseudo-Receptors for Enzyme Delivery

As noted above, effective enzyme therapy for lysosomal disorders requires internalization of the enzymes into target cells and subsequent trafficking to lysosomal compartments. Great efforts have therefore been devoted to enhance the interaction between recombinant enzymes and cell receptors that mediate clathrin-dependent uptake and subsequent lysosomal delivery [3,24,27,28]. In contrast, little attention has been paid toward targeting of lysosomal enzymes to molecules other than endocytic receptors, likely because of skepticism that such an approach could lead to efficient lysosomal delivery.

However, binding of molecules to certain (nonreceptor) surface molecules can promote internalization of the resulting ligand/pseudo-receptor complex by novel mechanisms. This is the case of Intercellular Adhesion Molecule 1 (ICAM-1), a 110 kDa transmembrane glycoprotein from the immunoglobulin superfamily [34]. Binding of ICAM-1 to its natural ligands (e.g., white blood cells) does not prompt uptake by endocytic routes, but rather results in proteolytic shedding of this molecule from the plasma membrane following ligand release. In a similar manner, monoclonal antibodies recognizing the ICAM-1 extracellular domain avidly bind to this molecule without significant internalization [35]. Strikingly, however, certain artificial multimeric ligands, such as those generated by conjugation of biotinylated anti-ICAM with streptavidin or by coating the surface of carrier beads with these antibodies [36], are readily internalized into ICAM-1-expressing cells by a novel endocytic mechanism [37,38] (Figure 8.1C).

The pathway by which these multimeric ligands trigger ICAM-1-mediated uptake has been studied in detailed using human endothelial cells [37,39]. Interestingly, endocytosis of anti-ICAM "nano" particles by these cells ("CAM-mediated endocytosis") was shown to be unique and distinct from classical mechanisms, including clathrin- or caveoli-mediated pathways, macropinocytosis, or phagocytosis. Importantly, however, uptake by this unique pathway also resulted in lysosomal delivery.

Targeting of lysosomal enzymes to ICAM-1 may provide several advantages over conventional lysosomal delivery methods. First, ICAM-1 is widely distributed among many cell types and up-regulated in certain pathological states (such as inflammation) common to lysosomal diseases [34,40,41]; second, enzyme uptake and lysosomal delivery is independent of clathrin-mediated mechanisms, which may be impaired in some lysosomal storage disease cells [42]; and thirdly, targeting via ICAM-1 may favor transcytosis to adjacent tissues through pathways similar to those employed by ICAM-1 recognizing leukocytes and pathogens [34,43]. ICAM-1 targeting of lysosomal enzymes may be achieved by several means, including the use of chimeric proteins containing the enzyme (catalytic) domain and an anti-ICAM antibody domain, and/or the use of nanocarriers containing both domains. These strategies are briefly discussed below.

Fusion-Independent Alternatives to Improve Delivery of Lysosomal Enzymes: Immunotargeting

Immunotargeting of drugs and protein therapeutics is a growing research field and is based on coupling the cargo of interest to an antibody that recognizes a particular antigen present on the surface of cells [44]; thus, the antibody acts as an affinity carrier providing active targeting (Figure 8.1C). This strategy has proven to be highly efficient and can maximize delivery to the target sites and reduce the dose needed to achieve therapeutic effects. Selection of the antigen/antibody to be targeted will determine delivery to the proper cell types, tissues, and organs affected by a given disease condition. Thus, similar to peptide-tag based systems, immunoconjugates can be addressed to either cell-specific determinants or broadly distributed antigens, to better fit particular disease requirements.

Immunoconjugates can be directly formed utilizing bi-specific antibodies that act as heterofunctional crosslinkers with two distinct binding specificities, one for the recombinant enzyme cargo and the other for the target surface receptor [45]. In addition, conjugation of a given targeting antibody to a recombinant enzyme cargo can be achieved by utilizing chemical crosslinkers or biotin-(strept)avidin strategies to produce both single bi-functional entities or even multimeric conjugates [46–48]. Fine tune sizing of such multimeric conjugates is a critical parameter dictating subcellular addressing, and thus the conjugates can be designed to maximize internalization by endocytic pathways and delivery to lysosomal compartments within the cell [49]. Multimeric conjugates of recombinant lysosomal enzymes and targeting antibodies may greatly enhance the affinity of these conjugates to the target surface antigen, facilitating endocytic uptake by the cells, and providing a means for delivery of multiple enzyme copies per individual binding event.

The same targeting concept can be applied to liposomes and polymer nanocarrier systems, where antibodies can be utilized as targeting entities for efficient, site-specific delivery of the loaded therapeutics [50]. Particularly, immunotargeting of biodegradable nanocarriers may represent a highly promising strategy for lysosomal enzyme replacement therapy and offer several inherent advantages. For example, nanocarriers can typically incorporate large quantities of the therapeutic cargo, and they also offer the advantage of having a biodegradable structure [51]. In addition, immunotargeted nanocarriers can be designed to increase the circulation time of a given recombinant enzyme, e.g., by coupling poly(ethylene) glycol (PEG) to the polymer structure, which hinders recognition by the immune system and protects from rapid degradation [50]. Furthermore, final enzyme release from the nanocarrier can be controlled by manipulating the polymer degradation rate in environments with a particular acidic pH.

A remarkable advantage of immunotargeting-based vehicles is that the immunoconjugates can be designed to target any surface determinant (either receptors or noninternalizable molecules) by generating determinant-specific antibodies. In addition, antibodies to a given surface molecule can be produced to recognize distinct accessible epitopes on the antigen, and this modulation can provide fine-tuning not only of the binding ability of the immunoconjugate, but also may provide control over the specific cell response to the bound "ligand," both in terms of internalization (mechanism, kinetics, etc.) and trafficking (transcytosis, lysosomal destination, processing, and stability).

Lysosomal Delivery of Recombinant Acid Sphingomyelinase (ASM) by ICAM-1-Targeted Nanocarriers

An example of the applicability of immunotargeting for enzyme replacement therapy of lysosomal diseases is highlighted by the use of recombinant acid sphingomyelinase (ASM, the enzyme deficient in types A and B Niemann-Pick disease), whose lysosomal delivery was greatly enhanced in several cell types when targeted by multimeric nanocarriers immunotargeted to the "noninternalizable" surface molecule, ICAM-1 [52]. AntiICAM/ASM nanocarriers were efficiently targeted (on the range of one order of magnitude greater) to control endothelial cells, endothelial cells with imipramine-induced deficiency of ASM, and skin fibroblasts from patients diagnosed from type B Niemann-Pick disease. Despite the presence of sugar targeting residues (e.g., M6P) on the recombinant enzyme, binding of anti-ICAM/ASM nanocarriers to the cells was not competed by excess M6P, but was abolished by excess free anti-ICAM, indicating that these nanocarriers are recognized by cells in a glycosylation-independent manner.

Furthermore, internalization of anti-ICAM/ASM nanocarriers by the cells was rapid ($t_{1/2}\approx15$ min) and efficient (90% internalization), and occurred by a nonclassical uptake mechanism (CAM-mediated endocytosis) [37]. Bypassing clathrin-mediated pathways, such as those mediating internalization of glycosylated enzymes and fusion proteins targeted through IGF-II or RAP, may represent an important advantage given that endocytosis by clathrin-coated pits has been observed to be deficient in cells with certain lysosomal enzyme deficiencies such as NPD [42]. Highlighting this notion, anti-ICAM/ASM nanocarriers delivered significantly higher amounts (about one order of magnitude increase) of recombinant enzyme intracellularly when compared with untargeted nanocarriers or free ASM [52], and showed higher therapeutic efficiency. Currently, the utility of ICAM/ASM nanocarriers is being validated in animal models of ASM-deficient Niemann-Pick disease.

Apart from providing a highly efficient delivery system, targeting to ICAM-1 may offer additional advantages derived from the intrinsic characteristics of this molecule [34,40,43]. For instance, ICAM-1 is highly expressed in the luminal surface of endothelial cells, and thus is easily accessible from the circulation. It is involved in transmigration of white blood cells and also serves as a pseudo-receptor for pathogens that cross the blood brain barrier; therefore, ICAM-1 might provide transendothelial delivery to peripheral tissues. Additionally, ICAM-1 is expressed in a variety of other cells types, including epithelial cells, glial cells, fibroblasts, myoblasts, and the monocyte-macrophage system (among others), and thus may provide effective targeting to different tissues and organs affected by lysosomal disorders. ICAM-1 surface density is high on most cells types ($1\text{--}2 \times 10^5$

molecules/cell), and is further enhanced by pathological factors (e.g., inflammation), including those pertinent to lysosomal disorders, therefore providing powerful targeting primarily to sites of disease. Targeting to ICAM-1 also blocks recognition by leukocyte beta 2 integrins and certain pathogens, thus affording secondary benefits such as anti-inflammatory effects and hindering of such opportunistic infections, respectively.

Therefore, the efficacy of anti-ICAM nanocarriers for glycosylation- and clathrin-independent lysosomal delivery, and other secondary benefits expected from their utilization, indicate that anti-ICAM nanocarriers may represent a unique platform to improve classical delivery approaches and could evolve into a highly efficient system for enzyme replacement therapy of lysosomal disorders.

New *In Vivo* Delivery Methods

To date, enzyme replacement therapy has relied on the intravenous infusion of the recombinant enzymes to deliver the therapeutic proteins. However, systemic delivery often leads to infusion reactions that require careful medical monitoring and slow infusion rates, and occasional termination of therapy. In addition, systemic delivery also results in inefficient delivery to many target organs (such as muscle, lung, and brain), requiring large amounts of enzyme to be infused at high cost. Furthermore, the stability of many recombinant enzymes in the circulation remains unknown. For these reasons, it is important to investigate alternative delivery methods for recombinant enzymes, enzyme-peptide fusion proteins, or the immunotargeting conjugates described above. Such delivery methods could include (but not be limited to) subcutaneous administration, aerosolization, and/or intrathecal injection. For example, intrathecal injection of alpha-L-iduronidase into MPS I dogs resulted in a mean 23- to 300-fold increase in enzyme levels in total brain and meninges, respectively [53]. Analysis of the stored glycosaminoglycan substrate revealed a 57% decrease in meningeal levels accompanied by histological improvement of lysosomal storage in all CNS cell types. In addition, Bae et al. [54] compared the intravenous vs. subcutaneous administration of acid sphingomyelinase in acid sphingomyelinase deficient (NPD) mice. They observed a slow release of the recombinant enzyme into the circulation following subcutaneous administration and a longer half-life, and achieved similar clinical efficacy as compared to intravenous delivery.

Thus, these alternative delivery approaches may overcome some of the obstacles inherent to systemic delivery, and may be clinically useful for some lysosomal storage disorders. Clearly, pursuit of these approaches will require new formulations that permit enzyme preparations suitable for such delivery (e.g., aerosols) and must be fully evaluated in animal models prior to their clinical use. However, the high cost of systemic enzyme delivery together with the targeting limitations and immunological complications, warrant serious investigation of such alternative delivery approaches.

New Technologies and Gaucher Disease

Despite more than a decade of success, enzyme replacement therapy is burdened with certain inherent limitations. For example, this approach requires frequent infusions of enzyme, is extremely expensive, and provides only variable success in certain target organs. In addition, intravenous enzyme delivery does not significantly impact the neurodegenerative disease forms of these disorders. Thus, there is considerable room for improvement that requires the continued development and evaluation of new enzyme therapy technologies. Despite the high cost of drug development, it is imperative that such new technologies be implemented to improve the therapeutic effectiveness of enzyme therapy, reduce costs, and open new disease markets (e.g., for neurological disorders).

For example, although Cerezyme, the recombinant form of beta glucosidase currently used for enzyme therapy in Type 1 Gaucher disease, effectively targets macrophages through mannose receptors, enzyme delivery to other cell types lacking such receptors may be limited. Thus, the alternative enzyme targeting strategies discussed above (e.g., chimeric enzymes or antibody targeted conjugates/nanocarriers) might be used to bypass the natural, mannose receptor-mediated uptake mechanisms, providing enhanced enzyme delivery and clinical efficacy in organ systems where the current therapy has variable and limited effects (e.g., bones). The ICAM-1 targeting method may be particularly relevant for Gaucher disease since, like Niemann-Pick disease, inflammation is an important component of this disorder, leading to up-regulation of ICAM-1 expression. Moreover, enzyme uptake via clathrin-mediated pathways may be similarly defective in the two disorders. However, such immunotargeting approaches are not limited to ICAM-1, and could be employed for other surface molecules expressed on relevant cell types in the disease. Such approaches could also be used to target molecules used for transcytosis across the blood brain barrier to enhance enzyme delivery into the CNS for the neurological lysosomal disease forms such as Types 2 and 3 Gaucher disease.

In addition to enhanced enzyme targeting and uptake, the development of new enzyme administration methods could also be particularly important. These alternative administration approaches could reduce the clinical and financial burden of frequent, life-long intravenous infusions, and might provide improved delivery to important target organs. For example, intravenous enzyme delivery is not effective for the neurological components of Type 2 and 3 Gaucher disease, and thus alternative delivery approaches, such as intrathecal enzyme administration, should be evaluated. Subcutaneous enzyme pumps, which provide a slow delivery of recombinant protein into the circulation, could also be studied since this approach may provide a different biodistribution pattern and such pumps might only need to be replenished on a monthly or bimonthly basis.

Unfortunately, as with many successful drug approaches, a degree of inertia exists in the enzyme therapy field and there is a general reluctance to invest (intellectually and financially) in modifying a proven (and regulatory approved) technology. However, it is precisely the great success of enzyme therapy, highlighted by Type 1 Gaucher disease, that requires scientists and physicians to continue to explore and improve this technology. While other approaches, such as small molecule-based substrate reduction and enzyme enhancement, as well as gene therapy, hold great promise, only enzyme therapy has the proven track record of long-term efficacy and safety, and with continued modification and enhancement, this approach could be even more effective in the future and available to an even wider patient population.

ACKNOWLEDGMENTS

The authors thank Dr. Vladimir Muzykantov (Department of Pharmacology, University of Pennsylvania School of Medicine) for his valuable guidance on immunotargeting- and nanocarrier-based therapeutic systems and enriching comments on this chapter. This work was supported by NIH grants HD28607 and DK54830 to E.S., as well as by a fellowship from Fundación Ramón Areces (Spain) and a scientist development grant from American Heart Association 0435481N to S.M.

REFERENCES

1. de Duve, C., From cytases to lysosomes, *Fed. Proc.*, 23, 1045, 1964.
2. Sabatini, D.D. and Adesnik, M.B., The biogenesis of membranes and organelles, in *The Metabolic and Molecular Bases of Inherited Diseases*, Scriver, C.R., Beaudet, A.L., Sly, W.S., and Valle, D., Eds., McGraw-Hill, New York, 2001, 433.
3. Neufeld, E.F., The uptake of enzymes into lysosomes: an overview, *Birth Defects Orig. Artic. Ser.*, 16, 77, 1980.
4. Kornfeld, S., Lysosomal enzyme targeting, *Biochem. Soc. Trans.*, 18, 367, 1990.
5. Barton, R.W. and Neufeld, E.F., The Hurler corrective factor. Purification and some properties, *J. Biol. Chem.*, 246, 7773, 1971.
6. Porter, M.T., Fluharty, A.L., and Kihara, H., Correction of abnormal cerebroside sulfate metabolism in cultured metachromatic leukodystrophy fibroblasts, *Science*, 172, 1263, 1971.
7. O'Brien, J.S. et al., Sanfilippo disease type B: enzyme replacement and metabolic correction in cultured fibroblasts, *Science*, 181, 753, 1973.
8. Cantz, M. and Kresse, H., Sandhoff disease: defective glycosaminoglycan catabolism in cultured fibroblasts and its correction by beta-N-acetylhexosaminidase, *Eur. J. Biochem.*, 47, 581, 1974.
9. Hers, H.G., Alpha-glucosidase deficiency in generalized glycogen storage disease (Pompe's disease), *Biochem. J.*, 86, 11, 1963.
10. Scriver, C.R. et al., The *Metabolic and Molecular Bases of Inherited Diseases*, 8th ed., McGraw-Hill, New York, 2001.
11. Krivit, W., Stem cell transplantation in patients with metabolic storage diseases, *Adv. Pediatr.*, 49, 359, 2002.
12. Malatack, J.J., Consolini, D.M., and Bayever, E. The status of hematopoietic stem cell transplantation in lysosomal storage diseases, *Pediatr. Neurol.*, 29, 391, 2003.
13. Johnson, W.G. et al., Intravenous injection of purified hexosaminidase A into a patient with Tay-Sachs disease, in *Enzyme Therapy in Genetic Diseases*, Desnick, R.J., Bernlohr, R.W., and Krivit, W., Eds., The National Foundation, New York, 1973, 120.
14. Brady, R.O. et al., Demonstration of a deficiency of glucocerebroside-cleaving enzyme in Gaucher's disease, *J. Clin. Invest.*, 45, 1112, 1966.

15. Desnick, R.J., Thorpe, S.R., and Fiddler, M.B, Toward enzyme therapy for lysosomal storage diseases, *Physiol. Rev.*, 56, 57, 1976.

16. Pentchev, P.G. et al., Gaucher disease: isolation and comparison of normal and mutant glucocerebrosidase from human spleen tissue, *Proc. Natl. Acad. Sci. U.S.A.*, 75, 3970, 1978.

17. Doebber, T.W. et al., Enhanced macrophage uptake of synthetically glycosylated human placental beta-glucocerebrosidase, *J. Biol. Chem.*, 257, 2193, 1982.

18. Barton, N.W. et al., Therapeutic response to intravenous infusions of glucocerebrosidase in a patient with Gaucher disease, *Proc. Natl. Acad. Sci. U.S.A.*, 87, 1913, 1990.

19. Barton, N.W. et al., Replacement therapy for inherited enzyme deficiency — macrophage-targeted glucocerebrosidase for Gaucher's disease, *N. Engl. J. Med.*, 324, 1464, 1991.

20. Heiber, V.C., Cloning of a cDNA complementary to rat preputial gland beta-glucuronidase mRNA, *Biochem. Biophys. Res. Commun.*, 104, 1271, 1982.

21. Suzuki, K. et al., Are animal models useful for understanding the pathophysiology of lysosomal storage diseases, *Acta Pediatr. Suppl.*, 92, 54, 2003.

22. Achord, D.T. et al., Human beta-glucuronidase: *in vivo* clearance and *in vitro* uptake by a glycoprotein recognition system on reticuloendothelial cells, *Cell*, 15, 269, 1978.

23. Furbish, F.S. et al., Uptake and distribution of placental glucocerebrosidase in rat hepatic cells and effects of sequential deglycosylation, *Biochim. Biophys. Acta*, 673, 425, 1981.

24. Murray, G.J., Lectin-specific targeting of lysosomal enzymes to reticuloendothelial cells, *Methods Enzymol.*, 149, 25, 1987.

25. Zhu, Y. et al., Dexamethasone-mediated up-regulation of the mannose receptor improves the delivery of recombinant glucocerebrosidase to Gaucher macrophages, *J. Pharmacol. Exp. Ther.*, 308, 705, 2004.

26. Xia, H., Mao, Q., and Davidson, B.L., The HIV Tat protein transduction domain improves the biodistribution of beta-glucuronidase expressed from recombinant viral vectors, *Nat. Biotechnol.*, 19, 640, 2001.

27. LeBowitz, J.H. et al., Glycosylation-independent targeting enhances enzyme delivery to lysosomes and decreases storage in mucopolysaccharidosis type VII mice, *Proc. Natl. Acad. Sci. U.S.A.*, 101, 3083, 2004.

28. Prince, W.S. et al., Lipoprotein receptor binding, cellular uptake, and lysosomal delivery of fusions between the receptor-associated protein (RAP) and alpha-L-iduronidase or acid alpha-glucosidase, *J. Biol. Chem.*, 279, 35037, 2004.

29. Morgan, D.O. et al., Insulin-like growth factor II receptor as a multifunctional binding protein, *Nature*, 329, 301, 1987.

30. Li, Y. et al., Receptor-associated protein facilitates proper folding and maturation of the low-density lipoprotein receptor and its class 2 mutants, *Biochemistry*, 41, 4921, 2002.

31. Czekay, R.P. et al., Endocytic trafficking of megalin/RAP complexes: dissociation of the complexes in late endosomes, *Mol. Biol. Cell*, 8, 517, 1997.

32. Green, I. et al., Protein transduction domains: are they delivering? *Trends Pharmacol. Sci.*, 24, 213, 2003.

33. Brooks, H., Lebleu, B., and Vives, E., Tat peptide-mediated cellular delivery: back to basics, *Adv. Drug Deliv. Rev.*, 57, 559, 2005.

34. Muro, S., Output: VCAM-1 and ICAM-1, in *The Endothelium: A Comprehensive Reference*, Aird, W., Ed, Cambridge University Press, In press.

35. Murciano, J.C. et al., ICAM-directed vascular immunotargeting of antithrombotic agents to the endothelial luminal surface, *Blood*, 101, 3977, 2003.

36. Muro, S., Muzykantov, V.R., and Murciano, J., Characterization of endothelial internalization and targeting of antibody-enzyme conjugates in cell cultures and in laboratory animals, in *Methods in Molecular Biology*, Bioconjugation protocols, Niemeyere, C.M., Ed., Humana Press, Totowa, 2004.

37. Muro, S. et al., A novel endocytic pathway induced by clustering endothelial ICAM-1 or PECAM-1, *J. Cell. Sci.*, 116, 1599, 2003.

38. Muro, S. et al., ICAM-1 recycling in endothelial cells: a novel pathway for sustained intracellular delivery and prolonged effects of drugs, *Blood*, 105, 650, 2005.

39. Muro, S., Koval, M., and Muzykantov, V., Endothelial endocytic pathways: gates for vascular drug delivery, *Curr. Vasc. Pharmacol.*, 2, 281, 2004.

40. DeGraba, T. et al., Profile of endothelial and leukocyte activation in Fabry patients, *Ann. Neurol.*, 47, 229, 2000.

41. Muro, S. and Muzykantov, V.R., Targeting of antioxidants and anti-thrombotic drugs to endothelial cell adhesion molecules, *Current Pharm. Desig.*, 11, 2383, 2005.

42. Dhami, R. and Schuchman, E.H., Mannose 6-phosphate receptor-mediated uptake is defective in acid sphingomyelinase-deficient macrophages: implications for Niemann-Pick disease enzyme replacement therapy, *J. Biol. Chem.*, 279, 1526, 2004.

43. Hopkins, A.M., Baird, A.W., and Nusrat, A., ICAM-1: targeted docking for exogenous as well as endogenous ligands, *Adv. Drug. Deliv. Rev.*, 56, 763, 2004.

44. Muzykantov, V.R., Targeting pulmonary endothelium, in *Biomedical Aspects of Drug Targeting*, Muzykantov, V.R., and Torchilin, V., Eds., Kluwer Academic Publishers, Boston, 2003.

45. Cao, Y. and Suresh, M.R., Bispecific antibodies as novel bioconjugates, *Bioconjug. Chem.*, 9, 635, 1998.

46. Muzykantov, V.R., Conjugation of catalase to a carrier antibody via a streptavidin-biotin cross-linker, *Biotechnol. Appl. Biochem.*, 26, 103, 1997.

47. Wilbur, D.S. et al., Development of new biotin/streptavidin reagents for pretargeting, *Biomol. Eng.*, 16, 113, 1999.

48. Jeong Lee, H. and Pardridge, W.M., Drug targeting to the brain using avidin-biotin technology in the mouse; (blood-brain barrier, monoclonal antibody, transferrin receptor, Alzheimer's disease), *J. Drug. Target.*, 8, 413, 2000.

49. Wiewrodt, R. et al., Size-dependent intracellular immunotargeting of therapeutic cargoes into endothelial cells, *Blood*, 99, 912, 2002.

50. Dziubla, T. et al., Nanoscale anti-oxidant therapeutics, in *Oxidative Stress, Disease and Cancer*, Singh, K.K., Ed., Imperial College Press, London. In press .

51. Dziubla, T., Karim, A., and Muzykantov, V.R., Polymer nanocarriers protecting active enzyme cargo against proteolysis, *J. Control Release*, 102, 427, 2005.

52. Muro, S., Schuchman, E., and Muzykantov, V., Lysosomal enzyme delivery by ICAM-1 targeted nanocarriers bypassing glycosylation- and clathrin-dependent endocytosis, *Mol. Ther.*, 13, 135, 2005.

53. Kakkis, E. et al., Intrathecal enzyme replacement therapy reduces lysosomal storage in brain and meninges of the canine model of MPS I, *Mol. Genet. Metab.*, 83,163, 2004.

54. Bae, J.S. et al. Comparative effects of recombinant acid sphingomyelinase administered by different routes in Niemann-Pick disease mice, *Exp. Anim.*, 53, 417, 2004.

CHAPTER 9

Gaucher Disease Animal Models

Ying Sun, You-Hai Xu, and Gregory A. Grabowski

CONTENTS

INTRODUCTION

Animal models for lysosomal storage diseases have greatly facilitated patho-physiologic and natural history studies. These analogues of the human diseases have led to characterization of visceral and central nervous system (CNS) involvement, and they have provided useful tools for therapeutic endeavors. For lysosomal diseases lacking spontaneously occurring animal models, targeted disruption or "knock-in" techniques have created such models.[1-5] Several animal models have approximated the human phenotypes, but Farber disease (ceramidase deficiency) and Gaucher disease (GCase deficiency) have been notable exceptions.[3,6] The ceramidase null mouse, Farber disease, has an early embryonic lethal phenotype because of aggressive apoptosis of the fertilized zygote.[3] The glucocerebrosidase (GCase) null mouse (Gaucher disease analogue) was created by disruption of the *gba* locus and displays neonatal lethality.[6] Until recently, efforts have failed to establish viable models of GCase deficiency and phenotypes resembling human Gaucher disease. Here, we review these efforts and potential limitations of this approach for Gaucher disease.

MOUSE MODELS

The *gba* Knockout Model

The *gba* locus was disrupted by insertion of a *neo*-cassette in exons 9 and 10. Homozygotes died within 24 h of birth with irregular respiration, poor feeding, and decreased movement.[6] Tissue GCase activity was less than 4% of control. Glucosylceramide accumulation was present in lung, brain, and liver, and macrophages.[6,7] The cause of neonatal death was a marked transepidermal water loss because of altered skin permeability.[8] A similar finding is present in humans with complete GCase deficiency and death in the neonatal period.[9-11] The neonatal lethality limited the utility of this mouse model for studies of pathophysiology but did facilitate understanding the role of GCase in skin permeability.[8]

RecNcil and L444P/L444P Mouse Models

A single insertion mutagenesis procedure, SIMP, was used to generate RecNcil (L444P/A456P) and L444P point mutant mice.[4] These mice were generated with the idea that a less complete deficiency would be compatible with survival. The homozygous RecNcil- or L444P-SIMP mice had red wrinkled skin, similar to the *gba* knock-out, and died within 24–48 h after birth. GCase activity in L444P-SIMP mouse liver, brain, and skin was ~20% of wild type, but was only 4–9% in tissues from RecNcil-SIMP mice.[4] Glucosylceramide (GC) was biochemically detected in the brain, liver, and skin (~10–15 fold) of RecNcil homozygotes but not of L444P/L444P mice. Compared to the *gba* null mouse, no GC accumulation was found in macrophages of RecNcil or L444P homozygotes.[4,8] In humans, Gaucher fibroblasts from L444P homozygotes have low residual activity (3–13% of WT),[12] and do not manifest skin defects or neonatal lethality. This may reflect differences in skin barrier formation during the fetal development of mouse and human. In the mouse, a competent skin permeability barrier forms later than in humans, about 1–2 days before birth compared with 30–36 wk gestation in humans.[13,14] This epidermal permeability barrier formation is mediated by a mixture of nonpolar lipids (enriched in ceramides) in the intercellular spaces of the stratum corneum (SC).[15] The severe deficiencies of GCase activity in human or mouse epidermis altered the ultrastructure of extracellular lamellar structures in the stratum corneum without an apparent alteration of lamellar body contents.[8] The short lifespan of the *gba* null or RecNciI- or L444P-homozygous mice has significantly limited the utility of these models.

Gba[L444P/L444P]-*Ugcg*[+/+]Mouse Model

An additional mouse with *Gba*[L444P/L444P] was made by intercrossing L444P het-erozygotes into a mouse line bearing a glucosylceramide synthase (GCS) knock-out. Following extensive intercrosses, mice with L444P homozygosity and GCS heterozygosity were obtained, as well as L444P/L444P and wild-type GCS in the GCS strain background. These mice were respectively designated as *Gba*[L444P/L444P]*Ugcg*[+/KO] and *Gba*[L444P/L444P]*Ugcg*[+/+].[16,17] About 50% of *Gba*[L444P/L444P]*Ugcg*[+/+] could pass through weaning and survived about 1 yr. but without significant GC accumulation in liver, spleen, brain, or macrophages.[16] These mice did display lymph node hypertrophy and elevated TNFα/IL-1 mRNA expression,[16] a possible deregulated immune response. The most likely explanation for the differences between the different *Gba*[L444P/L444P] with wild-type GCS is a strain difference or genetic background.

Viable GCase Point-Mutated Mouse Models

Mice bearing the following mutations have been characterized: V394L/V394L, N370S/N370S, D409H/D409H, and D409V/D409V.[18] These mice express differential levels of GCase residual activities in various tissues leading to differences in phenotypes. The phenotypic, histological, and biochemical abnormalities of these mice are summarized below.

Similar to the *gba* null and "L444P-SIMP" homozygotes, the intermediate point-mutant N370S+*neo* and V394L+*neo* homozygotes died within 24 h of birth with the skin permeability barrier defects.[18] However, viable mice for all the above point mutations, except N370S, were obtained when the *neo* was excised.[18] Additional viable mice were obtained by intercrosses with the *gba* null and each point mutant line. The residual GCase activities for homozygotes were 2.5 to 10% of WT in various tissues and cultured fibroblasts in the following order: WT>V394L>N370S (fibroblasts only)>D409H>D409V.[18] Compared with other tissues, the residual activity of CNS GCase from all variants was increased (21 to 28% of WT). A similar tissue differential pattern of GC accumulation was observed in human GD.[19] The basis for this differential GCase residual activity in brain has not been defined. Also, GC accumulation was not present in the brain at any age for any variants. These results suggest a potential differential and posttranslational processing or turnover of GCase protein in various tissues and a need for further studies.

The gross phenotypes were concordant with the GCase residual activity in the corresponding point mutated Gaucher mice. The V394L/V394L or V394L/null mice displayed little histological changes in liver, spleen, or lungs throughout their normal life-span. In mice >1.5 yrs., about twofold increases in GC accumulation were present in these tissues as were occasional storage cells. Similarly, the D409H or D409V homozygotes had some storage of GC by biochemical and histological analyses in these same tissues. With age the D409V/null mouse develops extensive (10- to 40-fold) GC storage and Gaucher-like cells in liver, spleen, and particularly lungs. Thus, these mice with various point mutations in *gba* survive and have some of the histological and biochemical findings of humans with Gaucher disease type 1. In general, the tissue lesions of GCase deficiency in these mice were less severe than the corresponding tissues from patients with Gaucher disease. None of the mice had gross CNS defects. These observations suggested different fluxes and thresholds of glycosphingolipid (GSL) substrate and metabolites or altered pathways of GC metabolism in the mouse compared to humans. However, further study is needed to determine if alternative routes are present for the efflux of accumulated GC in our mutant mice. In addition, GCase deficiencies may produce large GC contents that may favor the GSL shunting to synthesis of oligosaccharide compounds, e.g., LacCer or gangliosides. The accumulation of LacCer, GM3, GM2, GM1, GD3, and gluco-sylsphingosine in spleen, liver, and CNS was observed in GD.[20–22] Also, significant levels of GSL accumulate as secondary metabolites in other lysosomal storage diseases (LSDs).[23] Similar observations were made in a mouse model of Niemann-Pick C disease, in which GC, LacCer, sphingomyelin, and cholesterol were increased to 2~100-fold wild-type levels in several tissues, i.e., liver, spleen, lung, kidney, and thymus.[24] The changes in influx and efflux of GSL substrate and metabolites may modulate the GC to levels below the threshold for accumulation in tissues, and therefore reduce the severity of Gaucher disease phenotype. Analysis of GSL metabolism and turnover will be useful for understanding the molecular and biochemical events causal to GD. Determination of the secondary cellular and biochemical pathways that are affected by substrate accumulation, and intervention in these pathways, is potentially important for pathphysiologic and therapeutic studies.[25]

The lethality of the N370S/N370S in mice was unexpected. As is described elsewhere in this volume and by others,[26] N370S homozygosity is associated with less severe phenotypes in human Gaucher disease. These results were unexpected for several reasons: 1) the mouse N370S enzyme, expressed in the baculovirus system, had properties that were nearly identical to its human counterpart.[18] 2) The residual GCase activity in skin from N370S homozygote humans is relatively high. However, the skin appearance of N370S/N370S pups was similar to that in the GCase-null or L444P-SIMP homozygote pups, in spite of the high level of *in vitro* residual activity of cultured N370S/N370S fibroblasts (~13% of WT). This implies that mouse N370S GCase may not effectively hydrolyze the skin GC. Skin GC contains additional hydroxyl groups and very long chain fatty acids.[27,28] The contents of long and very long chain fatty acids in sphingolipids in mouse epidermis are higher than those in humans [Table 9.1]. The total glycolipid content of mouse epidermus is greater than in that from humans (3.6 and 2.4 μmol of glucose/100 g tissue, respectively).[29] In RecNciI (type 2) mice,

Table 9.1   Fatty Acid Composition of Human and
Mouse Epidermis

	Mouse [28]	Human [68]	
	SC-Cer2[a]	SP[b]	GSL
14:0		2.6	5.1
15:0		1.1	3.4
16:0	17.4	14.6	8.2
17:0	0.12	2.0	2.4
18:0	0.59	6.4	4.3
18:1		2.8	17.9
20:0	0.66	1.6	7.7
21:0	0.24	1.3	1.7
22:0		8.9	4.3
23:0	1.68	1.6	1.7
24:0	38.11	18.8	10.0
24:1		9.5	2.0
25:0	7.29	2.0	5.2
26:0	47.33	5.8	5.4
27:0	0.57		
28:0	1.53	0.7	5.9
30:0	0.11		
24:0:OH			2.6
26:0:OH			5.6
Not identified			8.6
Units	%	%	%
% of total lipids by weight			

[a] SC-Cer2: stratum corneum ceramide 2
[b] SP: sphingomyelin

Cer-16 [C16-AS (C16-α-hydroxylated N-acyl fatty acid sphingosine=AS)] levels remained unaffected. However, the levels of C24-/C26-AS or ω-OH fatty acid and ceramides (Cer-OS) were reduced to <40% or 10% of wild-type control, respectively.[30] MALDI-TOF analyses revealed the presence of very long chain fatty acid (C32:1, C30:0, and C34:1) in GC-OS.[30] The above suggest a potential for less efficient hydrolysis of GCs containing long chain fatty acid by mouse N370S than by its human counterpart. Importantly, this skin defect was specific for the N370S enzyme and was not present in the other, lower residual enzyme activity, enzyme variants. Further study of lipid metabolism in these point-mutated mice *in vivo* or *ex vivo* will be necessary to elucidate this pathophysiology. The above findings point to potential factors that could affect the efficiency of variant GCase functions in human and mouse.

Comparative study of GSL composition in human and mice could reveal the biochemical basis of the differential phenotypes from the same *gba* mutations. The available data are difficult to compare, but some salient features are evident. The nonhydroxy FA composition in mouse brain ceramide is different from that in human cerebellar/cerebral GC. The ratio of C_{14}-C_{20} to C_{22}-C_{24} in mouse brain was 81–85.9 to 14.3–19.8. These ratios were 47–48 to 50–53 in human brain GC.[31–33] In normal human liver these ratios were 12 to 87.[33] The specificity of variant GCases activities has not been well studied toward GCs of specific sphingoid base and fatty acid acyl chain compositions. Such analyses using physiologic assays may provide insights into the differential tissue effects of the variant GCases and their resultant pheno-types. Such viable GCase point-mutation models could prove useful for study of variant GCase activity, as well as development of therapy and molecular pathophys-iologic studies (see below).

OTHER MODELS RELATED TO GAUCHER DISEASE

Chemically Induced Mouse Model

The first mouse model for Gaucher disease was generated by administration of conduritol B epoxide (CBE) to wild-type mice.[34] CBE is a covalent inhibitor of GCase.[35,36] Complete inhibition of GCase was achieved in adult mice and GC accumulated in Kupffer cells when GC-loaded liposomes (PC and PS) were admin-istered.[34] The GC accumulation lasted for 5 days following a single CBE injection that provides a therapeutic window. Importantly, chronic injection of CBE to wild-type mice produced CNS damage that could not be reversed by withdrawal of CBE.[34] This model has been used for limited enzyme or gene therapy studies.[37]

Chimeric Mouse Model of Gaucher Disease

The chimeric mouse model was generated by infusing the hematopoietic stem cells from *gba* null fetuses into irradiated normal mice. These chimeric mice man-ifested total deficiency of GCase in peripheral blood cells. No Gaucher cells were present in these mice and only limited amounts of GC stored in the tissues primarily involved in Gaucher disease.[38]

GCase Activator Mouse Models

GCase requires a protein activator, saposin C, and phospholipids to achieve the maximal activity. Saposin C is one of the four saposins (A, B, C, and D) proteolyt-ically derived from their precursor, prosaposin (see Chapter 4). Deficiency of saposin C has been reported in two human patients with GC accumulation and a Gaucher-like phenotypes.[39,40] The saposin C-deficient animals are not available. Deficiency of saposin A in mice leads to a late-onset, chronic form of globoid cell leukodys-trophy.[41] The deficiency of saposin D resulted in kidney dysfunction and cerebellar Purkinje cell loss with a life span up to 8 months.[42] Both saposin A and D deficiencies in mice have mild phenotypes relative to defects in their cognate enzymes, galacto-sylceramidase and ceramidase, respectively. Prosaposin, or total saposin, deficiency was generated in Dr. Kunihiko Suzuki's lab by gene targeting.[43] Partial saposin deficiency mice were created in our lab by expressing the prosaposin transgene at subnormal levels against the background of total saposins deficiency.[44]

Total Saposin Deficiency Mice

A complete deficiency of mouse prosaposin and saposins led to severe neuro-logical phenotype with a life-span between 20 to 30 days and neonatal death in common. The lipid accumulation profile, the phenotype and pathology of total saposin deficiency mice closely resemble the human prosaposin deficiency.[45,46] Each saposin has activator functions of specific cognate enzymes at different steps in GSL metabolism pathway. Thus, the phenotype directly due to deficiency of saposin C cannot be determined from these mice. The major accumulated GSL was LacCer in brain, liver, and kidney. GC, ceramide and gangliosides were increased in brain. In the liver, GC, ceramide, and globotriaosylceramide were significantly increased. No Gaucher cells were found in any tissue in these mice. Interestingly, GCase activity was decreased in liver, kidney, and fibroblast of the total saposin deficiency mice, but not significantly altered in brain. The turnover rates of mouse and human GCase were increased in fibroblasts with total saposin deficiency.[47] Prosaposin deficiency cells and tissues from both human and mice were used to show that saposin C functions to protect GCase from proteolysis in lysosome.[47]

Partial Saposin Deficiency in Mice

Partial prosaposin deficiency in mice was developed by introducing a low level expressing (5–45%) mouse prosaposin transgene into deficient mice (termed NA). Such mice survived up to 220 days and had delayed neurological phenotype with development of ataxia and tremor after 3 months. This was accompanied by extensive Purkinje cell loss in cerebellum. LacCer and GC accumulated in liver but were reduced in NA brains relative to those from prosaposin deficient mice. Gaucher cells were not found in any tissue. Decreased GCase activities were evident in cells from the NA mice.[44,47]

Partial Prosaposin Deficiency Combined with GCase Point-Mutations

A variant Gaucher mouse by cross-breeding the NA mice into homozygotes for GCase point mutations (V394L or D409H).[48] The striking finding was the presence of large numbers of storage macrophages in liver, lung, and spleen; none were present in age matched the NA, or V394L or D409H homozygotes. Upon ultrastructural examination, the storage macrophages in lung contained characteristic braided struc-tures of GC as seen in human Gaucher cells. GC accumulation was more than 2- to 10-fold increased in visceral tissues and CNS compared to NA deficient mice.[48] Reduction of stability of mutant GCase against the background of subnormal saposin levels, most likely saposin C, accounts for the GC accumulation. The NA model provides insight into the threshold level (~10–15%) of saposin C needed for variant GCase function *in vivo*. The visceral manifestations in this model closely mimicked the Gaucher phenotype. These viable animals of Gaucher disease should prove useful for therapeutic and pathogenic studies.

Canine Model of Gaucher Disease

An 8-month-old Australia Sydney Silky dog was described to have glucocere-broside storage with progressive neurological impairment.[49] Gaucher cells were found in liver, lymph nodes, and cerebellum, but not in spleen. By electron micro-scopy, the membrane bounded lamellae bodies were present in neurons of dorsal thalamic area and hippocampus. There were losses of granule cells in cerebellum. Curiously, Gaucher cells were seen in the granule cell layer, but no storage materials were present in Purkinje or glial cells. GCase activity was negligible in liver, lymphocytes and granulocytes, about 80% remained in brain.[50] This is the only naturally occurring animal model of Gaucher disease but is not available.

CURRENT ISSUES IN MOUSE MODELS OF GAUCHER DISEASE

Global Gene Expression in Gaucher Disease

Initial studies were conducted to investigate the correlation of neuropaothlogy with gene expression profiles in CNS Gaucher disease patients and mice.[51,52] In *gba* disrupted mice, mRNA microarray analysis showed a 3.4-fold decrease in bcl-2 RNA expression in E19.5 brain stem and cerebellum that was confirmed by RT-PCR and Western analyses. In GCase point-mutated mice, about 44–59% of the differ-entially expressed genes in lung, liver, and spleen were shared between V394L/V394L (4L/4L) and D409/null (9V/null) mice in newborns and 4-, 12-, 18-, and 28-week-old mice (Figure 9.1) (unpublished observations). The genes include those for cytokines, the TNF family, macrophage markers, apoptosis, and the MHC family, indicating pro-inflammation components in the Gaucher disease. These stud-ies establish a starting point for understanding the molecular basis of the progressive pathogenesis of Gaucher disease.

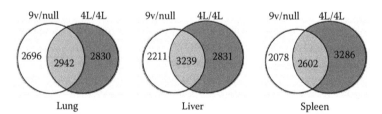

Figure 9.1   Venn diagram shows the number of mRNAs that were altered and shared by V394L/V394L and D409V/null mice in lung, liver, and spleen tissues at the age of newborn, 4-, 12-, 18-, and 28-weeks. Duplicated tissue RNA pooled samples (each pooled sample from three individual mice) from V394L/V394L, D409V/null and WT mice were applied to Affymetrix MOE430 2.0 GeneChip and analyzed with GeneSpring 7.2 software. Following quantile normalization across all gene chips and a t-test (p<0.05), an RNA expression change (>±1.8-fold) was detected in 5772, 6070, or 5888 genes in lung, liver, or spleen of V394L/V394L mice, or 5638, 5450, or 4680 genes in D409V/null mice, respectively. Among them 44–59% of the genes (2602 to 3239, light gray) was shared by samples of either genotypes.

Inflammatory and Immune System Studies

The pathologic hallmark of Gaucher disease is the presence of lipid-storage macrophages, Gaucher cells, in visceral organs.[53] These cells develop by ingesting sources of GC, and can lead to fibrosis scars in tissues. The exact malfunction of such macrophages in Gaucher disease is largely unexplored. Studies from affected patients showed increased levels of pro-inflammatory genes (i.e., TNF-α, IL-6, IL-8, and IL-1β) and anti-inflammatory cytokine genes (i.e., CD14).[54-56] These anti-inflammatory mediators are markers for alternatively activated macrophages.[57-61] These results suggest a central role of GC in altering macrophage function as an initial step in the pathophysiology of Gaucher disease.

Therapeutic Studies

Available Gaucher mouse models will be valuable resources for enzyme and gene therapy studies. The models can be used to investigate the tissue or cell type specific distribution, modification of GCase enzyme for enhanced efficacy. Pro-inflammatory effects in disease progression have been suggested in several lysosomal storage disease including gangliosidosis,[62] mucopolysaccharidoses,[63] and lysosomal acid lipase deficiency[64] as well as Gaucher disease.[65] Irrespective of the underlining mechanism, anti-inflammatory agents could be tested for beneficial effects in mouse models. The CNS is a specific challenge in treating Gaucher disease, since enzyme therapy does not alter the CNS impairment in types 2 and 3 patients. Additional alternative approaches could be tested for efficacy, e.g., receptor mediated transcytosis[66,67] and a lipid-mediated delivery system.[67]

SUMMARY

In the past decade, a series of Gaucher disease mouse models have been developed. The studies from such models have provided information about pathophysiology that is not obtainable in humans. The availability of newer mouse models should provide enhanced approaches to the therapy and understanding of the pathogenesis of Gaucher disease.

REFERENCES

1. Clarke, L.A. et al., Murine mucopolysaccharidosis type I: targeted disruption of the murine alpha-L-iduronidase gene, Hum. Mol. Genet., 6, 503, 1997.
2. de Geest, N. et al., Systemic and neurologic abnormalities distinguish the lysosomal disorders sialidosis and galactosialidosis in mice, Hum. Mol. Genet., 11, 1455, 2002.
3. Li, C.M. et al., Insertional mutagenesis of the mouse acid ceramidase gene leads to early embryonic lethality in homozygotes and progressive lipid storage disease in heterozygotes, Genomics, 79, 218, 2002.

4. Liu, Y., et al., Mice with type 2 and 3 Gaucher disease point mutations generated by a single insertion mutagenesis procedure, Proc. Natl. Acad. Sci. U.S.A., 95, 2503, 1998.

5. Marathe, S. et al., Creation of a mouse model for non-neurological (type B) Niemann-Pick disease by stable, low level expression of lysosomal sphingomyelinase in the absence of secretory sphingomyelinase: relationship between brain intra-lysosomal enzyme activity and central nervous system function, Hum. Mol. Genet., 9, 1967, 2000.

6. Tybulewicz, V.L.J. et al., Animal model of Gaucher's disease from targeted disruption of the mouse glucocerebrosidase gene, Nature, 357, 407, 1992.

7. Orvisky, E. et al., Glucosylsphingosine accumulation in mice and patients with type 2 Gaucher disease begins early in gestation, Pediatr. Res., 48, 233, 2000.

8. Holleran, W.M. et al., Consequences of beta-glucocerebrosidase deficiency in epidermis. Ultrastructure and permeability barrier alterations in Gaucher disease, J. Clin. Invest., 93, 1756, 1994.

9. Sidransky, E. et al., The clinical, molecular, and pathological characterisation of a family with two cases of lethal perinatal type 2 Gaucher disease, J. Med. Genet., 33, 132, 1996.

10. Tayebi, N. et al., Prenatal lethality of a homozygous null mutation in the human glucocerebrosidase gene, Am. J. Med. Genet., 73, 41, 1997.

11. Finn, L.S. et al., Severe type II Gaucher disease with ichthyosis, arthrogryposis and neuronal apoptosis: molecular and pathological analyses, Am. J. Med. Genet., 91, 222, 2000.

12. Grace, M.E. et al., Analysis of human acid beta-glucosidase by site-directed mutagenesis and heterologous expression, J. Biol. Chem., 269, 2283, 1994.

13. Hanley, K. et al., Glucosylceramide metabolism is regulated during normal and hormonally stimulated epidermal barrier development in the rat, J. Lipid Res., 38, 576, 1997.

14. Hanley, K. et al., Hypothyroidism delays fetal stratum corneum development in mice, Pediatr. Res., 42, 610, 1997.

15. Yardley, H.J. and Summerly, R., Lipid composition and metabolism in normal and diseased epidermis, Pharmacol. Ther., 13, 357, 1981.

16. Mizukami, H. et al., Systemic inflammation in glucocerebrosidase-deficient mice with minimal glucosylceramide storage, J. Clin. Invest., 109, 1215, 2002.

17. Yamashita, T. et al., A vital role for glycosphingolipid synthesis during development and differentiation, Proc. Natl. Acad. Sci. U.S.A., 96, 9142, 1999.

18. Xu, Y.H. et al., Viable mouse models of acid beta-glucosidase deficiency: the defect in Gaucher disease, Am. J. Pathol., 163, 2093, 2003.

19. Nilsson, O. and Svennerholm, L., Accumulation of glucosylceramide and glucosylsphingosine (psychosine) in cerebrum and cerebellum in infantile and juvenile Gaucher disease, J. Neurochem., 39, 709, 1982.

20. Nilsson, O. et al., The occurrence of psychosine and other glycolipids in spleen and liver from the three major types of Gaucher's disease, Biochim. Biophys. Acta, 712, 453, 1982.

21. Gonzalez-Sastre, F., Pampols, T., and Sabater, J., Infantile Gaucher's disease: a biochemical study, Neurology, 24, 162, 1974.

22. Kaye, E.M. et al., Type 2 and type 3 Gaucher disease: a morphological and biochemical study, Ann. Neurol., 20, 223, 1986.

23. Walkley, S.U., Secondary accumulation of gangliosides in lysosomal storage disorders, Semin. Cell Dev. Biol., 15, 433, 2004.

24. Pentchev, P.G. et al., A lysosomal storage disorder in mice characterized by a dual deficiency of sphingomyelinase and glucocerebrosidase, Biochim. Biophys. Acta, 619, 669, 1980.

25. Futerman, A.H. and van Meer, G., The cell biology of lysosomal storage disorders, Nat. Rev. Mol. Cell. Biol., 5, 554, 2004.

26. Sibille, A. et al., Phenotype/genotype correlations in Gaucher disease type I: clinical and therapeutic implications, Am. J. Hum. Genet., 52, 1094, 1993.

27. Wertz, P.W., Epidermal lipids, Semin. Dermatol., 11 (2), 106–13, 1992.

28. Uchida, Y. et al., Epidermal sphingomyelins are precursors for selected stratum corneum ceramides, J. Lipid Res., 41, 2071, 2000.

29. Ohashi, M., A comparison of the ganglioside distributions of fat tissues in various animals by two-dimensional thin layer chromatography, Lipids, 14, 52, 1979.

30. Doering, T., Proia, R.L., and Sandhoff, K., Accumulation of protein-bound epidermal glucosylceramides in beta-glucocerebrosidase deficient type 2 Gaucher mice, FEBS Lett., 447, 167, 1999.

31. Joseph, K.C. et al., Fatty acid composition of cerebrosides, sulphatides and ceramides in murine leucodystrophy: the quaking mutant, J. Neurochem., 19, 307, 1972.

32. Joseph, K.C. and Hogan, E.L., Fatty acid composition of cerebrosides, sulphatides and ceramides in murine sudanophilic leucodystrophy: the Jimpy mutant, J. Neurochem., 18, 1639, 1971.

33. Nilsson, O. and Svennerholm, L., Characterization and quantitative determination of gangliosides and neutral glycosphingolipids in human liver, J. Lipid Res., 23, 327, 1982.

34. Kanfer, J.N. et al., The Gaucher mouse, Biochem. Biophys. Res. Commun., 67, 85, 1975.

35. Grabowski, G.A. et al., Human acid b-glucosidase. Use of conduritol B epoxide derivatives to investigate the catalytically active normal and Gaucher disease enzymes, J. Biol. Chem., 261, 8263, 1986.

36. Radin, N.S. and Vunnam, R.R., Inhibitors of cerebroside metabolism, Meth. Enzymol., 72, 673, 1981.

37. Marshall, J. et al., Demonstration of feasibility of *in vivo* gene therapy for Gaucher disease using a chemically induced mouse model, Mol. Ther., 6, 179, 2002.

38. Beutler, E. et al., A chimeric mouse model of Gaucher disease, Mol. Med., 8, 247, 2002.

39. Schnabel, D., Schroder, M., and Sandhoff, K., Mutation in the sphingolipid activator protein 2 in a patient with a variant of Gaucher disease, FEBS Lett., 284, 57, 1991.

40. Rafi, M.A. et al., Mutational analysis in a patient with a variant form of Gaucher disease caused by SAP-2 deficiency, Somat. Cell. Mol. Genet., 19, 1, 1993.

41. Matsuda, J. et al., A mutation in the saposin A domain of the sphingolipid activator protein (prosaposin) gene results in a late-onset, chronic form of globoid cell leukodystrophy in the mouse, Hum. Mol. Genet., 10, 1191, 2001.

42. Matsuda, J. et al., Mutation in saposin D domain of sphingolipid activator protein gene causes urinary system defects and cerebellar Purkinje cell degeneration with accumulation of hydroxy fatty acid-containing ceramide in mouse, Hum. Mol. Genet., 13, 2709, 2004.

43. Fujita, N. et al., Targeted disruption of the mouse sphingolipid activator protein gene: a complex phenotype, including severe leukodystrophy and wide-spread storage of multiple sphingolipids, Hum. Mol. Genet., 5, 711, 1996.

44. Sun, Y. et al., Prosaposin: threshold rescue and analysis of the "neuritogenic" region in transgenic mice, Mol. Genet. Metab., 76, 271, 2002.

45. Hulkova, H. et al., A novel mutation in the coding region of the prosaposin gene leads to a complete deficiency of prosaposin and saposins, and is associated with a complex sphingolipidosis dominated by lactosylceramide accumulation, Hum. Mol. Genet., 10, 927, 2001.

46. Paton, B.C. et al., Additional biochemical findings in a patient and fetal sibling with a genetic defect in the sphingolipid activator protein (SAP) precursor, prosaposin. Evidence for a deficiency in SAP-1 and for a normal lysosomal neuraminidase, Biochem. J., 285, 481, 1992.

47. Sun, Y., Qi, X., and Grabowski, G.A., Saposin C is required for normal resistance of acid b-glucosidase to proteolytic degradation, J. Biol. Chem., 278, 31918, 2003.

48. Sun, Y. et al., Gaucher disease mouse models: point mutations at the acid b-glucosidase locus combined with low-level prosaposin expression lead to disease variants, J. Lipid Res., 46, 2102, 2005.

49. Hartley, W.J. and Blakemore, W.F., Neurovisceral glucocerebroside storage (Gaucher's disease) in a dog, Vet. Pathol., 10, 191, 1973.

50. Van De Water, N.S., Jolly, R.D., and Farrow, B.R., Canine Gaucher disease — the enzymic defect, Aust. J. Exp. Biol. Med. Sci., 57, 551, 1979.

51. Hong, Y.B., Kim, E.Y., and Jung, S.C., Down-regulation of Bcl-2 in the fetal brain of the Gaucher disease mouse model: a possible role in the neuronal loss, J. Hum. Genet., 49, 349, 2004.

52. Myerowitz, R. et al., Global gene expression in a type 2 Gaucher disease brain, Mol. Genet. Metab., 83, 288, 2004.

53. Beutler, E. and Grabowski, G.A., Gaucher Disease, in The Metabolic and Molecular Basis of Inherited Disease, 8th ed., Scriver, C.R., Beaudet, A.L., Sly, W.S., and Valle, D., Eds., McGraw-Hill, New York, 2001, 3635.

54. Michelakakis, H. et al., Plasma tumor necrosis factor-a (TNF-a) levels in Gaucher disease, Biochim. Biophys. Acta, 1317, 219, 1996.

55. Hollak, C.E. et al., Elevated levels of M-CSF, sCD14 and IL8 in type 1 Gaucher disease, Blood Cells Mol. Dis., 23, 201, 1997.

56. Barak, V. et al., Cytokines in Gaucher's disease, Eur. Cytokine Netw., 10, 205, 1999.

57. Gratchev, A. et al., Alternatively activated macrophages differentially express fibronectin and its splice variants and the extracellular matrix protein betaIG-H3, Scand. J. Immunol., 53, 386, 2001.

58. Frings, W., Dreier, J., and Sorg, C., Only the soluble form of the scavenger receptor CD163 acts inhibitory on phorbol ester-activated T-lymphocytes, whereas membrane-bound protein has no effect, FEBS Lett., 526, 93, 2002.

59. Goerdt, S. et al., Alternative versus classical activation of macrophages, Pathobiology, 67, 222, 1999.

60. Gordon, S., Alternative activation of macrophages, Nat. Rev. Immunol., 3, 23, 2003.

61. Mantovani, A. et al., Macrophage polarization: tumor-associated macrophages as a paradigm for polarized M2 mononuclear phagocytes, Trends Immunol., 23, 549, 2002.

62. Jeyakumar, M., et al., Central nervous system inflammation is a hallmark of pathogenesis in mouse models of GM1 and GM2 gangliosidosis, Brain, 126, 974, 2003.

63. Ohmi, K. et al., Activated microglia in cortex of mouse models of mucopolysaccharidoses I and IIIB, Proc. Natl. Acad. Sci. U.S.A., 100, 1902, 2003.

64. Lian, X., Yan, C., Yang, L., Xu, Y., and Du, H., Lysosomal acid lipase deficiency causes respiratory inflammation and destruction in the lung, Am. J. Physiol. Lung Cell Mol. Physiol., 286, L801, 2004.

65. Boven, L.A. et al., Gaucher cells demonstrate a distinct macrophage phenotype and resemble alternatively activated macrophages, Am. J. Clin. Pathol., 122, 359, 2004.
66. Urayama, A. et al., Developmentally regulated mannose 6-phosphate receptor-mediated transport of a lysosomal enzyme across the blood-brain barrier, Proc. Natl. Acad. Sci. U.S.A., 101, 12658, 2004.
67. Pardridge, W.M., Blood-brain barrier drug targeting: the future of brain drug development, Mol. Interv., 3, 90, 2003.
68. Gray, G.M. and Yardley, H.J., Lipid compositions of cells isolated from pig, human, and rat epidermis, J. Lipid Res., 16, 434, 1975.

CHAPTER 10

Type 1 Gaucher Disease — Clinical Features

Pramod K. Mistry and Ari Zimran

CONTENTS

INTRODUCTION

The history and physical examination of a patient with advanced type 1 Gaucher disease (GD) creates a lasting impression. The history of chronic disability in these often highly accomplished individuals, massive organomegaly, cytopenia, bone disease with additional rare complications are unmistakable hallmarks of the disease. The dramatic clinical phenotype underscores a single gene's power to alter the structure and function of a single cell type that triggers the development of a multisystem disease. While the clinical consequences of this genetic disease have been even more clearly defined in the last decade, the molecular mechanisms that induce these clinical transformations are largely not elucidated. The clinical features and the natural history of type 1 GD have been described comprehensively in recent reviews.[1-4] During the past decade, the introduction of enzyme replacement therapy (ERT) as the standard of care and insights into pathobiology of the disease have led to renewed interest in precise delineation of the spectrum of Gaucher phenotype, its natural history, and impact of therapy in modifying the phenotype. Early descriptions of Gaucher phenotypes were skewed by ascertainment bias in European patient populations, especially those of Ashkenazi Jewish ancestry and the fact that in most patient series, severely affected symptomatic patients comprised the majority.[5-7] However, two major factors have led to an appreciation of the expanding spectrum of phenotypic diversity of GD. First, enzyme replacement therapy (ERT) has become the standard of care for type 1 GD throughout the world[8]; second, the International Gaucher Registry (ICGG) has been instrumental in capturing data on the phenotypes of new patient populations.[3,9] This chapter will briefly review the classic phenotypic features of nonneuronopathic type 1 GD, focus on new understanding of diversity of phenotypes, including the changing natural history in the era of interventional therapeutics, emergence of new phenotypes due to differential responsiveness of affected tissues to ERT and the increasing awareness that accumulating glucosylceramide in macrophages may trigger a unique form of macrophage activation and inflammatory response that leads to development of the disease. For neuronopathic Gaucher disease, genetic epidemiology and laboratory biomarkers, the reader is referred to the relevant chapters.

GAUCHER PHENOTYPE(S)

The broad clinical classification of GD into three distinct types has some practical importance in predicting prognosis and response to ERT. However it is an imperfect representation of a continuum of phenotypes that can present major diagnostic challenges especially in children with severe systemic disease but apparent absence of CNS involvement.[10] Such affected children can be classified confidently as nonneuronopathic (type 1) if they harbour at least one N370S GBA allele.[11] In all other genotypes presenting in childhood, designation as type 1 disease has to be tentative since neurologic symptoms may emerge later in the natural history of the disease. For GD beyond childhood, the classification of nonneuronopathic disease becomes increasingly more certain while rate of progression and severity of systemic disease becomes increasingly less predictable. Recent data suggests that adults with type 1 GD may on rare occasions manifest variants of Parkinsonism.[12,13] The association is not specific to any GBA-alleles and causal relationship has not been proven. However, the association of Parkinsonism variants with GD in rare patients is important to recognize since its responsiveness to standard anti-Parkinsonian therapies may differ from that in patients who do not have Gaucher disease.[12]

A severity score index (SSI) has been developed by Zimran et al. to represent the overall Gaucher phenotype as a semiquantitative index of disease severity that incorporates age of presentation, extent of different organ involvement, spleen status, and subjective pain criteria.[5] While it is an imperfect tool for precise ascertainment of the severity of a protean Gaucher phenotype or the evaluation of the candidacy for enzyme therapy and its response, SSI has been used effectively in genotype/phenotype correlation studies,[14] evaluation of natural history of GD,[14–16] as well as evaluation of biomarkers of disease activity.[17,18] Unfortunately, SSI has precluded validation in part because currently there is no methodology to quantify total body burden of Gaucher cells and their activity or glucosylceramide content of tissues, *in vivo*. Severity of disease in individual organs can be quantified to follow disease progression or response to therapy in these individual compartments; i.e., volumetric measurement of liver and spleen[6,19] or MRI-based bone marrow burden (BMB) score developed by Maas et al.[20,21] However, focusing on individual organs in this manner provides no information regarding the rest of the Gaucher phenotype since there is poor concordance of severity of visceral and/or hematologic disease with the extent of involvement of skeletal or other disease compartments among patients with same GBA genotypes and even among affected siblings.[22]

ROLE OF MACROPHAGE ACTIVATION AND INFLAMMATION IN DEVELOPMENT OF GAUCHER DISEASE

The hallmark of GD is widespread accumulation of tissue macrophages engorged with lysosomes that are laden with glucosylceramide. Although organomegaly may assume vast proportions, the glycolipid content in tissues constitutes less than 1% of the increased visceral mass: approximate glucosylceramide content of tissues in μmol/g is 35 in spleen, 35 in lymph nodes of asplenic patients, 12 in the liver of patients with intact spleen, 25 in liver of asplenic patients and 5 in the lungs.[23] Abnormal accumulation of glucosylceramide within Gaucher cells appears to trigger an altered macrophage phenotype and a low grade chronic inflammation that is occasionally interspersed with florid acute inflammatory state of Gaucher crises. Together this leads to the development of the diverse manifestations of GD such as hypertrophy of primary and secondary lymphoid organs, high incidence of monoclonal gammopathy, hyper-metabolic state, development of fibrosis in affected organs including pericellular fibrosis around the storage cells and the prolonged systemic inflammatory state following the Gaucher bone crises.[7,24,25] Thus the latter represents the most florid acute inflammatory presentation of GD, and in all other settings there is low grade chronic inflammation, which leads to other diverse manifestations of the disease.

NATURAL HISTORY

The progressive nature of type 1 GD has been demonstrated in natural history studies in patient populations from The Netherlands and Japan.[15,16] Disease progression was most marked in patients who presented in childhood before age 4 and to a lesser degree in those presenting between age 4 and 25. However, among those presenting after age 25, there was no increase in SSI during the observation period. It should be noted that during a longer period of observation in the Dutch study, disease severity progressed regardless of age at presentation. In another natural history study, wherein many of the patients were N370S homozygous, there was no progression of disease during 2–13 years (mean 5.6 yrs) of follow up.[5] In fact, N370S allele is markedly under-represented in Ashkenazi Jewish patient populations compared to control population, and the extent of under-representation has led to the estimate that up to two thirds of N370S homozygous individuals may never come to medical attention, pointing to low expressivity of this GBA allele.[26,27] However, within each genotype category, there is considerable variability in disease severity, which precludes reliable prognosis in individual patients.

Insight into the natural history of type 1 GD in pre-ERT era is derived from University of Pittsburgh Gaucher Registry compiled by Robert E. Lee.[7] Among 235 patients with type 1 GD, 35 died due to a variety of causes including malignancies, pulmonary hypertension, sepsis, bleeding complications, and portal hypertension. The patients who died prematurely with mean age at death among patients with malignancies of 59 years in contrast to those without cancers who died at mean age of 46 from other causes, including sepsis, pulmonary hypertension, cirrhosis, and bleeding complications.

MODE OF PRESENTATION OF TYPE 1 GAUCHER DISEASE

Just as GD is so diverse in its overall phenotypic expression, there is extreme variability in its initial clinical presentation, which may include any sub-specialty spanning the entire spectrum from pediatric to geriatric medicine. It is striking to learn repeatedly from patients how signs and symptoms of GD remained unrecognized for months or even years, occasionally even in patients who exhibited florid manifestations.[28] Patients may experience fatigue and weakness for a long time before seeking medical evaluation. Bruising, epistaxis, and bone pain in childhood may not arouse concern until other typical manifestations are noticed.[4] Similarly, low platelets during pregnancy may be attributed to pregnancy-related thrombocytopenia and fractures in older patients to osteoporosis. Further complexity is added by the changing spectrum of the disease in the era of ERT. The presentation of the first generation of patients on ERT was characterized by preponderance of those with severe and often irreversible complications of disease such as avascular necrosis (AVN), hip replacements, and vertebral fractures[29]; many of these patients previously underwent splenectomy, which may have altered the natural history of the disease. In contrast, the new generation of patients is characterized by presentation before irreversible complications occur and who therefore have the prospect of achieving complete reversal of phenotype by ERT.

In the ICGG registry, pediatric presentation is characterized by growth failure in a third of the patients, moderate to severe hepatosplenomegaly in the majority (>90%) and anemia in only a third of the patients, in contrast to symptomatic thrombocytopenia in over 90%. Bone pain was reported in a third of the patients (~10% of them had bone crises), although evidence of skeletal involvement on imaging was present in over 90% of the children. With increasing awareness of GD and earlier administration of ERT, most physicians managing a large number of patients have noticed a dramatic decline in the incidence of bone crises as a pre-senting feature of the disease.[30]

In patients presenting as young adults or at an older age, hepatosplenomegaly is less striking compared to pediatric patients.[7] There is no correlation between the degree of splenomegaly and the symptoms.[31] Splenomegaly is always more pro-nounced compared to the liver[32]; if there is disproportionate hepatomegaly, presence of a primary hepatic co-morbidity should be sought. It should be noted that even among apparently asymptomatic patients, osteoporosis, osteonecrosis or lytic lesions may be found. Among older patients, nonclassical presentation of GD may occur such as infectious complications after incidental surgery, osteoporotic fractures, cholelithiasis or benign monoclonal gammopathy and multiple myeloma.

Rarely, an adult patient may harbour massive splenomegaly that results in inca-pacitating abdominal discomfort, exacerbated by recurrent splenic infarcts and peri-splenitis. Massive splenomegaly descending into the pelvis may result in urinary frequency and dyspareunia.[5] In adults as with children, thrombocytopenia is usually more significant than anemia at presentation.[9,33] In a patient with intact spleen, the finding of significant anemia without thrombocytopenia should lead to a search for co-morbidity such as iron deficiency, vitamin B_{12} deficiency, or hemolysis.[4] Older patients with mild or no reduction of platelets may present with severe bleeding complications after incidental surgery due to recently described platelet dysfunction or an acquired coagulopathy.[34]

Another distinctive presentation of GD is sudden development of severe sple-nomegaly following infectious mononucleosis.[35–36] Similarly, patients known to have GD may develop marked progression of splenomegaly during Epstein Barr Virus (EBV) infection. In both settings, there is regression of splenomegaly when the intercurrent infection resolves. In all other settings, splenomegaly seldom regresses spontaneously although the rate of progression is very slow. Thus a rapid enlargement of spleen should prompt a search for a co-morbidity such as infection, lympho-proliferative disorder, or haemolytic anemia.

THE SPLEEN IN GAUCHER DISEASE

Normal spleen size is estimated to be 0.2% of body weight, and in GD it may increase to over 100 multiples of normal. Massive splenomegaly is most frequently seen in children and it may be associated with cachexia, growth failure, and hyper-metabolic state.[7,32] Recurrent infarction and fibrosis occurs as splenomegaly progresses. With greater splenomegaly there is increasing likelihood of focal defects (on imaging) that may represent areas of infarction, fibrosis, vascular malformations, focal collections of Gaucher cells or extramedullary hematopoesis.[37]

Among the less typical presentations of splenomegaly, we have observed a young girl with "wandering spleen" located in the mid-lower abdomen.[38] This 18-year-old patient elected to undergo splenectomy to prevent future obstetrical complications. Another unusual presentation was of a 48-year-old man presenting with two months' history of severe left flank pain, and found to have a large subcapsular splenic hematoma[39]; on further history, he recalled experiencing blunt trauma to the abdomen. In another patient, splenic rupture occurred after severe infectious mononucleosis at age 25. It is remarkable to note the rarity of splenic rupture in GD despite the frequent occurrence of gross organomegaly.

Rarely, Gaucher-related splenomegaly may be aggravated by infectious, haemolytic, or malignant disorders. Two examples include the development of brucellosis in a Turkish patient with type 1 GD,[40] and an American patient who developed Littoral cell angioma, an uncommon vascular lesion mimicking traumatic organ injury or malignancy.[41] When masses in the spleen are noted in patients not known to have GD, a prolonged diagnostic work-up is undertaken before the correct diagnosis is eventually made.

The spleen can be viewed as the central hub of GD activity as evidenced by presence of these space-occupying lesions in large spleens and the fact that splenectomy is associated with correction of hyper-metabolic state, unequivocal progression of type 3 Norrbottnian GD, worse hepatic disease, overt pulmonary disease, and worse skeletal disease. It is also important to keep in mind that GD patients who perish from life-threatening infections are usually asplenic.[1] Thus, splenectomy and progressive disease thereafter may be viewed as leading to re-orientation of monocyte-macrophage system towards the liver and the lung with progressive replacement of the bone marrow.

THE LIVER IN GAUCHER DISEASE

It is envisaged that every patient has an accumulation of Gaucher cells within the liver even if there is no hepatomegaly that may progress to mild-severe hepatomegaly with or without elevated liver function tests. A minority of patients go on to develop severe hepatic fibrosis with portal hypertension and liver failure. Some have associated hepatopulmonary syndrome (HPS).[42] In the ICGG registry, approximately half of the patients have hepatomegaly. The mean increase of liver volume was 2.0 multiples of the normal expected volume relative to the spleen enlargement which was 19.8 multiples of the normal volume.[9] Although the degree of hepatomegaly is never as massive as splenomegaly, patients who underwent splenectomy in the pre-ERT era may develop massive enlargement of the liver extending into the pelvis and occupying the left upper quadrant. The extent of splenomegaly and hepatomegaly are usually correlated and when there is disproportionate hepatomegaly or liver function abnormalities, search for a primary hepatic co-morbidity should be undertaken. In a series of 25 severely affected patients reported by James et al. in the 1980s, liver enzyme was elevated in nearly half the patients and 10% had evidence of portal hypertension.[43] In this series there was also evidence of extramedullary hemopoesis in 25% of patients. The pattern of hepatic fibrosis leading to portal hypertension appears to be unique to GD with an unusual type of confluent central fibrosis lacking typical features of cirrhosis such as regenerative nodules or bile ductular proliferation.[44] This type of presentation of GD is rare and in our combined series, we estimate it is only ~1%.

The liver takes up the largest amount of infused enzyme and is therefore the most responsive organ to ERT, even in the presence of advanced disease, although fibrosis is irreversible.[44]

Several cases of hepatocellular carcinoma (HCC) have been reported in type 1 GD.[45–46] Hence, alpha-feto protein and triphasic contrast-enhanced CT scan of the liver with or without liver biopsy are mandatory for further evaluation of any suspicious lesions.[12,47] It should be noted that HCC occurs in the setting of severe disease while another malignancy associated with GD, namely, multiple myeloma, may occur also in the setting of mild disease and low SSI.[7,29,31]

When evaluating the liver in GD, it is important to keep in mind possible co-morbidity in the form of chronic hepatitis C (HCV). Among GD patients infected with HCV, there was evidence of increased T cell response to the virus compared to patients with GD not infected with HCV, chronic HCV non-GD patients as well as healthy controls[48]: HCV-infected GD patients manifested increased peripheral NKT cells and elevated serum IFN levels.

Recent studies have defined a billiary manifestation of type 1 GD in the form of increased prevalence of cholelithiasis. Rosenbaum and Sidransky reviewed the clinical records of 66 type 1 patients evaluated at the NIH — 21 of them (31.8%) had either gallstones or (in 13 of them) a history of cholecystectomy.[49] In the same study, they also reported 20% prevalence of gallbladder involvement in 16 out of 80 Ashkenazi Jewish patients from Northern Israel. Ben Harosh-Katz et al. also found a higher incidence of cholelithiasis — 25.4% (82 patients out of 323) among adult patients evaluated at a national referral Gaucher Clinic in Israel[50]; in her study she also looked at the possibility of finding a correlation between gallbladder disease and the UDPGT1 polymorphism (Gilbert syndrome). Such an association was statistically significant only with the splenectomized patients. The cause of increased prevalence of cholelithiasis in GD is unknown.

HEMATOLOGIC MANIFESTATIONS

The hematological features of type I GD have been reviewed comprehensively by Zimran and associates.[51] Anemia and/or thrombocytopenia are common presenting features of GD either as an incidental finding during routine blood count or in association with typical symptoms of fatigue and/or bleeding tendency. Leucopenia occurs commonly, and, in the presence of concurrent impairment of neutrophil chemotaxis, contributes to susceptibility to infections. Cytopenia in patients with intact spleen is generally attributed to hypersplenism, but marrow replacement by Gaucher cells may also contribute. It is important to keep in mind that vitamin B_{12} deficiency may be common in GD,[52] but macrocytosis is common also due to liver involvement. Some patients have been found to have concurrent autoimmune hemolytic anemia or idiopathic thrombocytopenia. Searching for another etiology is particular important in patients who show unexpected decline in blood count, despite either stable course over years in the untreated patient, or following earlier satisfactory response to ERT.

It is now recognized that abnormal platelet function (aggregation and/or adhesion) may be another cause for bleeding in GD unrelated to actual platelet count or coagulation profile.[53] ERT corrects platelet dysfunction, similar to its effect on reversing abnormal neutrophil chemotaxis.[54]

Two recent studies have defined rheologic abnormalities in GD. In Jerusalem, 60 patients with GD and matched controls were studied and found to have elevated fibrinogen and accelerated erythrocyte sedimentation rates.[55] These findings were associated with enhanced erythrocyte and leukocyte adhesiveness and aggregation, and were consistent with the notion that low-grade chronic inflammation of GD may result in impaired microcirculation and tissue hypoxemia. A second study from London investigated several hematological and hemorheological parameters in GD patients stratified according to spleen status compared to healthy controls with and without spleen. Only the splenectomized patients with GD showed significant hemorheological abnormalities, pointing to a specific effect of the storage disease on rheologic properties of blood.[56]

THE BONES IN GAUCHER DISEASE

Overall Spectrum of Skeletal Disease in Type 1 Gaucher Disease

It is somewhat arbitrary to separate bone marrow disease and skeletal disease since the two are intimately linked, and it seems that bone marrow cells, in concert with storage cells, orchestrate much of the diverse skeletal pathology seen in GD. However, it is important to think of the mineral skeleton as a distinct disease compartment to stage overall disease severity, determine prognosis, and set therapeutic goals. The skeletal manifestations of GD are associated with significant morbidity as revealed in analysis of ~1500 type 1 GD patients in the ICGG registry.[57] As expected, radiological bone manifestations were more common among patients receiving ERT. The prevalence of common skeletal findings in ERT-treated compared to ERT-naive was Erlenmeyer flask deformity (61% vs. 56%), bone marrow infiltration (59% vs. 53%), osteopenia (50% vs. 34%), avascular necrosis (34% vs. 12%), infarction (35% vs. 16%), new fractures (26% vs. 10%), lytic lesions (18% vs. 10%), and joint replacements (14% vs. 6%). In this study, history of bone crises was present in 29% of ERT-patients compared to 5% of patients not receiving ERT. Skeletal disease had major impact on quality of life as evidenced by presence of bone pain in 66% of patients receiving ERT compared to 49% naïve to ERT. Furthermore, 20.8% of the patients had some degree of disability: 12% walked with some difficulty, 7% required an orthopaedic aid to walk, 1% required a wheelchair and 0.5% were bedridden. These data from the ICGG registry indicate that bone involvement confers pain and discomfort in most patients and it is associated with variable degree of disability.

Skeletal Disease Severity: Staging Disease Severity and Its Determinants

A skeletal disease severity score has been developed to depict severity of skeletal disease and encompass the extreme diversity of disease manifestations in this compartment. Hermann and colleagues suggested five stages of skeletal disease severity: stage 1, modelling deformity; stage 2, osteopenia; stage 3, medullary expansion; stage 4, osteolysis; and stage 5, avascular necrosis.[58] An attractive feature of this score is that it may reflect the sequential natural history of bone involvement in GD. However, it cannot be used longitudinally to assess response to therapy.

Another determinant of severity of skeletal disease is the spleen status. Several studies in pre-ERT era showed that bone disease is more severe in asplenic patients, leading to the conclusion that splenectomy results in increased accumulation of storage cells in extra-splenic sites and an accelerated natural history of the disease in these compartments.[40] However, this concept has been challenged.[1,7] Fortunately, ERT has virtually eliminated the need for splenectomy and a concomitant dramatic decline of incidence of avascular necrosis (AVN) has been noted among patients on ERT.

Gaucher Bone Crises

Gaucher bone crises have been virtually eliminated in patients who have access to ERT, and those currently known to have AVN are most likely to suffer these episodes prior to starting ERT.[30] When a Gaucher bone crisis occurs, it is the most incapacitating manifestation of GD, frequently requiring opiate analgesics. Differential diagnosis from septic osteomyelititis may be difficult. Typically, technetium-99m bone scan within 2–3 days of onset of bone crises is negative in Gaucher bone crises; in septic osteomyelitis it is positive. It is wise to avoid biopsy due to risk of introducing infection, chronic osteomyelitis, and development of sinus tracts. Osteonecrosis can affect medullary bone as well as cortical. Medullary osteonecrosis may have typical presentation as described above but may also be entirely asymptomatic. These lesions may cause major disability from pathological fractures and collapse of osseous end plates. In children, the disorder is sometimes misdiagnosed as Legg-Calve-Perthes disease. In the ICGG registry, a third of patients had a history of osteonecrosis.[9] While specific genotypes, prior splenectomy, trauma, strenuous physical exertion, and pregnancies appear to be predisposing factors for the development of AVN, it is not possible to accurately predict in individual patients, the future risk for development of AVN. Therefore, there is an urgent need for development of reliable biomarkers and/or prognostic models that predict development of AVN and advanced skeletal disease which will assist in evaluation of the candidacy for ERT in individual patients.

Osteopenia and Osteoporosis

Osteopenia and osteoporosis are common among patients with type 1 GD, and the severity of this complication correlates strongly with other clinical indicators of disease severity, including the N370S/84GG genotype, prior splenectomy, and hepatomegaly, as well as with the overall severity of skeletal disease as assessed by skeletal radiography.[59] The most striking effect of ERT in ameliorating osteopenia is seen in the pediatric age group. In adults, ERT alone is not sufficient to ameliorate osteopenia in many patients. A randomized control trial of patients receiving ERT demonstrated that adding bisphosphonate (alendronate) at dose 40 mg per day, but not placebo, resulted in greater improvement in bone density.[60]

Mass Lesions of the Bone (Gaucheromas)

One of the most unusual lesions to occur in GD is a mass lesion contiguous with the bone, often referred to as a "Gaucheroma," although not all such lesions are composed predominantly of Gaucher cells. Recently, four patients from Germany were described with such lesions from their total of 70 patients; in each, there was destruction or protrusion of the cortex with extraosseous extension into soft tissues.[16] It is important to understand that surgical manipulation or biopsy of these lesions may trigger dangerous hemorrhage.

Another unusual type of musculoskeletal mass complication that mimics acute bone crisis, AVN, or septic arthritis of the hip joint, is ileopsoas hematoma, which developed spontaneously in one patient or following extraneous physical activity.[61] These hematomas cause severe pain and may be associated with signs of an acute inflammatory response. As with many unusual complications, there should be a high index of suspicion; in two such cases, we performed an ultrsonographic-guided drainage of the hematomas, plus bed rest and pain control, and both young patients had uneventful recovery.

Orthopedic Procedures in Type 1 Gaucher Disease

In general, patients with skeletal complications benefit from a conservative approach to overall management such as avoidance of surgical drainage during an acute episode of bone crisis and avoidance of splenectomy. There is an important place for joint replacement surgery that produces outstanding results in GD despite bleeding risk, generalized osteopenia, and marrow infiltrated with lipid-laden cells, which may contribute to loosening of prosthesis and decreased immunity. It is important to recognize that these outstanding results are possible through developing core expertise in multidisciplinary centers of excellence.[62] With appropriate haematological preparation, prior ERT and choice of suitable prosthesis, excellent results have been achieved in all patients including young patients and with noncemented prostheses.

Pathophysiology of Bone Disease

The pathophysiology of localized bone involvement (osteonecrosis, lytic lesions, and endosteal scalloping) and generalized bone disease (osteoporosis) is not well understood. The pathways to development of these lesions from accumulating glucoerceramide-laden macropages in the bone marrow have not been elucidated. It is believed that marrow expansion due to infiltrating cells causes vascular occlusion and compression, and increased intraosseous pressure; however, there is no data to support this concept. The mechanisms by which Gaucher cells displace normal bone marrow cells, and cause marrow fibrosis, oedema and ischaemia are not known. Data suggest that serum levels of cytokines that affect osteoclastic function may be up-regulated in patients with GD, although the results are conflicting (see Chapter 14).

THE LUNGS IN GAUCHER DISEASE

The lungs represent a quantitatively important site for accumulation of Gaucher cells. However, overt pulmonary manifestations are rare in type 1 GD although in types 2 and 3 forms of the disease, alveolar consolidation ("lipid pneumonia") or infiltrative and fibrotic parenchymal disease, occur respectively in a significant number of patients. A large study from Israel of type 1 GD reported an unexpectedly high prevalence of sub-clinical abnormalities of pulmonary function tests in the form of reduced diffusion capacity (in 42% of 95 patients) or small airway obstruction (68%).[63] An intriguing finding on cardiopulmonary exercise testing in several patients with type 1 GD was of reduced peripheral oxygen consumption.[64] It is tempting to speculate if there may be a mitochondrial defect and/or if this contributes to the highly prevalent symptom of fatigue in type 1 GD. Severe lung involvement in type 1 GD is extremely rare and when it occurs it reflects alveolar, interstitial or vascular infiltration with Gaucher cells as well as pulmonary fibrosis (see Figure 12.26, Figure 12.27, Figure 12.28, and Figure 12.29 in Chapter 12), leading to dyspnea and hypoxemia. This type of advanced lung disease occurs in only ~1% of the patients in large referral centers; the predictors of severe pulmonary parenchymal disease are asplenia, severe GBA alleles, and overall severe disease with high SSI.[65] While there are reports of satisfactory response of pulmonary disease to ERT,[66] this compartment does not respond in many patients due to pulmonary fibrosis.[67] The delivery of mannose-terminated enzyme to lung tissue is less compared to the liver, spleen, and bone marrow and novel delivery methods may have a role in treatment of this manifestation.

In type 1 disease, Gaucher cells are known to accumulate intra-vascularly in the small capillaries and arterioles (pulmonary intravascular macrophages, PIMs).[68] Two distinct pulmonary vascular disorders have been well-described in the literature despite their rarity because of their life-threatening presentations: severe pulmonary hypertension (PH)[7,65,69] and hepatopulmonary syndrome (HPS).[42] In the pre-ERT era, development of HPS led to rapid demise of the patient, unless orthotopic liver transplantation was an option[70] but with ERT, there is an extraordinary reversal of this manifestation.[42] Therefore, presence of HPS in type 1 GD is an absolute indication for emergency administration of ERT and it is life-saving. Initial investigations include demonstration of severe hypoxemia, which partially reverses on breathing 100% oxygen and presence of platypnea and orthodeoxia. Confirmatory diagnostic investigations include Doppler echocardiogram with agitated saline bubble contrast to demonstrate appearance of bubble contrast in left atrium, 4 to 5 heartbeats after its appearance in the right atrium. Shunt fraction may be calculated using technicium labeled macroaggregated albumin and pulmonary angiography should be considered.

In two patients, ERT dramatically reversed HPS, dependency on oxygen as well as hepatomegaly, but it unmasked underlying PH.[42] Pulmonary hypertension was a rare but well documented complication of GD prior to the availability of ERT.[7] Two large scale studies have been performed to determine prevalence of PH in type 1 GD using Doppler echocardiography.[65,69] Mild to moderate, asymptomatic PH (right ventricular systolic pressure >25 <50 mm Hg) was present in ~ 7% of the patients in 2 different referral centers. This type of PH does not appear to be clinically significant or progressive. Severe, clinically significant PH (RVSP >50 mm Hg) occurred in only ~1% of patients. Development of this severe phenotype was confined exclusively to asplenic patients. Indeed, severe PH occurs at higher frequency among non-GD patients who have been splenectomized for hematological or other indications.[71] It is important to recognize that PH can develop silently over many years. Therefore it is recommended that all asplenic patients have a baseline Doppler echocardiogram and if abnormal it should be repeated at 1 to 2 year intervals. Once PH is suspected, right heart catheter is mandatory to confirm the diagnosis and obtain a baseline before starting specific vasodilator therapies. The role of ERT in management of patients with severe PH has been a subject of appropriate vigorous debate based on a concern that it may exacerbate the condition. As a result, there is divergence of approaches to management of these patients with respect to ERT. At Yale and in other parts of the U.S., where we currently manage 10 such patients, we use ERT to aggressively manage the underlying disease and use vasodilators to manage the pulmonary pressures. At present, eight patients have had stabilization or reduction of their pressures; two patients died from overwhelming septicemia but after significant amelioration of PH. In Israel, in addition to measures to manage PH, the management plan includes a trial of cessation of ERT. In general, there is agreement that because development of severe PH is confined entirely to asplenic patients, the new generation of patients is unlikely to be at risk of this complication because ERT has virtually eliminated the need to perform splenectomy. The basis of these unusual pulmonary vascular abnormalities, intrapulmonary vascular dilations on the one hand and plexogenic arteriopathy at the other end of the spectrum, is not known and it is the subject of active investigation.

THE KIDNEY IN GAUCHER DISEASE

There are only a few reports of renal dysfunction manifesting as proteinuria associated with Gaucher cells in glomeruli. Recently, a large study was undertaken to determine whether there was high prevalence of sub-clinical renal manifestations in type 1 GD. The study included 161 patients (26 were children) who underwent blood pressure recording, renal sonography, serum chemistries, urinalysis, urine electrolytes, and total protein excretion. In addition, tubular proteinuria and glomerular filtration rate (GFR) were estimated.[72] GFR was found to be significantly greater in GD compared to controls, and it was associated with more severe disease. Significant proteinuria was found only in patients with co-morbidities such as diabetes mellitus or multiple myeloma; no patient had decreased renal function. The finding of increased GFR, which was more marked in patients with severe disease, suggests that subtle renal abnormalities in the form of hyperfiltration are more common than expected in type 1 GD.

MALIGNANCY IN TYPE 1 GAUCHER DISEASE

The possibility that patients with GD may have increased risk of developing cancers, especially hematological malignancies, has been the topic of discussion in more than 200 publications since 1965. Most are single case reports of patients with GD developing malignancy. In a postmortem series of 35 patients, Lee noted 19 (54%) malignant tumors.[7] In this report, malignancy seemed to be associated with milder forms of Gaucher disease in that the mean age at death of patients with cancers was 59 years compared to 46 years for the mean age at death of patients without cancers.

Determination of any risk for cancer development in Gaucher disease is of profound importance in management of patients and counseling as well as understanding pathogenetic mechanisms. Attempts to assess potential risk of malignancy in type 1 GD has been plagued by ascertainment bias. For example, Shiran et al. reported a 20.8% incidence of cancer among 48 patients with Gaucher disease compared with 6.8% in a healthy population.[73] However this study was conducted by a search through the archives of a general medical center. In fact, half of the patients presenting with malignancies were diagnosed upon the finding of Gaucher cells in bone marrow smears. The control population was taken from the clinic of a family practitioner in the community. Thus, the estimate 14.7-fold increase of cancer risk may have been somewhat exaggerated due to ascertainment bias. Three recent studies have re-examined the topic. Two studies, one from the ICGG registry[74] and another from the world's largest single referral clinic of Gaucher disease in Jerusalem,[75] failed to demonstrate increased incidence of all malignancies among a 2742 and 383 patient population, respectively, aged >18 years with the exception of multiple myeloma. In the publication from ICGG registry, relative risk of multiple myeloma was estimated to be 5.9 compared to an age- and sex-matched U.S. population. The increased risk of multiple myeloma has been confirmed in a new

Dutch-German collaborative report of ~150 non-Jewish patients with type 1 Gaucher disease.[46] In addition, this study found an overall higher incidence of all cancers in their combined cohorts, and the authors emphasized an apparent dramatic increase of risk of hepatocellular carcinoma (HCC).

Our conclusions and recommendations on this topic at the present time are as follows:

1. There is unequivocal evidence of increased risk of multiple myeloma in type 1 GD. We recommend that protein electrophoresis and immunoglobulin measurements be part of annual evaluation of all adult patients. In patients with known MGUS, any change in status (i.e., new onset of anemia, immune-paresis, bone pain, pathologic fractures) should prompt further investigations to search for multiple myeloma (skeletal series, bone marrow aspirate, serum light chains, and immunofixation). The increased risk of multiple myeloma in GD has a plausible basis in the form of chronic B cell stimulation via IL 6 secreted by Gaucher cells.
2. There is currently no compelling evidence of increased risk of HCC due to type 1 GD per se. However, every patient should have full HCV and HBV testing and if there is hepatic co-morbidity, and especially evidence of cirrhosis/fibrosis, close surveillance for HCC should be performed.
3. Although there are many case reports of lymphoproliferative disorders in type 1 GD, studies from ICGG and Jerusalem Gaucher clinic failed to demonstrate increase risk. We recommend that a high index of suspicion be maintained in this regard when evaluating patients.
4. Cancer surveillance should be a normal part of routine multidisciplinary Gaucher care.

THE BRAIN IN TYPE 1 GAUCHER DISEASE

Definition of type 1 GD is predicated on the absence of neurologic involvement and any detectable neurological signs. It is generally accepted that presence of at least one N370S GBA allele absolutely precludes development of neuronopathic Gaucher disease.[2] Nevertheless, in a few autopsy reports perivascular accumulation of Gaucher cells in cerebral cortex white matter has been reported.[7] The clinical significance, if any, of this finding is not known. In the last decade there is increased recognition of the association between GD and early-onset Parkinsonism, as well as between carriership of GD and idiopathic Parkinson disease. This is discussed in great detail in Chapter 12.

OBSTETRIC AND GYNECOLOGICAL ASPECTS

Gynecological and obstetric aspects are significant issues in comprehensive multi-disciplinary management of type 1 GD since women in reproductive years comprise a significant proportion of patients. Thrombocytopenia, anemia, and occasionally massive organomegaly as well as joint deformities if present, are a cause for concern. In the pre-ERT era, there were sporadic reports of infertility, delayed menarche and menorrhagia, which today may be reversed by ERT. There is no evidence of premature menopause.

The obstetrical aspects are important since pregnancy may alter the natural history of GD, and there is always concern about the effect of GD of the mother on the fetus and pregnancy outcome. Two large studies from the Jerusalem Clinic addressed these issues.[77,78] The first study reports observations between 1990 and 1995 of 102 spontaneous pregnancies: 25 (24.5%) ended in spontaneous first-trimester abortions and 72 continued beyond the 22nd week of gestation. Nine patients (27.7%) were diagnosed as having GD during their first pregnancies. Worsening of thrombocytopenia and anemia occurred in most women, but antepartum blood transfusion was not required. Early postpartum hemorrhage and fever occurred more frequently after cesarean as well as vaginal deliveries. Seven of 102 women developed Gaucher bone crisis during the third trimester or early postpartum periods. There was a tendency for bone crises to recur in subsequent pregnancies.[110] A subsequent study reported observations on 66 pregnancies in 43 women during 1997 to 2003.[72] The live birth rate was 78.3% among ERT-treated compared to 86.0% among ERT-naïve patients. Most untreated women with milder disease had uncomplicated pregnancies, whereas patients on ERT due to their more severe disease had higher incidence of bleeding complications and infections postpartum, but few had spontaneous abortions. These observations confirm previous anecdotal reports of disease progression during pregnancy[47] and, therefore, close hematologic consultation and monitoring during pregnancy is recommended. If epidural anesthesia is performed, we consider administration of single-donor platelet transfusion to patients with platelet count $<70,000/mm^3$ or severe platelet dysfunction immediately prior to the procedure.

In the past decade, we have accumulated enough experience to encourage our patients to continue with ERT during pregnancy. Imiglucerase is designated a category B drug by the FDA, i.e., a drug not likely to pose a threat to the fetus from the evidence in animal studies, but no well-controlled studies have been performed in pregnant women. The benefits of continued ERT in these patients are clear with reduced risk of disease progression during pregnancy and amelioration of bleeding or infectious complications during the postpartum period.

CONCLUSIONS

The wide spectrum of type 1 GD phenotypes has been defined using large patient populations rather than anecdotal reports. This has enabled evidence-based recognition of the prevalence of specific syndromes within the vast phenotypic continuum such as Gaucher/pulmonary hypertension, Gaucher/hepatic fibrosis and portal hypertension, Gaucher/multiple myeloma and Gaucher/Parkinsonism. Each of these phenotypic paradigms offers outstanding opportunities to dissect environmental and/or genetic modifiers of the Gaucher phenotype. These efforts, in concert with recognition of the need to develop a multi-dimensional disease severity index that includes all of the domains of the disease, including perhaps the temporal evolution of the phenotype in individual patients, will refine patient management and develop therapeutic goals.

ACKNOWLEDGMENTS

We acknowledge the support and assistance of the National Gaucher Foundation (U.S.) and Gaucher Association (U.K.). PKM is supported by the National Institutes of Health, NIDDK grant 1 K24 DK066306-01.

REFERENCES

1. Cox, T.M. and Schofield, J.P., Gaucher's disease: clinical features and natural history, Baillieres Clin Haematol, 10(4), 657, 1997.
2. Beutler, E. and Grabowski, G.A., *Gaucher Disease*, Scriver CR, B.A., Sly W.S. et al., Eds., New York: McGraw-Hill, 2001.
3. Grabowski, G.A. et al., *Gaucher Disease: Phenotypic and Genetic Variation*, Scriver CR, B.A., Sly W.S. et al., Ed., McGraw-Hill, New York, 2006.
4. Zimran, A. et al., Survey of hematological aspects of Gaucher disease, Hematology, 10(2), 151, 2005.
5. Zimran, A. et al., Gaucher disease. Clinical, laboratory, radiologic, and genetic features of 53 patients, Medicine (Baltimore), 71(6), 337, 1992.
6. Sibille, A. et al., Phenotype/genotype correlations in Gaucher disease type I: clinical and therapeutic implications, Am J Hum Genet, 52, 1094, 1993.
7. Lee, R.E., The pathology of Gaucher disease, Prog Clin Biol Res, 95, 177, 1982.
8. Pastores, G.M. et al., Therapeutic goals in the treatment of Gaucher disease, Semin Hematol, 41(4 Suppl 5), 4, 2004.
9. Charrow, J. et al., The Gaucher registry: demographics and disease characteristics of 1698 patients with Gaucher disease, Arch Intern Med, 160(18), 2835, 2000.
10. Sidransky, E., Gaucher disease: complexity in a "simple" disorder, Mol Genet Metab, 83, 6, 2004.
11. Grabowski, G.A., Gaucher disease: gene frequencies and genotype/phenotype correlations, Genet Test, 1(1), 5, 1997.
12. Neudorfer, O. et al., Occurrence of Parkinson's syndrome in type I Gaucher disease, QJM, 89(9), 691, 1996.
13. Sidransky, E., Gaucher disease and parkinsonism, Mol Genet Metab, 84(4), 302, 2005.
14. Zimran, A. et al., Prediction of severity of Gaucher's disease by identification of mutations at DNA level, Lancet, 2(8659), 349, 1989.
15. Ida, H. et al., Type 1 Gaucher disease: phenotypic expression and natural history in Japanese patients, Blood Cells Mol Dis, 24(1), 73, 1998.
16. Maaswinkel-Mooij, P. et al., The natural course of Gaucher disease in The Netherlands: implications for monitoring of disease manifestations, J Inherit Metab Dis, 23(1), 77, 2000.
17. Aerts, J.M. and Hollak, C.E., Plasma and metabolic abnormalities in Gaucher's disease, Baillieres Clin Haematol, 10(4), 691, 1997.
18. Deegan, P.B. et al., Clinical evaluation of chemokine and enzymatic biomarkers of Gaucher disease, Blood Cells Mol Dis, 35(2), 259, 2005.

19. Barton, N.W. et al., Therapeutic response to intravenous infusions of glucocerebrosi-
    dase in a patient with Gaucher disease, Proc Natl Acad Sci U.S.A., 87(5), 1913, 1990.
20. Maas, M. et al., Quantification of bone involvement in Gaucher disease: MR imaging
    bone marrow burden score as an alternative to Dixon quantitative chemical shift MR
    imaging — initial experience, Radiology, 229(2), 554, 2003.
21. Maas, M. et al., Quantification of skeletal involvement in adults with type I Gaucher's
    disease: fat fraction measured by Dixon quantitative chemical shift imaging as a valid
    parameter, AJR Am J Roentgenol, 179(4), 961, 2002.
22. Goker-Alpan, O. et al., Divergent phenotypes in Gaucher disease implicate the role
    of modifiers, J Med Genet, 42(6), e37, 2005.
23. Svennerholm, L., Hakansson, G., Mansson, J.E., and Nilsson, O., Chemical differ-
    entiation of the Gaucher subtypes, Prog Clin Biol Res, 95, 231, 1982.
24. Mankin, H.J., Rosenthal, D.I., and Xavier, R., Gaucher disease. New approaches to
    an ancient disease, J Bone Joint Surg Am, 83-A(5), 748, 2001.
25. Barton, D.J. et al., Resting energy expenditure in Gaucher's disease type 1: effect of
    Gaucher's cell burden on energy requirements, Metabolism, 38(12), 1238, 1989.
26. Grabowski, G.A., Gaucher disease: Gene frequencies and genotype/phenotype cor-
    relations., Genet Testing, 1, 5, 1997.
27. Beutler, E. et al., Gaucher disease: gene frequencies in the Ashkenazi Jewish popu-
    lation, Am J Hum Genet, 52(1), 85, 1993.
28. Sidransky, E., Tayebi, N., and Ginns, E.I., Diagnosing Gaucher disease. Early recog-
    nition, implications for treatment, and genetic counseling, Clin Pediatr (Phila), 34(7),
    365, 1995.
29. Grabowski, G.A., Leslie, N., and Wenstrup, R., Enzyme therapy for Gaucher disease:
    the first 5 years, Blood Rev, 12(2), 115, 1998.
30. Weinreb, N.J. et al., Effectiveness of enzyme replacement therapy in 1028 patients
    with type 1 Gaucher disease after 2 to 5 years of treatment: a report from the Gaucher
    Registry, Am J Med, 113(2), 112, 2002.
31. Gielchinsky, I. et al., Is there a correlation between degree of splenomegaly, symptoms
    and hypersplenism? A study of 218 patients with Gaucher disease, Br J Haematol,
    106(3), 812, 1999.
32. Elstein, D., Abrahamov, A., Hadas-Halpern, I., and Zimran, A., Gaucher's disease,
    Lancet, 358(9278), 324, 2001.
33. Grabowski, G.A. et al., Pediatric non-neuronopathic Gaucher disease: presentation,
    diagnosis and assessment. Consensus statements, Eur J Pediatr, 163, 58, 2004.
34. Gillis, S. et al., Platelet function abnormalities in Gaucher disease patients, Am J
    Hematol, 61(2), 103, 1999.
35. Eapen, M., Hostetter, M., and Neglia, J.P., Massive splenomegaly and Epstein-Barr
    virus-associated infectious mononucleosis in a patient with Gaucher disease, J Pediatr
    Hematol Oncol, 21(1), 47, 1999.
36. Kolodny, E.H. et al., Phenotypic manifestations of Gaucher disease: clinical features
    in 48 biochemically verified type 1 patients and comment on type 2 patients, Prog
    Clin Biol Res, 95, 33, 1982.
37. Neudorfer, O. et al., Abdominal ultrasound findings mimicking hematological malig-
    nancies in a study of 218 Gaucher patients, Am J Hematol, 55, 28, 1997.
38. Dweck, A. et al., Wandering spleen in a young girl with Gaucher disease, Isr Med
    Assoc J, 3(8), 623, 2001.
39. Aharoni, D., Hadas-Halpern, I., Elstein, D., and Zimran, A. Huge subcapsular splenic
    hematoma in a patient with Gaucher disease, Isr Med Assoc J, 2(1), 61, 2000.

40. Turfaner E. et al., Gaucher disease and brucella: just a mere coincidence? Genet Couns, 14, 363, 2003.

41. Gupta, M.K. et al., Littoral cell angioma of the spleen in a patient with Gaucher disease, Am J Hematol, 68, 61, 2001.

42. Dawson, A. et al., Pulmonary hypertension developing after alglucerase therapy in two patients with type 1 Gaucher disease complicated by the hepatopulmonary syndrome, Ann Intern Med, 125(11), 901, 1996.

43. James, S.P., Stromeyer, F.W., Stowens, D.W., and Barranger, J.A., Gaucher disease: hepatic abnormalities in 25 patients, Prog Clin Biol Res, 95, 131, 1982.

44. Lachmann, R.H. et al., Massive hepatic fibrosis in Gaucher's disease: clinico-pathological and radiological features, QJM, 93(4), 237, 2000.

45. Xu, R. et al., Hepatocellular carcinoma in type 1 Gaucher disease: a case report with review of the literature, Semin Liver Dis, 25(2), 226, 2005.

46. de Fost, M. et al., Increased incidence of cancer in adult Gaucher disease in Western Europe, Blood Cells Mol Dis, Oct. 2005 (Epub ahead of print).

47. Patlas, M. et al., Multiple hypoechoic hepatic lesions in a patient with Gaucher disease, J Ultrasound Med, 21(9), 1053, 2002.

48. Margalit, M. et al., Immunomodulation by β-glucosylceramide, Presented at Int. Symp. on Immune Mediated Diseases, Moscow, September 3–7, 2005.

49. Rosenbaum, H. and Sidransky, E., Cholelithiasis in patients with Gaucher disease, Blood Cells Mol Dis, 28(1), 21, 2002.

50. Ben Harosh-Katz, M. et al., Increased prevalence of cholelithiasis in Gaucher disease association with splenectomoy but not with gilbert syndrome, J Clin Gastro 38(7), 586, 2004.

51. Zimran, A. et al., Survey of hematological aspects of Gaucher disease, Hematology, 10, 151, 2005.

52. Gielchinsky, Y. et al., High prevalence of low serum vitamin B12 in a multi-ethnic Israeli population, Br J Haematol, 115(3), 707, 2001.

53. Gillis, S. et al., Platelet function abnormalities in Gaucher disease patients, Am J Hematol, 61, 103, 1999.

54. Zimran, A. et al., Correction of neutrophil chemotaxis defect in patients with Gaucher disease by low-dose enzyme replacement therapy, Am J Hematol, 43, 69, 1993.

55. Zimran, A. et al., Rheological determinants in patients with Gaucher disease and internal inflammation, Am J Hematol., 75, 190, 2004.

56. Bax, B.E. et al., Haemorheology in Gaucher disease, Eur J Haematol, 75(3), 252, 2005.

57. Bembi, B. et al., Bone complications in children with Gaucher disease, Br J Radiol, 75(Suppl 1), A37, 2002.

58. Hermann, G., Pastores, G.M., Abdelwahab, I.F., and Lorberboym, A.M., Gaucher disease: assessment of skeletal involvement and therapeutic responses to enzyme replacement., Skeletal Radiol, 26, 687, 1997.

59. Pastores, G.M. et al., Bone density in Type 1 Gaucher disease, J Bone Miner Res, 11, 1801, 1996.

60. Wenstrup, R.J. et al., Gaucher disease: alendronate disodium improves bone mineral density in adults receiving enzyme therapy, Blood, 104(5), 1253, 2004.

61. Jmoudiak, M. et al., Iliopsoas hematoma in a young patient with type I Gaucher disease, Isr Med Assoc J, 5(9), 673, 2003.

62. Itzchaki, M. et al., Orthopedic considerations in Gaucher disease since the advent of enzyme replacement therapy, Acta Orthop Scand, 75(6), 641, 2004.

63. Kerem, E. et al., Pulmonary function abnormalities in type I Gaucher disease, Eur Respir J, 9(2), 340, 1996.
64. Miller, A., Brown, L.K., Pastores, G.M., and Desnick, R.J., Pulmonary involvement in type 1 Gaucher disease: functional and exercise findings in patients with and without clinical interstitial lung disease, Clin Genet, 63(5), 368, 2003.
65. Mistry, P.K. et al., Pulmonary hypertension in type 1 Gaucher's disease: genetic and epigenetic determinants of phenotype and response to therapy, Mol Genet Metab, 77(1–2), 91, 2002.
66. Banjar, H., Pulmonary involvement of Gaucher's disease in children: a common presentation in Saudi Arabia, Ann Trop Paediatr, 18(1), 55, 1998.
67. Goitein, O. et al., Lung involvement and enzyme replacement therapy in Gaucher's disease, Qjm, 94(8), 407, 2001.
68. Amir, G and Ron N., Pulmonary pathology in Gaucher's disease, Hum Pathol, 30, 666–70, 1999.
69. Elstein, D. et al., Echocardiographic assessment of pulmonary hypertension in Gaucher's disease, Lancet, 351(9115), 1544, 1998.
70. Lachmann, R.H. et al., Massive hepatic fibrosis in Gaucher's disease: clinico-pathological and radiological features, Qjm, 93(4), 237, 2000.
71. Hoeper, M.M. et al., Pulmonary hypertension after splenectomy? Ann Intern Med, 130(6), 506, 1999.
72. Becker-Cohen, R. et al., A comprehensive assessment of renal function in patients with Gaucher disease, Am J Kidney Dis, 46(5), 837, 2005.
73. Shiran, A., Brenner, B., Laor, A., and Tatarsky, I., Increased risk of cancer in patients with Gaucher disease, Cancer, 72, 219, 1993.
74. Rosenbloom, B.E. et al., Gaucher disease and cancer incidence: a study from the Gaucher Registry, Blood, 105(12), 4569, 2005.
75. Zimran, A. et al., Incidence of malignancies among patients with type I Gaucher disease from a single referral clinic, Blood Cells Mol Dis, 34(3), 197, 2005.
76. de Fost, M. et al., Increased incidence of cancer in adult Gaucher disease in Western Europe, Blood Cells Mol Dis, 2005.
77. Granovsky-Grisaru, S. et al., Gynecologic and obstetric aspects of Gaucher's disease: a survey of 53 patients, Am J Obstet Gynecol, 172(4 Pt 1), 1284, 1995.
78. Elstein, Y. et al., Pregnancies in Gaucher disease: a 5-year study, Am J Obstet Gynecol. 190, 435, 2004.

CHAPTER 11

Neuronopathic Gaucher Disease

Raphael Schiffmann and Ashok Vellodi

CONTENTS

DEFINITION

Neuronopathic Gaucher Disease (NGD) is best defined as a confirmed diagnosis of Gaucher disease in the presence of neurological symptoms and signs, for which there is no other cause.

EPIDEMIOLOGY

Although no systematic study exists, it is estimated that about 6% of Gaucher patients have neuronopathic Gaucher disease, 5% have the chronic form (type 3) and 1% have the acute form (type 2) of the disease.[1] The true prevalence of NGD is not known. In Japan, up to 40% of a group of 59 Gaucher patients had NGD with similar number of type 2 and type 3 patients.[2] Interestingly, the N370S allele that is typically associated exclusively with the nonneuronopathic form of the disease or the 84gg mutation has not been described in the Japanese population.[3] No NGD patients were found in Romania[4] and only 5% were found in a Spanish cohort.[5]

Clinical Classification and Its History

Primary neurological involvement in Gaucher disease was first described in the early 20th century.[6,7] Classically, three forms of the disease are recognized: type I or nonneuronopathic, type 2 or acute neuronopathic, and type 3 or subacute or chronic neuronopathic.[8–10] The latter was also sometimes subdivided into 3A (patient with mild systemic disease and progressive myoclonic encephalopathy), 3B (neurologic deficit dominated by supranuclear gaze palsy and severe systemic disease), 3C (mild neurological disease with aortic valve calcifications), and the Norrbottnian variant.[10,11] However, it is increasingly recognized that these divisions are somewhat artificial and in fact any combination of the above abnormalities may occur in any given patient.[12,13] The only phenotype that seems to be distinct may be type 3C that is always caused by a homozygous mutation for the D409H (1342G-->C) mutation.[14,15] Therefore, NGD is better described as a continuum of characteristic neurological abnormalities.

CLINICAL FEATURES

Although, as stated above, NGD is a continuum, to reflect the existing literature we shall use here the classic clinical classification of this disorder.

Clinical Syndromes

Type 2 (Acute) Neuronopathic Gaucher Disease

Type 2 NGD usually refers to children who display neurological abnormalities before age 6 months and die by age 2–4 years despite enzyme replacement therapy (ERT), with an average age of death of 9 months.[16,17] These patients commonly present in early infancy with evidence of brainstem dysfunction consisting of supranuclear gaze palsy and hepatosplenomegaly followed by progressive deterioration associated with convergent strabismus caused by bilateral 6th nerve palsy, dysphagia, stridor with breath holding spells, pyramidal signs (cortical thumbs, retroflexion of the neck), failure to thrive, and cachexia.[12,17–20] The clinical course is a rapid and relentless deterioration. The child eventually succumbs to further brainstem involvement with either stridor leading to laryngeal obstruction and apnea, or to dysphagia provoking aspiration. These patients also uniformly have interstitial lung disease and repeated respiratory infections that compound the neurological deficits.[17,19] In the original description of 67 patients by Frederickson and Sloan, the average age at death was 8 months. Survival of a typical type 2 Gaucher patient beyond 2 years was rare.[21] The EWGGD Task Force divided type 2 patients into those who do not exhibit signs of pyramidal tract involvement, irritability or cognitive impairment (type 2A) and those who are more severely affected (type 2B).[22] Type 2B patients show marked evidence of pyramidal tract involvement, which is a poor prognostic sign, and, in addition, suffer from irritability and cognitive impairment. It is particularly challenging to differentiate between patients with mild type 2B disease and patients with type 3 Gaucher disease who have long-term survival rates with ERT.

Lethal Neonatal Variant

Congenital Gaucher disease leading to a form of "collodion baby" is associated with the virtual absence of residual GBA activity, an association that was first reported in 1988.[23] These newborns may present as hydrops foetalis or are dead at parturition or die within the first few days of life, at least partly due to the excessive water evaporation caused by the abnormal lipid composition of the epidermis.[24] The phenotype appears to be analogous to that of the null-allele mouse, which was created by the targeted disruption of the GBA gene,[25] and may be more common than the classical type 2 NGD.[26]

Type 3 (Subacute, Chronic Neuronopathic) Gaucher Disease

Currently, the minimum definition of type 3 NGD requires the presence of supranuclear gaze palsy.[22,27] This neurological abnormality consists of saccadic eye movements slowing affecting predominantly the horizontal eye movement, but vertical saccades are almost always slow as well (see below).[28,29] This finding may be isolated or include a developmental delay, hearing impairment and other brainstem deficits, and seizures.[11] Patients, however, often present not with a neurological abnormality, but with enlarged liver or spleen, anemia, or a failure to thrive. Most type 3 NGD patients present in the first 5 years of life.[30] However, when the systemic (nonneurologic) manifestations of the disease are relatively mild, they will come to medical attention during childhood, adolescence, and even adulthood because of seizure, myoclonic movements, or other neurological abnormalities.[1] The natural history of the disease is variable. It was once dominated by the visceral complications of hypersplenism, respiratory distress, portal hypertension, infection, and bone disease. However, these problems are now effectively treated by ERT, which significantly prolongs the life expectancy of these patients but has no visible effect on the brain.[30–32]

Historically, a number of syndromes of type 3 NGD has been recognized such as the genetic isolate of patients originated in Norrbotten region of Northern Sweden (hence, the term Norrbottnian variant).[33] Another form that was recognized in the past exhibited a relatively mild systemic burden of disease and progressive myoclonic encephalopathy, with seizures, dementia, and death.[34] However, there are patients with both severe systemic disease and supranuclear gaze palsy who ultimately develop progressive myoclonic encephalopathy.[30] The most recently described syndrome consists of the association of heart valve and aortic calcification, supranuclear gaze palsy, mild hepatosplenomegaly, and bone disease. This variant, sometimes referred to as the type 3C variant, was first described in 1995 by two different groups in Israel and Spain.[14,35] It is almost always associated with homozygosity for the *D409H* (G1342C) mutation.

We shall now describe in greater detail the typical features seen in patients with type 3 NGD.

Saccadic Paresis and Other Ocular Abnormalities

The most consistent clinical feature is an abnormality of horizontal gaze, first described in the late 1970s.[36] This has been mistakenly referred to as oculomotor apraxia, but should more accurately be called supranuclear saccadic gaze palsy.[37] The abnormality consists of a combination of saccade (quick eye movement) slowing sometimes associated with failure to initiate the saccade (saccade initiation failure or SIF).[28,29] Supranuclear saccadic gaze palsy can be difficult to detect clinically, especially in infancy, but is readily revealed as missed quick-phases during induced optokinetic and vestibular nystagmus.[28] The abnormality can often be observed when the child turns around while walking or when reading is associated with horizontal head jerks that represent an attempt to compensate for the saccadic deficit. Vertical movements are usually affected as well. Older children learn to compensate for their poor saccades by a combination of synkinetic blinking, looping, and head thrusting but may result in significant functional disability while driving or crossing a street, and even difficulty in reading books. If the vertical component of the saccade is involved, they may also have difficulties in looking up to and back down from the blackboard.

Saccadic velocity and amplitude can be measured and quantified in a number of ways. Both voluntary and reflexive saccades can be assessed,[28] for example, by asking a patient to follow an object on a video screen or by having the child look at a large revolving screen that induces optokinetic nystagmus.[28] The relationship between saccadic amplitude and velocity can be depicted and clear differences between type 3 NGD patients and healthy controls can be noted (Figure 11.1). In patients with type 2 NGD and in other progressive forms of type 3 NGD, supranuclear saccadic paresis progresses to nuclear paresis and, finally, to complete ocular paralysis.[30]

Auditory Dysfunction

The other consistent feature in NGD is an abnormality of brainstem auditory evoked potentials.[38] Waveforms are poor and show progressive deterioration over time. Apart from reflecting brainstem dysfunction, the significance of this finding is unclear, as many patients have normal peripheral hearing.[39] They do, however, appear to have a central auditory processing disorder. In a recent study using Mismatch Negativity (MMN), a standard electrophysiological technique for detecting an auditory processing disorder, 5 of 6 children with NGD had an abnormal result.[40]

Myoclonus and Seizures

As mentioned above, some NGD patients develop a progressive encephalopathy, characterized by recurrent and sometimes continuous myoclonus.[41] These myoclonic movements are multifocal, usually of neocortical origin, and highly resistant to medical therapy.[42] Somatosensory evoked potentials (SEP) in such patients show a greatly increased amplitude of the potential, a so-called "giant potential," which indicates a decreased cortical inhibitory input.[43] It was recently demonstrated that the SEP amplitude is also elevated in type 3 NGD patients who do not have myoclonus (Figure 11.2).[44] In this group, there is an inverse correlation between the SEP amplitude and the patient's intelligence quotient (IQ).[44] Therefore, it is likely that the development of overt myoclonus in Gaucher disease is the extreme manifestation of a general cortical pathogenic process that occurs in all neuronopathic Gaucher patients. Subcortical myoclonus associated with loss of neurons in the cerebellar dentate nucleus and Purkinje cells has also been described.[45] Myoclonus and seizures in type 3 NGD are associated with an abnormal EEG that shows diffuse slow background and multifocal spikes and spike and wave discharges that are usually bilateral (Figure 11.3). Patients with type 2 NGD also often develop progressive myoclonic encephalopathy. Frequently, patients with progressive myoclonic encephalopathy exhibit mild and even trivial systemic disease, sometimes with little or no storage material in organs outside the brain.[46] This apparent paradox has never been satisfactorily explained and suggests a complex pathogenesis of myoclonus in Gaucher disease.

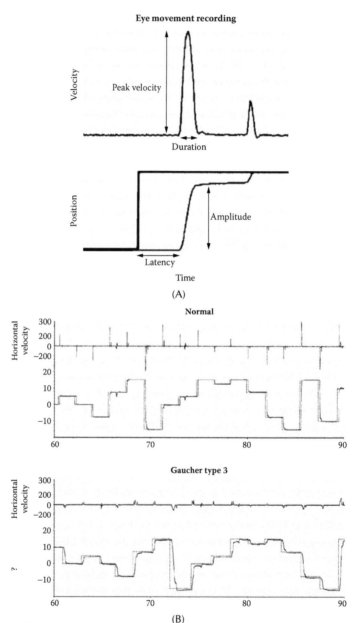

Figure 11.1   Measurement of voluntary saccadic eye movements using an infrared system. (A) The characteristics of a saccades are its velocity, amplitude, duration, and latency. The patient is following an object on a video screen. The position of the object and eye in space can be followed. (B) Recording of saccadic eye movements of different amplitudes. In each case, saccadic velocity is depicted in the upper tracing and the position of the object (interrupted line) and the eyes (continuous line) in the lower tracing. The velocities of the normal subject are often above 300 degrees per second, while the patient with type 3 Gaucher disease rarely goes above 100 degrees per second. The patient has normal latency (normal saccadic initiation) but the saccades are slow thus prolonging their duration.

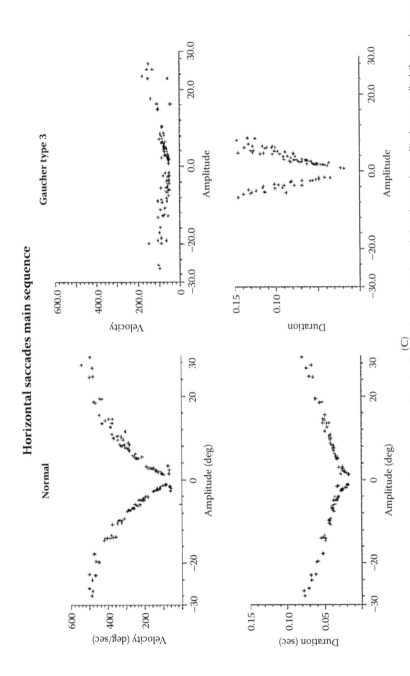

Figure 11.1 (Continued). (C) The relationships between saccadic velocity and the amplitude and duration and amplitude are called the main sequence. The difference between the patient and the normal control can be easily seen and thus quantify the slowness of the saccades. (Courtesy of Dr. Edward Fitzgibbon.)

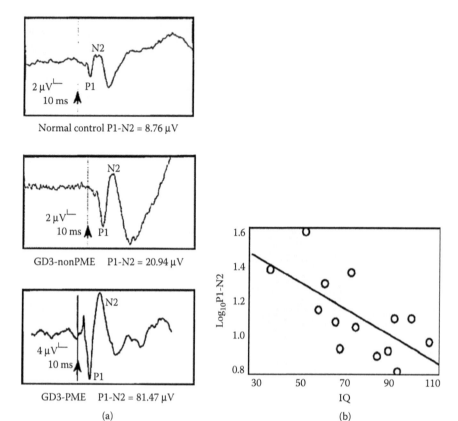

Normal control P1-N2 = 8.76 μV

GD3-nonPME   P1-N2 = 20.94 μV

GD3-PME   P1-N2 = 81.47 μV

(a)   (b)

Figure 11.2 (a) Panels show stretch-evoked somatosensory evoked potentials (SEP) from (top to bottom) a 9-year-old healthy boy; a 4-year-old boy with type 3 Gaucher disease without progressive myoclonus encephalopathy (GD3-non-PME); and a 33-year old woman with type 3 Gaucher disease with progressive myoclonic encephalopathy (GD3-PME). Arrow on each panel indicates time of sensory stimulus (tap). (b) Scatter plot shows the linear regression of log10 P1-N2 SEP with IQ in all GD3-non-PME patients. Open circles represent individual patients. (From Garvey, M.A. et al., *Neurology*, 56, 391, 2001. With permission.)

Type 3 NGD patients who are otherwise stable neurologically exhibit an increasing incidence of benign partial complex seizures not associated with increased SEP amplitude or clinical myoclonus (unpublished observations). It is likely that this epileptic activity reflects a stable brain abnormality possibly related to the hippocampal pathology in Gaucher disease (see pathology) and is not a grave prognostic sign.

Gaucher Disease and Parkinsonism

Over the years there have been reports of rare patients with typical or atypical parkinsonism in type 1 (nonneuronopathic form) Gaucher disease suggesting that neurological involvement can occur as a late complication.[47–51] In some instances,

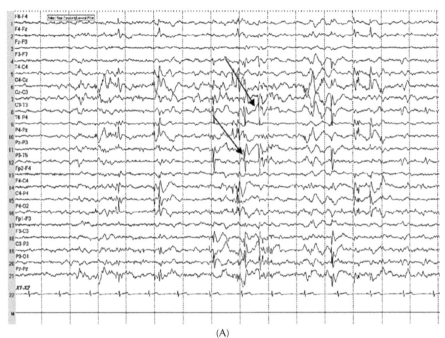

(A)

Figure 11.3 Typical EEG of a patient with type 3 Gaucher disease and cortical myoclonus. A. Diffusely slow background with bilateral spike and wave epileptogenic discharges (arrows). B. In the same patient, bilateral, synchronous spike and wave discharges (arrow). These abnormalities indicate a generalized cortical functional abnormality.

patients had myoclonus that may be seen in type 3 Gaucher disease.[47,48] On occasion, we and others have observed parkinsonian features such as rigidity and hypokinesia in patients who otherwise have typical type 3 NGD.[52] Recently, a number of lines of evidence suggest that the clinical association of parkinsonism or Parkinson disease and Gaucher disease may not be purely coincidental.

In a patient with parkinsonism, dementia and supranuclear gaze palsy, and Lewy bodies, the cardinal neuropathologic feature of Parkinson disease, were found in CA2–CA4 regions of the hippocampus. This hippocampal region is typically affected by gliosis or neuronal loss in all forms of Gaucher disease.[53] This region is selectively affected only in dementia with diffuse Lewy body disease.[54] On the other hand, α-synuclein-positive Lewy bodies are not typically found in these brain regions in Parkinson disease or a related disorder called dementia with diffuse Lewy body disease. These inclusions stained strongly for ubiquitin and α-synuclein. Therefore, the presence of the pathological hallmark of parkinsonism and Parkinson disease in a neuropathological distribution typical of Gaucher disease suggests that the two processes are pathogenetically related. Alterations in the glucocerebrosidase gene were found in 12 brain samples (21%) of 57 randomly chosen subjects carrying the pathological diagnosis of Parkinson disease.[55] There were eight (14%) with mutations (N370S, L444P, K198T, and R329C) and four with probable polymorphisms (T369M and E326K). Subsequently, three studies in patients with Parkinson disease

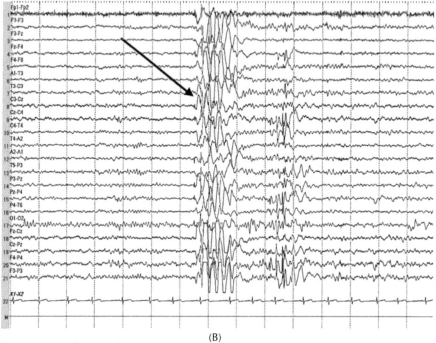

(B)

Figure 11.3 (Continued).

were published. In a clinic-based case series from Israel, 31% of 99 Ashkenazi Jewish patients with Parkinson disease were found to have any of 6 mutations in the glucocerebrosidase gene.[56] All but three of these had only one mutated allele, while three subjects were homozygous for the N370S mutation. Among 74 Alzheimer disease patients and 1543 healthy Ashkenazi Jewish controls, 4.1% and 6.2% had glucocerebrosidase mutations respectively.[56] Similar findings were obtained by other groups in Ashkenazi Jewish patients in the U.S.,[57] and in non-Jewish patients with Parkinson disease,[58] but not in Norway.[59]

Taken together, these findings suggest that mutations in the glucocerebrosidase gene are a new risk factor for Parkinson disease, and that one mutated allele is sufficient to confer this risk. However, the spectrum of Parkinson disease or parkinsonism phenotypes that are indeed associated with these mutations are not yet fully established. It is therefore important that in the future these patients be examined carefully and comprehensively documented with particular attention to their saccadic eye movements. Since carriers of mild mutations of the glucocerebrosidase gene have normal enzyme activity and therefore normal level of the substrates of this enzyme, the pathogenetic mechanism is less likely to be related to the glycolipid abnormality of Gaucher disease. A more likely hypothesis is that the mutated glucocerebrosidase is misfolded in the cell and causes stress that in rare individuals compounds existing but yet unknown cellular susceptibilities (Figure 11.4). Alternatively, it is possible that glucocerebrosidase participates in a yet unknown cellular process or acquires a new function when mutated. Finally, and myoclonic dystonia

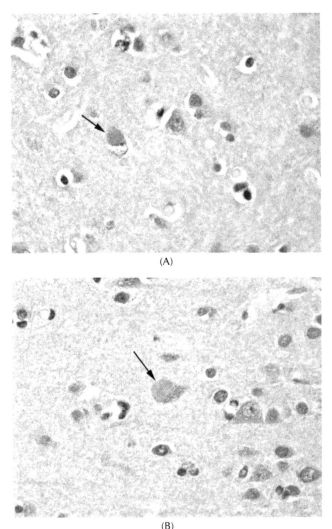

(A)

(B)

Figure 11.4 **(See color insert following page 304.)** Staining pattern for glucocerebrosidase using the 8E4 monoclonal antibody of cortical neurons in a patient with Gaucher disease and parkinsonism-dementia syndrome that was found to have diffuse Lewy body disease on autopsy. The antibody stains a structure resembling an inclusion that is otherwise is α–synuclein positive (arrows). Initial magnification ∞100. (From Wong, K. et al., *Mol. Genet. Metab.*, 82, 192, 2004.)

without seizures or other cortical abnormalities was observed in a Mexican patient (genotype *D409H/R131L*) with mild systemic disease and a normal SEP (unpublished data).

Visceral Disease

It should be emphasized that NGD patients often have very aggressive visceral disease.[11,27] Presentation with severe visceral disease at an age younger than 2 years is frequently associated with type 3 NGD. Pulmonary involvement, at least radio-logically, is seen in at least 50% of these patients and may become symptomatic.[30] Shortness of breath on exercise, coughing, and wheezing are sometimes quite severe and may be associated with vomiting. Pulmonary function tests show abnormalities of diffusion, occasionally with a restrictive pattern.

NEUROPATHOLOGY

The neuropathology of Gaucher disease is described in detail in Chapter 13. It can be summarized as follows: in general, there are certain features common to all forms of Gaucher disease, including type 1. These include perivascular lipid-laden macrophages (Gaucher cells) and perivascular or region-specfic gliosis such as in CA2–CA4 of the hippocampus, layer 3 and 5 for parietal cortex and layer 4b of the occipital cortex.[53] Neuronal loss occurs as well in these regions in patients with acute NGD and in patients with the chronic NGD that manifest neurological dete-rioration such as progressive myoclonic encephalopathy.[53] Interestingly, ERT is associated with reduction in the number of perivascular Gaucher cells.[27]

GENETICS

Unlike type 1 Gaucher disease, which has a particularly high prevalence in Ashkenazi Jews, NGD is panethnic. However, founder effects have been described in several parts of the world. The most striking example is seen in northern Sweden.[60] Molecular studies show that the two clusters are compatible with a single founder, who arrived in northern Sweden in or before the 16th century.[61] Another founder effect has been reported in the Mappila Muslims of Kerala state, in southern India.[62]

The phenotype–genotype correlation in NGD is often not evident. It has been recognized, however, that patients who are either homozygous or compound-het-erozygous for certain mutations, such as *L444P* and *D409H*, are at high risk of developing CNS involvement, while patients with the *N370S* mutation may be afforded protection from the classical neurological involvement.[1,13] Homozygosity for *L444P* is particularly common in Japan, where *L444P* is the most common allele, accounting for 41% of all disease alleles.[2]

Sixteen consecutive patients with progressive myoclonic encephalopathy had a total of 14 different genotypes.[63] However, mutations V394L, N188S, and G377S are often associated with progressive myoclonic encephalopathy, although each has been seen with other forms of GD.[63] On the other hand, mutation R463C and homozygosity for L444P rarely are associated with myoclonic seizures. A common hypothesis is that the neuroprotective effect of some mutations is dose-dependent, and that there is a threshold of enzyme activity, below which the likelihood of developing NGD is high.[64] However, a threshold effect in NGD has never been demonstrated. Variable levels of ER retention and ER-associated degradation in the proteasomes likely explains at least part of phenotypical variability.[65]

PATHOGENESIS

Type 1 GD is caused by accumulation of glucosylceramide in one cell type, the macrophage.[66] This glycolipid is derived from phagocytosis and the breakdown of senescent erythrocytes and leucocytes. In contrast, in NGD disease there is also cerebral dysfunction resulting from neuronal and possibly also macroglial and microglial accumulation of glycosphingolipid.[53,67,68] Smaller residual enzyme was thought to lead to the neuronopathic forms. In both human and mouse glucocerebrosidase mutants, enzyme activity is generally highest in the brain.[69]

Although the precise mechanism of CNS damage has not yet been elucidated, agonist-induced calcium release via the ryanodine receptor was significantly enhanced in brain microsomes from patients with type 2 NGD compared with the type 3 NGD, type 1 forms, and controls, and this alteration correlated with levels of glucocerebroside accumulation.[68,70] These findings suggested that defective calcium homeostasis may be a mechanism responsible for cerebral pathophysiology in the most severe form of NGD. Cultured neurons were more sensitive to glutamate-induced neuronal toxicity and to toxicity induced via various other cytotoxic agents.[71] Immunohistochemical studies indicate that brain regions containing cells that are selectively harmed in NGD have the highest level of glucocerebrosidase expression.[53] Therefore, a major role of glucocerebrosidase in the brain appears to be to keep intracellular glycosphingolipid concentrations below toxic levels. Additional mechanisms that cause cerebral damage and dysfunction in NGD likely exist. Excessive amounts of glycolipid in various compartments such as the endoplasmic reticulum or the plasma membrane and not necessarily in lysosomes are seen with special stains.[53,68]

The deacylated form of glucosylceramide, glucosylsphingosine, is often invoked as an offending metabolite in NGD.[72,73] However, recent findings indicate that its levels in brain microsomes appear far too low for it to play a major role in the Ca^{2+} release in Gaucher disease,[70] but it does induce Ca^{2+} release, but via a different mechanism than glucocerebroside.[70] Glucosylsphingosine is thought to cause Ca^{2+} release by mobilizing Ca^{2+} as an inositol triphosphate receptor agonist from thapsigargin-sensitive Ca^{2+} stores. Therefore, it is unlikely that glucosylsphingosine plays a pathogenic role in NGD, at least via an effect on the regulation of intracellular Ca^{2+} homeostasis. Interestingly, glucosylsphingosine was not elevated in the brain of an adult patient with type 3 Gaucher disease, parkinsonism and dementia.[74]

DIAGNOSIS

The diagnosis of NGD is purely clinical and depends on the finding of supranuclear gaze palsy with or without other neurological manifestations. NGD cannot be diagnosed solely on the basis of low residual GBA activity or a particular genotype. Because there are commonly few signs of apparent neurological involvement, a diagnosis of type 3 NGD is often missed in the early stages of the disease or in mild forms. Patients should be observed for the presence of head jerks while reading or while turning around when walking. Clinical features, such as early onset of disease, aggressive visceral disease, falling or low IQ scores in a patient with a "high-risk" genotype, can also suggest type 3 NGD.

Neuropsychometric Assessment

Neuropsychometric tests can be administered to assess IQ (Wechsler Intelligence Scale for Children, 3rd revision), language (Clinical Evaluation of Language Fundamentals, revised), visuospatial ability (Bender–Gestalt test), and reading ability (Wechsler Objective Reading Dimension).[22,27,30] It is important to use universally available techniques in order that global comparisons may be made.

A particular diagnostic challenge is the differentiation between type 2 and type 3 NPD.[12] The striking difference in prognosis between the two forms of NGD and the resultant implications for treatment mean that the clinical distinction at presentation is critical. Severe forms of type 2 disease (2A) are relatively easy to identify, but milder cases can be diagnosed with certainty only retrospectively when death occurs before age 4 years. The skin of type 2 patients has an increased ratio of epidermal glucosylceramide to ceramide, as well as extensive ultrastructural abnormalities.[16,75] It is suggested that these changes might form a reliable basis for discrimination between type 2 and type 3 disease.[16] Skin biopsies of infants with type 1 or 3 GD have been studied to date.

Imaging Techniques

Results of brain imaging tests such as computed tomography (CT) and magnetic resonance imaging (MRI) are usually normal. In patients with progressive disease such as myoclonic encephalopathy, diffuse cerebral atrophy is seen in relatively advanced stages.[27,76] Rarely, cerebellar calcification and involvement of the basal ganglia on CT and other abnormalities are seen.[22,77]

TREATMENT

Type 2 Neuronopathic Gaucher Disease

Only a handful of type 2 patients have been treated with ERT, and it is quite clear that it does not result in significant clinical benefit.[17,78] It should not, therefore, be used in this form of NGD. However, early in the course of disease, it may be difficult to make a definitive diagnosis of type 2 NGD in some patients. In such cases, it is reasonable to provide ERT until an early and rapid neurological decline indicates the correct diagnosis and ERT may be withdrawn.

Type 3 Neuronopathic Gaucher Disease

Currently, ERT is the mainstay of treatment for type 3 patients. Rarely, splenectomy or bone marrow transplantation (BMT) may be indicated.

Enzyme Replacement Therapy

ERT has been shown to be effective in ameliorating the visceral manifestations of Gaucher disease in type 3 patients.[22,27,30,79] Its effect on the nonneurological manifestations of type 3 NGD is similar to the effect on type 1 GD with normalization of blood counts and regression of liver and spleen size. ERT should be started as soon as the patient is diagnosed to prevent irreversible skeletal complications. The usual recommended dose is at least 60 IU/kg of body weight every 2 weeks. This dose should not be reduced, even if the visceral signs improve. Occasionally, patients require higher doses or more frequent administration to control the systemic disease or when more rapid disease reversal is needed.[22,27,30] The systemic effect of ERT has markedly modified the natural history of most patients with type 3 NPD. Quality of life has markedly increased and early therapy prevents bone crises, bone fractures, liver cirrhosis, and portal hypertension.[1] Its effect on the lung disease in these patients is less evident with almost no cases of regression of the radiological interstitial abnormalities and the appearance of respiratory symptoms in some patients on ERT.[80] Lymph nodes probably represent an area of refractory disease.[17,81]

Despite demonstration that ERT is associated with reduction of perivascular lipid-laden macrophages (Gaucher cells) in the brain,[27] ERT does not seem to have any effect in patients with myoclonic seizures, supranuclear gaze palsy, or cognitive deficit, nor in patients with parkinsonism or other movement disorders.[30] There is therefore an increasing consensus that ERT has no measurable effect on the neurological manifestations of type 3 NPD patients.[30,31,82–84] Although it has been demonstrated that neurons can be protected from damage if the enzyme can be delivered to them,[85] the major obstacle thus far has been the inability of intravenously infused glucocerebrosidase to cross the blood–brain barrier in appreciable amounts.[27,78]

Splenectomy

In patients with very aggressive disease presenting with massive hepatosplenomegaly in infancy, respiratory function may be compromised to such an extent that rapid "debulking" is necessary. On occasion, severe anorexia and early satiety may be caused by the compression of the stomach by the large spleen. In these rare cases, splenectomy may be indicated, and partial, rather than total, splenectomy is recommended, as total splenectomy may be associated with worse neurological prognosis.[79]

Bone Marrow Transplantation (BMT)

BMT has been carried out in a small number of patients.[86–89] There is currently no clear role for BMT in NGD. As noted above, it is clear that successful engraftment will cure the manifestations of visceral disease because they are caused by enzyme-deficient macrophages. Bone marrow-derived perivascular lipid-laden macrophages (Gaucher cells) are common in the brain of all GD patients including in type 1 Gaucher disease. Clinical observation does not indicate a therapeutic benefit of BMT in NGD. We have observed the development or progression of typical neurological disease manifestations in patients who successfully received a transplant. Three patients underwent BMT at the Westminster Children's Hospital, London, U.K., at the ages of 1.25, 3.25, and 11.9 years. BMT did not reverse neurological deficit or prevent continued deterioration. The clinical course of patients post-BMT is similar to the one commonly observed in type 3 NGD patients on ERT.[30] We have observed two additional patients who developed progressive myoclonic encephalopathy leading to death, a few years after a successful BMT, and another patient who developed supranuclear gaze palsy after BMT. The procedure itself is not without risk. Furthermore, a total splenectomy is required pre-BMT.[90] Therefore, in the event of graft rejection, the patient will be left asplenic and is consequently at risk of rapidly progressive neurological deterioration.[79]

Treatment of Lung Involvement

As mentioned above, lung involvement is common and often symptomatic in NGD patients. ERT has little if any effect on it and is particularly ineffective in reversing the deadly lung abnormalities in type 2 NGD. Anecdotal evidence suggests that oral and long-term inhaled corticosteroid therapy is often effective. It is usually possible, over a period of time, to gradually wean patients from corticosteroids who are receiving ERT (Unpublished data).

FUTURE DIRECTIONS

Enzyme Delivery to the Central Nervous System

The blood-brain barrier is the major obstacle to use of ERT in NGD. The observation that GBA replacement reverses the neuronal functional abnormalities suggests that ERT is inefficient in the treatment of type 3 NGD, not that neurons are incapable of taking up enzyme.[85] Direct enzyme infusion of imiglucerase into rat and monkey brain and brainstem resulted in neuronal uptake.[91,92] Efforts are being made to modify glucocerebrosidase to enable it to cross the blood-brain barrier. For example, conjugation with the TAT proteintransduction domain has been shown in some cases to allow a large protein to enter the brain.[93]

Substrate Reduction Therapy

Clinical studies in patients with type 1 Gaucher disease have demonstrated the efficacy of substrate reduction therapy using N-butyldeoxynojirimycin (OGT 918 or miglustat or Zavesca®) in ameliorating visceral disease to a degree.[94–96] Based on its mode of action in the catabolic process of glycosphingolipids and its ability to cross the blood-brain barrier, substrate reduction therapy is a logical candidate for the treatment of type 3 NGD, and studies are now underway to assess this mode of therapy in this patient group. Miglustat level in the spinal fluid is about 40% of the plasma level in type 3 GD. Thus far no dramatic response has been observed, but final analysis is expected in 2006. Until then, we do not recommend the routine use of miglustat in NGD.

Chemical Chaperones

Since all Gaucher patients have residual enzyme activity, enhancing the enzyme activity using small molecules that cross the blood-brain barrier and inducing normal folding and appropriate posttranslation modification of the mutated glucocerebrosidase is an attractive option. Currently, such molecules are active site competitive inhibitors that activate the enzyme in much lower concentrations,[97] but noncompetitive activator may be developed in the future.

Other Potential Therapeutic Approaches

The new understanding of the neuronal mechanism by which glucosylceramide causes neuronal dysfunction and death suggests the use of inhibitors of the ryanodine receptor that block intracellular Ca^{2+} release such as dantrolene, similar to that attempted in epilepsy and other neurodegenerative diseases in which Ca2+-mobilization is altered.[98,99] The use of delivery vehicles such as liposomes and various modes of gene transfer are under investigation as well.[100-102]

SUMMARY

The diagnosis of NGD may be difficult because the neurologic abnormalities can be subtle. In addition, the symptoms and signs combine to form a wide variety of syndromes. There is considerable variation in the natural history, making the assessment of treatment problematic. The association of glucocerebrosidase gene mutation with Parkinson disease and other Lewy body-related disorders further enlarge the pathogenetic scope and phenotypic spectrum. ERT in sufficient doses has a marked effect on the nonneurological manifestations of the disease, especially in patients with severe forms if initiated early in life. However, it has no demonstrable effect on the neurological abnormalities. Substrate reduction therapy, the development of clinically viable direct enzyme delivery systems, and the improved understanding of the pathophysiology of NGD offer grounds for cautious optimism regarding the future treatment of the neurological symptoms of Gaucher disease.

REFERENCES

1. Charrow, J. et al., The Gaucher registry: demographics and disease characteristics of 1698 patients with Gaucher disease. Arch Intern Med, 160, 2835, 2000.
2. Ida, H. et al., Clinical and genetic studies of Japanese homozygotes for the Gaucher disease L444P mutation. Hum Genet, 105, 120, 1999.
3. Eto, Y. and Ida, H., Clinical and molecular characteristics of Japanese Gaucher disease. Neurochem Res, 24, 207, 1999.

4. Drugan, C. et al., Gaucher disease in Romanian patients: incidence of the most common mutations and phenotypic manifestations. Eur J Hum Genet, 10, 511, 2002.
5. Alfonso, P. et al., Mutation prevalence among 51 unrelated Spanish patients with Gaucher disease: identification of 11 novel mutations. Blood Cells Mol Dis, 27, 882, 2001.
6. Rusca, C.L., Sul morbo del Gaucher. Hematoogica (Pavia), 2, 441, 1921.
7. Oberling, C. and Woringer, P., La maladie de Gaucher chez le nourisson. Rev Franc de Pediatr, 3, 475, 1927.
8. Lachiewicz, P.F., Gaucher's disease. Orthop Clin North Am, 15, 765, 1984.
9. Glew, R.H. et al., Enzymic differentiation of neurologic and nonneurologic forms of Gaucher's disease. J Neuropathol Exp Neurol, 41, 630, 1982.
10. Brady, R.O., Barton, N.W., and Grabowski, G.A., The role of neurogenetics in Gaucher disease. Archives of Neurology, 50, 1212, 1993.
11. Patterson, M.C. et al., Isolated horizontal supranuclear gaze palsy as a marker of severe systemic involvement in Gaucher's disease. Neurology, 43, 1993, 1993.
12. Goker-Alpan, O. et al., Phenotypic continuum in neuronopathic Gaucher disease: An intermediate phenotype between type 2 and type 3. J Pediat, 143, 273, 2003.
13. Sidransky, E., Gaucher disease: complexity in a "simple" disorder. Mol Genet Metab, 83, 6, 2004.
14. Chabas, A. et al., Neuronopathic and non-neuronopathic presentation of Gaucher disease in patients with the third most common mutation (D409H) in Spain. J Inh Metab Dis, 19, 798, 1996.
15. Bohlega, S. et al., Gaucher disease with oculomotor apraxia and cardiovascular calcification (Gaucher type IIIC). Neurology, 54, 261, 2000.
16. Sidransky, E. et al., Epidermal abnormalities may distinguish type 2 from type 1 and type 3 of Gaucher disease. Pediatr Res, 39, 134, 1996.
17. Bove, K.E., Daugherty, C., and Grabowski, G.A., Pathological findings in Gaucher disease type 2 patients following enzyme therapy. Hum Pathol, 26, 1040, 1995.
18. Kaye, E.M. et al., Type 2 and type 3 Gaucher disease: a morphological and biochemical study. Ann Neurol, 20, 223, 1986.
19. Takahashi, T. et al., Enzyme therapy in Gaucher disease type 2: an autopsy case. Tohoku J Exp Med, 186, 143, 1998.
20. Sinclair, G., Choy, F.Y.M., and Humphries, L., A novel complex allele and two new point mutations in type 2 (acute neuronopathic) Gaucher disease. Blood Cells Mole Dis, 24, 420, 1998.
21. Frederickson, D.S. and Sloan, H.R., Glucosylceramide lipidoses — Gaucher's disease, in The Metabolic Basis of Inherited Disease, Stanbury, J.B., Fredrickson, D.S., Eds., McGraw-Hill, New York, 1972, 730–759.
22. Vellodi, A. et al., Management of neuronopathic Gaucher disease: a European consensus. J Inher Metabol Dis, 24, 319, 2001.
23. Lui, K. et al., Collodion babies with Gaucher's disease. Arch Dis Child, 63, 854, 1988.
24. Soma, H. et al., Identification of Gaucher cells in the chorionic villi associated with recurrent hydrops fetalis. Placenta, 21, 412, 2000.
25. Tybulewicz, V.L. et al., Animal model of Gaucher's disease from targeted disruption of the mouse glucocerebrosidase gene. Nature, 357, 407, 1992.
26. Stone, D.L. et al., Is the perinatal lethal form of Gaucher disease more common than classic type 2 Gaucher disease? Eur J Hum Genet, 7, 505, 1999.
27. Schiffmann, R. et al., Prospective study of neurological responses to treatment with macrophage-targeted glucocerebrosidase in patients with type 3 Gaucher's disease. Ann Neurol, 42, 613, 1997.

28. Harris, C.M., Taylor, D.S., and Vellodi, A., Ocular motor abnormalities in Gaucher disease. Neuropediatrics, 30, 289, 1999.

29. Harris, C.M. et al., Morbus Gaucher. Auditory brainstern response and eye movement in diagnosis of Morbus Gaucher type 3. Monatsschrift Kinderheilkunde, 151, 854, 2003.

30. Altarescu, G. et al., The efficacy of enzyme replacement therapy in patients with chronic neuronopathic Gaucher's disease. J Pediatr, 138, 539, 2001.

31. Germain, D.P., Gaucher disease: clinical, genetic and therapeutic aspects. Pathologie Biologie, 52, 343, 2004.

32. Erikson, A., Bembi, B., and Schiffmann, R., Neuronopathic forms of Gaucher's disease. Baillieres Clin Haematol, 10, 711, 1997.

33. Erikson, A., Gaucher disease — Norrbottnian type (III). Neuropaediatric and neuro-biological aspects of clinical patterns and treatment. Acta Paediatr Scand Suppl, 326, 1, 1986.

34. Frei, K.P. and Schiffmann, R., Myoclonus in Gaucher disease. Adv Neurol, 89, 41, 2002.

35. Abrahamov, A. et al., Gaucher's disease variant characterised by progressive calcification of heart valves and unique genotype. Lancet, 346, 1000, 1995.

36. Tripp, J.H. et al., Juvenile Gaucher's disease with horizontal gaze palsy in three siblings. J Neurol Neurosurg Psychiatry, 40, 470, 1977.

37. Steinlin, M., Thunhohenstein, L., and Boltshauser, E. Congenital Oculomotor Apraxia — Presentation, Developmental Problems and Differential-Diagnosis. Klinische Monatsblatter Fur Augenheilkunde, 200, 623, 1992.

38. Bamiou, D.E. et al., Audiometric abnormalities in children with Gaucher disease type 3. Neuropediatrics, 32, 136, 2001.

39. Campbell, P.E., Harris, C.M., and Vellodi, A., Deterioration of the auditory brainstem response in children with type 3 Gaucher disease. Neurology, 63, 385, 2004.

40. Musiek, F.E. and Lee, W.W., Auditory middle and late potentials, in *Contemporary Perspectives in Hearing Assessment*, Musiek, F.M., Rintelmann, W.F., Eds., Allyn and Bacon, Nedham Heights, 1999.

41. Zupanc, M.L. and Legros, B., Progressive myoclonic epilepsy. Cerebellum, 3, 156, 2004.

42. Rapin, I. Myoclonus in neuronal storage and Lafora diseases. Advances in neurology, 43, 65, 1986.

43. Manganotti, P. et al., Hyperexcitable cortical responses in progressive myoclonic epilepsy: a TMS study. Neurology, 57, 1793, 2001.

44. Garvey, M.A. et al., Somatosensory evoked potentials as a marker of disease burden in type 3 Gaucher disease. Neurology, 56, 391, 2001.

45. Verghese, J., et al., Myoclonus from selective dentate nucleus degeneration in type 3 Gaucher disease. Arch Neurol, 57, 389, 2000.

46. Wenger, D.A. et al., Biochemical studies in a patient with subacute neuropathic Gaucher disease without visceral glucosylceramide storage. Pediatr Res, 17, 344, 1983.

47. Sack, G.H., Jr., Clinical diversity in Gaucher's disease. Johns Hopkins Med J, 146, 166, 1980.

48. Neudorfer, O. et al., Occurrence of Parkinson's syndrome in type I Gaucher disease. Qjm, 89, 691, 1996.

49. Neil, J.F., Glew, R.H., and Peters, S.P., Familial psychosis and diverse neurologic abnormalities in adult-onset Gaucher's disease. Arch Neurol, 36, 95, 1979.

50. Davison, C., Disturbances of lipid metabolism and the central nervous system. J Mt Sinai Hosp, 9, 389, 1942.

51. Bembi, B. et al., Gaucher's disease with Parkinson's disease: clinical and pathological aspects. Neurology, 61, 99, 2003.

52. Tayebi, N. et al., Gaucher disease and parkinsonism: a phenotypic and genotypic characterization. Mol Genet Metab, 73, 313, 2001.

53. Wong, K. et al., Neuropathology provides clues to the pathophysiology of Gaucher disease. Mol Genet Metab, 82, 192, 2004.

54. Dickson, D.W. et al., Hippocampal degeneration differentiates diffuse Lewy body disease (DLBD) from Alzheimer's disease: light and electron microscopic immuno-cytochemistry of CA2-3 neurites specific to DLBD. Neurology, 41, 1402, 1991.

55. Lwin, A. et al., Glucocerebrosidase mutations in subjects with parkinsonism. Mol Genet Metab, 81, 70, 2004.

56. Aharon-Peretz, J., Rosenbaum, H., and Gershoni-Baruch, R., Mutations in the glu-cocerebrosidase gene and Parkinson's disease in Ashkenazi Jews. N Engl J Med, 351, 1972, 2004.

57. Clark, L.N. et al., Pilot association studyof the beta-glucocerebrosidase N370S allele and Parkinson's disease in subjects of Jewish ethnicity. Mov Disord, 20, 100, 2005.

58. Sato, C. et al., Analysis of the glucocerebrosidase gene in Parkinson's disease. Mov Disord, 20, 367, 2005.

59. Toft, M. et al., Glucocerebrosidase gene mutations are not associated with Parkinson's disease in the Norwegian population. Mov Disord, 20, Suppl., 10, S37, 2005.

60. Dreborg, S., Erikson, A., and Hagberg, B., Gaucher disease — Norrbottnian type. I. General clinical description. Eur J Pediatr, 133, 107, 1980.

61. Dahl, N., Hillborg, P.O., and Olofsson, A., Gaucher disease (Norrbottnian type 3): probable founders identified by genealogical and molecular studies. Hum Genet, 92, 513, 1993.

62. Feroze, M., Arvindan, K.P., and Jose, L., Gaucher's disease among Mappila Muslims of Malabar. Indian J Pathol Microbiol, 37, 307, 1994.

63. Park, J.K. et al., Myoclonic epilepsy in Gaucher disease: genotype-phenotype insights from a rare patient subgroup. Pediatr Res, 53, 387, 2003.

64. Zhao, H. et al., Gaucher disease: In vivo evidence for allele dose leading to neurono-pathic and nonneuronopathic phenotypes. Am J Med Genet, 116, 52, 2003.

65. Ron, I. and Horowitz, M., ER retention and degradation as the molecular basis underlying Gaucher disease heterogeneity. Hum Mol Genet, 14, 2387, 2005.

66. Cox, T.M. and Schofield, J.P., Gaucher's disease: clinical features and natural history. Baillieres Clin Haematol, 10, 657, 1997.

67. Brady, R.O., Kanfer, J.N., and Shapiro, D., Metabolism of glucosylcerebroside II. Enzymatic deficiency in Gaucher's disease. Biochem Biophys Res Commun, 18, 221, 1965.

68. Pelled, D. et al., Enhanced calcium release in the acute neuronopathic form of Gaucher disease. Neurobiol Dis, 18, 83, 2005.

69. Xu, Y.H. et al., Viable mouse models of acid beta-glucosidase deficiency: the defect in Gaucher disease. Am J Pathol, 163, 2093, 2003.

70. Lloyd-Evans et al., Glucosylceramide and glucosylsphingosine modulate calcium mobilization from brain microsomes via different mechanisms. J Bio Chem, 278, 23594, 2003.

71. Korkotian, E. et al., Elevation of intracellular glucosylceramide levels results in an increase in endoplasmic reticulum density and in functional calcium stores in cultured neurons. J Biol Chem, 274, 21673, 1999.

72. Schueler, U.H. et al., Toxicity of glucosylsphingosine (glucopsychosine) to cultured neuronal cells: a model system for assessing neuronal damage in Gaucher disease type 2 and 3. Neurobiol Dis, 14, 595, 2003.
73. Orvisky, E. et al., Glucosylsphingosine accumulation in tissues from patients with Gaucher disease: correlation with phenotype and genotype. Mol Genet Metab, 76, 262, 2002.
74. Tayebi, N. et al., Gaucher disease with parkinsonian manifestations: does glucocerebrosidase deficiency contribute to a vulnerability to parkinsonism? Mol Genet Metab, 79, 104, 2003.
75. Stone, D.L. et al., Type 2 Gaucher disease: the collodion baby phenotype revisited. Arch Dis Child Fetal Neonatal Ed, 82, F163, 2000.
76. Yoshikawa, H. et al., Uncoupling of blood flow and oxygen metabolism in the cerebellum in type 3 Gaucher disease. Brain Dev, 13, 190, 1991.
77. Chang, Y.C. et al., MRI in acute neuropathic Gaucher's disease. Neuroradiology, 42, 48, 2000.
78. Migita, M. et al., Glucocerebrosidase level in the cerebrospinal fluid during enzyme replacement therapy — unsuccessful treatment of the neurological abnormality in type 2 Gaucher disease. Eur J Pediatr,162, 524, 2003.
79. Erikson, A., Remaining problems in the management of patients with Gaucher disease. J Inher Metab Dis, 24 Suppl. 2, 122, 2001.
80. Lee, S.Y. et al., Gaucher disease with pulmonary involvement in a 6-year-old girl: report of resolution of radiographic abnormalities on increasing dose of imiglucerase. J Pediatr, 139, 862, 2001.
81. Lim, A.K., Vellodi, A., and McHugh, K., Mesenteric mass in a young girl — an unusual site for Gaucher's disease. Pediatr Radiol, 32, 674, 2002.
82. Eto, Y. and Ohashi, T. Novel treatment for neuronopathic lysosomal storage diseases-cell therapy/gene therapy. Curr Mol Med, 2, 83, 2002.
83. Ono, H. et al., Neurological features in Gaucher's disease during enzyme replacement therapy. Acta Paediatr, 90, 229, 2001.
84. Michelakakis, H. et al., Early-onset severe neurological involvement and D409H homozygosity in Gaucher disease: outcome of enzyme replacement therapy. Blood Cells Molecules and Diseases, 28, 1, 2002.
85. Pelled, D., Shogomori, H., and Futerman, A.H., The increased sensitivity of neurons with elevated glucocerebroside to neurotoxic agents can be reversed by imiglucerase. J Inher Metabol Dis, 23, 175, 2000.
86. Tsai, P. et al., Allogenic bone marrow transplantation in severe Gaucher disease. Pediatr Res, 31, 503, 1992.
87. Erikson, A. et al., Clinical and biochemical outcome of marrow transplantation for Gaucher disease of the Norrbottnian type. Acta Paediatr Scand, 79, 680, 1990.
88. Ringden, O. et al., Long-term follow-up of the first successful bone marrow transplantation in Gaucher disease. Transplantation, 46, 66, 1988.
89. Kapelushnik, J. et al., Bone marrow transplantation from a cadaveric donor. Bone Marrow Trans, 21, 857, 1998.
90. Hobbs, J.R. et al., Beneficial effect of pre-transplant splenectomy on displacement bone marrow transplantation for Gaucher's syndrome. Lancet, 1, 1111, 1987.
91. Zirzow, G.C. et al., Delivery, distribution, and neuronal uptake of exogenous mannose- terminal glucocerebrosidase in the intact rat brain. Neurochem Res, 24, 301, 1999.
92. Lonser, R.R. et al., Convective perfusion of glucocerebrosidase for neuronopathic Gaucher disease. Ann Neurol 2005, 57, 542, 2005.

93. Lee, K.O. et al., Improved intracellular delivery of glucocerebrosidase mediated by the HIV-1 TAT protein transduction domain. Biochem Biophys Res Commun, 337, 701, 2005.

94. Cox, T.M. et al., The role of the iminosugar N-butyldeoxynojirimycin (miglustat) in the management of type I (non-neuronopathic) Gaucher disease: a position statement. J Inher Metab Dis, 26, 513, 2003.

95. Cox, T. et al., Novel oral treatment of Gaucher's disease with N-butyldeoxynojirimycin (OGT 918) to decrease substrate biosynthesis. Lancet, 355, 1481, 2000.

96. Heitner, R. et al., Low-dose N-butyldeoxynojirimycin (OGT 918) for type I Gaucher disease. Blood Cells Mol Dis, 28, 127, 2002.

97. Ishii, S. et al., Transgenic mouse expressing human mutant alpha-galactosidase A in an endogenous enzyme deficient background: a biochemical animal model for studying active-site specific chaperone therapy for Fabry disease. Biochim Biophys Acta, 1690, 250, 2004.

98. Pal, S. et al., Epileptogenesis induces long-term alterations in intracellular calcium release and sequestration mechanisms in the hippocampal neuronal culture model of epilepsy. Cell Calcium, 30, 285, 2001.

99. Mattson, M.P. et al., Calcium signaling in the ER: its role in neuronal plasticity and neurodegenerative disorders. Trends Neurosci, 23, 222, 2000.

100. Matzner, U. and Gieselmann, V. Gene therapy of metachromatic leukodystrophy. Expert Opin Biol Ther, 5, 55, 2005.

101. Hartung, S.D. et al., Correction of metabolic, craniofacial, and neurologic abnormalities in MPS I mice treated at birth with adeno-associated virus vector transducing the human alpha-L-iduronidase gene. Mol Ther, 9, 866, 2004.

102. Cheng, S.H. and Smith, A.E. Gene therapy progress and prospects: gene therapy of lysosomal storage disorders. Gene Ther, 10, 1275, 2003.

CHAPTER 12

Pathologic Anatomy of Gaucher Disease: A Pictorial Essay

Robert E. Lee

CONTENTS

INTRODUCTION

This disease was initially described by Dr. Gaucher, a 28-year-old intern in the Hospital Cochin in Paris. This institution has been active in patient care since 1784 (Figure 12.1). Dr. Gaucher chronicled the story of a 34-year-old woman patient in a thesis for a Doctorate of Medicine degree from the University of Paris in 1882 [1]. The patient was first noted to have splenomegaly at age 7 years. Throughout her life she had multiple hospital admissions for bleeding as well as abdominal pain. The autopsy revealed a very cachetic individual weighing only 31 kg with a markedly distended abdomen resembling a term pregnancy because of the enlargement of both the spleen (4770 g) and the liver (3880). If one allows for the enlarged spleen and liver, her cachexia was very marked at 24 kg.

For a patient with splenomegaly in the 1880s, the various diseases that had to be considered included typhoid fever, malaria, cirrhosis, and malignancy. Gaucher was able to eliminate all of these. Because of its localization and lack of spread, he termed the disorder a primary epithelioma of the spleen. The liver of his patient was enlarged, dark yellow, and cirrhotic. No mention was made of liver infiltration with the same cells that involved the spleen. Three drawings of the histology of spleen were included that accurately illustrated the light microscopic features of the patient's spleen. The reasoning for his conclusions shows an excellent understanding of pathologic mechanisms, and the recognition of this disease as a new disease testifies to the skill of French pathology as well to the skill of young Dr. Gaucher.

The purpose of this chapter is to illustrate the anatomic features of this common lipid storage disease and to point out the effects that these deposits have on normal function.

SPLENIC PATHOLOGY

Splenomegaly is the constant sign of this disease. Commonly when the physician detects splenomegaly, there is also hepatomegaly. In our series of 335 Gaucher patients, the largest spleen was 6665 g in a 17-year-old girl who was initially diagnosed at age 4 years. The only exception to this observation was in a 30-year-old woman who had a preoperative diagnosis of hemolytic anemia. Her spleen was 260 g and histologically suggested Gaucher disease. The diagnosis of "Idiopathic Hypertrophy of the Spleen" was made by Gaucher because the disease seemed to affect mostly the spleen along with abnormal splenic histology. The hematologic effects of splenic involvement most commonly manifest as thrombocytopenia, but anemia and leukopenia may additionally be present. Abdominal pain results from the splenic infarcts or pressure on abdominal viscera. The shape of the various types of splenic infarcts in this disorder are mostly round and not the wedge-shaped, subcapsular infarcts that result from spleni cartery emboli [5]. In addition there can be splenic nodules that may be partly necrotic or composed of benign elements (Figure 12.6 and Figure 12.7). The splenic site of accumulation of Gaucher cells is the red pulp cords (Figure 12.3). Because of their size these storage cells have a limited ability to pass from the red pulp through the endothelium into a splenic vein and then into the portal venous system (Figure 12.4). This blockage is not perfect because Gaucher cells primarily populate the liver sinusoids, bone marrow, lungs, and brain.

Probably the majority of the Gaucher cells are formed in the spleen, since this is a main site of blood cell turnover. Dr. Gaucher recognized that the large histiocytes were foreign to the spleen and illustrated them in drawings that were done at a time when light microscopy was just beginning. Electron microscopy reveals that a Gaucher cell has numerous lysosomes in which the lipid material is stored as long, narrow, and twisted [4] tightly arranged 60 Å bilayers (Figure 12.10, Figure 12.11, and Figure 12.12). Dr. Gaucher recognized the unique involvement of these "epi-thelial cells" in the spleen and beautifully illustrated them in the early 1880s when light microscopes were just beginning to find a place in medicine (Figure 12.4 and Figure 12.5). In every patient that we have studied of the more than 500 patients in our Gaucher Registry, Gaucher cells are identical. Many Gaucher cells, especially in older patients, contain abundant cytoplasmic iron. This presumably comes from the iron that comes from red cell destruction after erythrophagocytosis. It is microscopically best visualized by routine iron stains of Gaucher tissues. When the Gaucher tissue is minced and purified using density gradient centrifugation, the stored material is composed of long narrow 60 Å thick bilayers that are twisted in a right handed direction (Figure 12.13 and Figure 12.14). There are many other lipid storage or other storage diseases that may have a few storage cells that are similar to Gaucher cells by light microscopy. None are identical to the Gaucher cell [4,6–8] The sphingolipid storage disease that most closely resembles Gaucher disease is Krabbe Disease [3]. Here the galactocerebroside is stored exclusively in the nervous system and deposits take two forms, twisted resembling Gaucher deposits, and a second form that is straight or gently curved.

LIVER PATHOLOGY

The sinusoidal infiltration may not be recognized in the youngest patients, but biopsies in adults will reveal typical macrophages in the sinusoids that are quite different from adjacent hepatocytes. Abnormalities in liver function are not a prominent feature initially despite the hepatomegaly. However, when the patient has extensive liver involvement, liver failure with diffuse cirrhosis may develop. Parenchymal necrosis can also develop. At this stage there will be many of the associated abnormal physical and laboratory findings that are associated with liver failure. A number of our patients with liver failure have had liver transplants.

BONE/BONE MARROW INVOLVEMENT

Every patient with Gaucher disease will have Gaucher cells when the marrow is properly sampled and examined microscopically. Because of the focal nature of the deposits, a single exam may not locate the typical cells. When these cells are stained with a Romanovsky stain, such as a Wright stain, the classic wrinkled appearance of the cytoplasm is present (Figure 12.8). In the adult patient, frequently iron stains show individual positive Gaucher cells (Figure 12.15). By electron microscopy the iron is in the form of dense particles of ferritin that are free in the cytoplasm. Most of this iron resides within the same lysosomes that contain the storage deposits (Figure 12.16). In children this excess iron is not present. By iron kinetic studies no significant abnormalities of iron or red cell kinetics were detected in 12 Gaucher patients who had this excess of marrow iron [2]. It is assumed that the iron in the Gaucher cells is available for erythropoiesis and is possibly present because of erythrophagocytosis, a feature that is also noted in Gaucher cells. Beside the iron stain for identifying Gaucher cells, the PAS-diastase stain is helpful, but none of the usual stains are specific.

Fractures are a common painful complication that occur due to bone replacement and vascular interruption by the lipid deposits. Generally the bones have a thin cortex and necrosis is usually present at the fracture site. In severe cases, skeletal necrosis is widespread at autopsy. Fracture sites commonly involve the femoral head (Figure 12.18) as well as the long bones of the extremities [5]. Other sites can be the ribs, pelvis, and spine. A typical change seen in young patients is the widening of the distal femur, termed a flask-shaped deformity, that is associated with weakening due to thinning of the cortex.

LUNG INVOLVEMENT

Four types of involvement are noted histologically:

1. Intra alveolar collections of Gaucher cells. This distribution is usually found in the youngest patients (Figure 12.28).
2. Clusters of Gaucher cells in the interstitial space as well as in the pleura (Figure 12.27).
3. Periarterial deposits of Gaucher cells. These patients frequently have pulmonary hypertension (Figure 12 and Figure 12.26).
4. Gaucher cells in the alveolar capillaries is an unusual finding and may represent a terminal feature.

Lung involvement is usually found in the more severely affected patients and the most severe type is with periarterial deposits.

GAUCHER CELLS IN THE BRAIN

The patients with the most numerous Gaucher cells are the youngest, and the most severely handicapped are classified as having clinical Type II involvement. In these instances, the Gaucher cells are located free in the gray matter in one or more of the cerebral lobes or the cerebellum. Other patients who have little or no neurologic signs or symptoms may have perivascular collections of Gaucher cells involving vessels at the junction of the white and gray matter. From the history alone, I have never been able to predict which patients will have perivascular Gaucher cells. The majority of Gaucher adults have no Gaucher cells in the brain. In only one instance in a pair of brothers (Cases 339 and 340) have I seen Gaucher cells involving the spinal cord. These were the youngest patients; the older died at birth and the younger was aborted at 20 weeks.

I have been tempted to believe that Gaucher cells reach the infant brain before the blood-brain barrier is established. They then migrate from the gray matter to a perivascular position at the junction of the gray and white matter. In the absence of glucocerebrosidase, they may reside there for the life of the patient.

MOLECULAR PATHOLOGY

The form that Gaucher deposits take is as an elongated and twisted set of bilayers where each bilayer is 60 Å thick. The direction of the twist, determined by shadowing, is in a right-handed direction (Figure 12.10, Figure 12.11, and Figure 12.12). This elongated and insoluble deposit is located in the lysosome and probably accounts for the typical wrinkled cytoplasm shown by light microscopy of the Gaucher cell (Figure 12.8). Because of the hollow appearance of these deposits, they are commonly referred to as Gaucher tubules (Figure 12.10, Figure 12.11, and Figure 12.14). The reason that the deposits have this appearance is that the aqueous stains used for electron microscopy do not penetrate the interior or hydrophobic portion of the deposits. The tight bilayers are only revealed by freeze fracture techniques that enable one to see that the basic structure is a set of twisted bilayers.

MALIGNANT DISEASE IN THE PITT REGISTRY

In our series of 535 patients, we noted the frequency of malignant tumors: 61 tumors are present with the most frequent tumor in 8 patients being myeloma; 6 had colon carcinoma; 5 patients had lymphoma; 4 each had prostate, lung, and carcinomas; and 4 patients had chronic lymphatic leukemia. Studies of natural killer activity were carried out with 7 Gaucher homozygotes, 10 heterozygotes, and 106 normal patients. Only 2 Gaucher patients had low K cell activity.

SUMMARY

The anatomic and molecular features of Gaucher disease have been presented based on 43 years of experience and observation with the hope that effective therapy may arise from research in this field.

REFERENCES

1. Gaucher, Philip-Charles-Ernest, On primary epithelioma of the spleen: idiopathic hypertrophy of the spleen without leukemia, Thesis for Doctorate of Medicine Paris Faculty of Medicine, 1882.
2. Lee, R.E., Balcerzak, S.P., and Westerman, M.P., Gaucher's disease: a morphologic study and measurements of iron metabolism. *Am. J. Med.*, 42, 891, 1967.
3. Lee, R.E., Worthington, C.R., and Glew, R.H., The bilayer nature of the membranous deposits occurring in Gaucher's disease, *Lab. Invest.*, 24, 261, 1971.
4. Yunis, E.J. and Lee, R.E., The ultrastructure of Globoid (Krabbe) Leukodystrophy, *Lab Invest.*, 21, 415, 1969.
5. Lee, R.E. and Ellis, L.D., The storage cells of chronic myeloid leukemia, *Lab Invest.*, 24, 261, 1971.

6. Lee, R.E., Pathology of Gaucher Disease, Presented at the International Symposium on Gaucher Disease, July 22–24, 1981, NY, NY, in *Gaucher Disease: a Century of Delineation and Research*, R.J. Desnick et al., Ed.. Alan R. Liss Inc., 1982, New York.

7. Krause, J.R., Bures, C., and Lee, R.E., Acute leukemia in Gaucher's disease, *Scan. J. Hematology*, 23, 115, 1979.

8. Lee, R.E., Robinson, D., and Glew, R.H., Gaucher's disease: modern enzymatic and anatomic methods of diagnosis, *Arch. Pathol. Lab. Med.*, 105, 102, 1981.

9. Zidar, B.L. et al., Pseudo-Gaucher cells in the bone marrow of a patient with Hodgkin's disease, *Am. J. Clin. Path.*, 87, 533, 1987.

Figure 12.1   The Hospital Cochin is located in the center of Paris at 27 Faubourg Saint-Jacques Street and has been serving patients since 1784. Gaucher, pictured here, cared for S.V., a 34-year-old woman with a long illness and diagnosed a previously unrecognized Primary Epithelioma of the Spleen. Her illness was the topic of the thesis he submitted to the University of Paris for his Doctor of Medicine Degree in 1882.

Figure 12.2 **(See color insert following page 304.)** This close up gross photo of a Gaucher spleen is mostly white because the normally red pulp has been infiltrated by the white lipid contained in the Gaucher cells. The lymphoid centers that are normally grey and round are obscured by the Gaucher cell infiltrate.

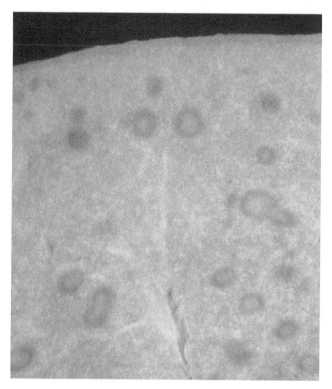

Figure 12.3 This gross view of the spleen was photographed with a dissecting microscope and shows several gray nodules of lymphoid tissue termed white pulp or Malpighian corpuscles. The red pulp has changed from red due to the presence of numerous Gaucher cells containing glucocerebroside, a white lipid.

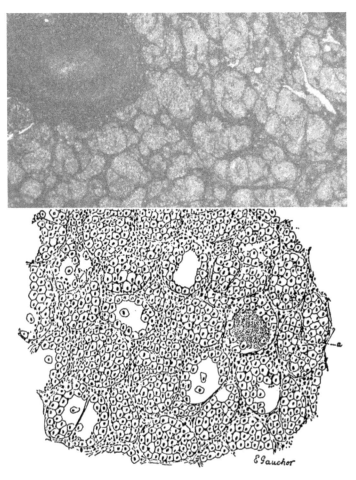

Figure 12.4   Above: a section of spleen with numerous rounded collections of Gaucher cells that are located in the red pulp. A single white pulp center is present, also termed a Malpighian corpuscle. Below: is Gaucher's illustration showing the identical features as he sketched in 1882.

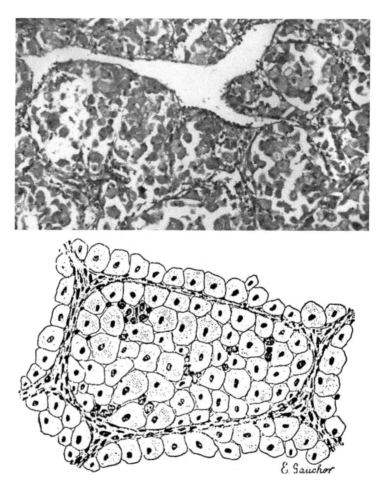

Figure 12.5   Above: A section of spleen with numerous Gaucher cells located in red pulp area. These cell collections compress a single venous sinus that will eventually flow into the portal vein of the liver. Below: Gaucher's signed illustration of a single splenic red pulp sinus.

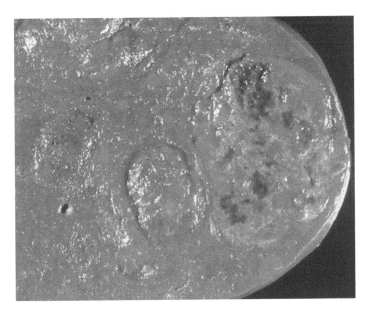

Figure 12.6   **(See color insert.)** A Gaucher patient was 37 years old when he had a chole-cystectomy. A liver biopsy was performed at that time and Gaucher cells were seen in the hepatic sinusoids. Fourteen year later, his spleen was removed and the spleen was enlarged to 1370 gms with parenchymal nodules that measured up to 4.5 cms. It is not rare that Gaucher spleens have these nodules. They can be composed of fibroblasts, erythroid cells of the marrow, and dilated capillaries. Later in life, this man developed carcinoma of the prostate.

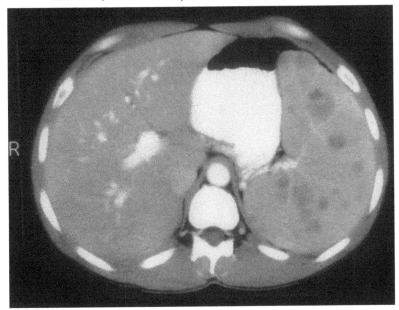

Figure 12.7   A 38-year-old man was examined for splenomegaly and pancytopenia. A CT scan with contrast revealed multiple splenic nodules. A bone marrow biopsy confirmed the diagnosis of Gaucher disease.

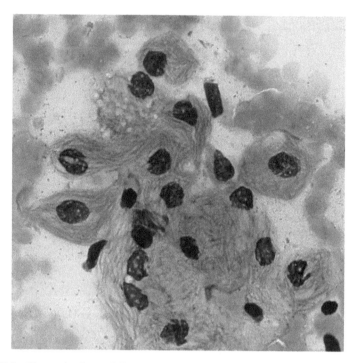

Figure 12.8 **(See color insert.)** These are Gaucher cells aspirated from bone marrow and stained with a Romanofsky stain. The diagnostic appearance relates to the large amount of cytoplasm with a wrinkled appearance and a benign appearing nucleus. This approach to the diagnosis via the marrow has been one of the early accomplishments of hematology.

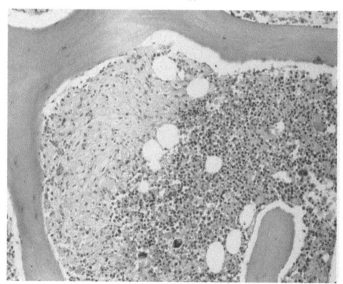

Figure 12.9 **(See color insert.)** Gaucher cells are clustered next to hyperplastic marrow. Excess iron is deposited in cells adjacent to a bony trabeculum.

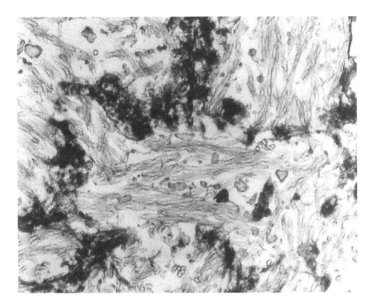

Figure 12.10   This is a typical electron microscopic appearance of the cytoplasm of a Gaucher cell. The wrinkled appearance of the cytoplasm seen by light microscopy is explained by the elongated and insoluble deposits that reside in the lysosomes. This patient presented with splenomegaly, anemia, and low platelets and was diagnosed at age 65 years. At age 71 he was found to have chronic lymphatic leukemia. He died with severe pneumonia. The autopsy: spleen 2100 g, liver 2675 g.

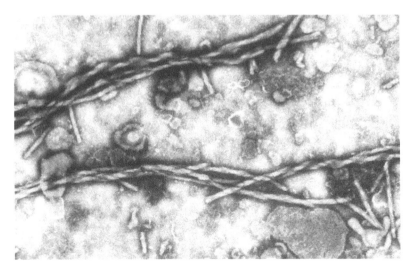

Figure 12.11   This is a the diagnostic appearance of a negative stain of Gaucher deposits showing long and twisted forms as viewed by electron microscopy. The stains used in this procedure are a heavy metal salt containing either uranyl, lead, or osmium.

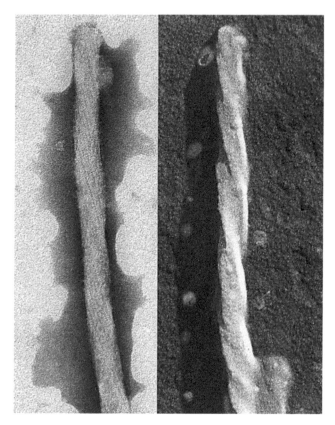

Figure 12.12   Left: a single negatively stained Gaucher deposit with uranyl acetate. Right: The shadowed deposit reveals that the deposit has a right-handed twist. The carbon-platinum shadow originated on the right.

Figure 12.13 Left: a discontinuous gradient with 3 densities using 0.3M, 1.5M, and 2.0M sucrose. A white layer resulted from prolonged centrifugation and was further purified using a continuous gradient. Right: A single white band exists in the continuous gradient between 0.7M and 0.85M sucrose that was free of other cellular membranes.

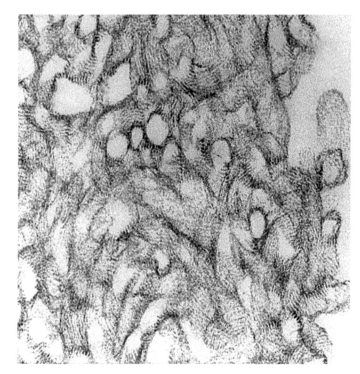

Figure 12.14 These are purified deposits that were prepared for electron microscopy by fixation and plastic embedding and were stained by osmium. The stain only involves the outer surface. Until we could look further into the deposits, we wrongly assumed that the deposits were hollow.

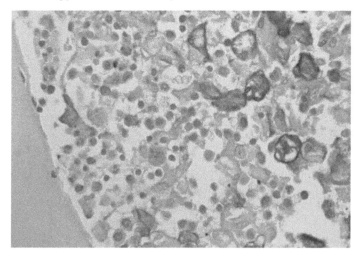

Figure 12.15 **(See color insert.)** This woman was diagnosed when she was pregnant at age 24 when the obstetrician palpated the enlarged spleen. More commonly the softness of the postpartum abdomen has revealed an enlarged spleen. The marrow was stained for iron to reveal an excess with varying amounts of iron in the Gaucher cells. Excess iron stores are not found in young Gaucher patients.

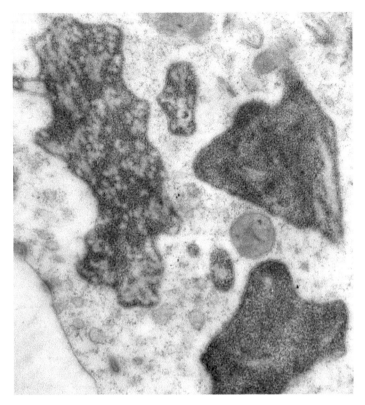

Figure 12.16   Electron microscopy of the Gaucher cell contains excess iron. The iron is in the form of ferritin and exists in the same lysosomes as the lipid deposits. Another possibility is that Gaucher cells hold on to iron more avidly than normal phagocytes.

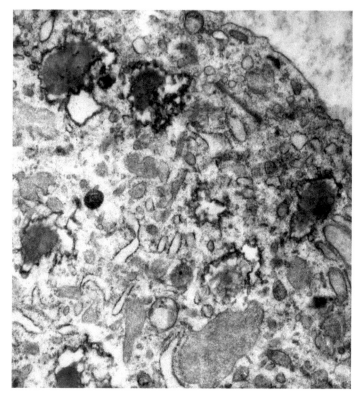

Figure 12.17   A Gaucher cell contains several wrinkled red cells within lysosomes. The excess iron in Gaucher cells may be partly explained by erythrophagocytosis.

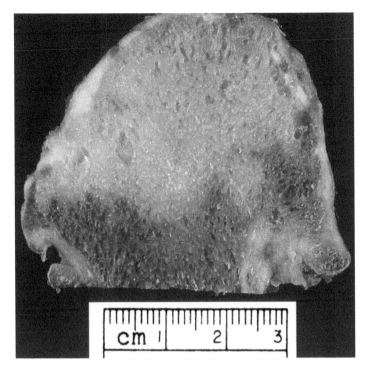

Figure 12.18   This deformed femoral head was removed from a 43-year-old man who was diagnosed at age 24 years while a medical student. The yellow discoloration and alterations in the cartilage are due to the necrosis. His spleen had not been removed.

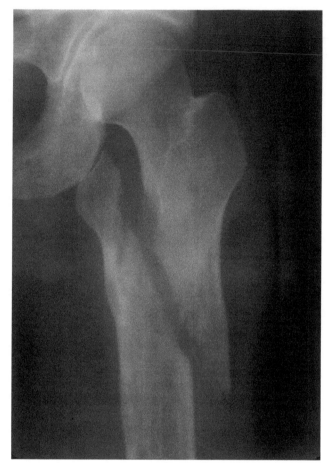

Figure 12.19 A woman sustained an intertrochanteric fracture of the left femur. She was diagnosed with Gaucher disease at age 7, splenectomized at 28 and had this fracture at 36. She subsequently developed a fracture of the tibia and, just before death at age 42, of the proximal head of the humerus.

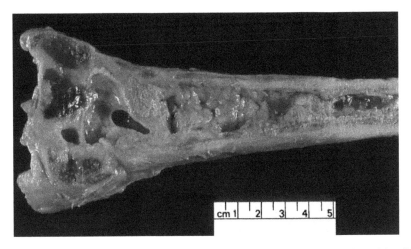

Figure 12.20   This is the result of an old cortical fracture in the proximal portion of the tibia. The marrow cavity contains markedly necrotic tissue.

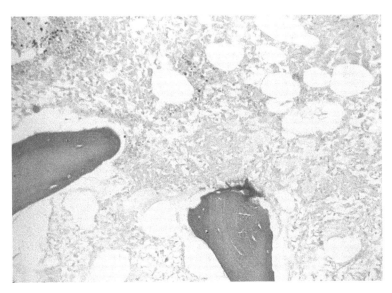

Figure 12.21   A 25-year-old woman had extensive disease that at autopsy involved liver, kidneys, and heart. Her diagnosis was made at age 8 and a splenectomy was performed at age 17 years. A biopsy of the left femur revealed necrosis of both bone and marrow.

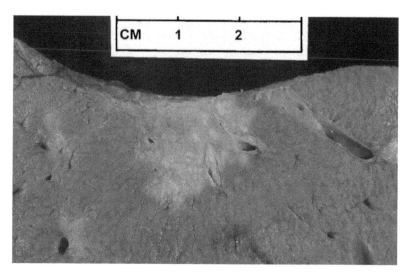

Figure 12.22 A 10-year-old boy received a liver transplant for progressive liver failure. His native liver seen here was involved with foci of necrosis as well as cirrhosis. He is now well and 16 years' posttransplant.

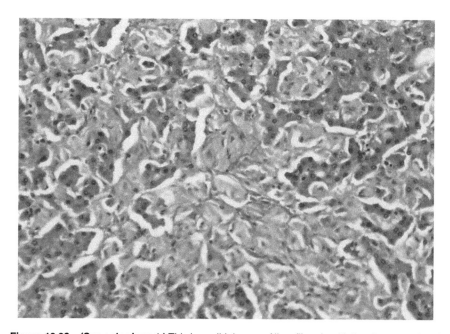

Figure 12.23 **(See color insert.)** This is a mild degree of liver fibrosis with the changes related to the presence of the Gaucher cells. Collagen is stained blue with a trichrome stain.

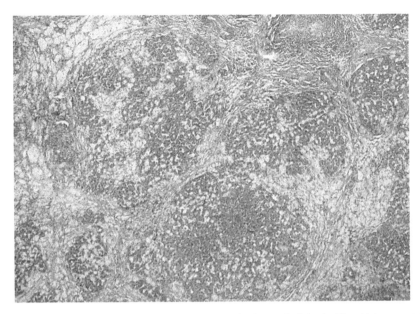

Figure 12.24 **(See color insert.)** This patient had advanced cirrhosis. The trichrome stain reveals the blue fibers of fibrosis. Fibrosis is one of the constants in Gaucher disease: wherever Gaucher cells are located, fibrous tissue is also present.

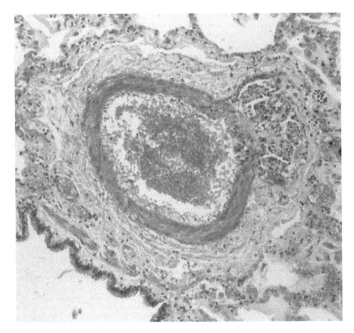

Figure 12.25 **(See color insert.)** Well-developed plexiform arteriopathy of the pulmonary artery was present at autopsy in an 18-year-old girl with evidence of pulmonary hypertension. She was diagnosed at age 2 years by splenectomy. The pale cells adjacent to the muscularis of the artery are Gaucher cells.

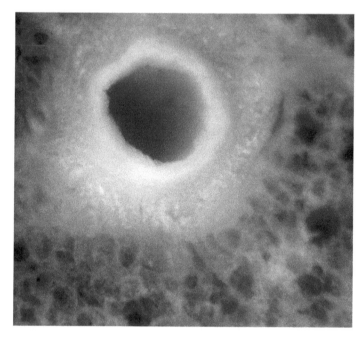

Figure 12.26   A pulmonary artery is surrounded by a collar of white tissue representing the Gaucher lipid, glucocerebroside, that is white when purified.

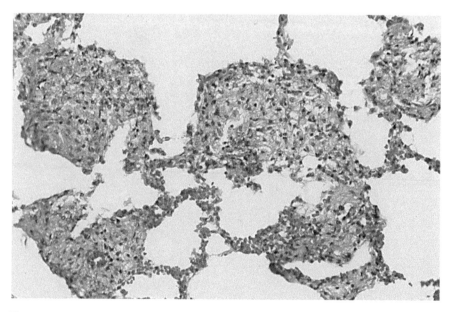

Figure 12.27   Clusters of Gaucher cells are located in the interstitial space of the lung in a 15-year-old girl who died immediately following splenectomy. Two siblings also had severe Gaucher disease.

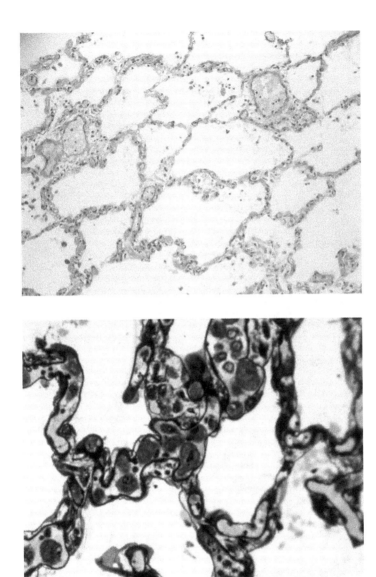

Figure 12.28   Above: This low power microscopic view of the lung appeared normal. At high power below: numerous Gaucher cells were found in alveolar capillaries.

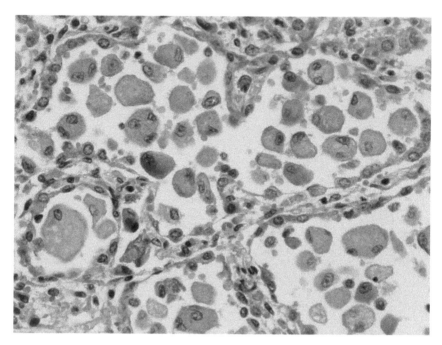

Figure 12.29   (See color insert.) These Gaucher cells were found in the alveoli of a 3-year-old boy who had marked cyanosis in his terminal course. His younger brother also had a similar course.

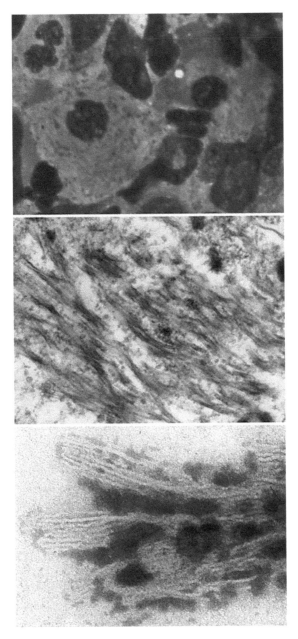

Figure 12.30 This group of three photos was made from marrow of a woman with chronic myeloid leukemia to show one of the variety of storage cells other than Gaucher cells. The upper photo shows a marrow storage cell with striated cytoplasm. The middle photo shows the microscopic appearance of the storage cells that is quite different to that of the Gaucher cells. Here the membranous bodies are curved and are not twisted. The lowest photo shows a cluster of negatively stained membranes that are curved, parallel, and not twisted.

CHAPTER **13**

Neuropathological Aspects of Gaucher Disease

Kondi Wong

CONTENTS

INTRODUCTION AND BACKGROUND

Neuropathology in Neuronopathic Gaucher Disease: The Classic Description

Neuropathology of Gaucher disease (GD) is the study of the central nervous system affected by a clinically and pathologically diverse and heterogenous disease directly and indirectly related to genetic mutations coding for the lysosomal enzyme glucocerebrosidase.[1] This results in variable functional and physiologic decreases in glucocerebrosidase activity associated with an accumulation of glycolipid substrates, primarily glucocerebroside and glucosylsphingosine.[2] As such, GD patients have widely varying degrees of severity in symptomatology, altogether different symptoms, rates of progression, anatomical regions affected, anatomical regions spared, and differing morbidity and mortality.

Typically, GD patients have been placed into three general categories based on their clinical presentation, time of onset, and rate of progression,[2,3] biochemical phenotypes,[4,5] functional enzyme activity levels, and other biochemical alterations. Genetic phenotyping of GD patients offers an additional level of categorization and sorting and has identified some mutations that are associated with and may portend a more severe or a milder course of disease, but is not always predictive of central nervous system (CNS) disease.[6] The crux of Gaucher disease type classification is whether the brain is affected or not, how severely, and the rate of progression of neuronopathic disease; the severity of CNS disease being indirectly proportional to survival. GD patients are most commonly affected by the type 1 form of disease with the central nervous system and brain unaffected clinically, pathologically and biochemically as classically described.[7,8] Patients with the neuronopathic forms of Gaucher disease have brain disease and differ in the onset of signs and symptoms, with the acute neuronopathic, type 2 form of GD patients developing disease in infancy. In patients with chronic, neuronopathic, type 3 GD, symptoms may commence at any age.[9]

NEURONOPATHIC GAUCHER DISEASE: BACKGROUND

A type of Gaucher disease affecting the central nervous system, the neuronopathic form[10] of Gaucher's disease (NPGD), was described by Philippe Gaucher.[11] The hallmark of neuronopathic Gaucher disease and one of the most consistently

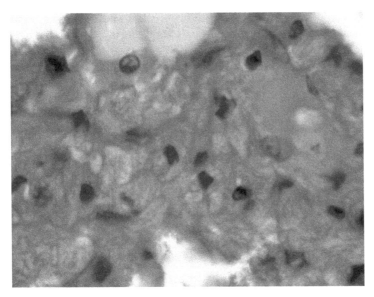

Figure 13.1 **(See color insert following page 304.)** Gaucher Cells: A collection of perivascular or periadventitial Gaucher cells stained with periodic acid-Schiff stain and
hematoxylin. The characteristic cytoplasm has been described as resembling
"crinkled tissue paper." (Periodic acid-Schiff with hematoxylin counter stain, 400
original magnification.)

identified pathologic features is the perivascular and periadvential accumulation of
lipid-laden macrophages, called Gaucher cells[2,12–15] (Figure 13.1). An extracortical,
discernible loss of neurons, sometimes associated with crumpled, shrunken-atrophic
neurons[12,14] has been reported involving the basal ganglia, nuclei of the midbrain,
pons and medulla, cerebellum, dentate nucleus, and hypothalamus.[14,16,17] A severe
neuronal loss and degeneration of pyramidal cell neurons of the hippocampal CA2
region, and to a slightly lesser extent, CA4 and CA3, have been described.[18] Remarkably, the hippocampal CA1 region, directly adjacent and juxtaposed to the hippocampal CA2 region, is largely spared of disease without notable neuronal loss or
degeneration.

Cerebral cortical pathology in neuronopathic Gaucher disease consisting of cortical laminar necrosis of the third and fifth cortical layers,[12,13,17] neuronal loss with
astrogliosis in layer IV, Gaucher cells in the visual cortex and occipital-temporal
lobes,[19,20] as well as parenchymal Gaucher cells lying free within cerebral cortex,
especially in the occipital lobes,[13,15,16] has been reported. Astrogliotic neuronal loss
of layer IV has been reported, more typical of the occipital lobe visual cortex than
other cerebral regions. More specifically, marked astrogliotic neuronal loss appears
to be confined to neurons of the calcarine visual cortex layer 4b. Laminar astrogliosis
of cerebral cortex layer 3 and 5 is a more diffuse process, affecting the parietal and
temporal lobes, including the cingulate gyrus and entorhinal cortex. The frontal and
occipital lobes have more sporadic and milder neuronal cell loss and astrogliosis of
cortical layer 3 and 5. Notably, sensorimotor cortex involvement by cortical layer 3
and 5 neuronal loss and astrogliosis has not been identified, and may be a brain

region relatively spared of pathological involvement in neuronopathic Gaucher disease. Finally, nonspecific gray and white matter gliosis[13,14,20] and microglial proliferation have been noted.[14,16] Neuronal storage of glucosylceramide (GlcCer)[13,14] has also been described, but the identification of GlcCer tubular structures in neurons is rare.[21]

Although by definition type 1 Gaucher disease does not involve the brain, there are reports of CNS involvement in patients who would otherwise be considered to have type 1 Gaucher disease. These patients are asymptomatic until well into adulthood and the pathological CNS findings are much milder.[8,18,22,23] Late onset Parkinson's disease or parkinsonian symptoms in patients diagnosed as having type 1 GD have been noted incidentally in early reports,[15,24–26] but recent reports causally linking clinical and pathologic Parkinsonism and dementia with Gaucher disease of all types has been suggested.[18,26–33]

CELLULAR NEUROPATHOLOGY IN GAUCHER DISEASE

Astrocytes and neurons are the main cell types affected in Gaucher disease. The most common astrocytic change is perivascular astrogliosis of gray and white matter. Perivascular astrogliosis consists of thickened astrocytic processes outlining the cerebral capillaries and/or perivascular fibrillary astrocytes with spherically radiating thin glial processes.

Fibrillary astrogliosis is a reaction of astrocytes to nonspecific injury and insults resulting in an increased expression of glial fibrilllar acidic protein intermediate filaments in thin glial processes radiating outward spherically with the astrocyte cell body and nucleus at the center. In fibrillary gliosis of the brain parenchyma, astrocytes label with glial fibrillary acidic protein (GFAP) and are evenly dispersed with varying cellularity and density of staining, thickness, and length of glial processes. Fibrillary astrogliosis accompanies perivascular astrogliosis but is seldom present without perivascular astrogliosis.

The third type of astrogliosis in Gaucher disease is the gemistocytic type. Gemistocytic astrogliosis is astrogliosis in which the astrocyte cell body is swollen or "stuffed" with GFAP intermediate filaments. Gemistocytic astrogliosis may form a near-solid mass of tissue as a gliotic, scarring response to severe injury; however, this pattern is not typical of Gaucher disease. Gemistocytic astrogliosis occurs in localized regions where there is the greatest injury response, i.e., CA2 hippocampal region in Gaucher disease, isolated scattered foci in white matter, striatum, brainstem, and cerebellum. In Gaucher disease of all types, perivascular gliosis is present in white matter centrum ovale, white matter tracts and interspersed gray matter of the striatum, cerebellar white matter, and the brainstem interlaced with white matter tracts and interconnected brainstem nuclei. Parenchymal fibrillary astrogliosis appears to accompany perivascular gliosis in localized areas where the density of astrogliotic staining is stronger and denser. Gemistocytic astrogliosis was only identified in neuronopathic localized areas of injury with neuronal cell loss.

Neuronal changes in Gaucher disease consist of neuronal loss, degenerative changes, and neuronal inclusion. Neuronal loss is difficult to quantify in most areas of the brain such as the cerebral cortex, and subcortical gray matter, thalami and striatum, except in the most severe cases. Loss of neurons can be most reliably and easily quantified in areas of the brain where the density of neurons can be reliably observed, such as the hippocampal CA regions and dentate, specific lamina of cerebral visual cortex, and certain midbrain and brainstem nuclei.

Degenerative changes in the brain in Gaucher disease consist of basophilia of the neuronal cytoplasm, loss of fine detail in the neuronal nucleus giving a smudged appearance, shrunken atrophy of the neuron cell body with pink hyalinization or basophilic hue to the cytoplasm. Degenerative changes of neurons in Gaucher disease may also result in formation of neuronal inclusions. On hematoxylin and eosin (H&E) stain, rare tubular or vesicular inclusions in neuronal cytoplasm suggestive of tubular structures in Gaucher cells may be seen. Cerebral neurons (primarily in cerebral cortical layer 5) with cortical-type Lewy bodies in Gaucher disease associated Lewy Body Dementia (Diffuse Lewy Body Disease) have pink, eosinophilic, reniform to round cytoplasmic inclusions. Likewise, brainstem Lewy bodies of the substantia nigra, neuromelanin containing neurons and brainstem-type Lewy bodies of the pyramidal cell neurons of CA2–4 of the hippocampus have large, round Lewy bodies. Brainstem-type Lewy bodies are larger, round, well defined, cytoplasmic bodies with a clear or pale halo, concentric rings of radiating spicules or lines of material and a central hyaline core when the Lewy body is sectioned at its midpoint. Cortical Lewy bodies are smaller, reniform, less well defined, may be amorphous, usually do not have the concentric structure, hyaline core and the halo.[34] Rarely, neuronal appearing cells having abundant cytoplasm engorged with vesicular and tubular inclusions are identified on H&E. Although such cells appear to be neurons with Gaucher tubular inclusions, it is difficult to convincingly prove the assumption, morphologically, antigenically or ultrastructurally.[20,35]

Other neuronal inclusions are best demonstrated with special staining techniques. A standard Alcian blue/Periodic acid-Schiff (AB/PAS) stain or one of the modifications of the AB/PAS stains, when properly performed,[18] stains the neuronal and glial cell cytoplasm pink whereas perivascular/periadvential Gaucher cells stain blue. Unusually, single and groups of neurons in the striatal nuclei take on a blue hue or may be brilliant sky-blue in coloration identical to the coloration of Gaucher cells (Figure 13.2B, Figure 13.2C, and Figure 13.2D). On occasion, some of the "Alcian blue neurons" may also contain tubular vesicular inclusions (Figure 13.2A, Figure 13.2D). The Alcian blue neurons are unequivocal neurons, identified by neuronal nuclei, cytoplasm, prominent extended dendritic and axonal projections (Figure 13.2D). They have only been identified in neuronopathic Gaucher disease patients.

Synuclein has become a valuable aid in the study of Gaucher disease because of the association of Gaucher Disease Associated Synucleinopathies corresponding to clinical Parkinsonism and/or clinical dementia of the Diffuse Lewy Body Dementia (DLBD) type and type I Gaucher disease. Interestingly, in a number of populations studied, a relatively high frequency of patients with idiopathic Parkinson's disease have one allele with a Gaucher gene mutation or polymorphism.[30,31,36,37]

Synuclein is more sensitive than standard hematoxylin and eosin (H&E) in detecting synuclein inclusions in DLBD and Parkinson's disease. In addition to labeling the cortical and brainstem Lewy bodies seen on H&E stained histologic

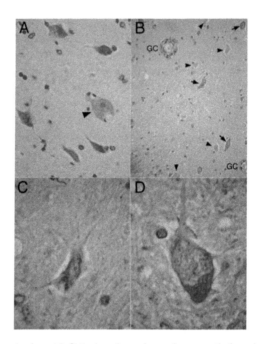

Figure 13.2   (See color insert.) Cytoplasmic and membrane pathology in type 2 Gaucher's
disease patient neurons: (A) One apparent globus pallidus neuron (arrowhead)
is filled and distended by tubular–filamentous structures. The cell is favored to
be neuronal rather than monocytic because of the centrally located nucleus that
is not displaced to the side, spacing of other neurons, the large size of the cell
without an accompanying entourage of perivascular Gaucher cells (H&E 200
original magnification). (B) Low power view of globus pallidus with blue perivascular Gaucher cells (GC), the usual pink (periodic acid–Schiff staining) neurons,
and blue (alcian blue staining) neurons that tinctorially stain the same as Gaucher's cells. Alcian blue 8GX/PAS at pH 2.5 normally stains neurons pink, but in
Gaucher's disease, isolated neurons, in rare instances, may exhibit blue staining
characteristics identical to Gaucher's cells that normally stain blue (alcian blue
8GX/PAS, pH 2.5, 100 original magnification). (C) High power view of an alcian
blue stained neuron with blue cytoplasm, blue neuronal membrane and neuronal
processes (alcian blue 8GX/PAS at pH 2.5, 400 original magnification). (D) Close-
up view of a neuron with proximal axonal process. Alcian blue stains the membrane and the internal cytoplasmic contents of the neuron. The internal cytoplasmic structures appear to have a convoluted tubular–filamentous structure (alcian
blue 8GX/PAS at pH 2.5, 400 original magnification).

sections, synuclein will be immunoreactive to aggregates of synuclein in the proximal axon, synuclein in early formative stages of Lewy bodies, and smaller cytoplasmic synuclein inclusions of varying morphology and size. In an anecdotal
observation, unusual, small punctate collections of synuclein or small vesicular rings
of synuclein immunoreactivity were seen in the cytoplasm of pyramidal cell neurons
of the CA4–2 of the hippocampus in one patient (not shown).

An antiglucocerebrosidase antibody is available and potentially useful for the
study of Gaucher disease patients. However, the specificity of the antibody is a
problem in that the mutated enzyme is variably and weakly labeled by the antibody
in most cases. However, the glucocerebrosidase antibody is useful for identifying

the pattern of glucocerebrosidase enzyme labeling in normal control patients. In one Gaucher disease patient in which the antibody labeled the mutated enzyme, a quantitative pattern of distribution in the hippocampus was split between the hippocampal CA4–2 pyramidal cell neuron region and the CA1 region.

The Bielschowsky stain is useful for confirming the lack of neuritic plaques in the cerebral cortex and hippocampal CA1 region, and the lack of significant numbers of neurofibrillary tangles, characteristic of Alzheimer's disease. The Bielschowsky stain is also useful for identifying dystrophic neurites in the hippocampal (HPC) CA2 region described as a feature of Diffuse Lewy Body Disease, independent of (unassociated with) Gaucher disease. When HPC CA2 region dystrophic neurites are identified in Gaucher disease patients, it is in the context of dementia and Diffuse Lewy Body Disease.

REGIONAL NEUROPATHOLOGY IN GAUCHER DISEASE

Hippocampus

The hippocampus is probably one of the most, if not the best, studied regions in the human brain. This is because the hippocampus is involved in many diseases and also largely in part because the well-characterized regions and lamina of the hippocampus allow assessment of neuronal and glial changes, neuronal degeneration, inclusions and neuronal loss before end-stage disease ensues. The underlying general pattern of hippocampal disease in Gaucher disease lies in the contrasting selective vulnerability of the CA2–4 regions as compared with the CA1 region. Neuropathologic changes on H&E staining, immunoreactivity with glial fibrillary acidic protein, glucocerebrosidase, synuclein, and studies with Ca^{2+} uptake autoradiography serve to probe the underlying cause of this pattern of disease.

Hippocampal Pyramidal Cell Neuron Loss and Gliosis in Gaucher Disease

The hippocampus is affected in Gaucher disease patients of all types. In type 1 Gaucher disease patients, the regions affected are gliotic with perivascular and fibrillary astrogliosis that may be mild and not apparent on H&E staining but readily demonstrable on GFAP immunostaining. In type 2 and 3 Gaucher disease patients, the same regions are affected and the same regions are relatively spared pathologic changes.

The pattern of Gaucher disease neuropathology in the hippocampus (HPC) is hippocampal, CA2, CA3, and CA4 regions affected with relative sparing of the CA1 region (Figure 13.3A and Figure 13.3B) in all GD patients. In type 2 and 3 GD patients, the CA2 region is most severely affected with severe neuronal loss and associated dense fibrillar and occasionally gemistocytic astrogliosis (Figure 13.3D). The CA3 and CA4 regions are also diseased with moderate neuronal loss and astrogliosis (Figure 13.3A and Figure 13.3B). In contrast, the CA1 region of the hippocampus, affected and targeted in the vast majority of brain diseases such as

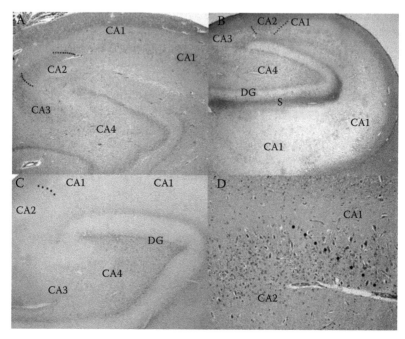

Figure 13.3   Hippocampal CA4-2 pyramidal cell neuron loss in neuronopathic Gaucher disease and astrogliosis in adult type 1 GD: A. Type 2 Gaucher disease. Hippocampal pyramidal neuron cell loss and astrogliosis are most severe in CA2 (region bracketed between rows of dots), moderate to severe in CA3, and moderate in CA4. The CA1 region of the hippocampus has minimal background gliosis and is largely spared of significant pathology. (H&E, 40 original magnification.) B. Type 1 Gaucher disease. Hippocampal pyramidal neuron cell loss is not readily discernible and is, at most, minimal. Hippocampal CA2 region is strongly and densely glial fibrillary acidic protein (GFAP) immunoreactive. CA3 and CA4 also have moderate GFAP immunoreactivity whereas the CA1 region only has background, scattered, perivascular immunoreactivity. DG denotes the dentate gyrus. S denotes the stratum lacunosum, a region that is often GFAP immunoreactive. (GFAP immunoperoxidase, 40 original magnification). C. Type 3 Gaucher disease. The patient's symptoms included developmental delay, supranuclear gaze palsy and mild dysphagia, but no myoclonus or seizures. The patient died during bone marrow transplantation. CA4 region has mild gliosis and GFAP immunoreactivity that extends into CA3. CA2 and CA1 do not have an appreciable increase in GFAP immunoreactivity. (GFAP immunoperoxidase, 40 original magnification). D. Type 2 Gaucher disease (Patient 10). Close-up of hippocampal CA2–CA1 interface. The few remaining hippocampal pyramidal cell CA2 neurons are basophilic and shrunken. Astrogliosis with eosinophilic astrocytes and glial processes is prominent. The CA1 region at the interface has mildly affected pyramidal cell neurons (slightly basophilic) and mild astrogliosis 150 μm from the CA2–CA1 interface. The CA1 region has no significant changes above background (not shown). (H&E, 100 original magnification.)

ischemia, Alzheimer's disease, hippocampal sclerosis, etc., and directly adjoining the CA2 region, did not have any detectable neuronal loss or astrogliosis.

In type 1 Gaucher disease patients, the same hippocampal regions are affected, but significant neuronal loss and degeneration is absent. The hippocampal CA2 region has marked astrogliosis and the CA3–4 regions have moderate astrogliosis of CA3–4. Again, the hippocampal CA1 region is relatively spared of disease with no increased astrogliosis of CA1 (Figure 13.2B), above the low level of background. This pattern of astrogliosis is demonstrable in long-lived, type 1 Gaucher disease patients without any neurologic deficits (Figure 13.6D).

The astrogliosis in regions of the hippocampus are different than in ischemia/hypoxia or in hippocampal sclerosis associated with seizures. In the latter disease, astrogliosis accompanies, or is a result of neuronal loss. In Gaucher disease, astrogliosis may be prominent with fibrillary and perivascular astrocytes. Gliotic GFAP immunoreactive astrocytic processes lining along a capillary as part of the blood brain barrier along one aspect, and extending another set of processes toward a viable, pyramidal cell HPC neuron, may also be present. Pyramidal neuronal cell loss, if present, is not readily identifiable in type 1 Gaucher disease.

Hippocampal Antiglucocerebrosidase Immunohistochemistry

A high-specificity glucocerebrosidase antibody has been used to label the hippocampal region neurons to explore the functional relationship between the region pattern of neuronal damage and level of glucocerebrosidase expression. Neuronal loss and neuronal degenerative changes in neuronopathic GD and astrogliosis in all GD types, appears to target the hippocampal CA4–2 region, but relatively spares the adjoining CA1 region. Glucocerebrosidase enzyme is mutated and with variable loss of enzyme activity in GD and differential glucocerebrosidase enzyme levels also appear to differ in the CA4–2 region as compared with the adjoining CA1 region.

The hippocampal CA2–4 region pyramidal cell neurons normally have strong, dense immunoreactivity to a highly specific glucocerebrosidase antibody. In hippocampal CA2–4, almost all the neurons have strong, dense immunoreactivity (Figure 13.4A, Figure 13.4B, and Figure 13.4C). In contrast, CA1 pyramidal cell neurons have sparser glucocerebrosidase immunoreactivity, scattered in a smaller percentage of neurons (Figure 13.4A, Figure 13.4B, and Figure 13.4D). Because of the specificity of the glucocerebrosidase antibody, the antibody does not reliably or consistently label mutated glucocerebrosidase enzyme in Gaucher disease patients. However, in one patient with type 1 GD, the mutated glucocerebrosidase enzyme was immunoreactive and demonstrated the same pattern of strong dense immunoreactivity in CA4–2 and sparser, scattered reactivity in CA1 (Figure 13.4B and Figure 13.4D).

In the hippocampal CA4–2 region, dense antiglucocerebrosidase labeling indicating high enzyme levels in CA4–2 may serve to tightly regulate GlcCer levels in the CA4–2 pyramidal cell neurons. Patients with Gaucher disease have decreased glucocerebrosidase activity and resultant elevated glucocylceramide in the CA4–2 region, may increases spontaneous calcium induced calcium release (CICR) by up to 300%, and cause ryanodine calcium channel dysfunction[38] as is further discussed in the next section.

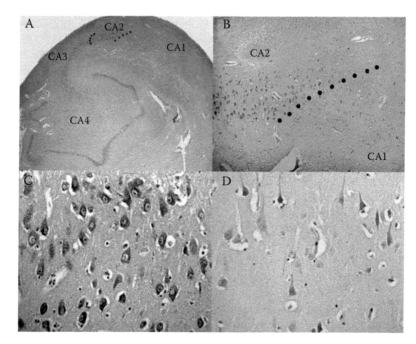

Figure 13.4 Antiglucosylceramidase immunoreactivity is most intense in CA4–2 region: A. Normal human control, hippocampal formation. Immunoreactivity to 8E4 antiglucosylceramidase antibody is strong in the pyramidal cell neurons of the CA2, CA3, and CA4 regions. CA1 has much weaker immunoreactivity. (8E4 immunostain, 40 original magnification.) B. Normal human control, hippocampal, CA2–CA1 interface. Strong immunoreactivity to 8E4 antiglucosylceramidase antibody is present in the pyramidal cell neurons of the CA2 region whereas the CA1 pyramidal cell neuron staining dense is sparse. (8E4 immunostain, 100 original magnification.) C. Type 1 Gaucher disease. Hippocampal CA2 region. CA2 pyramidal cell neurons have dense, granular, cytoplasmic 8E4 immunoreactivity. (8E4 immunostain, 200 original magnification.) D. Type 1 Gaucher disease. Hippocampal CA1 region. Some CA1 pyramidal cell neurons have mild to moderate 8E4-glucosylceramidase immunoreactivity, but other neurons lack significant immunoreactivity. (8E4 immunostain, 200 original magnification.)

Hippocampal ^{45}Ca^{2+} Uptake Autoradiography Elucidates Functional Link Between CA4–2 Neuropathology and Increased Risk of CA4–2 Neuronal Excitatory Cytotoxicity Induced by Excess Glucosylceramide

Glucosylceramide has other effects such as sensitization of ryanodine receptors (RyaR) and potentiation of CICR. The localization of RyaR receptors and CICR susceptibility in the hippocampus may play a key role in contributing to the hippocampal pattern of disease in GD.

Earlier studies had already found an IP3-sensitive pool and a CICR-sensitive pool in the hippocampus.[39–41] A set of similar studies to address the CA4–2 and CA1 regions demonstrated that the CA4–2 region neurons display predominantly CICR, whereas the CA1 region neurons are inositol triphosphase IP$_3$-sensitive and CICR-insensitive[18] (Figure 13.5A, Figure 13.5B, and Figure 13.5C). The CICR-sensitive

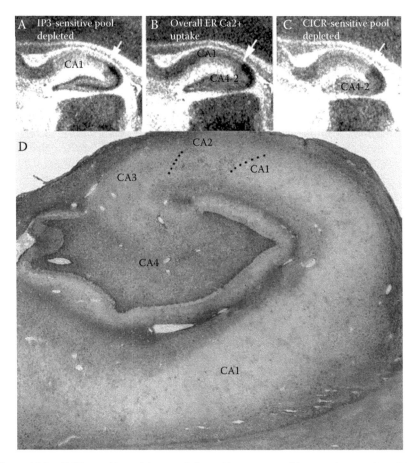

Figure 13.5   CICR sensitive calcium pools in rat hippocampus CA4–2 region matches CA4–2 region astrogliosis in type 1 GD: Selective enrichment of CICR in rat hippocampal subregions. Endoplasmic reticulum (ER) calcium pools were localized in sagittal rat brain sections using the $^{45}Ca^{2+}$ autoradiography. Darker areas reflect higher amounts of ER sequestered $^{45}Ca^{2+}$. The approximate boundary distinguishing the CA1 region from the CA4–2 regions is indicated by the white arrows. $^{45}Ca^{2+}$ uptake performed in the presence of 10 μM IP3 to selectively deplete the IP3-sensitive pools. CA1 region in Figure A is lighter density than CA1 in Figure B, indicating IP3-sensitive calcium depletion. B. Overall ER $^{45}Ca^{2+}$ uptake. C. $^{45}Ca^{2+}$ uptake performed in the presence of 10 mM caffeine to deplete the CICR-sensitive pools. CA4–2 region in Fig. C is lighter density than CA4–2 region in Figure B, indicating caffeine-sensitive calcium depletion of CICR regions. Figures A–C document that the CICR pools in the CA4–2 regions and the dentate gyrus are characterized by their relative resistance to IP3, but sensitivity to caffeine. D. Type 1 GD, nonneurologic, hippocampus: Marked astrogliosis in CA4–2 regions (CICR-sensitive region) contrasts with relatively spared CA1 region (IP3-sensitive region). Dense GFAP immunoreactive astrogliosis of abuts directly against the dentate granular cells neurons. (GFAP 200 original magnification.)

characteristic of CA4–2 hippocampal pyramidal cell neurons is a significant additional component of the overall selective vulnerability to cytotoxic injury in Gaucher disease.

Selective Vulnerability of Hippocampal CA4–2 Pyramidal Cell Neurons in Gaucher Disease

In concert with the native susceptibility of the hippocampal CA4–2 regions to injury through hyperexcitability states, the CA4–2 neurons are targeted due to differential localization of CICR-sensitive Ca^{2+} pools, ryanodine receptor binding sites, ryanodine receptors (RyaR), and glucocerebrosidase enzyme levels.

The neuroanatomical morphological distinction between the CA1 and CA4–2 regions were observed by Cajal who described the area now known as CA1, as "regio superior," large cell region, and CA3 as "regio inferior," small cell region.[42] The neural projections, degree of glutaminergic exposure, and function of hippocampal CA1 pyramidal cell neurons differ in marked contrast from CA4-2 neurons.

The CA1 region of the hippocampus receives input from a different entorhinal cortical layer (CA1 III, CA2–4 layer II) and the returning, recurrent axonal projection array[43] is comparatively weak and sparse[44,45] without any mossy fiber input.[45] CA1 hippocampal pyramidal cell neurons also lack branches of collateral projections distributed in CA1,[46, 47] and the massive association network, characteristic of CA3 is, for the most part, absent in CA1.[45]

In comparison, the CA4–2 region, neurons have significant, recurrent, excitatory glutaminergic projections and receive mossy fiber input from the dentate gyrus granule cells (excepting CA2).[48,49] For example, each hippocampal CA3 pyramidal cell neuron on the average contacts 30,000 to 60,000 neurons in the ipsilateral hippocampus and receives 25,000 excitatory glutaminergic synapses.[50,51] Mossy fibers, which have the second largest synaptic structures in the mammalian CNS, are excitatory, glutaminergic, and have a very large input into CA3, as well as CA4. The CA2 region receives recurrent, excitatory feedback from CA3 neurons.[50,51]

The predominant positive feedback characteristic in the CA4–2 regions,[45,52] in concert with the intrinsic bursting properties of hippocampal CA3 pyramidal cell neurons, produces a hyperexcitable state that predisposes the CA4–2 region to large synchronous discharges, initiated within the CA2 or CA3 regions.[43,53–55]

The end result is a coordinated susceptibility to glutaminergic excitatory injury by hyperexcitation that is usually mitigated and moderated by an equally strong GABAergic interneuron inhibition. But the strength of the opposing forces of excitation and inhibition produces a climate in which subtle alterations, such as small changes in firing properties of neurons, can tip the balance and override inhibitory control. Loss of inhibitor control results in uncontrolled excitatory electrical activity (seizures) that spread into CA1[45] and the pathologic changes of neuron loss and gliosis seen in hippocampal sclerosis.[56]

Increased Levels of Glc-Cer Increases Sensitivity of CA4–2 Specific RyaR Receptors and Potentiates Spontaneous Neuronal Discharges

Astrogliosis is a response to a variety of insults to the brain.[57] In hippocampal sclerosis associated with hypoxia or ischemia, the entire hippocampus can be gliotic with degenerative changes.[56] In mild to moderate cases, the mid to proximal portion of CA1 is gliotic and affected by seizure activity that originates in the CA3–2 excitatory, glutaminergic feedback loops.[45]

In Gaucher disease, a different set of disease related modulating factors is superimposed upon the functional anatomic susceptibility to aberrant excitatory disease states. This results in a different set of neuropathologic changes and alters the localization of pathology by targeting the CICR-susceptible neuronal population and augmenting the sensitivity of selected calcium channel receptors in CA4–2.[39–41,58] Biochemical selective vulnerability of CA4–2 is conferred by the localization of certain Ca^{2+} pools, receptors, and binding sites in the CA4–2. A potential key mechanism to cytotoxic injury in the CA4–2 region in Gaucher disease is the effect of increased intracellular glucocylceramide (GlcCer). GlcCer specifically modulates the ryanodine receptor (RyaR) and behaves in a messenger-like fashion, perhaps through a redox receptor mechanism.[39–41,58] At physiologically increased levels, GlcCer both increases the sensitivity of the ryanodine receptor to stimuli and augments the RyaR response to stimuli such as spontaneous CICR spikes or electrical discharges. In other words, an accelerated response occurs, greater in magnitude as well as a lower threshold response to stimuli results from increased cellular GlcCer. Thus, CICR spikes or electrical events have a greater likelihood of triggering a response and each response is more likely to be an accelerated response of greater magnitude resulting in a potentiation of CICR-sensitive regions.[38,59] In contrast, CA1 region pyramidal cell neurons belonging to the IP3-sensitive Ca^{2+} pools[39–41,58] co-localize with the sarcoplasmic/endoplasmic reticulum Ca^{2+}-ATPase (SERCA),[58,60] which is CICR-insensitive is not modulated by glucocylceramide. Thus, CA2–4 targeting and CA1 sparing are affected by the biochemical, neuronanatomical functional characteristics of hippocampal CA4–2 neurons in patients with Gaucher disease.

Hippocampal Pathology in GD Patients with Parkinsonism and Dementia; a Gaucher Disease Associated Synucleinopathy with Parkinsonism and Dementia

An initial four patients with type 1 Gaucher disease, Parkinsonism, and dementia were studied and described as an unusual variant of Gaucher disease.[18,30] As a category, the patients are probably best described as having a "Gaucher disease associated synucleinopathy with Parkinsonism and dementia." This is because the main neuropathologic finding is the neuronal deposition of synuclein inclusions in various typical as well as unusual locations. The synuclein inclusions are deposited in specific neuropathologic loci and pathologic patterns corresponding to Parkinson's disease, diffuse Lewy body disease, and recently described hippocampal CA4–2 lesions in Gaucher disease.

Half of the patients with type 1 GD and Parkinsonism with dementia[29] had numerous, hippocampal CA4–2, intraneuronal, synuclein-immunoreactive inclusions (Figure 13.6A, Figure 13.6B, Figure 13.6D, Figure 13.6E, and Figure 13.6F), morphologicaly indistinguishable from the typical brainstem-type Lewy bodies found in the substantia nigra of idiopathic Parkinson's disease patients. The Lewy bodies inclusions were absent in CA1 (Figure 13.6C). A number of hippocampal pyramidal cell neurons embedded in the stratum lacunosum moleculare near the most proximal CA1 region also had hippocampal "brainstem type" Lewy bodies (HPC-LB). One patient, in addition to CA4-2 HPC-LB, had cortical Lewy bodies characteristic of Diffuse Lewy Body Dementia (DLBD), and substantia nigra brainstem-type Lewy bodies characteristic of Parkinson's disease and DLBD.

Some type 1 GD and Parkinsonism with dementia patients lacked hippocampal "brainstem-type" Lewy Bodies (HPC-LB). One patient had marked neuronal loss of substantia nigra (SN) neurons accompanied by SN brainstem-type Lewy bodies. Another patient had, in addition to SN brainstem-type Lewy bodies, numerous cortical-type Lewy bodies in the temporal lobe entorhinal cortex and cingulate gyrus, consistent with Parkinson's disease and Diffuse Lewy Body Dementia (DLBD). This patient with diffuse Lewy body disease had, in addition, mild CA2 neuronal loss and dystrophic neurites in the CA2–3 region, findings characteristic of DLBD patients.

In summary, some clinical-pathologic patterns of disease may be discerned in Gaucher disease (GD) associated synucleinopathy with Parkinsonism and dementia patients. One group of patients has hippocampal Lewy Bodies, HPC-LB with or without other associated synucleiopathy. The group of patients lacking hippocampal Lewy bodies has Diffuse Lewy Body Dementia. Notably, three of these patients had the homozygous N370S mutations, a mutation that is thought to be neuroprotective. The coincidental crossroads of finding a rare neuroanatomical disease location, CA2, typical of a rare disease (GD), affected by neuronal inclusions characteristic of Parkinson's disease matching the pattern of glucocerebrosidase-staining density, is highly suggestive of a common pathogenic mechanism.

Epidemiological studies have also shown a strong correlation in a number of populations studied. The epidemiology of Gaucher disease associated Parkinson's disease is addressed elsewhere, therefore, only a few notable points regarding the subtle pathologic mechanism of disease will be addressed here.

In a number of studies, a surprisingly high frequency of patients, 14%, 21%, and over 31% of patients with sporadic, idiopathic Parkinson's disease had alterations in the glucocerebrosidase sequence consisting mostly of heterozygote carriers, a few known polymorphisms and a few homozygous patients.[30,31,36] The statistically significant association between the Gaucher gene and Parkinson's disease suggests the presence of a native, biochemical interaction between the glucocerebrosidase enzyme and/or glucosylceramide and synuclein metabolism that when perturbed, results in synucleinopathies such as Parkinson's disease and Diffuse Lewy body disease. The association notably involves genetic carriers and polymorphisms. In this context, the detection, presence, or absence of gross genetic abnormalities in the glucocerebrosidase gene is a crude measure of the fine biochemical interactions necessary to avoid metabolic disequilibrium leading to synuclein aggregation. The

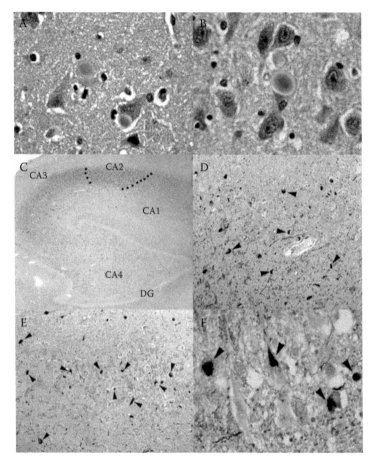

Figure 13.6 Brain stem-type Lewy bodies in hippocampal pyramidal cell neurons of the CA4–2 region: A. Type 1 Gaucher disease, hippocampal CA2 region. Large brain stem-type, intracytoplasmic, Lewy bodies are present in CA2 pyramidal cell neurons. (H&E, 400 original magnification.) B. Type 1 Gaucher disease, hippocampal CA3 region. Large brain stem-type, intracytoplasmic, Lewy bodies are present in CA3 pyramidal cell neurons. Brain stem-type Lewy bodies differ from "cortical Lewy bodies" in that the former have well formed, concentric, laminar layers of radiating fibrillar material, often a central strongly eosinophilic hyaline core, and a well-defined "halo" or lightly staining, outer fibrillar ring. (H&E, 400 original magnification.) C. Type I Gaucher disease, hippocampal formation. Synuclein immunoreactive, intracytoplasmic, pyramidal cell neuron inclusions labeling formative and mature Lewy bodies are abundant in the hippocampal CA4–2 regions. Except for a few inclusions at the CA2–CA1 interface and pyramidal cells in the stratum lacunosum, CA1 is devoid of synuclein immunoreactive Lewy body inclusions. (Synuclein immunoperoxidase, 40 original magnification.) D. Type 1 Gaucher disease, hippocampal CA4 region. Synuclein immunoreactive, intracytoplasmic, pyramidal cell neuron, inclusions (arrowheads) are abundant in the hippocampal CA4 region. (Synuclein immunoperoxidase, 200 original magnification.) E. Type I Gaucher disease, hippocampal CA2 region. Synuclein immunoreactive, intracytoplasmic, pyramidal cell neuron, inclusions (arrowheads) are also plentiful in the hippocampal CA2 region. (Synuclein immunoperoxidase, 200 original magnification.) F. Type 1 Gaucher disease, hippocampal CA2 region, close-up. Synuclein immunoreactive inclusions fill the cytoplasm of pyramidal cell neurons. Some of the inclusions (far left arrowhead) are consistent with the morphology of a brain stem-type Lewy body. (Synuclein immunoperoxidase, 400 original magnification.)

interactive biochemical pathways of disease are intact in patients without genetic sequence abnormalities and other triggers such as age, mitochondrial disease, cytotoxic exposures, etc., can be the causative factor of disequlibrium leading to synuclein-related disease. Thus, the lack of glucocerebrosidase gene mutation association with Parkinsons disease in a population[61] does not necessarily mean that the underlying mechanism of pathogenesis is void.

CEREBRAL CORTICAL PATHOLOGY

Overall, the cerebrum is least affected in Gaucher disease. In neuronopathic Gaucher disease, the white matter gliosis of the centrum ovale and accumulation of periadvential Gaucher cells is diminished in the cerebral cortex. Although the cerebral cortex pathology in a given patient, namely astrogliosis, Gaucher cells, and neuronal loss, is attenuated compared to pathologic changes in the adjacent white matter and in the brain stem, the cortex is not completely spared of disease.

The neuropathologic changes consist of parenchymal and perivascular fibrillary astrogliosis and focal mild to moderate neuronal cell loss. The cortical areas most consistently affected were the temporal lobe, entorhinal cortex, parietal lobe, cingulate gyrus, posterior parietal lobule and occipital. Cerebral cortical layer 5 was most consistently and prominently affected. Cortical layer 3 pathology was more variable and the changes were somewhat more diffuse, less linear and of a slightly milder degree (Figure 13.7A). Patients with type 1 Gaucher disease had only mild astogliosis in these areas, usually not apparent using the H&E stain, but demonstrable with glial fibrillary acid protein (GFAP) immunostaining. Patients with type 2 Gaucher disease had the same pattern of astrogliosis, but more easily visualized on standard H&E staining and more defined and linear on GFAP immunostaining. In addition, cortical layer 5, occasionally, had discernible, scattered, focal, mild to moderate neuron cell loss.

CALCARINE CORTEX

Neuronal loss with gliosis in the occipital lobe and visual cortex (Calcarine cortex) has been reported[19,20] to involve cerebral cortical layer IV. The finding of specific targeting of a single lamina in the calcarine cortex was the result of a chance meeting with the right expertise and the right question at the Armed Forces Institute of Pathology (AFIP).

At that time, there was an ongoing study and correlation of the functional neuroanatomic targeting of the pyramidal cell neurons of the hippocampus. The essential functional mechanism in the hippocampus targeted excitatory glutaminergic neurons with extensive glutaminergic positive feedback within a region of strong local inhibitory feedback. However, layer IV neuronal loss as previously reported[19,20] in the cerebral cortex, and especially the calcarine cortex, is in a strong inhibitory layer, not a glutaminergic excitatory projection layer. Thus, layer IV pathology did not fit into the hypothetical paradigm suggested by the study. Professor John Allman

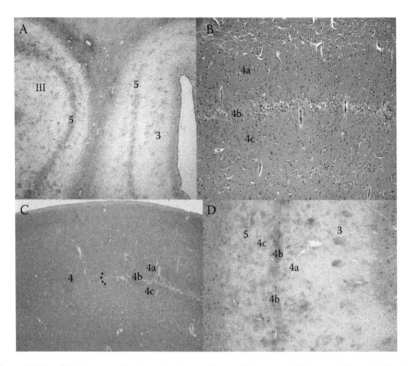

Figure 13.7 Calcarine cerebral cortex, layer 4b specific neuropathology: A. Type 1 Gaucher disease, Cerebral cortex, Parietal lobe: Laminar astrogliosis involves cortical layer 5 most consistently. Laminar astrogliosis accentuates the background perivascular astrogliosis along cortical layer 3, but may be diffuse and less organized in other areas (III). (GFAP, 40 original magnification.) B. Type 2 Gaucher disease, Calcarine cortex, V1, stria of Gennari: A precise, demarcated line of neurodegeneration and astrogliosis involves layer 4b, but spares adjacent layers 4a and 4c. Pathology involving of all of layer 4 in the calcarine cortex was not observed. (H&E, 40 original magnification.) C. Type 2 Gaucher disease (Patient 9), Calcarine cortex, Termination of V1, Stria of Gennari, Interface where 4a, 4b, and 4c merge (curved row of dots) into a single layer 4: Layer 4b (which corresponds to the stria of Gennari) has severe neurodegeneration and astrogliosis that abruptly terminates at the point where layer 4b ends. (H&E, 20 original magnification.) D. Type 1 Gaucher disease (Patient 4), Calcarine cortex, V1, stria of Gennari: In type 1 GD, neurodegeneration and neuronal cell loss are not identified. More subtle changes are present. A laminar region of astrogliosis along layer 4b, and to a lesser extent, cortical layers 3 and 5 is demonstrated by GFAP immunohistochemistry. (GFAP, 40 original magnification.)

of the California Institute of Technology, an expert on the human-primate visual cortex, was sojourning at the AFIP on a separate project. Upon presentation of the apparently incongruous data, he responded that there is a single lamina of the 12 or more lamina in calcarine cortex that fits the paradigm of strong glutaminergic, excitatory, recurrent with local inhibition and that lamina was layer 4b. After months of working with normal controls, 1 micron plastic sections, and various stains, and utilizing Professor Allman's expertise, the study was able to convincingly document that layer 4b, a thin lamina within the broader cortical layer 4, was found to be specifically targeted (layers 4a and 4c were spared) in Gaucher disease.[18]

In type 2 Gaucher disease patients, layer 4b neurons had severe neuronal cell loss and astrogliosis (Figure 13.7B and Figure 13.7C). In type 1 Gaucher disease patients, glial fibrillary acid protein immunohistochemistry demonstrated layer 4b astrogliosis and mild astrogliosis of layers 3 and 5 in addition to the mild layer 3 and 5 cerebral cortical astrogliosis (Figure 13.7D). Layer 4b pathology in type 2 and 3 GD was highly specific. The pattern of astrogliosis and neuronal cell loss (in type 2 GD) did not encroach upon layer 4a or 4c, directly adjacent to layer 4b, only several neurons cell body widths away from each other (Figure 13.7B). Layer 4b pathology immediately commenced and abruptly terminated at the laminar end-points, at the exact point where layer 4 split into layers 4a–4c, and where layers 4a–4c fused into layer 4 (Figure 13.7C).

CALCARINE CORTEX, LAYER 4B, GLUTAMINERGIC EXCITOTOXICITY, AND VISUOSPATIAL FUNCTION

The stria or line of Gennari, V1, and layer 4b are names of corresponding structures in the calcarine cortex.[62,63] As described previously, layer 4b has a large recurrent, excitatory, glutaminergic projection and feedback loop to the middle temporal visual area (MT) and strong local inhibition in the other layer 4 lamina, a characteristic that layer 4b has in common with the hippocampal CA4–2 region.[18] This pathologic mechanism of disease is supported in a study using audiologic and oculomotor testing that demonstrated an unusual pattern of audiologic and oculo-motor abnormalities "consistent with an excitotoxic mechanism predisposing nerve cells to glucocerebroside toxicity."[64]

The middle temporal visual area is present only in primates and humans, enabling visualization and comprehension of complex objects, i.e., a hand, as well as a visuospatial function, dynamic tracking of objects. Specifically, area MT is involved in tracking of moving objects against a moving background while the primate (or human) is also moving.[65–67]

Thus, injury leading to astrogliosis and neuron cell loss in layer could potentially produce detectable defects in visuospatial function. Such is the case in a neuropsy-chiatric cognitive study of 115 type 1 Gaucher disease patients that performed equally well, cognitively, compared with controls, except in a statistically significant visuospatial function.[68]

VASCULATURE, THE BLOOD BRAIN BARRIER AND THE GAUCHER CELL

Perivascular astrogliosis and perivascular Gaucher cells are the most widespread pathologic changes in the brain encompassing larger regions of the brain than specific regional disease. Because of the generalized, pervasive perivascular pathology, most of the pathology and damage to the brain is associated with the changes to the vasculature in Gaucher disease.

The hallmark of Gaucher disease, the perivascular or periadvential Gaucher cell is present in all forms of Gaucher disease; although they are more scarce in type 1 Gaucher disease, they are readily identified with a thorough search using periodic acid-Schiff stain or Alcian blue. Electron microscopy shows the perivascular Gaucher cell to lie external to (or adventitious to, thus the term "periadventitial" Gaucher cell) the capillary or blood vessel and outside of the brain parenchyma in an abnormal space within the blood brain barrier. Between the Gaucher cell and the bloodstream is the fibromuscular wall and/or endothelial cell. Between the Gaucher cell and the brain parenchyma is the astrocyte. In the abnormal space within the blood brain barrier, Gaucher cells lie amidst collagen fibers, periadvential fibroblasts, irregular basement membrane deposits and lipid material collections of lipid and membranous material. In this space, the fibroblasts appear to be laying down collagen fibrils and numerous basement membrane loops and reduplications are present in an apparent attempt to encompass the Gaucher cells. Vesicular lipid and membrane material collects in the extracellular blood brain barrier, pericapillary region and perhaps focally in the Virchow-Robin space, and astrocytes manifest increased deposition of glial fibrillary intermediate filaments in reactive astrogliosis.

Collections of periadventitial Gaucher cells are most numerous in the cerebral deep white matter, centrum ovale, and in regions where there is an admixture of gray and white matter such as the striatum and brain stem, as a whole. These regions also have the greatest degree of astrogliosis. The cerebral cortex and gray matter, in general, has far fewer and smaller collections of periadventitial Gaucher cells. At the cerebral gray-white junction, where the centrum ovale white matter meets the cerebral cortex, one can follow a capillary lined by groups of Gaucher cells traversing through the white matter leading into the cerebral cortex. However, at the gray-white junction, the periadventitial Gaucher cells, in most instances, will abruptly terminate at the junction as if a barrier were erected, preventing the Gaucher cells from following the vessel into the cerebral gray matter. In general, in a given patient, perivascular pathology is mildest in the cerebral cortex being (nearly absent in the sensorimotor cortex), mild to moderate in the white matter, and most pronounced in the striatum and brain stem regions.

The central point of these observations is that the blood brain barrier is not a homogeneous barrier throughout the brain. Given that the astrocyte confers blood brain barrier properties to the BBB endothelial cell,[69] and that closely approximated interactive region is disrupted by Gaucher cells and other material, it follows that the blood brain barrier is altered in Gaucher disease. In terms of therapeutic delivery of enzymes, medications, and molecular-genetic-therapeutics, one may need to take into the account the variation in delivery to different regions of the brain due to variable BBB permeability and receptor concentrations in different regions of the brain, and altered permabilities due to BBB effects in Gaucher disease.[70,71]

MEMBRANE, LIPID EFFECTS OF PHOSPHATIDYLCHOLINE AND GLUCOSYLCERAMIDE IN GAUCHER DISEASE

Gaucher disease is, in essence, a lipid perturbation. However, the abnormal balance of glucosylceramide and phosphtidylcholine metabolism appears to play a variety of possible roles in neuron cell loss, neurodegenerative changes, synuclein aggregation, perivascular disease, localization of periadventitial Gaucher cells and the blood brain barrier. Glucosylceramide is elevated in Gaucher disease, resulting in tubular endoplasmic reticulum, increased release of calcium stores, modulation of ryanodine receptor sensitivity (increasing susceptibility to excitotoxic neuronal injury and death), and increased phosphatidylcholine concentrations to levels detectable by Alcian blue staining.[18,38,59] One report even hypothesizes a mechanism of synuclein aggregation and accumulation secondary to phosphatidylcholine lipid and membrane lipid-vesicle alterations.[18]

A series of studies from one group shows that the physiologic increase in glucosylceramide concentration in Gaucher disease increases phosphatidylcholine synthesis in cultured neurons and macrophages.[72] This correlates with the neuroanatomic staining of Alcian blue positive Gaucher cell macrophages and, on rare occasions, neuronal plasma membrane, cytoplasmic and tubulovesicular structures staining, Alcian blue being a good stain for phosphatidyl-choline content[18] (Figure 13.2B, Figure 13.2C, and Figure 13.2D).

In addition, in type 2 Gaucher disease, there are small irregular deposits and vesicular collections of lipid and membranous material, not only amongst periadvential Gaucher cells, but also following along the brain capillaries in a pericapillary basement membrane location. Alcian blue-Periodic acid-Schiff (AB8GX-PAS) stains the basement membrane area blue, instead of the usual pink staining due to glycogen moieties of the basement membrane, consistent with increased phospholipid content of the basement membrane region. Juxtaposed against the basement membrane are the vascular endothelial cells of brain capillaries. Again, the phosphatidylcholine-glucosylceramide lipid axis plays a role as polar opposites with phosphatidylcholine selectively enriched in the apical plasma membrane (intravascular aspect) and glucosylceramide enriched in the basolateral (basement membrane-brain parenchyma aspect). The selective enrichment of phosphatidylcholine and glucosylceramide in blood brain barrier (BBB) endothelial cells is thought to be important for different membrane protein domains and for the maintenance of two membrane domains with different biophysical properties and permeabilities essential to the function of the BBB.[73] In the midst of this region of enriched glucosylceramide from the basolateral endothelial cell plasma membrane, and vesicular membranous lipid material along the basement membrane, lies the Gaucher cell, in an ideal location, engulfing material that cannot be digested.

REFERENCES

1. Brady, R.O., et al., Demonstration of a deficiency of glucocerebroside-cleaving enzyme in Gaucher's disease, *J. Clin. Invest.*, 45, 1112, 1966.

2. Beutler, E. and G.A. Grabowski, Gaucher disease, in *Metabolic & Molecular Basis of Inherited Disease*, W.S. Sly, Ed., McGraw-Hill, New York, 2001, 3635.

3. Brady, R.O. and J.A. Barranger, Glucosylceramide lipidosis, in *The Metabolic Basis of Inherited Disease*, J.B. Stanbury, J.B. Wyngaarden, and D.S. Fredrickson, Eds., McGraw-Hill, New York, 1983, 842.

4. Ginns, E.I., et al., Mutations of glucocerebrosidase: discrimination of neurologic and non-neurologic phenotypes of Gaucher disease, *Proc. Natl. Acad. Sci. U.S.A.*, 79, 5607, 1982.

5. Ginns, E.I., et al., Determination of Gaucher's disease phenotypes with monoclonal antibody, *Clin. Chim. Acta*, 131, 283, 1983.

6. Harris, C.M., et al., Auditory brainstem response and eye movement in diagnosis of Morbus Gaucher type 3, *Monatsschrift Kinderheilkunde*, 151, 854, 2003.

7. Brady, R.O., Sphingolipidoses, *Ann. Rev. Biochem.*, 47, 687, 1978.

8. Miller, J.D., R. McCluer, and J.N. Kanfer, Gaucher's disease: neurologic disorder in adult siblings, *Ann. Intern. Med.*, 78, 883, 1973.

9. Vellodi, A., et al., Management of neuronopathic Gaucher disease: a European consensus, *J. Inherit. Metab. Dis.*, 24, 319, 2001.

10. Niemann, A., Ein unbekanntes Krankheitsbild, *Jahrb. f Kinderheilk*, 1, 1914.

11. Gaucher, P.C.E., De L'epithelioma primitif de la rate hypertrophie idiopathique de la rate sans leucemie. *Medicine*, University of Paris, Paris, 1882.

12. Adachi, M., et al., Fine structure of central nervous system in early infantile Gaucher's disease, *Arch. Pathol.*, 83, 513, 1967.

13. Banker B.Q.M.J. and C.A.C., The cerebral pathology of infantile *Gaucher's Disease*, Academic Press, New York, 1962, 73.

14. Norman, R.M. and O.C. Lloyd, The neuropathology of infantile Gaucher's disease, *J. Pathol. Bacteriol.*, 72, 1956.

15. Lee, R.E., *The Pathology of Gaucher Disease*, Alan R. Liss, New York, 1982, 177.

16. Kaye, E.M., et al., Type 2 and type 3 Gaucher disease: a morphological and biochemical study, *Ann. Neurol.*, 20, 223, 1986.

17. Espinas, O.E. and A.A. Faris, Acute infantile Gaucher's disease in identical twins. An account of clinical and neuropathologic observations, *Neurology*, 19, 133, 1969.

18. Wong, K., et al., Neuropathology provides clues to the pathophysiology of Gaucher disease, *Mol. Genet. Metab.*, 82, 192, 2004.

19. Conradi, N., et al., Late-infantile Gaucher disease in a child with myoclonus and bulbar signs: neuropathological and neurochemical findings, *Acta Neuropathol.* (Berlin), 82, 152, 1991.

20. Conradi, N.G., et al., Neuropathology of the Norrbottnian type of Gaucher disease. Morphological and biochemical studies, *Acta Neuropathol.* (Berlin), 65, 99, 1984.

21. Conradi, N.G., H. Kalimo, and P. Sourander, Reactions of vessel walls and brain parenchyma to the accumulation of Gaucher cells in the Norrbottnian type (type III) of Gaucher disease, *Acta Neuropathol.* (Berlin), 75, 385, 1988.

22. Neil, J.F., R.H. Glew, and S.P. Peters, Familial psychosis and diverse neurologic abnormalities in adult-onset Gaucher's disease, *Arch. Neurol.*, 36, 95, 1979.

23. Soffer, D., et al., Central nervous system involvement in adult-onset Gaucher's disease, *Acta Neuropathol.* (Berlin), 49, 1, 1980.

24. Van Bogaert, L.F., Un cas de maladie de Gaucher de l'adulte avec syndrome de Raynaud, pigmentation, et rigidite du type extra-pyrajidal aux membres inferieurs, *Ann. Med.*, 45, 57, 1939.

25. McKeran, R.O., et al., Neurological involvement in type 1 (adult) Gaucher's disease, *J. Neurol. Neurosurg. Psych.*, 48, 172, 1985.

26. Neudorfer, O., et al., Occurrence of Parkinson's syndrome in type I Gaucher disease, *Qjm*, 89, 691, 1996.

27. Machaczka, M., et al., Parkinson's syndrome preceding clinical manifestation of Gaucher's disease, *Am. J. Hematol.*, 61, 216, 1999.

28. Varkonyi, J., et al., Gaucher disease associated with parkinsonism: four further case reports, *Am. J. Med. Genet. A*, 116, 348, 2003.

29. Tayebi, N., et al., Gaucher disease and parkinsonism: a phenotypic and genotypic characterization, *Mol. Genet. Metab.*, 73, 313, 2001.

30. Sidransky, E., Gaucher disease: complexity in a "simple" disorder, *Mol. Genet. Metab.*, 83, 6, 2004.

31. Aharon-Peretz, J., H. Rosenbaum, and R. Gershoni-Baruch, Mutations in the glucocerebrosidase gene and Parkinson's disease in Ashkenazi Jews, *N. Engl. J. Med.*, 351, 1972, 2004.

32. Bembi, B., et al., Gaucher's disease with Parkinson's disease: clinical and pathological aspects, *Neurology*, 61, 99, 2003.

33. Tayebi, N., et al., Gaucher disease with parkinsonian manifestations: does glucocerebrosidase deficiency contribute to a vulnerability to parkinsonism? *Mol. Genet. Metab.*, 79, 104, 2003.

34. Lowe, J., Lewy Bodies, in *Neurodegenerative Diseases*, D.B. Calne, Ed., Saunders, Philadelphia, 1994, 51.

35. Grafe, M. et al., Infantile Gaucher disease: a case with heuronal storage, *Ann. Neurol.*, 23, 300, 1988.

36. Lwin, A., et al., Glucocerebrosidase mutations in subjects with parkinsonism, *Mol. Genet. Metab.*, 81, 70, 2004.

37. Goker-Alpan, O., et al., Parkinsonism among Gaucher disease carriers, *J. Med. Genet.*, 41, 937, 2004.

38. Lloyd-Evans, E., et al., Glucosylceramide and glucosylsphingosine modulate calcium mobilization from brain microsomes via different mechanisms, *J. Biol. Chem.*, 278, 23594, 2003.

39. Verma, A., et al., Inositol trisphosphate and thapsigargin discriminate endoplasmic reticulum stores of calcium in rat brain, *Biochem. Biophys. Res. Commun.*, 172, 811, 1990.

40. Verma, A., et al., Rat brain endoplasmic reticulum calcium pools are anatomically and functionally segregated, *Cell. Regul.*, 1, 781, 1990.

41. Verma, A., D.J. Hirsch, and S.H. Snyder, Calcium pools mobilized by calcium or inositol 1,4,5-trisphosphate are differentially localized in rat heart and brain, *Mol. Biol. Cell.*, 3, 621, 1992.

42. Cajal, R.Y., Histologie du Systeme Nerveux de l'Homme et des *Vertebres*, Maloine, Paris, 1911.

43. Johnston, D. and T.H. Brown, Giant synaptic potential hypothesis for epileptiform activity, *Science*, 211, 294, 1981.

44. Radpour, S. and A.M. Thomson, Coactivation of Local Circuit NMDA Receptor Mediated epsps Induces Lasting Enhancement of Minimal Schaffer Collateral epsps in Slices of Rat Hippocampus, *Eur. J. Neurosci.*, 3, 602, 1991.

45. Johnston, D. and D.G. Amaral, Hippocampus, in *The Synaptic Organization of the Brain*, G.M. Shepard, Ed., Oxford University Press, New York, 1998, 429.

46. Tamamaki, N., K. Abe, and Y. Nojyo, Columnar organization in the subiculum formed by axon branches originating from single CA1 pyramidal neurons in the rat hippocampus, *Brain Res.*, 412, 156, 1987.

47. Amaral, D.G., C. Dolorfo, and P. Alvarez-Royo, Organization of CA1 projections to the subiculum: a PHA-L analysis in the rat, *Hippocampus*, 1, 415, 1991.

48. Miles, R. and R.K. Wong, Excitatory synaptic interactions between CA3 neurones in the guinea-pig hippocampus, *J. Physiol.*, 373, 397, 1986.

49. MacVicar, B.A. and F.E. Dudek, Local synaptic circuits in rat hippocampus: interactions between pyramidal cells, *Brain Res.*, 184, 220, 1980.

50. Ishizuka, N., W.M. Cowan, and D.G. Amaral, A quantitative analysis of the dendritic organization of pyramidal cells in the rat hippocampus, *J. Comp. Neurol.*, 362, 17, 1995.

51. Li, X.G., et al., The hippocampal CA3 network: an *in vivo* intracellular labeling study, *J. Comp. Neurol.*, 339, 181, 1994.

52. Traub, R.D. and R. Miles, Multiple modes of neuronal population activity emerge after modifying specific synapses in a model of the CA3 region of the hippocampus, *Ann. N.Y. Acad. Sci.*, 627, 277, 1991.

53. Johnston, D. and T.H. Brown, The synaptic nature of the paroxysmal depolarizing shift in hippocampal neurons, *Ann. Neurol.*, 16 Suppl., S65, 1984.

54. Johnston, D. and T.H. Brown, *Control Theory Applied to Neural Networks Illuminates Synaptic Basis of Interictal Epileptiform Activity*, Raven Press, New York, 1986, 263.

55. Johnston, D. and T.H. Brown, *Mechanisms of Neuronal Burst Generation*, Academic Press, London, 1984, 277.

56. Bruton, C.J., *The Neuropathology of Temporal Lobe Epilepsy*, Oxford University Press, New York, 1988.

57. Eng, L.F., A.C. Yu, and Y.L. Lee, Astrocytic response to injury, *Prog. Brain Res.*, 94, 353, 1992.

58. Miller, K.K., et al., Localization of an endoplasmic reticulum calcium ATPase mRNA in rat brain by *in situ* hybridization, *Neuroscience*, 43, 1, 1991.

59. Korkotian, E., et al., Elevation of intracellular glucosylceramide levels results in an increase in endoplasmic reticulum density and in functional calcium stores in cultured neurons, *J. Biol. Chem.*, 274, 21673, 1999.

60. Gunteski-Hamblin, A.M., J. Greeb, and G.E. Shull, A novel Ca2+ pump expressed in brain, kidney, and stomach is encoded by an alternative transcript of the slow-twitch muscle sarcoplasmic reticulum Ca-ATPase gene. Identification of cDNAs encoding Ca2+ and other cation-transporting ATPases using an oligonucleotide probe derived from the ATP-binding site, *J. Biol. Chem.*, 263, 15032, 1988.

61. Toft, M., et al., Parkinsonism, FXTAS, and FMR1 premutations, *Mov. Disord.*, 20, 230, 2005.

62. Polyak, S.L., *The Vertebrate Visual System*, University of Chicago Press, Chicago, 1957.

63. Brodemann, K., Vergleichende Lokalisation-slehre der Grosshirnrinde, Barth Press, Leipzig, 1909.

64. Campbell, P.E., et al., A model of neuronopathic Gaucher disease, *J. Inherit. Metab. Dis.*, 26, 629, 2003.

65. Fitzpatrick, D., J.S. Lund, and G.G. Blasdel, Intrinsic connections of macaque striate cortex: afferent and efferent connections of lamina 4C, *J. Neurosci.*, 5, 3329, 1985.

66. Spatz, W.B., Thalamic and other subcortical projections to area MT (visual area of superior temporal sulcus) in the marmoset Callithrix jacchus, *Brain Res.*, 99, 129, 1975.

67. Tigges, J., et al., Areal and laminar distribution of neurons interconnecting the central visual cortical areas 17, 18, 19, and MT in squirrel monkey (Saimiri), *J. Comp. Neurol.*, 202, 539, 1981.

68. Elstein, D., et al., Computerized cognitive testing in patients with type I Gaucher disease: effects of enzyme replacement and substrate reduction, *Genet. Med.*, 7, 124, 2005.
69. Cancilla, P.A., J. Bready, and J. Berliner, Brain endothelial-astrocyte interactions, in *The Blood-Brain Barrier: Cellular and Molecular Biology*, W.M. Pardridge, Ed., Raven Press, New York, 1993, pp. 25–46.
70. Pardridge, W.M., The blood-brain barrier: bottleneck in brain drug development, *NeuroRx*, 2, 3, 2005.
71. Pardridge, W.M., The blood-brain barrier and neurotherapeutics, *NeuroRx*, 2, 1, 2005.
72. Bodennec, J., et al., Phosphatidylcholine synthesis is elevated in neuronal models of Gaucher disease due to direct activation of CTP:phosphocholine cytidylyltransferase by glucosylceramide, *Faseb. J.*, 16, 1814, 2002.
73. Tewes, B.J. and H.J. Galla, Lipid polarity in brain capillary endothelial cells, *Endothelium*, 8, 207, 2001.

CHAPTER 14

Diagnosis and Laboratory Features

Carla E.M. Hollak and Johannes M.F.G. Aerts

CONTENTS

DIAGNOSIS

Since the discovery of the enzymatic defect in Gaucher disease [1,2], the golden standard for the diagnosis of Gaucher disease is the determination of deficient glucocerebrosidase activity in leucocytes or fibroblasts. An enzymatic assay using a fluorogenic substrate (4-methylumbelliferyl-β-D-glucopyranoside) is now the most widely applied method to confirm a diagnosis of Gaucher disease [3–5]. In addition to leukocytes and fibroblasts, urine can also serve as a reliable specimen to perform the test and carries the advantage that it is easy to obtain [6]. Apparently, a soluble form of glucocerebrosidase can be found in the urine, which exhibits identical properties as the membrane-associated tissue enzyme. This soluble form cannot be detected in the plasma, and therefore plasma and serum cannot serve as reliable sources for the determination of glucocerebrosidase deficiency. When appropriately applied on leukocytes, fibroblasts, or urine, the test is very accurate and will identify almost 100% of patients. The only rare exception is deficiency of the activator protein saposin C, which can clinically result in "Gaucher-like" cases [7,8]. Reduced activity of glucocerebrosidase in these cases *in vivo* can be missed with the *in vitro* assay, since sodium taurocholate is added as an activator in the glucocerebrosidase assay. The finding of Gaucher cells in tissue samples is not pathognomonic for Gaucher disease. In several diseases, especially disorders that are associated with high cell turnover such as chronic myeloid leukaemia or in chronic infections, the occurrence of pseudo-Gaucher cells have been reported [9,10]. Also Gaucher cells have to be distinguished from other diseases that exhibit pathological macrophages as a hall-mark such as the sea blue histiocyte syndrome or Niemann Pick disease. Thus, demonstration of deficient glucocerebrosidase activity is obligatory to confirm a diagnosis of Gaucher disease. For the detection of carriers of a glucocerebrosidase mutation, assessment of glucocerebrosidase activity is less accurate. Although most carriers have significantly lower activities, roughly 20% of them show overlap with healthy controls [11,12]. Therefore, mutation analysis should be carried out to confirm the presence or absence of carriership (see Chapter 3). The availability of an accurate test has facilitated prenatal testing. Determination of the level of gluco-cerebrosidase activity in cultured amniocytes or chorionic villus tissue can identify affected pregnancies [13,14], although for adequate prediction of the phenotype, additional mutation analysis is needed [14]. In these cases, values range from 0–15% of normal (for review see 5). More recently, prenatal detection of lysosomal storage disorders by determination of oligosaccharides and glycolipids in amniotic fluid using advanced mass spectrometric methods has been reported [15]. This technique is applied for screening purposes as discussed below.

LABORATORY FEATURES

The key pathological cell in Gaucher disease is the lipid-laden macrophage. It has been recognized for many years now that the pathology of Gaucher disease is not simply caused by the mass effect of storage cells, but also results from the secretion of various proteins such as hydrolases and cytokines. Such factors may originate from storage cells themselves or from surrounding activated macrophages [16]. Increased levels of some of these proteins may exert biological activity and hence influence the phenotypic expression of the disease. Identification of abnormal factors produced in Gaucher cell lesions will result in improved understanding of the symptoms and associated conditions of Gaucher disease. Apart from this, many of these proteins can be detected in the plasma and thus serve as disease markers. Monitoring of several of these markers has proven very useful in monitoring therapeutic intervention. The following paragraphs will summarize the abnormal plasma and tissue findings in Gaucher disease and their possible implications for the pathophysiology of the disease and their application as markers.

Glucosylceramide and Ceramide

The key pathological feature of Gaucher disease is the Gaucher cell, massively loaded with glucosylceramide. It is therefore not surprising that glucosylceramide levels are grossly elevated in tissues of Gaucher patients such as the spleen, liver, and bone marrow. The most striking elevation is found in the spleen with an increase of 10- to 1000-fold compared to control [17,4]. Elevated levels of glucosylceramide and, in addition the presumably more neurotoxic glucosylsphingosine (psychosine), has been established in the brain of neuronopathic Gaucher disease patients [18]. Elevated plasma levels of glucosylceramide have been described in several studies [19–21]. In contrast to the levels in tissue, the plasma levels are only twofold increased. In our laboratory an improved HPLC technique for accurate concomitant measurement of glucosylceramide and ceramide has recently been developed, which revealed a threefold increase in plasma glucosylceramide and a slightly reduced ceramide level in Gaucher type 1 patients as compared to controls (Groener, manuscript submitted for publication). The ratio of glucosylceramide over ceramide allows an almost complete discrimination between Gaucher patients and controls. Elevated glucosylceramide in erythrocytes of Gaucher patients has earlier been established [22]. Both plasma glucosylceramide, glucosylceramide/ceramide ratios and erythrocyte glucosylceramide concentration have been used as surrogate markers to monitor the effect of enzyme therapy and substrate reduction therapy. For the latter therapeutic intervention, the reduction in glycolipids has contributed to establish the proof of principle of this treatment, but it is limited as a tool to monitor therapeutic efficacy. Enzyme therapy usually results in normalization of glucosylceramide levels after a few months to years and thus quickly loses its value as a disease marker. In addition, no clear correlation between plasma or erythrocyte levels of glucosylceramide and clinical signs or response to therapy has been reported.

Lysosome-Associated Membrane Proteins and Saposins

Accumulation of storage material in lysosomal storage disorders (LSDs) results in an increased size and number of lysosomes, which is reflected in the plasma by an increase in lysosomal proteins. Lysosome-associated membrane proteins 1 and 2 (LAMP-1 and LAMP-2) are the most discriminating between controls and patients with LSDs [23,24]. In 92% of patients with Gaucher disease, the concentrations of both LAMP-1 and LAMP-2 were above the 95 percentile of the control range. In addition, elevated levels of saposins have been described in LSDs, including Gaucher disease. Saposins are activator proteins that play an important role in several lysosomal hydrolases [25], which is illustrated by the fact that genetic defects in saposins have been found to be associated with the storage of sphingolipids. The importance of saposin activity in Gaucher disease is further emphasized by the recent development of a mouse model of Gaucher disease in which mutations in the glucocerebrosidase gene in combination with low prosaposin expression resulted in further decreases in residual enzyme activities and larger accumulations of glucosylceramide in tissues [26]. Accumulations of saposins can be detected in the tissues of several LSD-affected individuals. Increases as much as 80-fold have been observed in the livers and brains of patients affected with several LSDs, including Gaucher disease, Niemann-Pick disease, fucosidosis, Tay-Sachs, and Sandhoff disease [27]. Although Chang and co-workers reported elevated levels of saposin A, C, and D in 94–96% of already diagnosed and in the majority of cases of fully symptomatic patients with Gaucher disease [28], it remains questionable whether this method is able to identify all Gaucher patients, especially in an early stage of disease [29]. Determination of elevated levels of the LAMPs in combination with saposins has been proposed as a useful tool to screen newborns for the presence of LSDs. This topic is discussed below.

Cholesterol

Reduced plasma concentrations of total cholesterol LDL and HDL-cholesterol have been described in patients with type 1 Gaucher disease [30,31]. In addition, the apoprotein levels that are part of the structure of the HDL and LDL particles (apo-B and apo-A1, respectively) were also decreased [30], while apoE was increased. One study reported the effect of enzyme therapy on lipoproteins. After a treatment period of 18 months, the levels of HDL-cholesterol and apo-A1 were partly restored, but no complete normalization was achieved. In contrast, LDL-cholesterol and apo-B levels remained unchanged and apo-E levels decreased, probably reflecting the disappearance of storage cells [32]. Of interest is the finding that abnormal HDL-cholesterol levels can be found also in Gaucher disease carriers [33].

Clotting Factors

The first report on deficiencies of coagulation factors was described by Vreeken in 1967 [34]. A Gaucher patient was screened for coagulation abnormalities prior to splenectomy and deficient activities of factor V and VIII were established, which normalized after splenectomy. Consumption of the coagulation factors by the large spleen as well as low grade intravascular coagulation was hypothesized to be the underlying mechanism. Several studies followed in which deficiencies of factors XI, XII, VII, X, V, and II were described [35,36]. In around 40% of adult type 1 Gaucher disease patients, APTT and PTT were prolonged and enzyme therapy can partly restore the prolonged clotting times [36]. Factor XI deficiency was reported as a separate phenomenon in earlier studies and this has been suggested to coincide with Gaucher disease, as factor XI deficiency also has a higher prevalence in the Ashkenazi Jewish population [37,38]. The cause of the occurrence of clotting factor deficiencies may be partly explained by increased clearance through the enlarged spleen. The extremely low levels of factor V in nonsplenectomized patients point towards this phenomenon. However, deficiencies can still be present after splenectomy, and especially in these patients, parameters of increased activation of coagulation as well as fibrinolysis, such as elevated D-dimer and plasmin-α2antiplasmin complexes, can be established [36]. Thus it seems that in Gaucher disease there is evidence for low grade intravascular coagulation and fibrinolysis possibly related to cytokines secreted by the Gaucher macrophages. Indeed, elevated levels of several pro-inflammatory cytokines in plasma of Gaucher disease patients have been described (see below). In addition, Rogowski and co-workers reported significant elevations in fibrinogen, erythrocyte sedimentation rate and C-reactive protein, indicative of a low-grade inflammatory profile [39].

Barone et al. reported the presence of lupus anticoagulant activity in 3 of 5 type 1 Gaucher patients. Enzyme therapy did not influence the presence of this activity, thus revealing another mechanism that can contribute to prolonged APTT in Gaucher disease [40].

Hydrolases

The activities of several common lysosomal hydrolases have been reported to be elevated in Gaucher disease. As early as 1956, increased activity of tartrate resistant acid phosphatase (TRAP) in plasma of type 1 Gaucher patients was reported [41]. Further studies showed that the increased activity was due to elevations in isoenzyme 5B [42]. The finding of increased TRAP activity was followed by reports of the presence of several lysosomal enzymes in plasma as well as tissue samples [43,44]. Ockerman and Kohlin reported elevated levels of β-galactosidase, hexosaminidase, β-glucuronidase and α-fucosidase in plasma of six Gaucher patients, while α-mannosidase levels were normal. Acid phosphatase and beta-glucuronidase showed the most prominent elevations, being about twofold to sixfold elevated. Elevated acid phosphatase has been used as a diagnostic marker prior to the identification of the genetic defect in Gaucher disease. Later it was suggested that the levels of β-hexosaminidase A and B could be used to screen for Gaucher disease in Ashkenazi Jews that underwent testing for Tay-Sachs disease carriership, since Gaucher patients had clear elevations in beta-hexosaminidase B, while beta-hexosaminidase A was relatively low [45]. However, further studies highlighted the marked heterogeneity of these enzyme activities, making this test unreliable for screening purposes [46]. Galactocerebrosidase activities were reported to be elevated in Gaucher spleen and brain of an infantile case of Gaucher disease and its role in the pathophysiology of brain disease was considered [47]. The source of these elevated hydrolases is most likely the Gaucher cells. Immunohistochemical studies have shown that TRAP can be localized to the storage cells and surrounding activated macrophages in the spleen [48,16]. Differential gene expression techniques applied to Gaucher spleen samples have identified increased levels of cathepsins S, C, and K and again with immunohistochemical staining these hydrolases were found to originate from Gaucher cells [49]. Two other commonly elevated hydrolase activities in Gaucher patients deserve specific mention. It is known that serum angiotensin-converting enzyme (ACE) levels are often increased, ranging from near normal to more than 10-fold the control median [50–53]. Increased levels of lysozyme activity have also been documented [51,54].

Although many attempts have been made to relate the elevated levels of hydrolases to specific pathology in Gaucher disease or disease severity in general, no direct relationship has been established. Especially cathepsin K, which is reported to have a pathogenetic role in osteolysis, and TRAP, which is known to be secreted by osteoclasts, have been implicated in the pathophysiology of Gaucher bone disease. However, no evidence exists that these hydrolases can be used as plasma markers for severity of skeletal disease. Also, neither of these hydrolases is sensitive enough to be used as a plasma biomarker for the pathological Gaucher cell. In this respect, the discovery of chitotriosidase deserves to be discussed separately.

Chitotriosidase

The need for a very sensitive and specific plasma biomarker of Gaucher cells prompted a search in our laboratory that led to the discovery of a very marked abnormality in the serum of symptomatic patients with Gaucher disease. Serum from such individuals showed a 1000-fold increased capacity to degrade the fluorogenic substrate 4-methylumbelliferyl-β-D-chitotrioside [54]. The corresponding enzyme was named chitotriosidase. The chitotriosidase protein was subsequently isolated to homogeneity and its cDNA was cloned [55,56].

Chitotriosidase was found to be the human analogue of chitinases from lower organisms. *In situ* hybridization and histochemistry of bone marrow aspirates and sections of spleen from Gaucher patients revealed that chitotriosidase is very specifically produced by storage cells. This is also supported by the close linear relationship between levels of chitotriosidase and glucosylceramide in different sections of spleen from Gaucher patients. As glucosylceramide is the best possible quantitative measure for storage cells, it may be deduced that chitotriosidase production is directly proportional to the number of Gaucher cells. In a culture model of Gaucher cells, chitotriosidase accounts for almost 10% of the total level of secreted protein. In sharp contrast, common tissue macrophages and dendritic cells produce little chitotriosidase. These observations help to explain the very specific, gross elevation in chitotriosidase levels in the blood of Gaucher patients. A relationship between the body burden on storage cells in Gaucher patients and their plasma chitotriosidase levels has been noted. The plasma chitotriosidase level does not reflect one particular clinical symptom of Gaucher disease, suggesting rather that it reflects the sum of the enzyme secreted by Gaucher cells in various body locations [54].

The molecular properties of chitotriosidase have been intensively studied. The enzyme is specifically produced by phagocytes. Neutrophils produce a 50 kDa enzyme that is stored in their specific granules and only released upon an appropriate stimulus [57]. Tissue macrophages largely constitutively secrete newly synthesized 50-kDa chitotriosidase, but about one third is directly routed to lysosomes and proteolytically processed to the 39-kDa unit that remains catalytically active [58]. The 3-D structure of the 39 kDa catalytic unit of chitotriosidase has been resolved by crystallography and x-ray analysis [59,60]. Detailed investigation of the catalytic mechanism has rendered an explanation for the apparent substrate inhibition [54], a phenomenon that prohibits accurate quantification of chitotriosidase levels by enzyme assay using conventional substrates like 4-methylubelliferyl-chitotriose or

4-methylumbelliferyl-chitobiose [61]. The inevitable use of sub-saturating substrate concentrations reduces the reproducibility and sensitivity of the assay and has hampered its wide-scale application outside expert laboratories. We discovered that the inhibition of enzyme activity at excess substrate concentration can be fully explained by transglycosylation of substrate molecules by chitotriosidase. The novel insight into transglycosidase activity of chitotriosidase led to the design of a new substrate molecule, 4-methylumbelliferyl-(4-deoxy)chitobiose. With this substrate, which is not an acceptor for transglycosylation, chitotriosidase shows normal Michaelis-Menten kinetics, resulting in major improvements in sensitivity and reproducibility of enzymatic activity measurements. The novel convenient chitotriosidase enzyme assay should facilitate the accurate monitoring of Gaucher disease patients receiving costly enzyme replacement therapy [61]. A significant additional advantage of the assay with the novel substrate is the ~fivefold increase in sensitivity of detection.

Intriguingly, one in every three individuals from various ethnic groups carries one abnormal chitotriosidase gene with a 24-bp duplication that prevents production of enzyme [62]. About 6% of the population is homozygous for this mutant allele and consequently completely lacks chitotriosidase activity. The recent discovery of the existence of a second chitinase in man, named acidic mammalian chitinase (AMCase), has raised the possibility that the common deficiency in chitotriosidase might be partly compensated for by the latter enzyme [63].

Analogous to the function of homologous chitinases in plants, the physiological role of chitotriosidase is most likely in innate immunity toward chitin-containing pathogens. Recombinant chitotriosidase exerts activity towards chitin-containing fungal pathogens both *in vitro* and *in vivo*. It prevents hyphal switch and causes hyphal tip lysis in *Candida albicans*. In neutropenic mouse models of systemic *Candidiasis* and *Aspergillosis*, administration of recombinant chitotriosidase markedly promoted survival [57]. An increased risk for nematode infection has indeed been described for chitotriosidase-deficient individuals [64]. Very recently, it has been reported that chitotriosidase deficiency may be associated with susceptibility to infection with Gram-negative bacteria in children undergoing therapy for AML [65].

Cytokines and Soluble Proteins

Our recent histochemical investigation of Gaucher spleen has revealed that mature lipid-laden Gaucher cells are surrounded by a mixture of phenotypically diverse macrophages producing a variety of cytokines [16]. Michelakakis and co-workers were the first to report elevated levels of TNF-α in plasma of type 2 and 3 Gaucher patients, and to a lesser extent in samples from type 1 Gaucher patients [66]. Allen et al. could not confirm the finding of elevated plasma TNF-α in type 1 Gaucher disease, but did observe increases in IL-6 and IL-10 [67]. In another study, we did not detect consistent elevated TNF-α or IL-6 in plasma of type 1 Gaucher patients studied [68]. The same study, however, revealed that IL-8 can be markedly increased in plasma of type 1 Gaucher patients. Very recently, analysis of plasma levels of TNF-α and genotyping for the -308 G-->A polymorphism in the promoter of the TNF-α gene was performed in 17 patients with type I Gaucher disease. A significant correlation was found between serum TNF-α levels and TNF-α genotypes for homozygotes vs. heterozygotes patients (p = 0.02), with patients homozygous for the polymorphism having the lower levels of serum TNF-α [69]. Polymorphisms in promoter regions of the TNF-alpha (and IL-6) gene may help to explain the heterogeneous findings on plasma elevations of these pro-inflammatory cytokines in type 1 Gaucher patients. As mentioned earlier, a low-grade inflammatory profile has been identified in Gaucher disease. In comparison to a multivariable database from healthy controls, patients with Gaucher disease had significant elevations in fibrinogen, accelerated erythrocyte sedimentation rate and C-reactive protein [39]. The authors pointed out that these parameters were not influenced by enzyme therapy, which might imply that these abnormalities are not directly related to the stored glycolipids but could be the result of cytokine or chemokine release.

Increased levels of macrophage colony stimulating factor (M-CSF), twofold to fivefold above normal, have been observed in plasma of most Gaucher patients [68]. In addition, plasma of many Gaucher patients contains up to sevenfold increased concentrations of the monocyte/macrophage activation marker soluble CD14 (sCD14). This finding supports the idea that activation of monocytes/macrophages occurs in symptomatic Gaucher patients.

Another marker for macrophage activation is soluble CD163 (sCD163). Increased concentrations have been measured in patients with infection and myeloid leukemias. The sCD163 plasma levels in type 1 Gaucher patients were found to be far above the levels in normal subjects [70,71].

Chemokine CCL18

About one in every 16 individuals, including Gaucher patients, completely lacks chitotriosidase activity. Obviously, measurement of plasma chitotriosidase levels is of no value for monitoring storage cells in deficient Gaucher patients. This prompted us to search for additional plasma biomarkers of Gaucher cells. To identify novel factors derived from Gaucher cells, we analyzed plasma samples of patients with Gaucher disease before therapeutic intervention and compared them to control plasma using surface-enhanced laser desorption/ionization (SELDI) mass spectrometry (MS) [72]. In plasma of a symptomatic patient, a peptide of 7856 Da was found to be markedly increased in relation to Gaucher disease manifestation. Subtractive hybridization studies previously revealed that RNA encoding for a protein with identical mass is up-regulated in a Gaucher spleen [49]. This protein, named pulmonary and activation-regulated chemokine (PARC, systematic name CCL18), is a member of the C-C chemokine family. It has been speculated that CCL18 plays a role in the recruitment of T and B lymphocytes toward antigen-presenting cells, a crucial step in the initiation of adaptive immune responses [73,74]. Abnormalities concerning serum immunoglobulins and other manifestations of disturbed B-cell function, like monoclonal or oligoclonal gammopathies, occur frequently in patients with Gaucher disease (see below). However, we did not observe any correlation between the plasma levels of, and the presence of, a monoclonal gammopathy. Because all symptomatic Gaucher patients show at least 10-fold increased plasma levels of CCL18, it can be envisioned that this constitutes a risk factor for the development of disturbed B-cell function.

In the case of Gaucher patients who are deficient in chitotriosidase activity, monitoring of plasma CCL18 seems a good and reliable alternative that can aid in the clinical management of Gaucher patients (see below). Immunohistochemical analysis of spleen sections from Gaucher patients has indeed revealed that Gaucher cells are the prominent source of CCL18 as well as chitotriosidase. Fractional corrections in plasma CCL18 levels and of chitotriosidase are similar in Gaucher patients following therapy. Corrections in bone marrow fat fraction, an indirect assessment of Gaucher cell infiltration of the marrow, are also paralleled by corrections in plasma CCL18 (or chitotriosidase).

Ferritin

The presence of high levels of iron and ferritin in Gaucher cells has been documented for a very long time. Most Gaucher patients show increased plasma ferritin levels [53,75]. It is presently unclear whether the high ferritin reflects increased storage of iron in the body or merely reflects an aspecific acute phase response. In other macrophage disorders, such as adult Still's disease, high levels of ferritin and low levels of glycosylated ferritin are found during the acute phase of the disorder and hence believed to reflect macrophage activation [76]. This phenomenon may also play a role in Gaucher disease. The observation that enzyme therapy reduces the abnormally high levels of ferritin [77] points in this direction. As in many other disorders, increased levels of ferritin may cause an imbalance in the immune system in an unfavorable way. Decreased phagocytosis as well as altered T-cell subsets have been found in these conditions, which may contribute to increased growth rates of cancer cells and infectious organisms [78,79].

Immunoglobulins

Immunoglobulin abnormalities in Gaucher disease are common. An increase in the gammaglobulin fraction on protein electrophoresis was already noted in 1950 by Goldfarb [80]. Subsequently several groups observed a high prevalence of both polyclonal and monoclonal gammopathies. Pratt et al. described hypergammaglobulinemia and monoclonal gammopathy in 10 of 16 type 1 Gaucher patients, aged 9 to 70 years, with the highest prevalence in patients over 50 years of age [81]. Marti and Allen have studied immunoglobulin abnormalities in adult patients and found a prevalence of 44% and 46% for polyclonal gammopathies respectively. Monoclonal gammopathies were found in 5 and 9% and were mainly IgG kappa [67,82]. The presence of a monoclonal gammopathy is associated with an increased risk for the development of multiple myeloma and amyloidosis. Indeed, several cases of multiple myeloma have been reported [83]. Recently, large cohort studies have confirmed that the risk for the development of multiple myeloma is increased in Gaucher disease [84–86]. Although very rare, AL amyloidosis has also been observed in type 1 Gaucher disease [87] and this disorder is always fatal. The cause for the high prevalence of immunoglobulin abnormalities has been sought in the chronic

258

stimulation of the B-cell system by several cytokines such as IL6 or PARC/CCL18 as described above. Indeed, Allen and co-workers reported an association between levels of IL6 and the presence of a monoclonal gammopathy [67]. The effect of enzyme therapy on the abnormal immunoglobulin pattern has been the subject of a few studies so far. In a case report, Deibener noted a linear decrease in monoclonal protein and β2-microglobulin in a 39-year-old male treated with enzyme therapy [88]. Very recently Brautbar and co-workers reported that polyclonal gammopathy may respond to enzyme supplementation, while monoclonal gammopathies remain unchanged [89].

Parameters of Bone Metabolism

Bone pathology in Gaucher disease is incompletely understood. The mass effect of Gaucher cell infiltration in the bone marrow is probably only one factor in a complex interplay of cytokines, local release of hydrolases and disturbed balance between osteosynthesis and resorption. For example, it has been clearly shown that many patients exhibit signs and symptoms of osteopenia, but the precise mechanism is still unclear [90,91]. Biochemical markers of bone turnover may help to elucidate the pathophysiological mechanisms involved and may also provide a tool for therapeutic intervention, since increased bone resorption could be effectively treated with bisphosphonates [92].

The parameters of bone turnover that have been studied so far include markers of bone formation such as osteocalcin, bone specific alkaline phosphatase and carboxyterminal propeptide of type I procollagen (PICP), as well as bone resorption such as type I collagen carboxy-terminal telopeptide (ICTP) in serum, and several urinary markers, including hydroxyproline, pyridinoline, and deoxypyridinoline cross-links and calcium. Conflicting results have been reported from studies in Gaucher disease. Drugan et al. reported decreased levels of osteocalcin as well as ICTP, compatible with low osteoblastic and osteoclastic activities. This was supported by findings in a study that addressed the influence of vitamin D supplementation on bone density in addition to enzyme therapy. Low levels of bone specific alkaline phosphatase as well as osteocalcin were found and although these levels increased with vitamin D treatment, no effect on bone density was recorded [93]. On the other hand, Fiore and co-workers reported enhanced pyridinoline end deoxypyridinoline cross-link excretion in the urine, indicating increased bone resorption [94]. Another study showed that, again, markers of bone formation were low, with low levels of PICP, while bone resorption parameters were enhanced, including high levels of ICTP [95]. These investigators also followed the effect of enzyme therapy on markers of bone turnover and observed no significant changes, except for increased calcium excretion in the urine, which does point to increased resorption. In contrast, they found improved bone density upon treatment with enzyme as measured by DEXA and concluded that the biochemical abnormalities are not useful in monitoring the effect of treatment on bone metabolism [96]. Others, however, have reported that enzyme therapy does not result in improved bone density [97,98]. It may be concluded that the only consistent finding is that markers of bone formation are low, compatible with decreased osteoblast activity. The conflicting results with respect to bone resorption markers could also be the result of differences in disease activity. For example, in a young male with type 1 Gaucher disease experiencing an acute bone crisis, clearly enhanced bone resorption has been biochemically established [99], while in the chronic situation osteoclast activity is normal or reduced. It seems that bone disease in Gaucher is not a systemic disorder of metabolism that affects bone uniformly, as was already suggested in the past [90]. If bone resorption is especially enhanced in bone crises, the use of bisphosphonates may specifically be of value during episodes of acute bone distress. Apart from the aforementioned case report [99], there is no evidence to support this. Wenstrup and co-workers showed that alendronate, 40 mg once daily, improved bone density in combination with imiglucerase, although the small treatment groups and absence of fractures in both treatment arms did not allow a conclusion with respect to a clinical effect [98].

CLINICAL APPLICATIONS OF BIOCHEMICAL MARKERS

Screening

Screening for Gaucher disease can be divided into two groups: carrier screening, which aims to identify couples at risk for giving birth to children with the disease, and screening that aims to identify people with Gaucher disease at an early stage, so that their condition can be more adequately managed. Carrier screening for Gaucher disease can only be done by identification of the glucocerebrosidase mutation. Large carrier screening programs for Ashkenazi Jews have been implemented for years, following the successful screening programs for Tay-Sachs disease [100]. For the identification of affected individuals, screening programs have largely focussed on methods to identify newborns with LSDs, including Gaucher disease. For these purposes several markers have been investigated. In contrast to their earlier results [23,24] Meikle and co-workers reported that in a large group of Danish patients with LSDs, LAMP-1 and saposin C were not sensitive indicators of the presence of a LSD [29]. However, using tandem mass spectrometry, several LSDs could be identified based on glycosphingolipid and oligosaccharide markers. For the detection of Gaucher disease, the measurement of glucosylceramide was used, which displayed a limited sensitivity of 60%. This technique was also applied on amniotic fluid samples and a correct identification of two Gaucher cases was made based on the elevated levels of glycosphingolipids [15]. Obviously, the advantages of these methods are that a wide variety of LSDs can be captured using only one panel of markers. It is, however, clear that for every suspected case, confirmation of a prenatal screenings test by enzymatic assay or mutation analysis should be performed. As previously stated, both methods are feasible using amniotic fluid or chorionic villi to confirm a diagnosis of Gaucher disease. Chamoles and co-workers have shown that dried blood spots on filter paper are reliable sources to identify patients with Gaucher disease by performing the enzymatic assays of both glucocerebrosidase and chitotriosidase [101]. In their study, the diagnosis was correctly established in all patients based on the determination of glucocerebrosidase deficiency in the blood spots, while purely based on chitotriosidase activity the diagnosis was missed in 4 of 54 Gaucher patients: in 3 because of deficient activity and in 1 because of a normal value in an asymptomatic individual. There is no doubt that prenatal screening or carrier screening in high risk populations can be justified, although counseling can be very difficult in the light of the phenotypic heterogeneity in type 1 Gaucher disease [102]. For population screening, however, including screening of newborns with the use of blood spots, strict ethical rules should be followed. Although effective treatment is available, therapeutic intervention is not always needed and rarely necessary in early childhood and thus a rationale for early recognition of the disorder is at least questionable. Furthermore, an early diagnosis in an individual who may remain completely asymptomatic carries a high socio-economic and psychological burden, which should be borne in mind.

Assessment of Disease Burden and Monitoring of Therapy

Since many of the biochemical markers are derived directly from the Gaucher cell, it is very tempting to use plasma levels as biomarkers to assess the disease severity. This has become of particular interest in the light of new therapeutic possibilities such as enzyme therapy and substrate reduction therapy. The plasma levels could indicate disease severity and thus aid in the decision to initiate therapy and they could be monitored during therapeutic intervention to help in assessing the efficacy of the treatment. To deserve the qualification "biomarker," the plasma abnormalities should be closely related to the clinical expression of the disease and hence predict clinical benefit or risk based on scientific evidence [103]. Early observations already point to the variability of elevations in several hydrolases without a clear correlation with clinical features [46]. Whitfield and co-workers found that there was a general relationship between the glucosylceramide/lactosylceramide ratio and the phenotype, with higher levels in more severely ill patients. In addition, generally lower LAMP-1 and saposin C levels were found in milder patients, but these levels were also influenced by age [104]. As markers for the response to treatment, many plasma abnormalities have been followed, including glucosylceramide, chitotriosidase, CCL18, ACE, TRAP, hexosaminidase, ferritin, sCD163, cathepsin K, and neopterin [Groener, submitted, 49,54,77,105,106]. In practice, the plasma abnormalities that showed the highest activities compared to controls and were the least laborious to measure have been most frequently applied. These are chitotriosidase, CCL18, ACE, and TRAP. The clinical usefulness of these biomarkers has been the subject of several studies. First of all it was documented that in patients after cessation of treatment, several parameters increased rapidly, indicating recurrence of disease activity [107]. The plasma markers included ACE, ferritin, acid phosphatise, and chitotriosidase. Clearly chitotriosidase was the most sensitive marker in predicting deterioration of clinical symptoms. Others reported that chitotriosidase did not correlate with the clinical course in two patients who developed neutralizing antibodies towards imiglucerase [108]. Cabrera-Salazar and co-workers followed ACE, TRAP, and chitotriosidase levels in 18 Gaucher patients during treatment with enzyme therapy. They found that the markers correlated with each other and also that there was a relation between the overall severity score (SSI) and chitotriosidase levels. It was also convincingly demonstrated that chitotriosidase showed the most impressive response to treatment [109]. Increase of the dose of enzyme also influenced the plateau that was reached for chitotriosidase. Similar conclusions were drawn by Vellodi et al., who retrospectively studied acid phosphatase, ACE, and chitotriosidase in a cohort of 28 paediatric patients. Chitotriosidase levels showed the steepest decrease and exhibited the lowest residual variance, making it the preferred biomarker [110]. Deegan and co-workers followed 20 adult patients during enzyme therapy using the same markers with the addition of CCL18 [111]. They argued that the measurement of CCL18 was superior to chitotriosidase for several reasons: CCL18 was detectable in all patients, while few patients had deficient chitotriosidase activity and CCL18 more closely reflected the visceral and platelet response to therapy. The authors plotted percent reductions of biomarkers against reductions in visceral volumes and found the closest relationship for reduction in splenic volume in 10 patients with CCL18. However, for excess liver volume, which included 17 patients, chitotriosidase was superior. Overall, in the assessment of response, again chitotriosidase showed the steepest decline. Since in this study the bone marrow compartment was not taken into account, it might well be that the chitotriosidase levels still provide the best indicator of overall Gaucher cell burden. In addition, Deegan and co-workers showed that platelet response was inversely correlated to CCL18 levels, but the relationship with the other markers was not given. As Vellodi and co-workers have also argued, there is no golden standard to assess Gaucher cell burden and therefore care must be taken with the interpretation of marker studies [110]. We feel that chitotriosidase, when appropriately measured, preferably using the newly developed deoxy-substrate, provides the most sensitive tool for follow-up. In the case of chitotriosidase deficiency, CCL18 is the second best biomarker. Based on a comprehensive review of our biochemical and clinical data, we have applied rules for the clinical application of chitotriosidase as follows: a level of 15,000 nmol/ml/h in a patient who does not carry the chitotriosidase mutation, and 7,500 nmol/ml/h in patients that are heterozygous for the chitotriosidase mutation is always related to extensive disease and is viewed as a separate indication for the initiation of treatment. Also a decrease of at least 15% in chitotriosidase activity should be achieved in 12 months of treatment and if not, the dose should be increased [112]. Currently the worldwide Gaucher Registry has formulated recommendations for the follow-up of patients, including chitotriosidase as the most valuable biomarker [113]. This has also been done for pediatric patients [114]. Long term follow-up studies that address the occurrence of clinical complications and the influence of co-morbidities, medication, treatment interruptions and dose changes should further elucidate the value of these biomarkers.

ONGOING SEARCH FOR ADDITIONAL SURROGATE MARKERS OF GAUCHER DISEASE

Plasma levels of the hydrolase chitotriosidase and the chemokine CCL18 are useful tools in clinical management of Gaucher patients. Both parameters reflect the body burden on storage cells. However, they render no impression of particular clinical complications. At present, the need remains for a surrogate marker in plasma that reflects skeletal disease in Gaucher patients. Currently, plasma of Gaucher patients is intensively investigated for such a marker using advanced analysis of plasma proteins such as surface-enhanced laser desorption/ionization mass spectrometry. Candidate proteins have already been identified in our laboratory but their value as surrogate markers for skeletal disease requires confirmation by retrospective analysis of a large cohort of Gaucher patients.

REFERENCES

1. Patrick, A.D., Short communications: a deficiency of glucocerebrosidase in Gaucher's disease, *Biochem. J.,* 97, 17C, 1955.
2. Brady, R.O., Kanfer, J.N., and Shapiro, D., Metabolism of glucocerebrosides. II. Evidence of an enzymatic deficiency in Gaucher's disease, *Biochem. Biophys. Res. Commun.*, 18, 221, 1965.
3. Beutler, E. and Kuhl, W., The diagnosis of the adult type of Gaucher's disease and its carrier state by demonstration of deficiency of beta-glucosidase activity in peripheral blood leukocytes, *J. Lab. Clin. Med.*, 76, 747, 1970.
4. Barranger, J.A. and Ginns, E.I., Gaucher disease, in *The Metabolic Basis of Inherited Disease,* Scriver, C.R., Sly W.S. and Valle, D. Eds., McGraw-Hill, New York, 1989, 1677.
5. Beutler, E. and Grabowski, G., Gaucher disease, in *The Metabolic Basis of Inherited Disease,* Scriver, C.R., Sly W.S. and Valle, D. Eds., McGraw-Hill, New York, 1995, 2641.
6. Aerts, J.M. et al., Deficient activity of glucocerebrosidase in urine from patients with type 1 Gaucher disease, *Clin. Chim. Acta*, 158, 155, 1986.
7. Schnabel, D., Schroder, M., and Sandhoff, K., Mutation in the sphingolipid activator protein 2 in a patient with a variant of Gaucher disease, *FEBS Lett.*, 284, 57, 1991.
8. Diaz-Font, A. et al., A mutation within the saposin D domain in a Gaucher disease patient with normal glucocerebrosidase activity, *Hum. Genet.*, 117, 275, 2005.
9. Busche, G. et al., Frequency of pseudo-Gaucher cells in diagnostic bone marrow biopsies from patients with Ph-positive chronic myeloid leukaemia, *Virchows Arch.,* 430, 139, 1997.
10. Dunn, P., Kuo, M.C., and Sun, C.F., Pseudo-Gaucher cells in mycobacterial infection: a report of two cases, *J. Clin. Pathol.*, 58, 1113, 2005.
11. Daniels, L.B. and Glew, R.H., beta-Glucosidase assays in the diagnosis of Gaucher's disease, *Clin. Chem.*, 28, 569, 1982.
12. Wenger, D.A. and Roth, S., Homozygote and heterozygote identification, *Prog. Clin. Biol. Res.*, 95, 551, 1982.
13. Svennerholm, L. et al., Prenatal diagnosis of Gaucher disease. Assay of the beta-glucosidase activity in amniotic fluid cells cultivated in two laboratories with different cultivation conditions, *Clin. Genet.* 19, 16, 1981.
14. Zimran, A. et al., Prenatal molecular diagnosis of Gaucher disease, *Prenat. Diagn.,* 15, 1185, 1995.
15. Ramsay, S.L. et al., Determination of oligosaccharides and glycolipids in amniotic fluid by electrospray ionisation tandem mass spectrometry: in utero indicators of lysosomal storage diseases, *Mol. Genet. Metab.*, 83(3), 231, 2004.

16. Boven, L.A. et al., Gaucher cells demonstrate a distinct macrophage phenotype and resemble alternatively activated macrophages, *Am. J. Clin. Pathol.*, 122(3), 359, 2004.

17. Suzuki, K., Glucosylceramide and related compounds in normal tissues and in Gaucher disease, *Prog. Clin. Biol. Res.*, 95, 219, 1982.

18. Nilsson, O. et al., The occurrence of psychosine and other glycolipids in spleen and liver from the three major types of Gaucher's disease, *Biochim. Biophys. Acta*, 712(3), 453, 1982.

19. Hillborg, P.O. and Svennerholm, L., Blood level of cerebrosides in Gaucher's disease, *Acta Paediatr.*, 49, 707, 1960.

20. Dawson, G. et al., Role of serum lipoproteins in the pathogenesis of Gaucher disease, *Prog. Clin. Biol. Res.*, 95, 253, 1982.

21. Nilsson, O. et al., Increased cerebroside concentration in plasma and erythrocytes in Gaucher disease: significant differences between type I and type III, *Clin. Genet.*, 22(5), 274, 1982.

22. Erikson, A. et al., Enzyme replacement therapy of infantile Gaucher disease, *Neuropediatrics*, 24(4), 237, 1993.

23. Meikle, P.J. et al., Diagnosis of lysosomal storage disorders: evaluation of lysosome-associated membrane protein LAMP-1 as a diagnostic marker, *Clin. Chem.*, 43(8 Pt 1):1325, 1997.

24. Hua, C.T. et al., Evaluation of the lysosome-associated membrane protein LAMP-2 as a marker for lysosomal storage disorders, *Clin. Chem.*, 44(10), 2094, 1998.

25. O'Brien, J.S. and Kishimoto, Y., Saposin proteins: structure, function, and role in human lysosomal storage disorders, *FASEB J.*, 5(3), 301, 1991.

26. Sun, Y. et al., Gaucher disease mouse models: point mutations at the acid {beta}-glucosidase locus combined with low-level prosaposin expression lead to disease variants, *J. Lipid Res.*, 46(10), 2102, 2005.

27. Inui, K. and Wenger, D.A., Concentrations of an activator protein for sphingolipid hydrolysis in liver and brain samples from patients with lysosomal storage diseases, *J. Clin. Invest.*, 72(5), 1622, 1983.

28. Chang, M.H. et al., Saposins A, B, C, and D in plasma of patients with lysosomal storage disorders, *Clin. Chem.*, 46(2), 167, 2000.

29. Meikle, P.J. et al., Newborn screening for lysosomal storage disorders: clinical evaluation of a two-tier strategy, *Pediatrics*, 114(4), 909, 2004.

30. Ginsberg, H. et al., Reduced plasma concentrations of total, low density lipoprotein and high density lipoprotein cholesterol in patients with Gaucher type I disease, *Clin. Genet.*, 26(2), 109, 1984.

31. Le, N.A. et al., Abnormalities in lipoprotein metabolism in Gaucher type 1 disease, *Metabolism*, 37(3), 240, 1988.

32. Cenarro, A. et al., Plasma lipoprotein responses to enzyme-replacement in Gaucher's disease, *Lancet*, 353(9153), 642, 1999.

33. Pocovi, M. et al., Beta-glucocerebrosidase gene locus as a link for Gaucher's disease and familial hypo-alpha-lipoproteinaemia, *Lancet*, 351(9120), 1919, 1998.

34. Vreeken, J. et al., A chronic clotting defect with some characteristics of excessive intravascular coagulation in a patient with Gaucher's disease, *Folia Med. Neerl.*, 10(6), 180, 1967.

35. Billett, H.H., Rizvi, S., and Sawitsky, A., Coagulation abnormalities in patients with Gaucher's disease: effect of therapy, *Am. J. Hematol.*, 51(3), 234, 1996.

36. Hollak, C.E. et al., Coagulation abnormalities in type 1 Gaucher disease are due to low-grade activation and can be partly restored by enzyme supplementation therapy, *Br. J. Haematol.*, 96(3), 470, 1997.

37. Freedman, S. and Puliafito, C.A., Peripheral retinal vascular lesions in a patient with Gaucher disease and factor XI deficiency. Case report, *Arch. Ophthalmol.*, 106(10), 1351, 1988.

38. Seligsohn, U. et al., Coexistence of factor XI (plasma thromboplastin antecedent) deficiency and Gaucher's disease, *Isr. J. Med. Sci.,* 12(12), 1448, 1976.

39. Rogowski, O. et al., Automated system to detect low-grade underlying inflammatory profile: Gaucher disease as a model, *Blood Cells Mol. Dis.,* 34(1), 26, 2005.

40. Barone, R. et al., Haemostatic abnormalities and lupus anticoagulant activity in patients with Gaucher disease type I, *J. Inherit. Metab. Dis.*, 23(4), 387, 2000.

41. Tuchmann, L.R., Suna, H., and Carr, J.J., Elevation of serum acid phosphatase in Gaucher's disease, *J. Mt. Sinai Hosp. N.Y.,* 23, 227, 1956.

42. Lam, K.W. and Desnick, R.J., Biochemical properties of the tartrate-resistant acid phosphatase activity in Gaucher disease, *Prog. Clin. Biol. Res.* 95, 267, 1982.

43. Ockerman, P.A. and Kohlin, P., Acid hydrolases in plasma in Gaucher's disease, *Clin. Chem.,* 15(1), 61, 1969.

44. Ockerman, P.A. and Kohlin, P., Tissue acid hydrolase activities in Gaucher's disease, *Scand. J. Clin. Lab. Invest.,* 22(1), 62, 1968.

45. Nakagawa, S. et al., Changes of serum hexosaminidase for the presumptive diagnosis of type I Gaucher disease in Tay-Sachs carrier screening, *Am. J. Med. Genet.,* 14(3), 525, 1983.

46. Natowicz, M.R., Prence, E.M., and Cajolet, A., Marked variation in blood beta-hexosaminidase in Gaucher disease, *Clin. Chim. Acta,* 203(1), 17, 1991.

47. Moffitt, K.D. et al., Characterization of lysosomal hydrolases that are elevated in Gaucher's disease, *Arch. Biochem. Biophys.,* 190(1), 247, 1978.

48. Hibbs, R.G. et al., A histochemical and electron microscopic study of Gaucher cells, *Arch. Pathol.,* 89(2), 137, 1970.

49. Moran, M.T. et al., Pathologic gene expression in Gaucher disease: up-regulation of cysteine proteinases including osteoclastic cathepsin K, *Blood,* 96(5), 1969, 2000.

50. Lieberman, J. and Beutler, E., Elevation of serum angiotensin-converting enzyme in Gaucher's disease, *N. Engl. J. Med.*, 294(26), 1442, 1976.

51. Silverstein, E. and Friedland, J., Elevated serum and spleen angiotensin converting enzyme and serum lysozyme in Gaucher's disease, *Clin. Chim. Acta,* 74(1), 21, 1977.

52. Silverstein, E., Pertschuk, L.P., and Friedland J., Immunofluorescent detection of angiotensin-converting enzyme (ACE) in Gaucher cells, *Am. J. Med.,* 69(3), 408, 1980.

53. Zimran, A. et al., Gaucher disease. Clinical, laboratory, radiologic, and genetic features of 53 patients, *Medicine* (Baltimore), 71(6), 337, 1992.

54. Hollak, C.E. et al., Marked elevation of plasma chitotriosidase activity. A novel hallmark of Gaucher disease, *J. Clin. Invest.,* 93(3), 1288, 1994.

55. Renkema, G.H. et al., Purification and characterization of human chitotriosidase, a novel member of the chitinase family of proteins, *J. Biol. Chem.,* 270(5), 2198, 1995.

56. Boot, R.G. et al., Cloning of a cDNA encoding chitotriosidase, a human chitinase produced by macrophages, *J. Biol. Chem.*, 270(44), 26252, 1995.

57. van Eijk, M. et al., Characterization of human phagocyte-derived chitotriosidase, a component of innate immunity, *Int. Immunol.,* 17(11), 1505, 2005.

58. Renkema, G.H. et al., Synthesis, sorting, and processing into distinct isoforms of human macrophage chitotriosidase, *Eur. J. Biochem.,* 244(2), 279, 1997.

59. Fusetti, F. et al., Structure of human chitotriosidase. Implications for specific inhibitor design and function of mammalian chitinase-like lectins, *J. Biol. Chem.,* 277(28), 25537, 2002.

60. Rao, F.V. et al., Crystal structures of allosamidin derivatives in complex with human macrophage chitinase, *J. Biol. Chem.*, 278(22), 20110, 2003.
61. Aguilera, B. et al., Transglycosidase activity of chitotriosidase: improved enzymatic assay for the human macrophage chitinase, *J. Biol. Chem.*, 278(42), 40911, 2003.
62. Boot, R.G. et al., The human chitotriosidase gene. Nature of inherited enzyme deficiency, *J. Biol. Chem.*, 273(40), 25680, 1998.
63. Boot, R.G. et al., Identification of a novel acidic mammalian chitinase distinct from chitotriosidase, *J. Biol. Chem.*, 276(9), 6770, 2001.
64. Choi, E.H. et al., Genetic polymorphisms in molecules of innate immunity and susceptibility to infection with Wuchereria bancrofti in South India, *Genes Immun.*, 2, 248, 2001.
65. Lehrnbecher, T. et al., Common genetic variants in the interleukin-6 and chitotriosidase genes are associated with the risk for serious infection in children undergoing therapy for acute myeloid leukaemia, *Leukemia,* 19(10), 1745, 2005.
66. Michelakakis, H. et al., Plasma tumor necrosis factor-a (TNF-a) levels in Gaucher disease, *Biochim. Biophys. Acta*, 1317(3), 219, 1996.
67. Allen, M.J. et al., Pro-inflammatory cytokines and the pathogenesis of Gaucher's disease: increased release of interleukin-6 and interleukin-10, *QJM.*, 90(1), 19, 1997.
68. Hollak, C.E. et al., Elevated levels of M-CSF, sCD14 and IL8 in type 1 Gaucher disease, *Blood Cells Mol. Dis.,* 23(2), 201, 1997.
69. Altarescu, G. et al., TNF-alpha levels and TNF-alpha gene polymorphism in type I Gaucher disease, *Cytokine,* 31(2), 149, 2005.
70. Moller, H.J. et al., Soluble CD163: a marker molecule for monocyte/macrophage activity in disease, *Scand. J. Clin. Lab. Invest.* Suppl., 237, 29, 2002.
71. Moller, H.J. et al., Plasma level of the macrophage-derived soluble CD163 is increased and positively correlates with severity in Gaucher's disease, *Eur. J. Haematol.,* 72(2), 135, 2004.
72. Boot, R.G. et al. Marked elevation of the chemokine CCL18/PARC in Gaucher disease: a novel surrogate marker for assessing therapeutic intervention, *Blood,* 103(1), 33, 2004.
73. Adema, G.J. et al., A dendritic-cell-derived C-C chemokine that preferentially attracts naive T cells, *Nature,* 387, 713, 1997.
74. Hieshima, K. et al., A novel human CC chemokine PARC that is most homologous to macrophage-inflammatory protein-1 alpha/LD78 alpha and chemotactic for T lymphocytes, but not for monocytes, *J. Immunol.* 159, 1140, 1997.
75. Morgan, M.A. et al., Serum ferritin concentration in Gaucher's disease, *Br. Med. J. (Clin. Res. Ed.),* 286(6381), 1864, 1983.
76. Lambotte, O. et al., High ferritin and low glycosylated ferritin may also be a marker of excessive macrophage activation, *J. Rheumatol.* 30(5), 1027, 2003.
77. Poll, L.W. et al., Correlation of bone marrow response with hematological, biochemical, and visceral responses to enzyme replacement therapy of nonneuronopathic (type 1) Gaucher disease in 30 adult patients, *Blood Cells Mol. Dis.,* 28(2), 209, 2002.
78. Bassan, R., Montanelli, A., and Barbui T., Interaction between a serum factor and T lymphocytes in Gaucher disease, *Am. J. Hematol.,* 18(4), 381, 1985.
79. Walker, E.M. Jr. and Walker, S.M., Effects of iron overload on the immune system, *Ann Clin. Lab. Sci.,* 30(4), 354, 2000.
80. Goldfarb, R.A., Atlas, D.H., and Goberman, P., Electrophoretic studies in Gaucher's disease, *Am. J. Clin. Pathol.,* 20, 963, 1950.
81. Pratt, P.W., Kochwa, S., and Estren, S., Immunoglobulin abnormalities in Gaucher's disease. Report of 16 cases, *Blood,* 31(5), 633, 1968.

82. Marti, G.E. et al., Polyclonal B-cell lymphocytosis and hypergammaglobulinemia in patients with Gaucher disease, *Am. J. Hematol.,* 29(4), 189, 1988.
83. Harder, H. et al., Coincidence of Gaucher's disease due to a 1226G/1448C mutation and of an immunoglobulin G lambda multiple myeloma with Bence-Jones proteinuria, *Ann. Hematol.,* 79(11), 640, 2000.
84. Zimran, A. et al., Incidence of malignancies among patients with type I Gaucher disease from a single referral clinic, *Blood Cells Mol. Dis.,* 34(3), 197, 2005.
85. Rosenbloom, B.E. et al., Gaucher disease and cancer incidence: a study from the Gaucher Registry, *Blood,* 105(12), 4569, 2005.
86. de Fost, M. et al., Increased incidence of cancer in adult Gaucher disease in Western Europe, *Blood Cells Mol. Dis.,* 2005, 36, 53, 2005.
87. Kaloterakis, A. et al., Systemic AL amyloidosis in Gaucher disease. A case report and review of the literature, *J. Intern. Med.,* 246(6), 587, 1999.
88. Deibener, J. et al., Enzyme replacement therapy decreases hypergammaglobulinemia in Gaucher's disease, *Haematologica,* 83(5), 479, 1998.
89. Brautbar, A. et al., Effect of enzyme replacement therapy on gammopathies in Gaucher disease, *Blood Cells Mol. Dis.,* 32(1), 214, 2004.
90. Stowens, D.W. et al., Skeletal complications of Gaucher disease, *Medicine* (Baltimore), 64(5), 310, 1985.
91. Pastores, G.M. et al., Bone density in Type 1 Gaucher disease, *J. Bone Miner. Res.,* 11(11), 1801, 1996.
92. Delmas, P.D. et al., The use of biochemical markers of bone turnover in osteoporosis, *Osteoporos. Int.,* 11 Suppl. 6, S2, 2000.
93. Schiffmann, R. et al., Decreased bone density in splenectomized Gaucher patients receiving enzyme replacement therapy, *Blood Cells Mol. Dis.,* 28(2), 288, 2002.
94. Fiore, C.E. et al., Bone ultrasonometry, bone density, and turnover markers in type 1 Gaucher disease, *J. Bone Miner. Metab.,* 20(1), 34, 2002.
95. Ciana, G. et al., Bone marker alterations in patients with type 1 Gaucher disease, *Calcif. Tissue Int.,* 72(3), 185, 2003.
96. Ciana, G. et al., Gaucher disease and bone: laboratory and skeletal mineral density variations during a long period of enzyme replacement therapy, *J. Inherit. Metab. Dis.,* 28(5), 723, 2005.
97. Lebel, E. et al., Bone density changes with enzyme therapy for Gaucher disease, *J. Bone Miner. Metab.,* 22(6), 597, 2004.
98. Wenstrup, R.J. et al., Gaucher disease: alendronate disodium improves bone mineral density in adults receiving enzyme therapy, *Blood,* 104(5), 1253, 2004.
99. Harinck, H.I. et al., Regression of bone lesions in Gaucher's disease during treatment with aminohydroxypropylidene bisphosphonate, *Lancet,* 2(8401), 513, 1984.
100. Charrow, J., Ashkenazi Jewish genetic disorders, *Fam. Cancer,* 3(3–4), 201, 2004.
101. Chamoles, N.A. et al., Gaucher and Niemann-Pick diseases — enzymatic diagnosis in dried blood spots on filter paper: retrospective diagnoses in newborn-screening cards, *Clin. Chim. Acta,* 317(1–2), 191, 2002.
102. Grabowski, G.A., Gaucher disease: considerations in prenatal diagnosis, *Prenat. Diagn.,* 20(1), 60, 2000.
103. Colburn, W.A., Optimizing the use of biomarkers, surrogate endpoints, and clinical endpoints for more efficient drug development, *J. Clin. Pharmacol.,* 40 (12 Pt 2), 1419, 2000.
104. Whitfield, P.D., et al., Correlation among genotype, phenotype, and biochemical markers in Gaucher disease: implications for the prediction of disease severity, *Mol. Genet. Metab.,* 75(1), 46, 2002.

105. Barton, N.W., et al., Replacement therapy for inherited enzyme deficiency — macro-phage-targeted glucocerebrosidase for Gaucher's disease, *N. Engl. J. Med.,* 324(21), 1464, 1991.
106. Casal, J.A., et al., Relationships between serum markers of monocyte/macrophage activation in type 1 Gaucher's disease, *Clin. Chem. Lab. Med.,* 40(1), 52, 2002.
107. vom Dahl, S., Poll, L.W., and Haussinger, D., Clinical monitoring after cessation of enzyme replacement therapy in M. Gaucher, *Br. J. Haematol.,* 113(4), 1084, 2001.
108. Zhao, H., Bailey, L.A., and Grabowski, G.A., Enzyme therapy of Gaucher disease: clinical and biochemical changes during production of and tolerization for neutral-izing antibodies, *Blood Cells Mol. Dis.,* 30(1), 90, 2003.
109. Cabrera-Salazar, M.A., et al., Correlation of surrogate markers of Gaucher disease. Implications for long-term follow up of enzyme replacement therapy, *Clin. Chim. Acta,* 344(1–2), 101, 2004.
110. Vellodi, A., Foo, Y., and Cole, T.J., Evaluation of three biochemical markers in the monitoring of Gaucher disease, *J. Inherit. Metab. Dis.,* 28(4), 585, 2005.
111. Deegan, P.B., et al., Clinical evaluation of chemokine and enzymatic biomarkers of Gaucher disease, *Blood Cells Mol. Dis.,* 35, 259, 2005.
112. Hollak, C.E., Maas, M., and Aerts, J.M., Clinically relevant therapeutic endpoints in type I Gaucher disease, *J. Inherit. Metab. Dis.,* 24 Suppl. 2, 97, 2001.
113. Weinreb, N.J. et al., Gaucher disease type 1: revised recommendations on evaluations and monitoring for adult patients, *Semin. Hematol.,* 41, 15, 2004.
114. Baldellou, A. et al., Paediatric non-neuronopathic Gaucher disease: recommendations for treatment and monitoring, *Eur. J. Pediatr.,* 163, 67, 2004.

CHAPTER 15

Imaging in Gaucher Disease, Focusing on Bone Pathology

Mario Maas and Erik M. Akkerman

CONTENTS

AIM

The aim of this chapter is to illustrate the strengths and weaknesses of various imaging techniques in adult patients with Gaucher disease type 1. The focus is primarily on imaging of invasion of bone; invasion of organs is not extensively covered. Outside the scope of this chapter is imaging of complications of Gaucher disease in other compartments than the skeleton, i.e., pulmonary hypertension is not included.

BACKGROUND

Since imaging of pathology is very different in various places throughout the world, an imaging protocol that can fit everywhere is impossible to produce. However, a few guidelines will be provided. From a radiological point of view, it is very important to realize that Gaucher disease is a multi-compartment-involving disease. The invasion of liver and spleen is a separate entity compared to invasion of the skeletal system. In the opinion of the Gaucher Institute of The Netherlands, it is mandatory to reflect this multi-compartment in imaging protocols developed to evaluate overall Gaucher burden. Each compartment needs its own imaging protocol and its own evaluation. One cannot evaluate visceromegaly by looking at bone. The extent of bone disease is also not adequately evaluated by determining the concentrations of glucocerebroside in tissue or plasma [1].

EVALUATION OF SKELETAL INVOLVEMENT

The skeletal manifestations of Gaucher disease are very variable. This variation occurs within groups but also within an individual. However, skeletal disease is a major complication of GD type 1, afflicting up to 75% of patients [2]. Manifestations that are described from plain films include: osteopenia, osteosclerosis, bone infarction, subcortical necrosis, remodeling (Erlenmeyer flask appearance), fractures, and joint collapse. It is thought that these changes represent three basic types of pathophysiological processes: (1) infarction of bone and bone marrow, (2) focal replacement of normal marrow by infiltration with Gaucher cells, and (3) osteopenia that can be focal or diffuse [1]. Since this all concerns bone marrow infiltration, it is absolutely mandatory that this compartment is evaluated with a bone marrow evaluation tool. From a radiological point of view, Magnetic Resonance Imaging (MRI) is the noninvasive state-of-the-art evaluation of bone marrow.

MRI

Qualitative

Magnetic Resonance Imaging (MRI) is the most sensitive technique to evaluate bone marrow invasion [3]. In order to understand bone marrow changes due to Gaucher disease, it is important to understand MRI in healthy bone marrow.

In adult long bones, fat is the predominant contributor to the bone marrow signal pattern on MR images [3,4]. This is explained by the fact that the T1 relaxation time is short and T2 is relatively long. A typical signal pattern on T1-weighted images of yellow marrow is roughly similar to the subcutaneous fat [3,5]. Normal red cellular marrow shows low (hypo-intense) signal intensity when compared to yellow marrow on T1-weighted images. On fast spin-echo T2-weighted sequences, both fat and water will show a high (hyper intense) signal intensity [3,5].

Since the content of the bone marrow is responsible for its appearance on MRI, we first will elaborate on this. Virtually the entire fetal marrow space is dedicated to red marrow at birth. In the immediate postnatal period, conversion from red to yellow marrow begins in an orderly pattern [3]. This conversion begins in the terminal phalanges and progresses from peripheral (appendicular) towards central (axial, vertebral-sacral) marrow. In an individual long bone it progresses from dia- physis to metaphysis. The cartilaginous epiphysis and apophysis can be characterized as yellow marrow from the moment of ossification. By the time a person is 25 years old, marrow conversion is complete and the adult pattern is achieved [3]. Red marrow predominantly is concentrated in the axial skeleton and proximal parts of the appen- dicular skeleton. However, there is a great variation possible, i.e., hematopoietic marrow may occupy up to of the femoral shaft. "Yellow marrow" is composed predominantly of fat cells (15% water, 80% fat and 5% protein). "Red marrow" is considered hematopoietically active marrow. It contains approximately 40% water, 40% fat and 20% protein [3].

Especially on T1-weighted images, the contrast between bright fatty marrow and darker (hypo intense) red marrow as well as most pathological processes (also hypo intense) is enhanced. When general marrow screening is performed, the primary anatomic sites for evaluation are the spine, particularly the lumbar spine due to the larger vertebral bodies, and the pelvis and proximal femurs. Since these areas contain a large percentage of red marrow throughout life it is logical to focus on this.

The use of MRI in analyzing bone marrow in Gaucher disease has been described as early as 1986 [1,6]. In general the focal replacement normal marrow contained is replaced by glucocerebroside-loaded macrophages, Gaucher cells, leading to a lowering of signal intensities both on T1- and T2-weighted images [1,7–9]. The markedly shorter T2 relaxation time and long T1 relaxation time appears to be the principal cause [6]. The presence of fast-exchanging protons in the substrate has been proposed as the most likely cause of this [6].

In Gaucher disease both homogeneous and heterogeneous patterns of involve- ment are encountered [7,10]. Some patients may have a diffuse infiltration and others may have a patchy infiltration, with islands of preserved marrow [7]. Marrow involvement generally follows the distribution of red cellular marrow in Gaucher disease, progressing from axial to peripheral and in a long bone from proximal to distal with a tendency to spare epiphysis and apophysis [1,11,12].

The first MRI results are from the group from Massachusetts General Hospital [1]. They described the bone marrow abnormalities on T1- and T2-weighted images in 24 patients with type 1 Gaucher disease, using a 0.6 Tesla magnet. The abnor- malities on T1- and T2-weighted images consisted of lowering of signal intensity on both sequences: i.e., the fatty bone marrow, which provides a high signal intensity was displaced/changed into a low signal intensity, most likely caused by Gaucher cell infiltration. The typical appearance of Gaucher cell infiltration of the lumbar spine and legs is shown in one patient in Figure 15.1. Mark the difference in aspect between the homogeneous low signal intensity infiltration pattern of the lumbar spine and the patchy low signal intensity in the legs.

Islands of high signal intensity on T2-weighted images and low signal intensity on T1, representing fluidlike lesions, are frequently encountered in patients with no

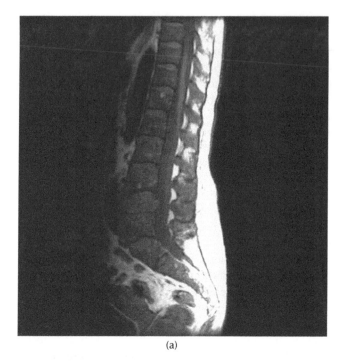

(a)

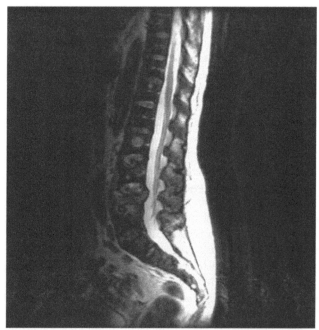

(b)

Figure 15.1 Sagittal T1 (a) and T2 (b) weighted images of the lumbar spine in a 59-year-old female Gaucher disease patient. Note the marked diffuse very low signal intensity of the axial bone marrow. A vertebral collapse is seen in L3 and L4.

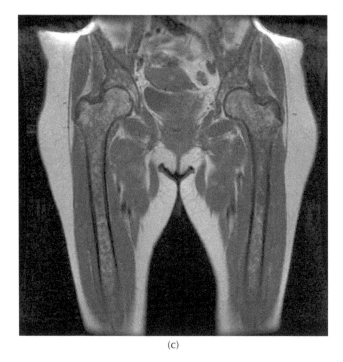

(c)

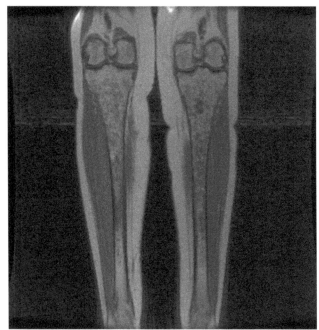

(d)

Figure 15.1 Coronal T1 weighted image of the femurs (c) and lower legs (d) of the same patient. The infiltration pattern is more patchy with infiltration of diaphysis and proximal epiphysis.

bone pain. These abnormalities with high signal on T2 may persist for a number of years, making an acute bone event unlikely. Accumulation of Gaucher cells in the bone marrow cavity produces intramedullary extravascular compression, with vascular obstruction and secondary ischemic events [7]. In others patients however, areas of high signal intensity on T2 or TSTIR will be interpreted as sign of active disease or active process within bone marrow [7]. This is thought to be the cause of these changes and this still needs clarification. One may also encounter islands of fat (high signal on T1- and T2-weighted images) in areas of older infarction in patients without bone pain, most likely representing evolving infarction [1].

A number of very important observations were described in Rosenthal's early MRI study [1]: (a) in every patient the axial bone marrow of the lumbar spine was involved when involvement of the lower extremity was variable; (b) the MR abnormalities were patchy or diffuse in the extremities when present, and diffuse only in the spine; (c) the distribution of abnormalities revealed a distinct pattern, with a proximal involvement prior to distal involvement, except for epiphyseal and apophyseal regions (femoral head and trochanteric region), which were often spared despite extensive involvement of metaphysis and diaphysis; (d) no correlation was found between abnormalities on plain film compared to MRI, while a strong correlation was observed between MRI abnormalities and clinical course of the patients. It was concluded that marrow infiltration in GD may remain silent for long periods of time before the appearance of clinically significant bone disease. These bones can be considered "bones at risk," especially since skeletal complications appeared to develop after extended bone marrow invasion.

Although conventional MRI evaluation is extremely important to establish the extent of disease in Gaucher disease patients, treating clinicians are lacking a quantitative measurement, something that is needed in treatment follow-up evaluation.

Quantitative (QCSI)

In order to quantify bone marrow changes in Gaucher disease, Quantitative Chemical Shift Imaging (QCSI) was explored [13]. The MR-signal for a normal MR-image originates from two types of hydrogen nuclei: nuclei in water molecules and nuclei in fat molecules [14]. The two types of nuclei have a slightly different MR frequency due to the so-called "chemical shift" effect. The Dixon method uses this frequency difference to separate the signals from water and fat, which makes it possible to quantify the fat signal fraction, F_f, hence, the Dixon Quantitative Chemical Shift Imaging (Dixon QCSI). The group from Massachusetts General Hospital evaluated this technique, in hematological bone marrow disorders in 1992 [15]. They investigated the use of QCSI in healthy volunteers, patients with leukemia or aplastic anemia on a 0.6 Tesla magnet. It was shown that the fat fraction, measured by this QCSI technique, was the single best discriminator between the groups. The same research group also explored the use of this technique to quantify longitudinal changes in bone marrow that occur during induction chemotherapy in patients with acute leukemia [16]. Results were correlated with bone marrow biopsy results. QCSI data showed sequential increase in fat fractions among responding patients, consistent with biopsy-confirmed clinical remission. It was concluded that QCSI proved useful in assessing treatment response in acute leukemia during early bone marrow regeneration and later in ascertaining remission or relapse. Furthermore, an additional described benefit of QCSI was the ability to sample a large portion of marrow.

The same group tested the QCSI technique in vertebral bone marrow in patients with Gaucher disease [13]. The measured fat fractions were correlated with quantitative analysis of marrow triglycerides and glucocerebrosides. An MR Spectroscopy performed in these surgical marrow specimens showed a single fat and water peak, thus validating use of QCSI. Glucocerebroside concentrations were higher in Gaucher marrow and inversely correlated with triglyceride concentrations. It was concluded that QCSI is a sensitive noninvasive technique for evaluating bone marrow infiltration in Gaucher disease, showing great promise as a noninvasive method to monitor bone marrow response to treatment. In order to validate bone marrow response data acquired by QCSI, an analysis of the lipids of normal and Gaucher bone marrow is performed [17]. In normal marrow, triglycerides were by far the most abundant lipid (278 ± 70 mg/gm wet weight); the concentration of glucocerebroside in normal marrow was 0.061 ± 0.06 mg/gm wet weight. Gaucher marrow had dramatically lower triglyceride levels of 82% (51 ± 53 mg/gm wet weight) and as expected marked elevation of glucocerebroside (7.1 ± 3.4 mg/gm wet weight) [18]. It was concluded that these data support a model of bone marrow alteration in Gaucher disease in which triglyceride-rich adipocytes are progressively replaced by Gaucher cells, leading to an overall reduction in total lipid content. This phenomenon provides an explanation for the changes found in QCSI measurements in Gaucher patients [17].

In the Academic Medical Center in Amsterdam, this QCSI technique has been explored [18–20]. Important issues that were addressed were the reproducibility of our protocol and the normal mean fat fraction within a healthy adult population [19]. The measured mean values in the lumbar spine were 0.37 (SD 0.08). The SD is due to repeating measurements, slice positioning and contour drawing were very small. It was concluded that reproducibility was excellent. Our set-up and protocol are described in more detail in order to enable other centers to implement this valuable technique.

We apply this method to measure F_f in the lumbar vertebrae (L3, L4, and L5). As shown in Figure 15.2, a coronal measurement slice (if necessary slightly tilted to transversal) is set out on a sagittal localizer image, such that the posterior part of the vertebrae of interest is optimally visualized.

In the original Two Point Dixon technique, which we use, two sets of acquisitions are done [14]. In one acquisition, the water signal (W) and the fat signal (F) are "in-phase" (I), which means that at any moment in time, the fat and water spins point in the same direction, and their signals add up in the image: $I = W + F$. In the second acquisition, fat and water have opposed phases (O): fat and water spins point in opposite directions, and the resulting image shows the magnitude of the difference between the fat and water signals: $O = |W - F|$. (See Figure 15.2.)

In order to separate the water and fat signals, we need to know which signal is stronger in each pixel. The phase-difference between the two acquisitions helps us to sort out regions with water dominant signals and regions with fat dominant signals. In the example of Figure 15.3, water dominant regions are gray in the

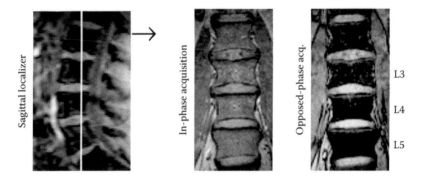

Figure 15.2 Localizer and Dixon acquisitions of the lumbar vertebrae L3, L4, and L5.

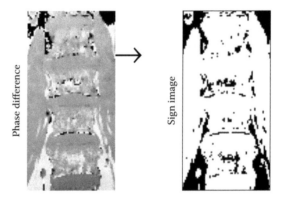

Figure 15.3 From the phase difference between the Dixon acquisitions the sign image is deduced.

phase-difference image and fat dominant regions are either white or black. A complication may be formed by inhomogeneities in the main magnetic field, which cause gray scale intensity variations across the phase-difference image. We developed an algorithm [18] that encompasses all these effects and is able to produce a sign image (S) (Figure 15.3) from the phase-difference which distinguishes water dominant regions (white, S = +1) from fat dominant regions (black, S = -1).

With the help of the sign image, water and fat images are obtained by simple algebraic manipulations. For the water image (W) we have: $W = I + S \cdot O$ (Figure 15.4), and for the fat image (F) we have: $F = I - S \cdot O$ (Figure 15.5).

The last step is to compute F_f, again by elementary algebra, as shown in Figure 15.6: $F_f = F / (F + W)$. In order to read the F_f values from the image easily, we display the F_f image with a color scale: the color strip next to the image tells which color corresponds to which F_f value from 0 (black) to 1 (white).

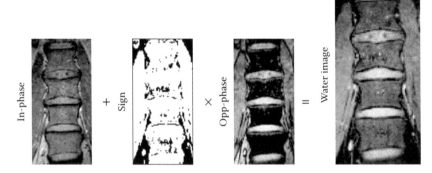

Figure 15.4 Calculation of the water image.

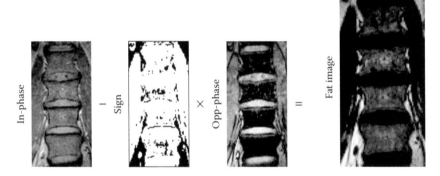

Figure 15.5 Calculation of the fat image.

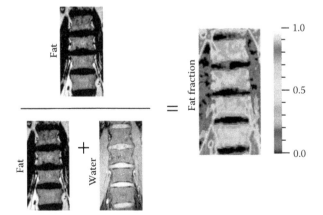

Figure 15.6 (See color insert following page 304.) Calculation of the fat fraction and display of the fat fraction with a color scale.

Clinical Application of QCSI

The important use of this technique is the ability to detect response of the bone marrow compartment to ERT. The Rosenthal group described this in 1995 in a paper in which both children and adults were evaluated [21]. Later our group described the effect of individualized doses of ERT in adult type 1 GD patients. [20]. We found an increase in fat fraction in almost all patients within the first year of treatment. Normalization was established within 4–5 years [20,21]. This technique is currently standard care in the Academic Medical Center, providing us with data on over 70 patients with longitudinal data in over 60 patients (Dutch and international). In Figure 15.7, an example is given of an adult type 1 Gaucher disease patient with regular QCSI measurements.

Although this technique was easily adapted from the Rosenthal group, it still is not used worldwide. Therefore semi-quantitative scoring systems are developed on several sites [2,7,8,10,22,23].

Semi-Quantitative

The first group to use a semi quantitative technique, which served as a basis for other alternatives, was the Boston group [10]. In their system, anatomical sites of involvement of bone marrow in the lower extremity were scored. The pattern of infiltration was basically centrifugal with sparing of epiphyses and apophyses. The highest numbered site of involvement as seen on MRI is assigned to the patient. In another study by Hermann et al., both sites of involvement (femurs and pelvis) and Signal Intensity (SI) on T1 and T2* were used [7]. Using SI three groups were made: group A, normal, high SI on T1 and T2, no increase on T2; group B, infiltrated, low SI on T1 and T2; group C marrow infiltration with active marrow process. The most severely involved group C showed clear correlation with clinical bone pain. Terk et al. [8] modified the Hermann classification by adding AVN and adding heterogeneity. A large number of patients was evaluated (n=62) on a 1.5 Tesla system. Correlation between AVN and higher grades of involvement of bone marrow was seen.

Semi-quantitative scoring systems are also described for detecting response to therapy [2,23]. The Terk group evaluated the use of SI changes on T1-weighted images of pelvis and femurs in 42 adult patients. Response to therapy was detected in 67% [22]. The German group of Poll et al. [2] evaluated response to therapy in 30 adult patients using MRI of the entire lower extremity and T1 and T2*. A combination of sites of involvement (Dusseldorf score) and Signal intensity was used. A response rate of 63% was found.

All visual scoring systems mentioned are restricted to evaluation of the peripheral bone marrow of the lower extremity. However, none of the published semi-quantitative scoring systems have evaluated axial bone marrow nor have they been compared to Dixon QCSI data. So far the only scoring system that produces this is the Bone Marrow Burden (BMB) score [23]. This BMB score is a combination of scoring systems of the peripheral skeleton already described with addition of the axial bone marrow component. It incorporates both the visual interpretation of the signal

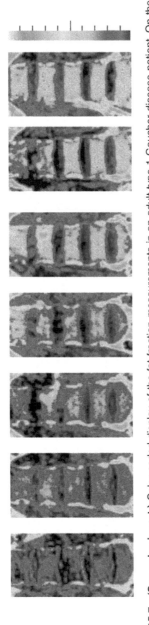

Figure 15.7 **(See color insert.)** Color coded display of the fat fraction measurements in an adult type 1 Gaucher disease patient. On the left (prior to Enzyme replacement therapy) the mean fat fraction is 0.08. The annual display shows a marked increase of fat fraction each year with a normal fat fraction of 0.44 on the latest measurement.

intensities and the geographic location of the disease on conventional MR images of the spine and femur. In the *lumbar spine* the SI of the marrow on T1-weighted sequences is compared to signal intensity of a healthy intervertebral disc [24]. The bone marrow SI on T2-weighted sequences was evaluated with the presacral fat as a normal iso-intense reference. In addition the infiltration pattern (patchy vs. diffuse) of the disease within the vertebral bodies is graded according to the distinction in pattern in evaluating hematological malignancies [2,24]. The patchy infiltration pattern in Gaucher disease consists of localized areas of abnormal marrow (low signal intensity on T1- and T2-weighted images) on a background of normal bone marrow (high signal intensity and T1- and T2-weighted images). In the diffuse pattern, the bone marrow is completely replaced. The diffuse pattern is considered to reflect a more advanced stage of the disease, as in multiple myeloma [1,25,26]. Furthermore, the absence of fat in the basivertebral vein region was added as a score, in a binominal fashion (present/absent). Since disappearance of fat at this site had been observed in our Gaucher disease patient group and from the literature, it was known to be an early sign of bone marrow invasion in malignancy [27], and it was included in the BMB.

In the *femurs* the marrow signal intensities on T1-weighted and T2-weighted images were graded compared to the signal intensity of subcutaneous fat, using a classification slightly modified from the one used in earlier studies [9,23]. The intensities were scored as hyperintense, slightly hyperintense, isointense, slightly hypo-intense and hypo-intense. Gaucher disease is expected to show hypo-intense signal intensity changes [1,8,10,11]. In severe cases, a mixed pattern of signal intensity is seen in the extremities with areas of high and low signal intensity especially on T2-weighted sequences. The femurs were divided according to three sites of involvement: proximal epiphysis/apophysis, (meta-) diaphysis, and distal epiphysis. This is a modification of the sites of involvement score [10,13].

Thirty patients were analyzed (16 male, and 14 female) with a mean age of 39.3 years (range 12–71). The bone marrow burden score BMB ranged from 3–14 points. The BMB is feasible with a good interobserver and intra-observer variability. There was a significant correlation between *QCSI* and BMB with = -0,78 (p<0.0001). For longitudinal follow-up, the measurements of 12 patients with a follow-up interval of 40 months were analyzed. When we analyzed our data in this way, we found 7 responders in the BMB Femurs group (58%) and 9 responders in the BMB overall group (75%). In our population we found more responders when a combination of visual scoring BMB of lumbar spine and femurs is used. One must consider that when no response is seen in the peripheral bone marrow, response may very well be present in the axial marrow.

Bone Marrow MRI in Pediatric Population with Gaucher Disease

In children, due to the normal age-related distribution of hematological bone marrow, with a low signal on T1-weighted images, the interpretation of MR images concerning Gaucher infiltration is severely hampered. Olsen et al. [28] performed routine control MRI of the spine in a group of 14 children with Gaucher disease (type 1 and 3) (age range 1.2 –14.1 years). Signal intensity of the lumbar disc was compared to bone marrow SI. A lower signal on T1 was considered abnormal above age 1 year. Lower or iso-intense signal intensity was abnormal in children older than 5 years. Therefore the use of lumbar spine axial bone marrow evaluation in children cannot be advised. For practical purposes, it might be wise to evaluate the bone marrow compartment in the lower extremities in children in order to detect extent of disease. Since there normally is rapid fatty replacement in the peripheral bone marrow, the appearance of low signal bone marrow both on T1- and T2-weighted images may serve as a marker for severe bone marrow involvement.

ROLE FOR IMAGING OTHER THAN MRI IN GAUCHER DISEASE

It was concluded that skeletal alterations as seen on MRI are far more extensive than would be suspected from either conventional radiographs or Computed Tomography (CT) [1,29]. Since there is no relevant data on using CT in bone marrow evaluation in Gaucher disease, this is not advised.

Conventional Radiology

Although advisable, in many centers worldwide is in not easy to have regular bone marrow evaluation with MRI in Gaucher disease patients. Since clinicians need to be informed about the status of the skeletal system concerning invasion with Gaucher cells, it often is thought that conventional plain radiography, being widely available and relatively inexpensive, is an acceptable alternative. A substantial Gaucher cell burden may be present without changes apparent on plain films. Gaucher cell manifestations on plain films may consist of osteoarticular and medullar bone infarction, focal lytic lesions, endosteal scalloping, Erlenmeyer flask deformity, pathological fracture, subperiosteal hemorrhage, periosteal new bone formation and generalized osteopenia [6,9,12]. Examples of classic Gaucher disease are shown in Figure 15.8.

Plain film can be used in specific situations such as in patients with localized pain to detect complications such as fractures or avascular necrosis. Likewise, the follow-up of arthroplasty is easily evaluated with plain radiography.

However there are many publications showing the very limited value of conventional radiography [1,8,11,12,29]. Plain radiography by no means has the capability to evaluate the extent of involvement of the skeletal system or to evaluate response to therapy. Conventional radiography is inadequate to assess the presence and extent of bone involvement [7]. Furthermore, there is no scientific evidence that a skeletal survey to evaluate focal or diffuse disease is advisable. The radiation dose of such a skeletal survey, especially of the axial skeleton, is high and therefore unadvisable.

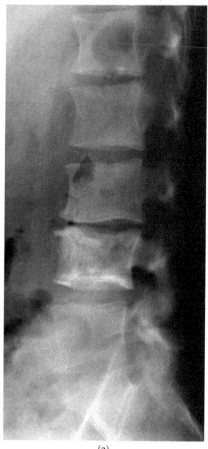

(a)

Figure 15.8 Conventional images of an adult type Gaucher disease male patient are shown. The classical image spectrum is present in this patient with a vertebral collapse, avascular necrosis of the right hip, Erlenmeyer Flask appearance of both femurs with marrow infarction, and both lytic and sclerotic lesions in the bones around the knee.

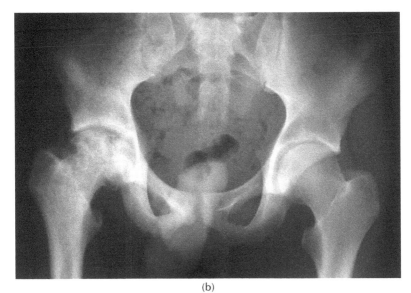

(b)

Figure 15.8 (Continued).

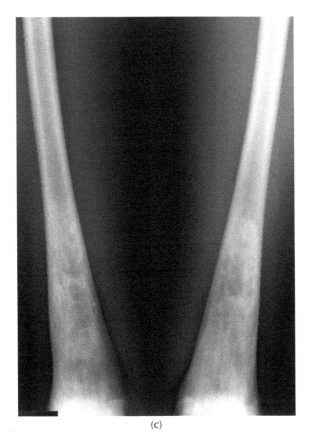

(c)

Figure 15.8 (Continued).

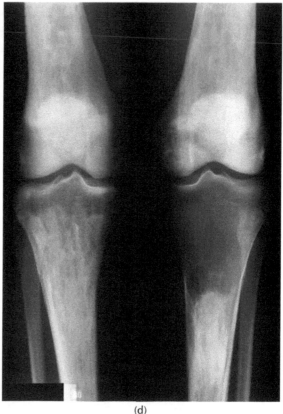

(d)

Figure 15.8   (Continued).

REFERENCES

1. Rosenthal, D.I. et al., Evaluation of Gaucher disease using Magnetic Resonance Imaging, *J. Bone Surg. Am.*, 68, 802, 1986.
2. Poll, L.W. et al., Magnetic resonance imaging of bone marrow changes in Gaucher disease during enzyme replacement therapy: first German long-term results, *Skeletal Radiol.,* 30, 496, 2001.
3. Vogler, B. III and Murphy, W.A., Bone marrow imaging, *Radiology*, 168, 679, 1988.
4. Dooms, G.C. et al., Bone marrow imaging: magnetic resonance studies related to age and sex, *Radiology,* 155, 429, 1985.
5. Steiner, R.M. et al., Magnetic resonance imaging of bone marrow: diagnostic value in diffuse hematologic disorders, *Magnetic Resonance Quarterly,* 6, 17, 1990.
6. Lanir, A. et al., Gaucher disease: assessment with MR imaging, *Radiology,* 161, 239, 1986.
7. Hermann, G. et al., MR imaging in adults with Gaucher disease type I: evaluation of marrow involvement and disease activity, *Skeletal Radiol.,* 22, 247, 1993.
8. Terk, M.R. et al., MR imaging of patients with type 1 Gaucher's disease: relationship between bone and visceral changes, *Am. J. Roentgenol.,* 165, 599, 1995.

9. O'Keefe, D. and Rosenthal, D.I., Computed tomography and magnetic resonance imaging in Gaucher's disease, in *MRI and CT of the Musculoskeletal System: a Text-Atlas*, Bloem, J.L. and Sartoris, D.J., Eds., William & Wilkins, Baltimore, 1992, 130.

10. Cremin, B.J., Davey, H. and Goldblatt, J., Skeletal complications of type I Gaucher disease: the magnetic resonance features, *Clin. Radiol.*, 41, 244, 1990.

11. Hermann, G. et al., Gaucher's disease type 1: assessment of bone involvement by CT and scintigraphy, *Am. J. Roentgenol.*, 147, 943, 1986.

12. Rosenthal, D.I. et al., Quantitative imaging of Gaucher disease, *Radiology*, 185, 841, 1992.

13. Johnson, L.A. et al., Quantitative chemical shift imaging of vertebral bone marrow in patients with Gaucher disease, *Radiology*, 182, 451, 1992.

14. Dixon, W.T., Simple proton spectroscopic imaging, *Radiology*, 153, 189, 1984.

15. Rosen, B.R. et al., Hematologic bone marrow disorders: quantitative chemical shift MR imaging, *Radiology*, 169, 799, 1988.

16. Gerard, E.L. et al., Compositional changes in vertebral bone marrow during treatment for acute leukemia: assessment with quantitative chemical shift imaging, *Radiology*, 183, 39, 1992.

17. Miller, S.P.F. et al., Analysis of the lipids of normal and Gaucher bone marrow, *J. Lab. Clin. Med.*, 127, 353, 1996.

18. Akkerman, E.M. and Maas, M., A region-growing algorithm to simultaneously remove dephasing influences and separate fat and water in two-point Dixon imaging. In: *Proceedings of the Society for Magnetic Resonance in Medicine and the European Society for Magnetic Resonance in Medicine and Biology*, 1995, 649.

19. Maas, M. et al., Dixon quantitative chemical shift MRI for bone marrow evaluation in the lumbar spine: a reproducibility study in healthy volunteers, *J. Comput. Assist. Tomogr.*, 25, 691, 2001.

20. Hollak, C.E. et al., Dixon Quantitative chemical shift imaging is a sensitive tool for evaluation of bone marrow responses to individualized doses of enzyme supplementation therapy in type 1 Gaucher disease, *Blood Cells Mol. Dis.*, 27, 1005, 2001.

21. Rosenthal, D. et al., Enzyme replacement therapy for Gaucher disease: skeletal responses to macrophage-targeted glucocerebrosidase, *Pediatrics,* 32, 33, 1995.

22. Terk, M.R., Dardashti, S., and Liebman, H.A., Bone marrow response in treated patients with Gaucher disease: evaluation by T1-weighted magnetic Resonance images and correlation with reduction in liver and spleen volume, *Skeletal Radiol.*, 29, 563, 2000.

23. Maas, M. et al., Quantification of bone involvement in Gaucher disease: MR imaging bone marrow burden score as an alternative to Dixon quantitative chemical shift MR imaging — initial experience, *Radiology,* 229, 554, 2003.

24. Moulopoulos, L.A. and Dimopoulos, M.A., Magnetic resonance imaging of the bone marrow in hematologic malignancies, *Blood*, 90, 2127, 1997.

25. Lecouvet, F.E. et al., Stage III multiple myeloma: clinical and prognostic value of spinal bone marrow MR imaging, *Radiology*, 209, 653, 1998.

26. Moulopoulos, L.A. et al., Multiple myeloma: spinal MR imaging in patients with untreated newly diagnosed disease, *Radiology*, 185, 833, 1992.

27. Algra, P.R., Bloem, J.L. and Valk, J., Disappearance of the basivertebral vein; a new MR imaging sign of bone marrow disease, *Am. J. Roentgenol.*, 157, 1129, 1991.

28. Olsen, Ø.E., Mchugh, K. and Vellodi, A., Routine magnetic resonance imaging of the spine in children with Gaucher disease: does it help therapeutic management? *Pediatr. Radiol.*, 33, 782, 2003.

29. Maas, M., Poll, L.W. and Terk, M.R., Imaging and quantifying skeletal involvement in Gaucher disease, *Br. J. Radiol.,* 75, A13, 2002.

CHAPTER 16

Radionuclide Evaluation of Gaucher Disease

Giuliano Mariani and Paola A. Erba

CONTENTS

BACKGROUND

Diagnostic nuclear medicine procedures are based on the possibility of mapping the *in vivo* distribution of agents labeled with radionuclides emitting gamma rays of suitable energy (radiopharmaceuticals), as resulting from specific biochemical processes defining the pathophysiologic profile of a certain tissue or organ. Such radiolabeled agents interact with cells/tissues at the molecular level, thus they are usually employed in micromolar mass amounts and can detect physiologic changes occurring in the nanomolar or even picomolar range. Depending on specific clinical applications, presence of disease is detected on scintigraphy either because a certain radiopharmaceutical abnormally accumulates in high concentration at the site of disease (positive indicator) or because the radiopharmaceutical accumulates physiologically in the normal tissue and therefore the site of disease shows up as an area with reduced uptake (negative indicator). In either instance, clinical information deriving from a diagnostic nuclear medicine procedure is more function-oriented than morphology-oriented, and anatomic definition of the disease site and surrounding normal tissues is generally rather poor.

Gaucher's disease is caused by a metabolic defect, and disease manifestations initiate first as functional abnormalities (e.g., reduced platelet, erythrocytes, and/or leukocyte counts as a consequence of bone marrow infiltration by Gaucher cells), then progress to anatomic abnormalities (e.g., hepato-splenomegaly, or skeletal changes detectable with conventional x-ray imaging). Different combinations of diagnostic morpho-functional procedures (e.g., ultrasound, conventional x-rays, nuclear medicine, transmission Computed Tomography, Magnetic Resonance) can therefore be employed in different stages of the disease, depending on the clinical condition, either for characterizing patients (staging) and/or for monitoring the efficacy of therapy.

Several radiopharmaceuticals and nuclear medicine imaging procedures have been employed for evaluating different aspects concerning both the skeletal and the nonskeletal complications of Gaucher's disease.

Although scintigraphy with bone-seeking agents is very sensitive for detecting early skeletal changes causing increased local turnover (primarily increased osteoblastic activity), its specificity is very low. Since skeletal changes observed in patients with Gaucher's disease are secondary to infiltration of the bone marrow by macrophages of the reticuloendothelial system turned into typical Gaucher cells, various scintigraphic approaches have been described to assess the degree of such infiltration, similarly to that used for assessing the severity of visceromegaly caused by the disease.

NUCLEAR MEDICINE PROCEDURES FOR ASSESSING NONSKELETAL COMPLICATIONS

Important advances in knowledge of the mechanisms involved in many hematologic diseases have been based on the use of radioactive tracers to explore various pathophysiologic steps leading to disease manifestations. In particular, radionuclides are used to label the various components of the blood with the aim of tracing their biological distribution, function, and life span *in vivo*, as well as assessing proliferation and differentiation of hematopoietic progenitor and precursor cells in the bone marrow. Other major applications of nuclear medicine to hematology include the determination of spleen size, splenic sequestration of blood cells, and assessment of the absorption, metabolism, and utilization of hematopoietic nutrients such as iron, vitamin B12, and folate.

Involvement of the bone marrow and spleen in patients with Gaucher's disease has been evaluated by means of radioisotopic techniques enabling measurement of blood volume, ferrokinetics, and spleen function. [1]

Blood Volume

Both the particulate fraction (red cell volume) and the liquid fraction (plasma volume) of blood can be measured accurately employing standardized procedures, which are currently based on widely available radionuclide-based techniques such as ^{99m}Tc-labelled red blood cells and ^{125}I-labelled human serum albumin, respectively. [2]

Ferrokinetics

To evaluate ferrokinetics, the plasma clearance $t_{1/2}$ and red cell iron utilization may be assessed using ^{59}Fe-labelled transferrin, while scintigraphy performed after ^{52}Fe-citrate administration depicts the process of erythropoiesis occurring in different sites (bone marrow, spleen, liver) and also allows estimation of the relative distribution of this process among such organs. [3]

The metabolic cycle of iron is intimately related to hemoglobin synthesis and erythropoiesis. Radionuclides can be used in tracing the movement of iron in the metabolic cycle to monitor the transport of iron, its uptake by hematopoietic tissue in the bone marrow and other organs, and the site, quantum, and nature of erythropoiesis. Radioactive iron (^{59}Fe) is used to assess different aspects of iron metabolism and the kinetics of erythropoiesis such as gastrointestinal absorption, distribution of ^{59}Fe, and uptake of radioactive iron by the bone marrow and other organs.

Iron absorption by the gastrointestinal tract is assessed by orally administering a standardized dose of radioactive iron (^{59}Fe, as ferric chloride), mixed with nonradioactive iron ($FeSO_4$, $7H_2O$) as the carrier and with a reducing agent (such as ascorbic acid). Feces passed by the patient are collected during the next 7 days and radioactivity excreted in the stools is measured with suitable equipment for gamma counting. Absorption is calculated as the difference between the intake of radioactive iron and its excretion in the stools, the normal value ranging between 10%–30% of the ingested test dose.

Intravenously injected ^{59}Fe provides the best attainable information on global iron kinetics, including its progress through the metabolic cycle and degree of total erythropoietic activity, as well as the relative extent of effective and ineffective erythropoiesis. Furthermore, iron is also taken up to some extent by the reticuloendothelial cells in the liver, spleen, and bone marrow (where it may be stored iron in the form of ferritin and hemosiderin), and also by circulating reticulocytes. The ferrokinetic data yield several types of semiquantitative and quantitative information such as the plasma iron clearance, plasma turnover or plasma iron transport rate, iron utilization by the newly formed red blood cells, and (based on body surface counting) local uptake and turnover of iron by various structures and organs. The whole study requires sequential blood sampling performed up to 10–15 days after the intravenous injection of a tracer dose of radioactive iron (^{59}Fe, as ferric chloride previously incubated *in vitro* with blood of the patient in order to achieve adequate binding to plasma transferring). Radioactivity in the blood samples is measured separately for the packed red cell component and for the plasma components and the resulting activity/time curves are analyzed to derive the parameters of interest.

The halftime of plasma iron clearance ($T_{1/2}$, time for plasma radioactivity to decrease to 50% its initial level) is mostly correlated with the mass of erythroid cell population in the bone marrow and, therefore, also with the total erythropoietic activity and, to some extent, with activity of the reticuloendothelial system in the liver, spleen, and bone marrow. A short ^{59}Fe $T_{1/2}$ would indicate increased total erythropoietic activity, usually associated with an increase of erythroid precursor cell mass (as seen in patients with megaloblastic anemia, myelodysplastic syndrome, thalassemia major, and iron-deficiency anemia). On the other hand, a long plasma ^{59}Fe $T_{1/2}$ is observed in patients with aplastic and hyperplastic anemias, characterized by reduced erythropoietic activity and depletion of erythroid precursor cell mass in the bone marrow with or without reduction of other cell lineages. [4]

In Gaucher's disease, iron turnover studies have consistently shown the occurrence of extramedullary hemopoiesis in the spleen and active medullary erythropoiesis associated with dyserythropoiesis, a feature whose exact cause has not yet been elucidated. Colonization of extramedullary sites by bone marrow-derived stem cells represents a possible mechanism of extramedullary hemopoiesis. In this regard, glycolipid-laden macrophages of Gaucher's disease may be less efficient in eliminating the "over-spilled" bone marrow-derived stem cells and probably allow extramedullary hemopoiesis to develop relatively early in the course of the disease. This phenomenon might explain why extramedullary hemopoiesis in these patients occurs in combination with a normal red cell volume.

Abnormalities of iron metabolism in adults with Gaucher's disease have been detected not only morphologically (staining of iron stored intracellularly in Gaucher cells), but also by evaluating iron kinetics and distribution as well as red cell survival. In particular, Lee et al. assessed iron absorption, iron kinetics, iron stores, and red cell survival in six Gaucher patients. [5] Plasma clearance of radioactive iron was rapid and associated with an increased overall plasma iron turnover. Iron-59 did not accumulate in the liver or the spleen but in the sacrum, especially over the first 2 hours (somewhat mirroring in an inverse fashion the decline in plasma radioactivity). At the 2-week measurement, radioactivity in the sacrum had considerably decreased, while the spleen showed slight ^{59}Fe accumulation. All the patients were characterized by low incorporation of iron into red cells, but the number and volume of red cells produced per day was mostly normal (except in a single patient whose production rate was increased). Iron absorption and the majority of patients had excess iron stores. These results are consistent with subsequent observations (performed in Gaucher patients without anemia) confirming abundant iron stores combined with near-normal erythrocyte survival and rapid plasma clearance of radioactive iron with subnormal uptake by the erythron. [6]

The most satisfactory explanation of the iron-erythrokinetic data described above postulates a moderately increased ineffective erythropoiesis. [7] This concept is consistent with the slight erythroid hyperplasia, high plasma iron turnover, rapid marrow uptake of radioiron, low incorporation of radioiron into red cells, and increased body iron stores. Ineffective erythropoiesis is not particularly severe, since erythroid hyperplasia is not pronounced, radioiron is not retained within the marrow (as determined by body surface counting), and fecal urobilinogen is not increased.

Finally, in the era before the introduction into clinical practice of enzyme replacement therapy (ERT) for Gaucher's disease, serial radioisotopic evaluations of blood volume, spleen function, and ferrokinetic provided an accurate guide as to when to undertake splenectomy, a procedure that is now not normally necessary because of the availability of ERT, and which should in any case be postponed as long as possible, given the more rapid progression of bone lesions after splenectomy. [8]

Spleen Function and Volume

The spleen has diverse important functions, some or all of which can be affected in several primary hematological disorders; on the other hand, disorders of the spleen may lead to hematological abnormalities. In both instances, evaluation of spleen functions and visualization of the spleen become important for diagnosis, prognosis (including staging), and therapeutic management.

One of the fundamental functions of the spleen is to filter out circulating blood of senescent red cells, abnormal red cells (culling), and intraerythrocytic inclusions (pitting), as well as extrinsic or foreign particles and morphologically abnormal red cells in hereditary spherocytosis, elliptocytosis, and sickle cell anemia. [9] In addition, the spleen is an important reservoir of blood cells, as it can sequester up to approximately 30% of the body's platelets and release them on demand. However, a pathologically enlarged spleen can sequester up to 90% of the body's platelets and thus cause severe thrombocytopenia. [10] On the other hand, the significant rise in platelet counts that usually follows splenectomy is sometimes only transient, because reticuloendothelial cells in other organs (including the liver) can increase their ability to sequester platelets. [10]

Several imaging techniques can be effectively used to visualize the spleen, including especially ultrasound imaging, magnetic resonance imaging (MRI), and computed tomography (CT). However, although these procedures yield excellent structural detail, they provide very little or no information about spleen function. On the contrary, radionuclide imaging of the spleen provides reliable information about spleen function.

Spleen scintigraphy is based on the intravenous injection of radiolabeled heat-damaged autologous red blood cells, which are rapidly removed by the spleen from circulating blood. Radiolabeling of heat-damaged erythrocytes (49.5°C for 20 minutes) can be performed using either ^{51}Cr, ^{111}In, or ^{99m}Tc, and scintigraphic imaging is usually performed 1 hour postinjection (but up to 3–4 hours) (see example in Figure 16.1). This procedure is very useful for mapping the spleen size (especially if tomographic acquisitions are recorded) and for detecting space-occupying lesions in the spleen, as well as for identifying possible ectopic or accessory spleens.

Spleen function can be assessed by evaluating the clearance of heat-damaged ^{51}Cr-labeled red cells from the circulation, whose $T_{1/2}$ ranges from 5 to 15 minutes in healthy subjects. The clearance rate is considerably prolonged in thrombocythemia and in other conditions associated with spleen atrophy. [11]

Scintigraphy with ^{99m}Tc-labeled, heat-damaged erythrocytes

 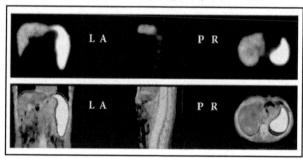

Figure 16.1 **(See color insert following page 304.)** Scintigraphy obtained after injection of autologous heat-damaged erythrocytes labeled with ^{99m}Tc-pyrophosphate (see text for details on preparation) in a 27-year-old man with hemolytic anemia and increased serum bilirubin (unconjugated). Planar imaging (left panel) shows markedly enlarged spleen with obvious sequestration of labeled red cells, much more intense than in the liver. Selected tomographic imaging obtained with a SPECT/CT gamma camera (right panel) confirms frank splenomegaly obvious in all three tomographic planes, frontal (left of panel), sagittal (center of panel) and transaxial (right of panel), with homogeneous distribution of radioactivity within the whole spleen. Bottom row shows the fusion images obtained by combining the scintigraphic information (orange to white color) with the anatomic CT information (gray background with different levels of opacity). Based on this scintigraphic evaluation, the patient underwent splenectomy, with ensuing recovery from the anemic condition.

^{99m}Tc-labeled colloids are used for combined imaging of the spleen, liver, and bone marrow, as they are rapidly cleared by macrophages of the reticuloendothelial system. [12] Considering that splenomegaly is a common feature in untreated Gaucher's patients and tends to recur after partial splenectomy, scintigraphy with ^{99m}Tc-radiocolloid allows both initial assessment of spleen volume and postoperative follow-up. In fact, recurrences are common if the patient does not initiate effective ERT, and spleen size should be assessed whenever symptoms of hypersplenism reappear after partial splenectomy. Radiocolloid scintigraphy can provide additional pathophysiologic information of some clinical relevance, such as splenic infarctions (detected as focal defects on the scan) explaining episodes of left upper quadrant pain. In this regard, SPECT considerably increases the diagnostic performance of liver and spleen scintigraphy over planar imaging. In this regard, Ohta et al. found that splenic lesions were clearly detected by all methods, but ^{99m}Tc-radiocolloid SPECT was superior to ultrasound imaging, CT and MRI for evaluating the reticuloendothelial function of a patient with Gaucher's disease. [13]

Liver Scintigraphy

Following the widespread use of ultrasound imaging, CT, and MRI over the past two decades, the primary focus of liver scintigraphy has shifted from the detection of focal hepatic lesions to tissue-specific characterization of such lesions, to evaluation of functional liver mass and to evaluation of hepatobiliary function.

Advances in technology and pharmacologic interventions during hepato-biliary scintigraphy (with development of new radiolabeled agents) have significantly improved the efficacy of scintigraphic imaging, thus expanding their clinical applications.

As mentioned earlier, ^{99m}Tc-radiocolloids represent a class of radiopharmaceuticals for liver/spleen and bone marrow imaging. Upon their systemic (intravenous) administration, these compounds are cleared by cells of the reticuloendothelial system, preferential organ distribution depending on size of the colloid. In this regard, approximately 85% of ^{99m}Tc-sulfur colloid (the radiocolloid most widely employed in the U.S.) is taken up in the liver (Kupffer's cells), 10% in the spleen, and 5% in the bone marrow. For comparison, spleen uptake of the smaller-sized ^{99m}Tc-phytate (the radiocolloid most widely employed outside the U.S.) is significantly lower than that of ^{99m}Tc-sulfur colloid.

Replacement of normal Kupffer's cells in the liver by Gaucher cells produces hepatomegaly and, if severe and long-standing, can lead to hepatic failure. James et al. described a consistent pattern of hepatomegaly and inhomogeneus uptake of ^{99m}Tc-phytate in patients with Gaucher's disease. [14] Varying degrees of scintigraphic liver involvement in three patients were observed also by Cheng and Holman, [15] while Israel et al. reported a whole spectrum of changes in the liver, varying from slight enlargement and inhomogeneity of uptake to markedly enlarged liver with frank focal defects. [16]

NUCLEAR MEDICINE PROCEDURES FOR ASSESSING SKELETAL COMPLICATIONS

Aseptic-Avascular Necrosis, Bone Infarction, Bone Crisis, and Osteomyelitis

Replacement of bone marrow by Gaucher cells results in bone destruction with ensuing propensity to fracture. One of the key features of untreated Gaucher's disease is reduced intramedullary blood flow due to the accumulation of glucocerebroside within the macrophages turned into Gaucher cells. In fact, increased intramedullary pressure interferes with the blood supply to the bone and can cause infarction and aseptic, avascular necrosis, which can occur slowly or develop acutely with a bone crisis. [17] As a result, bone pain and arthritic-type pain are common clinical manifestations of Gaucher's disease, particularly in patients with type 1 disease.

Imaging with bone-seeking agents such as ^{99m}Tc-labeled diphosphonates (the radiopharmaceuticals most commonly employed for skeletal scintigraphy) has been used during acute bone crises. [18] These agents concentrate predominantly in the mineral matrix of bone (newly formed hydroxyapatite crystals and amorphous calcium phosphate), while they do not localize to a significant degree in osteoblasts or in the osteoid component. Blood flow and extraction efficiency are the most important factors affecting the uptake of diphosphonates in the skeleton. Other factors affecting diphosphonate uptake in the bone include status of vitamin D, parathyroid hormone, corticosteroids, intraosseous tissue pressure, capillary permeability, acid-base balance, and sympathetic tone. In children, prominent uptake of the radiopharmaceutical is seen at the costochondral junctions, at the metaphyseal ends of the normal long bones, and in the facial bones. When the skeleton has reached maturation, this prominent uptake at the costochondral junctions and metaphyseal ends of long bones disappears, and skeletal accumulation of diphosphonates decreases with age, particularly in the extremities.

The pattern of scintigraphy with ^{99m}Tc-diphosphonate in patients with Gaucher's disease is variable and usually does not reflect the severity of bone loss or bone resorption as assessed by CT and ^{99m}Tc-radiocolloid scintigraphy. [19] Although experience with bone scan in Gaucher's disease is limited, dynamic changes in bone structure due to avascular necrosis, infarction, prior fracture and bone marrow expansion (Erlenmeyer flask) can be seen on a skeletal scan (Figure 16.2). [19,20]

Bone marrow cells and primitive osteoblasts only survive 6–12 hours after interruption of the blood supply, while mature osteoblasts and osteocytes survive up to 12–48 hours and adipocytes are more resistant and can survive as long as 2–5 days after beginning of ischemia. This sequence of events explains why changes on the bone marrow scintigraphy (reduced local uptake) appear earlier than changes on the bone scan, [21] also considering that ischemia per se does not directly affect the mineralized bone matrix or cartilage. In fact, the articular cartilage receives most of its nutrients by direct absorption from synovial fluid. However, increase of intracapsular pressure persisting for more than 5 days causes degeneration of the articular cartilage.

Recovery of bone and bone marrow infarct is initiated by neovascularization through the collateral circulation, advancing from the periphery of the area of necrosis or by recanalization of the occluded vessels. This granulation tissue provides all the elements necessary for the formation of bone matrix and new bone deposition by young osteoblasts. During the recovery phase, bone collapse often results from structural weakening and external stress. In the long-term, combined bone collapse and cartilage damage can result in significant deformity. [21]

Evolution of the scintigraphic patterns of avascular necrosis (typically, but not exclusively localized in the femoral head, see examples in Figure 16.3) is correlated with the sequence of pathophysiologic events summarized above. In Stage I (first 48 hours), the morphology of bone is preserved and plain x-ray examination is normal, while the bone scan varies from near-normal uptake (if ischemia is not severe) to absent uptake ("cold" area of necrosis caused by abrupt and severe interruption of blood supply). In Stage II (beginning of recovery within 1–3 weeks from the acute episode), hyperemia is frequent and there is diffuse demineralization of the area surrounding the necrotic tissue. Increased uptake is observed on the bone scan, starting at the boundaries between the site of necrosis and the normal tissue and progressing towards center of the infarct. This scintigraphic pattern (which usually involves the entire femoral head) lasts over several months, returning then to normal upon completion of the recovery process. However, in cases of bone collapse (Stage III), increased uptake may persist indefinitely. Stage IV is characterized by collapse of articular cartilage with degenerative changes on both sides of the hip joint, and results in increased periarticular uptake.

Avascular necrosis of the femoral head is not infrequent in patients with Gaucher's disease, especially if untreated. In this occurrence, evolution of the

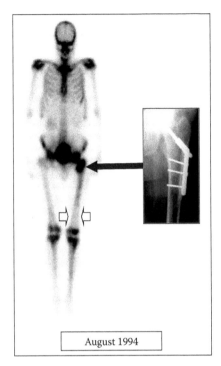 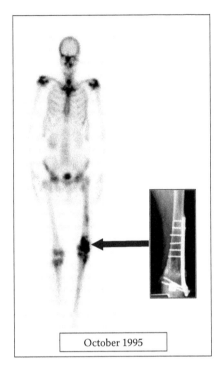

August 1994

October 1995

Figure 16.2 Whole-body skeletal scans obtained with [99mTc]-methylene diphosphonate ([99mTc]-MDP) on two separate occasions in a patient with Gaucher's disease. This patient (who was 67 years old at the time of the first scintigraphy) had been diagnosed with Gaucher's disease at the age of 36 years, when he underwent emergency splenectomy because of sudden rupture of the spleen. In December 1993 the patient suffered from spontaneous, pathological fracture of the left femoral neck that required surgical implantation of an osteosynthetic fixation screw-plaque (externally to the trochanteric region). When [99mTc]-MDP skeletal scintigraphy was performed in August 1994 (left panel), there was still increased uptake at the site of the prior pathological fracture (see inset with x-ray image of the region). In addition, the distal metaphyseal-diaphyseal region of the left femur was relatively photopenic, with a hollow-type pattern consistent with Erlenmeyer flask deformity (indicated by open arrows). It was possible to initiate enzyme replacement therapy only several months after this first observation. In the meantime, the patient suffered from a further pathological fracture, at the distal epiphyseal region of the left femur, at the site of thinning of the cortical bone visualized earlier by both x-ray examination and bone scintigraphy. This second fracture was also treated surgically with implantation of an osteosynthetic fixation screw-plaque. When the second bone scan was performed in October 1995, uptake of the bone-seeking agent had returned to nearly a normal pattern at the site of the first fracture, while markedly increased uptake was still visible at the site of the second fracture (see inset with x-ray image of the regions).

scintigraphic pattern mirrors the general pattern outlined above, [15,16,18,22] although increased uptake at an early stage has been reported. [15,16] This latter observation can most likely be explained by the occurrence of earlier subclinical, transient episode(s) of ischemia (similarly as observed in sickle cell anemia) or by coexistent fracture or osteomyelitis.

Bone scintigraphy with ^{99m}Tc-MDP

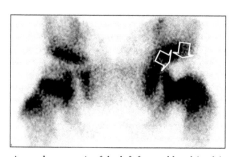
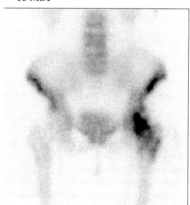
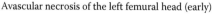

Avascular necrosis of the left femoral head (early)   Avascular necrosis of the
                                                             right femoral head (late)

Figure 16.3   Patterns of ^{99m}Tc-MDP uptake in the bone scan of two patients with different stages of avascular necrosis of the femoral head. Left panel shows the composite pin-hole images of both hips obtained in a teenage patient early after the acute phase of avascular necrosis of the left femoral head. Within a pattern of increased uptake of the bone-seeking agent in the growth plates (due to young age of the patient), uptake is absent in over half of the left femoral head (indicated by open arrows), reflecting poor arrival of the radiopharmaceutical due to reduced or absent blood flow. Right panel shows scintigraphy of the pelvis and hips obtained in an adult patient about 4 weeks after the acute phase of avascular necrosis of the right femoral head. At this stage, the entire right femoral head (including the trochanteric region) shows diffuse uptake of ^{99m}Tc-MDP, indicating activity of the recovery process.

As expected in any other more common clinical condition, both pathologic fractures and osteomyelitis occurring in patients with Gaucher's disease cause increased uptake on the bone scan. [23]

Diagnosis of osteomyelitis, and especially distinguishing aseptic pseudo-osteomyelitis (or bone crisis) from infectious osteomyelitis can be difficult in patients with Gaucher's disease. [17,24] Nevertheless, differential diagnosis of these two clinical conditions is crucial, because surgery (based, e.g., on an incorrect assumption of osteomyelitis) can by itself lead to chronic osteomyelitis and intractable draining sinuses. [25] Plain x-ray, CT, and MR imaging have either poor sensitivity (especially in the early phases of symptoms) or poor specificity for identifying infection in the bone infection. [26]

Three-phase bone scintigraphy can show typical changes of osteomyelitis (focally increased blood flow, increased capillary permeability and increased uptake on the 3-hour scan) as early as 24 hours after infection. [27] Especially at bone sites not previously affected by other pathologic conditions, this technique is both sensitive and specific (approximately 90% in both instances) and cost-effective for diagnosis of osteomyelitis 90%. [28] On the other hand, if bone has been affected by a previous pathology (particularly after orthopedic surgical procedures), three-phase bone scintigraphy is still highly sensitive, but its specificity drops dramatically (about

34%). [29] Thus, in case of equivocal findings on the three-phase bone scan, scintigraphy with autologous leukocytes labeled with ^{111}In-oxine or with ^{99m}Tc-hexamethyl propylene amine oxime (HMPAO) can be particularly advantageous (see Figure 16.4).

Differentiating bone infarct from osteomyelitis on clinical ground alone is also difficult, and findings on the bone scan can be quite variable. If scintigraphy is performed 1 week after the onset of symptoms, healing of the infarct may cause increased uptake rather than the typical pattern of a cold defect. On the other hand, osteomyelitis may also cause cold defects rather than increased uptake. [30] Addition of ^{67}Ga-citrate or ^{99m}Tc-colloid imaging to the bone scan enhances specificity and can resolve most of the diagnostic problems related to osteomyelitis in patients with sickle cell disease. [31] In particular, a normal bone marrow scintigraphy (^{99m}Tc-colloid) at the site where the bone scan shows increased uptake indicates osteomyelitis, while a radiocolloid photon-deficient area suggests healing infarct. Conversely, if the bone scan shows a photon-deficient area (as in an early phase of the episode), a positive ^{67}Ga scan indicates osteomyelitis (as will also, in a later phase of disease, a hot area on the bone scan with even higher uptake of ^{67}Ga). On the other hand, early infarcts are characterized by absent uptake of both ^{99m}Tc-diphosphonates and [67]Ga (congruent scintigraphic pattern), simply because the tracers cannot reach the area without adequate blood supply. [30]

The role of bone scintigraphy in the differential diagnosis of a bone crisis in patients with Gaucher's disease is controversial, in part reflecting uncertainties still remaining on the exact pathophysiologic mechanism(s) underlying this clinical condition. Employing three-phase bone scintigraphy, most authors have found decreased uptake on the delayed image at the onset of a crisis and increased uptake when the bone scan was performed 3 to 4 weeks later. [15,16,32,33] On the other hand, while a "case report" described a normal delayed image, [18] another described increased uptake at the onset of pain. [16] Nevertheless, bone scintigraphy was reported to be useful for demonstrating bone crisis in its early stage in patients with Gaucher's disease. [18]

In Gaucher patients who experience mild pain without local tenderness and swelling (episodes defined as "nonspecific bone pain" whose etiology is unknown), [34] the blood tests, plain x-rays, and bone scintigraphy are generally normal. Since the pathophysiologic mechanisms of a bone crisis include at least bone marrow infiltration and vascular insufficiency, the early scintigraphic pattern in a bone crisis of pseudo-osteomyelitis (within 1–3 days of onset) is reduced or absent uptake on the bone scan, often involving over one-third of the affected long bone. [15,16,18,32,33] Within approximately 6 months, uptake resumes a normal pattern (unless a fracture occurs in the meantime). On the repeat bone scan (e.g., 6 weeks after onset of the crisis), a photopenic area surrounded by a peripheral rim of increased uptake is generally observed on the delayed images. The photopenic area represents defective delivery of the tracer due to decreased medullary blood flow, and the rim reaction is thought to represent enhanced blood supply to the outer cortex resulting from an increased number of periosteal vessels penetrating the cortex. [35]

At the onset of a bone crisis, the angioscintigraphic component of a three-phase bone scintigraphy demonstrates decreased blood flow, as is similarly observed in

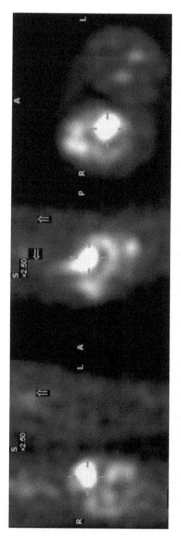
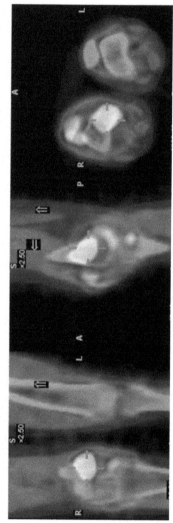

Figure 16.4 **(See color insert.)** Scintigraphy with autologous ^{99m}Tc-HMPAO-Leukocytes in a patient with osteomyelitis of the right knee presenting after arthroscopic meniscus surgery. Upper panel shows selected tomographic images obtained with a SPECT/CT gamma camera in different planes, frontal (left of panel), sagittal (center of panel) and transaxial (right of panel). Infection is clearly outlined by distribution of the labeled leukocytes, localized mostly in the medial epycondilus of the right femur, but also extending to the adjacent cartilages of the knee joint, and also to the underlying tibial plate (in its medial portion). Lower panel shows, for better topographic localization, the fusion images obtained by combining the scintigraphic information (orange to white color) with the antomic CT information (gray background with different levels of opacity).

patients who have a sickle-cell crisis. [36] The analogy with osteomyelitis of the clinical features in "pseudo-osteomyelitis" of patients with sickle-cell disease has been described both in Gaucher [37] and in non-Gaucher pediatric patients, [38] thus emphasizing the role of photopenic areas on bone scintigraphy for distinguishing early bone infarction from infection.

There are no reports on the bone scan pattern at the onset of acute osteomyelitis in patients with Gaucher's disease and only one article described a Gaucher disease patient who actually had osteomyelitis (showing focally increased uptake), evaluated, however, several weeks after the onset of symptoms. [25]

In the presence of repeated episodes of bone infarcts, scintigraphy with ^{99m}Tc-labeled bone-seeking agents does not facilitate the differential diagnosis between infarction and infection by imaging because of the presence of a spotty pattern of uptake. [39,40] In these conditions, a ^{67}Ga-citrate scan can be helpful for ruling out osteomyelitis, [32] as well as for confirming osteomyelitis in patients with concomitant bone disease. [41] While absent or mildly increased ^{67}Ga uptake favors the diagnosis of bone infarct vs. osteomyelitis, [42] increased uptake does not definitely confirm infection because ^{67}Ga itself behaves as a nonspecific bone-imaging agent and concentrates in areas with increased osteoblastic activity. [43] However, ^{67}Ga scintigraphy is useful to exclude osteomyelitis, complicating fractures and orthopedic surgery procedures. [44]

Nevertheless, because of possible equivocal or inconsistent scintigraphic patterns with combined imaging (^{99m}Tc-diphosphonates and ^{67}Ga) for discriminating bone infarcts from osteomyelitis (especially in case of prior multiple episodes), [31] alternative methods specific for bone infection are desirable. In this regard, scintigraphy with infection-imaging agents, such as autologous leukocytes labelled with ^{111}In-oxine (^{111}In-WBC) or with ^{99m}Tc-HMPAO (^{99m}Tc-HMPAO-WBC), provides highly helpful diagnostic information. [45–47]

Sensitivity of ^{111}In-WBC scintigraphy for detecting skeletal infection superimposed on traumatic, postsurgical, or neuropathic processes is virtually 100% for acute infections and approximately 60% for chronic infections, [48] with overall 88% sensitivity and 84% specificity. [49] This procedure is particularly helpful for excluding infection in previously violated bone sites such as in postsurgical and posttraumatic conditions. ^{99m}Tc-HMPAO-WBC scintigraphy has similar sensitivity/specificity yields as ^{111}In-WBC scintigraphy, being especially useful for exploring peripheral locations such as the extremities (see Figure 16.4).

Combined labeled leukocytes and bone scans have better accuracy than labeled-leukocyte scans alone in localizing abnormal foci. [50] Since labeled leukocytes physiologically concentrate in the active bone marrow (as well as in the liver and especially in the spleen), it can sometimes be difficult to discriminate this normal marrow uptake from abnormal uptake due to infection. Furthermore, prior surgical procedures can significantly alter the physiologic bone marrow distribution. In such conditions, ^{99m}Tc-colloid bone marrow scintigraphy can improve specificity. [51,52]

Scintigraphic identification of infectious foci can also be based on the use of radiolabeled antibodies (tagged with either ^{111}In or ^{99m}Tc) recently introduced in the clinical routine as simpler to prepare radiopharmaceuticals than labeled autologous leukocytes (Figure 16.4). In particular, both nonspecific polyclonal human IgG or antigranulocyte monoclonal antibodies can be employed, the former showing promising sensitivity of (95%) and specificity (83%) for diagnosing osteomyelitis [53]

In some patients with Gaucher's disease presenting with severe skeletal complications (especially avascular necrosis of the femoral head), total hip replacement can be necessary. One of the possible outcomes of such prosthetic surgery is the "painful hip," which can be the result of either loosening of the prosthetic implant, postsurgical osteomyelitis or, in Gaucher patients, simply one of the common features of the disease itself. In this condition, both three-phase bone scintigraphy and scintigraphy with ^{111}In-WBC or ^{99m}Tc-HMPAO-WBC provide useful information to discriminate loosening of the prosthetic implant from infection or other complications (see Figure 16.5). [54]

NUCLEAR MEDICINE PROCEDURES FOR ASSESSING BONE MARROW INFILTRATION

Indirect Method: Radiocolloid Scintigraphy

As alluded to above when describing the use of radiocolloid scintigraphy for evaluating liver and spleen involvement in Gaucher's disease, radiocolloids are particles included in a certain size range (between about 10 nm and 2000 nm) possessing surface characteristics that activate an efficient mechanism of clearance operated by the reticuloendothelial cells. The opsonized material activates a membrane-bound receptor on macrophages, the first step leading to phagocytosis of the radiolabeled compound. The efficiency of this clearing process varies with several factors besides the net surface charge and degree of opsonization such as antigenic properties, size and number of the particles, specific anatomical region, etc. [55,56] Following intravenous administration, radiocolloids are cleared from circulation by macrophages distributed throughout the reticuloendothelial system, i.e., in the bone marrow, in the liver (endothelial-lining Kupffer cells) and in the spleen.

Of interest to radionuclide evaluation of bone marrow involvement in Gaucher's disease, radiocolloid scintigraphy depicts physiological distribution of the bone marrow containing a normal population of macrophages. For proper evaluation of the abnormalities observed in patients with Gaucher's disease, it is important to take into account the physiologic evolution of bone marrow distribution in normal individuals. In an adult subject, radiocolloid scintigraphy is characterised by uptake predominantly in the axial skeleton (from skull to pelvis, including the ribs), while uptake in the appendicular bones is detectable only in the proximal femurs; this pattern of distribution reflects distribution of cellular red bone marrow, which in adults is restricted to the areas noted above. By contrast, newborns have a wide distribution of cellular red marrow in the bones of their appendicular and axial skeleton. Shift from high-cellular red marrow to low-cellular, high-fat yellow marrow occurs in the first decade of life and follows a predictable sequence. [57] Such marrow conversion proceeds from distal to proximal sites, distal bones being converted more rapidly than proximal bones. The process continues until the age of

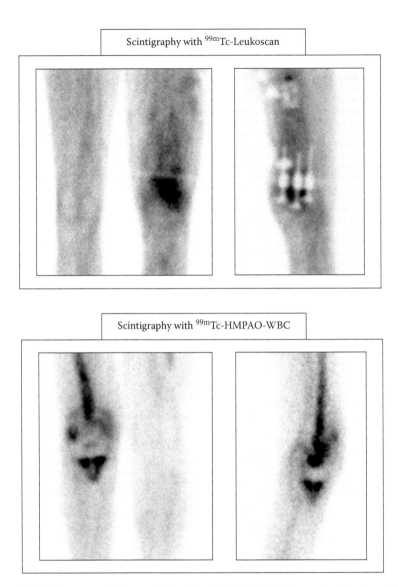

Figure 16.5 Upper panel: planar scintigraphy with 99mTc-labeled monoclonal antigranulocyte antibody obtained in a patient with osteomyelitis complicating a traumatic fracture of the left femur treated with external fixators. Uptake of the radiopharmaceutical is markedly increased (consistent with infection) at the site of the fracture, with a diffuse pattern better visible in the anterior view (left of panel), while lateral view of the left femur (left of panel) clearly shows artifacts caused by the metallic fixation device (photopenic areas within the region with increased uptake). Lower panel: planar scintigraphy with autologous 99mTc-HMPAO-Leukocytes in a patient with osteomyelitis complicating prosthetic implant of the right knee. Markedly increased localization of the labeled white blood cells is visible both at the femoral end of the prosthetic implant (including the entire distal portion of the femur) and under the tibial plate.

approximately 25 years, when most of the red bone marrow is retained in the vertebrae, sternum, ribs, pelvis, skull, and proximal shafts of the femora and humeri.

In certain pathophysiologic conditions, such as chronic anemia and chronic heart failure, as well as in marrow replacement disorders (such as Gaucher's disease), the yellow marrow often shifts back to red marrow. The spine and flat bones respond more quickly than other sites, followed by the proximal extremities and extending distally, thus mirroring in an inverse manner the physiologic conversion from red to yellow marrow. Colonization of extramedullary sites mediated by bone marrow-derived stem cells can also occur. No clinical or biochemical predictors have been identified for the rate of progression or future severity of the disease process, and osseous findings on plain x-ray do not always reflect the extent of visceral disease or the total burden of Gaucher cells within the marrow and, therefore, cannot be used to follow the progress of the disease with confidence. Combined bone marrow and liver/spleen scintigraphy is performed 30 min after intravenous injection of the radiocolloid (370 MBq, or 10 mCi) using a wide-field-of-view gamma camera to acquire anterior and posterior images of the whole body. Visual analysis of distribution of central bone marrow (sternum, lumbar spine, and pelvis) and peripheral bone marrow (appendicular skeleton with special attention to uptake in the proximal and distal femora and tibiae) and the pattern of pulmonary uptake are considered for the evaluation. It is to be emphasized that radiocolloid scintigraphy visualizes normal red bone marrow, thus pathologic involvement is demonstrated by reduced/absent uptake at the site of disease, the radiopharmaceutical belonging therefore to the group of negative indicators.

Several patterns of abnormal radiocolloid uptake in bone marrow have been described in patients with Gaucher's disease, [58] as follows: a) bone marrow expansion with a uniformly dense uptake, b) extensive peripheral bone-marrow expansion also involving the feet, c) bone-marrow expansion mixed in distribution and density, d) areas of uptake flanked by regions without uptake, and e) total absence of radiocolloid uptake by the marrow, or isolated uptake in the midshaft of the femur or tibia.

The extent of marrow involvement as evaluated by radiocolloid scintigraphy has been found to be directly correlated to the severity of skeletal disease as assessed by clinical and x-ray parameters. In particular, peripheral marrow extension alone was associated with minimal-to-mild radiographic and clinical bone disease, while increasing loss (either discrete or generalized) of uptake by peripheral marrow was associated with more severe bone disease. [58]

Lorberboym et al. reported increased uptake in the long bones of the lower extremities in the early stages of marrow expansion, whereas the most severely involved patients exhibited almost total absence of radiocolloid uptake. [59] A significant correlation between the extent of peripheral bone marrow expansion and the severity of hepatosplenomegaly was found, thus suggesting a possible role of bone marrow scintigraphy for monitoring the systemic burden of the disease. In their study, 83.3% of the patients on ERT showed a regression of the uptake in the proximal segments of the peripheral skeleton after therapy, indicating that this parameter may be useful in a long-term follow-up. Semiquantitative analysis of bone marrow redistribution identified significant depletion of central bone marrow activity in 70% of patients before therapy, with subsequent increased uptake in 73.7% of such patients following ERT. The lack of bone marrow changes in some of the patients on ERT might reflect an advanced stage of their disease. No significant difference was found in central marrow activity at baseline between splenectomized and nonsplenectomized patients. Although not quantitative, this parameter, in conjunction with peripheral marrow changes and CT/MRI findings in the spine, seems to be a good indicator of the bone marrow response to therapy in patients with Gaucher's disease.

Pulmonary infiltrates or respiratory symptoms are uncommon in Gaucher patients. [60] In this regard, animal studies show that macrophages migrate from bone marrow, liver, and spleen to the lungs and are trapped in the pulmonary capillary bed. These migrated macrophages retain their ability to phagocyte intravascular colloids, some radiocolloid uptake being thus possibly visualized on the scan. [61] In this regard, as many as 91% of Gaucher patients were reported to exhibit significant regression of pulmonary radiocolloid uptake following ERT, accompanied by central bone marrow repopulation in 70%. However, reduction of pulmonary uptake may also occur independently of central or peripheral bone marrow repopulation and may reflect changes in pulmonary macrophage function or number with treatment. These findings may indicate that the basic disease process of macrophage migration is receding.

Radiocolloid scintigraphy is therefore potentially useful in the characterization of patients with Gaucher's disease, especially if a baseline examination is available (preferably early in life), before development of significant bone involvement. When repeated over time, this imaging procedure has the potential of demonstrating the progression of bone marrow involvement in untreated patients (thus providing an additional parameter to consider for initiating ERT), or conversely of monitoring the efficacy of ERT. The major limitation of this technique is that it provides only indirect evaluation of bone marrow infiltration by Gaucher cells, as reduced or absent radiocolloid uptake cannot discriminate true bone marrow infiltration by Gaucher cells from substitution of bone marrow resulting from some other concomitant abnormality. This consideration is particularly important in Gaucher patients, since development of myeloproliferative disorders (causing impaired radiocolloid uptake) is not rare in this disease.

Direct Methods

Xenon-133

As glucocerebroside is a glycolipid, the use of Xenon-133 (a lipid-soluble, inert radioactive gas administered by inhalation) was envisioned as a means of deriving an index of lipidic infiltration of the bone marrow in patients with Gaucher's disease. [62,63] This radioactive gas has been used extensively Nuclear Medicine for lung ventilation studies and for measuring regional blood flow, especially in the brain and heart. [64,65] In addition, it has also been employed for determining triglyceride content in the liver. [66]

Early studies have shown that washout of ^{133}Xe from the bone is partly a function of the fatty content in the bone marrow, [67,68] and exhibits a biexponential clearance pattern related to the hematopoietic marrow (fast component) and to combined marrow fat and compact and cancellous bone (slow component). [67] Therefore, the potential of dynamic ^{133}Xe scintigraphy has been explored in Gaucher's disease, assuming that infiltration of the bone marrow with Gaucher cells would alter its cellular constituency, therefore also the patterns of tracer uptake and washout. [69]

^{133}Xe scintigraphy implies dynamic image acquisition during a 10-minute equilibration period after commencing ^{133}Xe inhalation, performed with an airtight mask (a procedure that can be problematic, especially in children). For quantitation purposes, uptake data are obtained during this period with gamma cameras placed over the lungs and the knees, respectively. The lung compartment serves as a depot for ^{133}Xe and as the numerical standard to which other body compartments are compared. Regions of interest (ROIs) are then defined over the entire right and left lung and the integral of the total lung counts accumulated over 10 minutes is considered as the administered dose of ^{133}Xe. Counts from ROIs drawn over the distal femoral and the proximal tibial metaphyses are normalized to account for intra-subject and inter-subject variability in bone size. Total ^{133}Xe uptake at 10 minutes in these ROIs is expressed as the ratio of counts to total lung counts. Increased ^{133}Xe uptake was found in the distal femoral metaphyses of Gaucher's patients, particularly those with immature skeleton. Xenon-133 uptake in the lower extremities increased as the normal medullary fat content, as measured by CT, decreased. An inverse correlation was also found between femoral tracer uptake and the spinal fat fraction as measured by MR, as well as (to a lesser but still significant extent) between tibial uptake and spinal fat fraction. [70] A typical dynamic ^{133}Xe inhalation study in an adult patient shows rapid uptake and more intense concentration of ^{133}Xe in the infiltrated bone marrow than in the noninvolved areas. During childhood, the ^{133}Xe flow pattern shows even more rapid uptake and washout of the tracer, with counts approaching background already about 5 minutes after discontinuing ^{133}Xe inhalation.

Analysis of dynamic data showed some peculiar patterns in the uptake of ^{133}Xe relative to soft tissue; patients with more severe bone marrow infiltration exhibit faster clearance of ^{133}Xe from the affected areas, especially in the second, slower component of the curve. However, this pattern reaches a plateau for the most severe degrees of infiltration. The $T_{1/2}$ value of ^{133}Xe clearance has been suggested as an index for monitoring the course of the disease, as it decreased significantly following ERT, with most of the reduction occurring in patients with rapidly growing skeletons. [71]

Nevertheless, exact interpretation of the ^{133}Xe uptake in bone marrow is still the object of debate, especially after treatment. In fact, while inhaled ^{133}Xe accumulates in the infiltrated bone marrow of patients with Gaucher's disease, it does not accumulate in liver and spleen (similarly infiltrated by Gaucher cells), thus suggesting that other factors are involved in the uptake process in addition to glucocerebroside accumulation in Gaucher cells. In this regard, glucocerebroside deposits per se do not seem to accumulate ^{133}Xe, as inferred by *in vitro* measurement of the partition coefficient, [72] and ^{133}Xe uptake is also influenced by cell density, as shown by

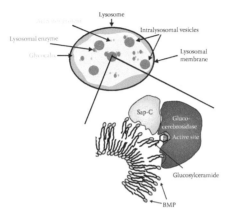

Figure 4.4 Model for the degradation of membrane-bound glucosylceramide by glucosylcera-mide--glucosidase (glucocerebrosidase) and Sap-C. (From Wilkening, G., Linke, T., and Sand-hoff, K., *J. Biol. Chem.*, 273, 30271, 1998. With permission.)

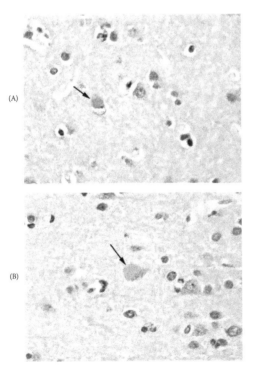

Figure 11.4 Staining pattern for glucocerebrosidase using the 8E4 monoclonal antibody of cortical neurons in a patient with Gaucher disease and parkinsonism-dementia syndrome that was found to have diffuse Lewy body disease on autopsy. The antibody stains a structure resembling an inclusion that is otherwise α–synuclein positive (arrows). Initial magnification ×100. (From Wong, K. et al., *Mol. Genet. Metab.*, 82, 192, 2004.)

Figure 12.2   This close up gross photo of a Gaucher spleen is mostly white   because the normally red pulp has been infiltrated by the white lipid contained in the Gaucher cells. The lymphoid centers that are normally grey and round are obscured by the Gaucher cell infiltrate.

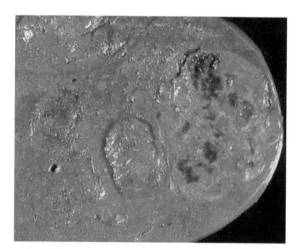

Figure 12.6   A Gaucher patient was 37 years old when he had a cholecystectomy. A liver biopsy was performed at that time and Gaucher cells were seen in the hepatic sinusoids. Fourteen year later, his spleen was removed and the spleen was enlarged to 1370 gms with parenchymal nodules that measured up to 4.5 cms. It is not rare that Gaucher spleens have these nodules. They can be composed of fibroblasts, erythroid cells of the marrow, and dilated capillaries. Later in life, this man developed carcinoma of the prostate.

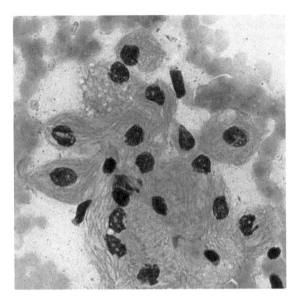

Figure 12.8 These are Gaucher cells aspirated from bone marrow and stained with a Romanofsky stain. The diagnostic appearance relates to the large amount of cytoplasm with a wrinkled appearance and a benign appearing nucleus. This approach to the diagnosis via the marrow has been one of the early accomplishments of hematology.

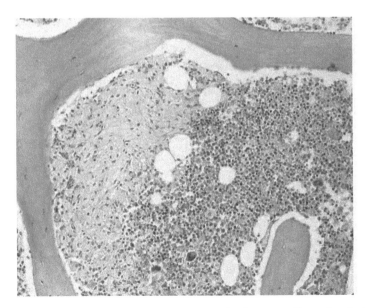

Figure 12.9 Gaucher cells are clustered next to hyperplastic marrow. Excess iron is deposited in cells adjacent to a bony trabeculum.

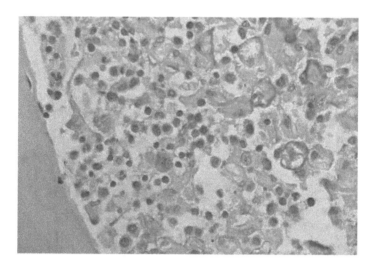

Figure 12.15 This woman was diagnosed when she was pregnant at age 24 when the obstetrician palpated the enlarged spleen. More commonly the softness of the postpartum abdomen has revealed an enlarged spleen. The marrow was stained for iron to reveal an excess with varying amounts of iron in the Gaucher cells. Excess iron stores are not found in young Gaucher patients

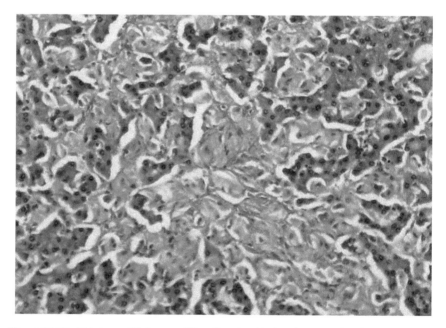

Figure 12.23 This is a mild degree of liver fibrosis with the changes related to the presence of the Gaucher cells. Collagen is stained blue with a trichrome stain.

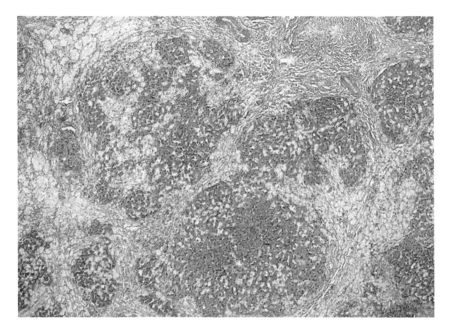

Figure 12.24 This patient had advanced cirrhosis. The trichrome stain reveals the blue fibers of fibrosis. Fibrosis is one of the constants in Gaucher disease: wherever Gaucher cells are located, fibrous tissue is also present.

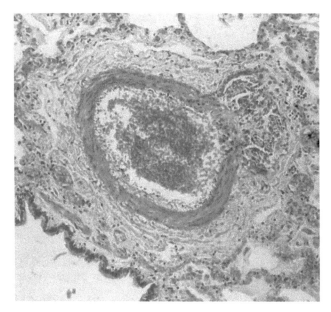

Figure 12.25 Well-developed plexiform arteriopathy of the pulmonary artery was present at autopsy in an 18-year old girl with evidence of pulmonary hypertension. She was diagnosed at age 2 years by splenectomy. The pale cells adjacent to the muscularis of the artery are Gaucher cells.

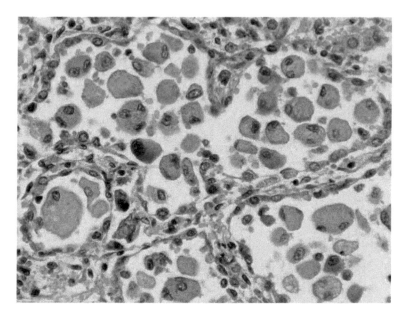

Figure 12.29 These Gaucher cells were found in the alveoli of a 3-year old boy who had marked cyanosis in his terminal course. His younger brother also had a similar course.

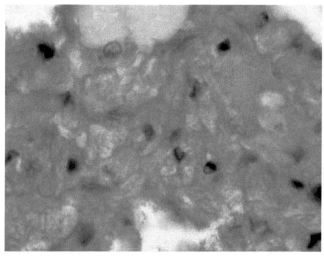

Figure 13.1 Gaucher Cells: A collection of perivascular or periadventitial Gaucher cells stained with periodic acid-Schiff stain and hematoxylin. The characteristic cytoplasm has been described as resembling "crinkled tissue paper." (Periodic acid-Schiff with hematoxylin counter stain, 400 original magnification.)

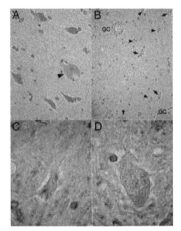

Figure 13.2 Cytoplasmic and membrane pathology in type 2 Gaucher's disease patient neurons: (A) One apparent globus pallidus neuron (arrowhead) is filled and distended by tubular–filamentous structures. The cell is favored to be neuronal rather than monocytic because of the centrally located nucleus that is not displaced to the side, spacing of other neurons, the large size of the cell without an accompanying entourage of perivascular Gaucher cells (H&E 200 original magnification). (B) Low power view of globus pallidus with blue perivascular Gaucher cells (GC), the usual pink (periodic acid–Schiff staining) neurons, and blue (alcian blue staining) neurons that tinctorially stain the same as Gaucher's cells. Alcian blue 8GX/PAS at pH 2.5 normally stains neurons pink, but in Gaucher's disease, isolated neurons, in rare instances, may exhibit blue staining characteristics identical to Gaucher's cells that normally stain blue (alcian blue 8GX/PAS, pH 2.5, 100 original magnification). (C) High power view of an alcian blue stained neuron with blue cytoplasm, blue neuronal membrane and neuronal processes (alcian blue 8GX/PAS at pH 2.5, 400 original magnification). (D) Close-up view of a neuron with proximal axonal process. Alcian blue stains the membrane and the internal cytoplasmic contents of the neuron. The internal cytoplasmic structures appear to have a convoluted tubular–filamentous structure (alcian blue 8GX/PAS at pH 2.5, 400 original magnification).

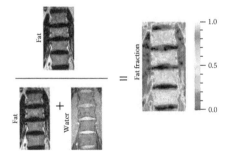

Figure 15.6 Calculation of the fat fraction and display of the fat fraction with a color scale.

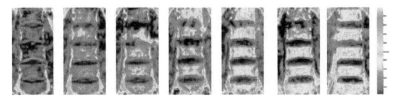

Figure 15.7 Color coded display of the fat fraction measurements in an adult type 1 Gaucher disease patient. On the left (prior to Enzyme replacement therapy) the mean fat fraction is 0.08. The annual display shows a marked increase of fat fraction each year with a normal fat fraction of 0.44 on the latest measurement.

Scintigraphy with ^{99m}Tc-labeled, heat-damaged erythrocytes

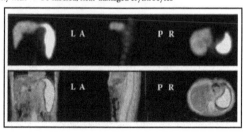

Figure 16.1 Scintigraphy obtained after injection of autologous heat-damaged erythrocytes labeled with ^{99m}Tc-pyrophosphate (see text for details on preparation) in a 27-year old man with hemolytic anemia and increased serum bilirubin (unconjugated). Planar imaging (left panel) shows markedly enlarged spleen with obvious sequestration of labeled red cells, much more intense than in the liver. Selected tomographic imaging obtained with a SPECT/CT gamma camera (right panel) confirms frank splenomegaly obvious in all three tomographic planes, frontal (left of panel), sagittal (center of panel) and transaxial (right of panel), with homogeneous distribution of radioactivity within the whole spleen. Bottom row shows the fusion images obtained by combining the scintigraphic information (orange to white color) with the anatomic CT information (gray background with different levels of opacity). Based on this scintigraphic evaluation, the patient underwent splenectomy, with ensuing recovery from the anemic condition.

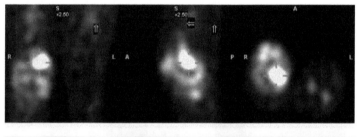

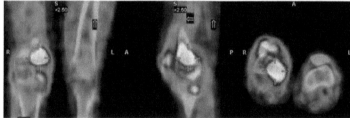

Figure 16.4 Scintigraphy with autologous ^{99m}Tc-HMPAO-Leukocytes in a patient with osteomyelitis of the right knee presenting after arthroscopic meniscus surgery. Upper panel shows selected tomographic images obtained with a SPECT/CT gamma camera in different planes, frontal (left of panel), sagittal (center of panel) and transaxial (right of panel). Infection is clearly outlined by distribution of the labeled leukocytes, localized mostly in the medial epycondilus of the right femur, but also extending to the adjacent cartilages of the knee joint, and also to the underlying tibial plate (in its medial portion). Lower panel shows, for better topographic localization, the fusion images obtained by combining the scintigraphic information (orange to white color) with the antomic CT information (gray background with different levels of opacity).

threefold to fivefold increases in blood flow to hypercellular compared to normal marrow. [67]

The above observations suggest that reduction of ^{133}Xe uptake following ERT may actually reflect a combination of growth- and cellularity-related changes in the bone and bone marrow consequent to treatment. Therefore, the effects of ERT on Gaucher cell infiltration of the bone marrow cannot be readily discerned by ^{133}Xe kinetic studies. Finally, the possibility of performing quantitative ^{133}Xe dynamic scintigraphy as described above is restricted to a few centers and is quite demanding from the technical point of view.

[99m]Tc-Sestamibi

Two lipophilic cationic agents, ^{99m}Tc-HMPAO (an intracellularly trapped complex commonly employed for brain perfusion scintigraphy and for labelling leukocytes) and especially ^{99m}Tc-hexakis-methoxy-isobutyl isonitrile (^{99m}Tc-Sestamibi, an indicator of cellular density and metabolic activity), have been found to accumulate in a stable manner in the bone marrow infiltrated by Gaucher cells. [20]

[99m]Tc-Sestamibi represents a novel class of radiopharmaceuticals with high potential for imaging cell biochemistry [73] and scintigraphy with this agent entails fewer technical problems than ^{133}Xe inhalation, in addition to being widely available, easy to administer and requiring widely available nuclear medicine imaging equipment. ^{99m}Tc-Sestamibi is a positively charged, lipophilic compound whose cellular uptake and retention are affected by regional blood flow, tissue viability, cell membrane potential, and mitochondrial uptake. Following systemic intravenous administration, it rapidly clears from the circulation and localizes intracellularly, normally remaining trapped within the mitochondria.

For scintigraphic evaluation of patients with Gaucher's disease, ^{99m}Tc-Sestamibi offers definite advantages over ^{133}Xe such as better nuclear characteristics (shorter physical half-life, lack of beta-emission, gamma emission energy with better imaging properties) and much longer retention than ^{133}Xe at the sites of uptake, thus allowing easier evaluation of bone marrow infiltrated by Gaucher cells (Figure 16.6). This latter feature is most likely linked to the well-known ability of ^{99m}Tc-Sestamibi to reveal sites with both increased cellular density and metabolic activity, as expected for the glucocerebroside-laden Gaucher's cells wherever they can be located (Figure 16.7). Therefore, ^{99m}Tc-Sestamibi scintigraphy provides a practicable manner of direct rather than indirect evaluation of bone marrow involvement, capable of yielding useful information equally well in ERT-naive patients as in patients on ERT. [74–82]

However, it must be emphasized that localization of ^{99m}Tc-Sestamibi (in the bone marrow as well as at other sites of increased cellularity) is not specific for Gaucher cells. [73] Thus, this radiopharmaceutical cannot be employed for diagnostic purposes but merely for assessing extent and severity of involvement after the diagnosis of Gaucher's disease has been ascertained according to current guidelines.

When evaluating patients with Gaucher's disease, scintigraphy with ^{99m}Tc-Sestamibi is performed after intravenous injection of a dose of radiopharmaceutical (370 MBq, or 10 mCi), approximately half the dose normally administered for other

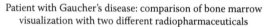

Patient with Gaucher's disease: comparison of bone marrow
visualization with two different radiopharmaceuticals

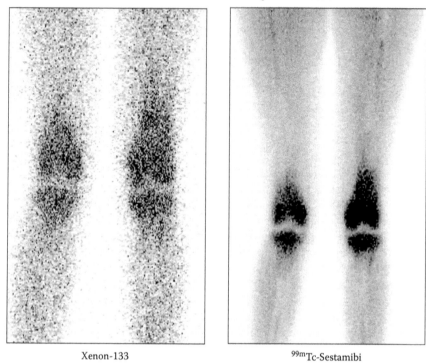

Xenon-133 ^{99m}Tc-Sestamibi

Figure 16.6   Comparison of scans obtained in the same patient with Gaucher's disease (a 47-
year-old woman with type 1 disease), respectively with Xenon-133 (left panel,
obtained as the summation of several 1-minute dynamic frames recorded during
inhalation of the radioactive gas) and with ^{99m}Tc-Sestamibi (right panel, obtained
as a single static image recorded 30 minutes after intravenous injection of the
radiopharmaceutical). Quality of the scintigraphic images obtained with ^{99m}Tc-
Sestamibi is clearly superior to those obtained with Xenon-133.

diagnostic purposes, after a 1-hour rest in order to reduce physiologic tracer uptake
in the skeletal muscles. Dynamic recording performed upon bolus intravenous injection of ^{99m}Tc-Sestamibi shows that the tracer starts to accumulate within the first
few minutes in the infiltrated bone marrow (Figure 16.8). Uptake of ^{99m}Tc-Sestamibi
reaches a plateau within 15–20 min after tracer injection, then remaining constant
for at least 2 hours, (see Figure 16.8). In one single patient, scintigraphic images
recorded 24 hours after tracer injection showed a decline in radioactivity uptake at
the involved bone marrow sites of only about 15%–20% of the peak activity (after
correction for the physical decay half-life of ^{99m}Tc).

The 30 min scan as recorded according to the standard protocol described earlier
by Mariani et al. [80] is therefore considered optimal and representative of the
highest radioactivity accumulation at the involved bone marrow sites. Focal uptake
of radioactivity in the bone marrow in the scans, normally not observed after
administration of ^{99m}Tc-Sestamibi, is considered as an indication of infiltration by

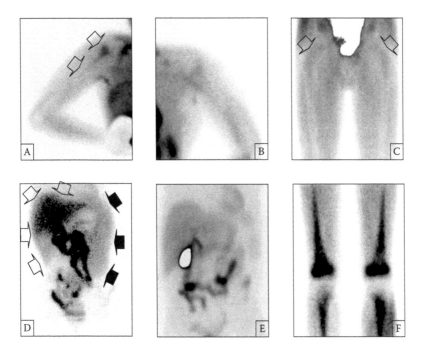

Figure 16.7 Patterns observed using scintigraphy with ^{99m}Tc-Sestamibi in different patients either with or without Gaucher's disease. (A) Open arrows indicate increased tracer uptake at the upper, proximal third of the right humerus in a patient with Gaucher's disease, indicating bone marrow involvement at this level. (B) For comparison, no uptake is detectable in the humerus of a control subject not affected by Gaucher's disease. (C) Open arrows indicate tracer uptake at the proximal thirds of both femurs in a patient with Gaucher's disease not showing the typical pattern of involvement in the more distal portions of the lower limbs observed in most of the patients. (D) Open arrows and solid arrows outline edges of the liver and spleen, respectively, in a patient with Gaucher's disease. Although physiologic hepatobiliary excretion of the radiopharmaceutical somewhat complicates interpretation of the scan in the abdominal area, both organs appear abnormally enlarged (especially the spleen, extending down to the iliac crest) and with increased tracer uptake. (E) For comparison, scan of the abdominal area is shown in a subject without Gaucher's disease, in whom the spleen only shows faint tracer uptake (radioactivity accumulation in the gallbladder shows up as a hypersaturated area). (F) Markedly increased accumulation of ^{99m}Tc-Sestamibi in the bone marrow of both femurs and tibiae in a patient affected by metabolic thesaurismosis different from Gaucher's disease.

Gaucher's cells. On the basis of the scintigraphic pattern, a combined semiquantitative score of bone marrow involvement was developed, modifying an index previously described for patients with bone marrow infiltration by multiple myeloma. [83] The main determinant in such a semiquantitative score is involvement of the knee area, i.e., the distal epiphysis of the femur and the proximal epiphysis of the tibia, extending progressively proximally in the femur and distally in the tibia.

Although ^{99m}Tc-Sestamibi accumulates also at other bone marrow sites infiltrated by Gaucher cells (as well as at other sites such as liver and spleen involved by the disease), attention is centered on the knees because this area is the most frequent,

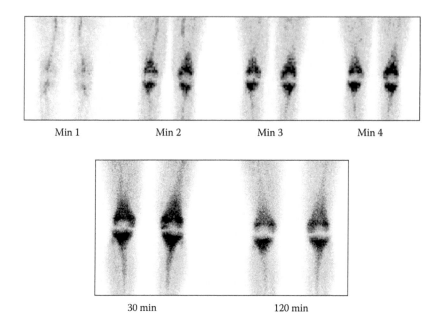

Min 1   Min 2   Min 3   Min 4

30 min   120 min

Figure 16.8 Upper panel: Single frames corresponding to the first 4 minutes of dynamic recording after intravenous bolus injection of ^{99m}Tc-Sestamibi in a patient with Gaucher's disease, showing obvious tracer accumulation in infiltrated bone marrow already within the first few minutes after injection. Lower panel: Static scans obtained in the same patient as in upper panel, respectively 30 and 120 minutes after injection of ^{99m}Tc-Sestamibi, showing virtually no change in radioactivity accumulation in extensively and intensely infiltrated bone marrow sites.

early site of one of the typical skeletal changes observed in Gaucher's disease. Scintigraphic evaluation of the knees with ^{99m}Tc-Sestamibi is especially favorable, because imaging of this area is not affected by physiologic hepatobiliary and renal excretion of radioactivity, which hampers visualization of bone marrow involvement in the proximal femurs, pelvic bones, and spine (see Figure 16.7).

The scintigraphic score developed for this purpose is the sum of separate visual assessments of both the extension and the intensity of uptake (examples in Figure 16.9), as follows: for extension (E), E0 = no evidence of bone marrow uptake, E1 = uptake at the distal femoral epiphyses, E2 = uptake at the distal femoral and proximal tibial epiphyses, E3 = uptake extending to the femoral diaphyses, E4 = uptake extending also to the tibial diaphyses, and E5 = uptake involving the entire femurs and tibias. For intensity (I), I0 = no evidence of bone marrow uptake, I1 = bone marrow uptake lower than in muscle, I2 = bone marrow uptake of the same intensity as in muscle, and I3 = bone marrow uptake higher than in muscle.

In a series of 74 patients with Gaucher disease, ^{99m}Tc-Sestamibi scintigraphy revealed some degree of bone marrow involvement in the majority of the patients. [80] The predominant scintigraphic pattern is represented by initial involvement of the distal femoral epiphyses in the mildest cases, progressively extending to the proximal tibial epiphyses, then to the diaphyseal portion of femurs and tibias, and, finally, in the more severe cases, to virtually the whole femurs and tibias. Intensity

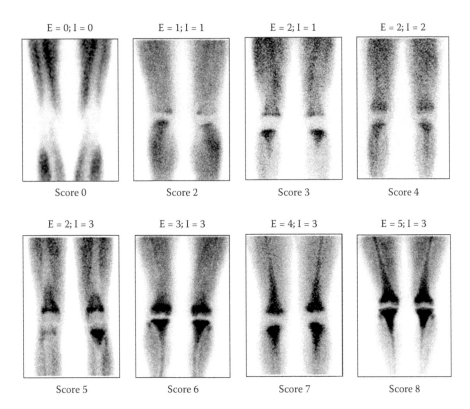

Figure 16.9   Examples of ^{99m}Tc-Sestamibi scans obtained in Gaucher patients with varying degrees of bone marrow involvement. Combined semiquantitative scintigraphic scores are indicated at the bottom, whereas scores for extension (E) and intensity (I) of radioactivity uptake in bone marrow are indicated at the top of each scan. Physiologic tracer uptake in skeletal muscles is minimized by keeping the patients at rest for at least 30 minutes before injecting the radiopharmaceutical.

of ^{99m}Tc-Sestamibi uptake varied from lower to higher than muscle uptake. Occasionally, involvement of the femoral heads and proximal femurs was observed in association with involvement of the distal femurs, without detectable radioactivity accumulation in the femoral diaphysis per se (see Figure 16.7). Visualization of the bone marrow was in general quite homogeneous at the sites of Gaucher's deposits, though with progressively decreasing intensity from the site with highest tracer accumulation. However, patients who had had prior episodes of osteonecrosis or fracture showed some inhomogeneity in distribution of radioactivity within the infiltrated bone marrow consisting of photopenic areas corresponding to the sites of the prior skeletal complications.

It should be noted that the pattern of bone marrow involvement visualized by scintigraphy with ^{99m}Tc-Sestamibi is somewhat discrepant with respect to the pattern visualized by MRI in patients with Gaucher's disease. [84] The most frequently reported MRI abnormality is reduced signal intensity of the bone marrow spaces on the T1- and T2-weighted images. This decrease in signal intensity, which seems to extend progressively from the proximal to the more distal portions of the long bones, is commonly attributed to progressive displacement of the hematopoietic, fat-rich marrow. However, a possible contributing factor to reduced MRI signal intensity is represented by edema, whose occurrence in the bone marrow infiltrated by Gaucher's cells can be linked to increased local production of proinflammatory cytokines, such as interleukin-6 and interleukin-10, in response to the presence of pathologic macrophages. [85] On the other hand, one cannot exclude the possibility that the ^{99m}Tc-Sestamibi pattern of initial bone marrow involvement centered electively around the knees (distal femoral and proximal tibial epiphyses) results simply from a scintigraphic summation effect on planar images. In fact, both the distal femoral and the proximal tibial epiphyses are the thickest portions of these long bones, so that they would appear as relatively hyperactive even if assuming homogeneous radioactivity accumulation in the bone marrow space rather than preferential accumulation at these sites. Nevertheless, the scintigraphic patterns observed in most of the patients evaluated with ^{99m}Tc-Sestamibi mimic the pattern of Gaucher's deposits visualized after injection of macrophage-targeted Alglucerase radiolabeled with Iodine-123. [86] Similar patterns of distribution have also been observed in patients with Gaucher's disease after intravenous administration of ^{99m}Tc-labeled low-density lipoprotein, which targets activated macrophages. [87]

The scintigraphic score was highly correlated, either positively or negatively, with the combined severity score index (SSI) described by Zimran et al. [88] (P = 0.002) as well as with some of the parameters contributing to SSI, as follows: hepatomegaly (P = 0.002), x-ray score according to Herman (P = 0.012), prior splenectomy (P = 0.022), bone pain (P = 0.044), hemoglobin concentration in blood (P = 0.022) (Figure 16.10). A highly significant negative correlation was also found between the scintigraphic score and prior ERT (P = 0.002), thus further confirming reliability of the ^{99m}Tc-Sestamibi scintigraphic score as an indicator that changes in response to treatment. The closest statistical correlation of the scintigraphic score was found with serum chitotrisidase (R = 0.79; P = 0.004), recently emerging as a powerful overall parameter of disease severity assumed to reflect the whole body burden of macrophages. [89] Therefore, imaging with ^{99m}Tc-Sestamibi appears to be a useful complement of topographic information concerning bone marrow sites involved by the disease with global disease severity indicated both by the combined Zimran SSI and by serum chitotriosidase. Thus, although based on localized evaluation of bone marrow involvement (lower limbs), the scintigraphic score described by Mariani et al. can be used to assess severity of overall bone marrow involvement (possibly also of overall disease severity) and permits comparison among various patients. [80]

Besides being highly correlated with other independent parameters of disease severity, the scintigraphic score also correlates with the response to ERT. The basic assumption is that radioactivity uptake in the infiltrated bone marrow actually reflects intracellular accumulation of ^{99m}Tc-Sestamibi within the glucocerebroside-laden macrophages that have been transformed into Gaucher's cells. In fact, clinical applications of scintigraphy with ^{99m}Tc-Sestamibi are based on its property to be actively concentrated in cells with increased metabolic expenditure such as myocardiocytes, tumor cells, and so forth. Increased resting energy expenditure is a well-known feature of patients with Gaucher's disease and has been attributed directly to the

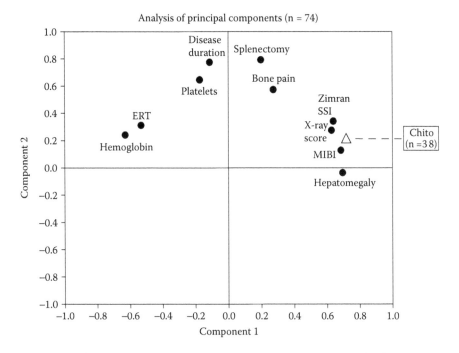

Figure 16.10   Results of multi-dimensional statistical analysis in patients with Gaucher's disease (irrespective of ERT treatment), represented as a "loading plot." Analysis of the principal components evaluates how the variables considered for each patient cluster together. The loading plot projects the original variables in the space of the principal components: coordinates of the variables along the principal components are the "loadings," that is, the statistical weights (or coefficients) of the original variables in the linear combinations that define the principal components themselves. In the loading plot, the most influential variables of a given component are those significantly farther from the center of the component, while any variable close to the center of the component has low loading, therefore virtually null statistical weight in the linear combination. Cosine of the angle defined by joining each pair of variables with the center of the two orthogonal axes corresponds to the correlation coefficient (R^2) of the two variables. Therefore, when any two variables form a 0° angle they are positively correlated (considering that cos 0° = 1). When the angle is 90° the two variables are not correlated at all (cos 90° = 0), and when the angle is 180° the two variables are correlated in a negative fashion (cos 180° = -1). Distance of each variable from the center of the orthogonal axes is proportional to the statistical weight of that variable on the principal components. ERT = enzyme replacement therapy; MIBI = semiquantitative score derived from scintigraphy with 99mTc-Sestamibi; Chito = serum chitotriosidase level (only available in 38 out of the entire group of 74 patients).

energy requirement of Gaucher's cells. [90,91] Therefore, the mechanism of accumulation of 99mTc-Sestamibi in the Gaucher's deposits is not solely based on lipophilicity of the compound (as is the case for the radioactive gas, 133Xe) but reflects the actual metabolic status of Gaucher's cells at the moment of tracer injection. This pattern of distribution is consistent with the observation of photopenic areas at marrow sites replaced by fibrous tissue because of prior acute episodes of necrosis or bone fractures.

Although still of a preliminary nature, more recent results validate the role of scintigraphy with ^{99m}Tc-Sestamibi as an additional method for monitoring the pattern of evolution of bone marrow involvement in these patients during an extended period of time. [81,82] These investigations were carried out to determine the trend of changes in the various parameters of disease (including the scintigraphic score obtained with ^{99m}Tc-Sestamibi) both in individual patients and in whole cohorts.

In particular, 39 patients with Gaucher disease underwent a baseline ^{99m}Tc-Sestamibi scan (prior to ERT in 28 patients, after an average 39 month-ERT in 11 patients), then at least one follow-up scan after 10–108 months (mean 26, median 18 months); at the end of follow-up 34 patients were on ERT (35 ± 15 U/kg body weight per month over 22 ± 18 months, cumulative dose 1122 ± 972 U/kg). Patients on longest follow-up also had additional ^{99m}Tc-Sestamibi scans at intermediate time points between baseline and last evaluation. The scintigraphic score was correlated with ERT dose/month, ERT cumulative dose, duration of ERT, and various clinical parameters defining the overall Zimran SSI. Inferential statistical assessment was based on stepwise backward and cross sectional time series analysis.

The pattern of change in the scintigraphic pattern of bone marrow infiltration as assessed by scintigraphy with ^{99m}Tc-Sestamibi following effective ERT was generally characterized by a gradual shift in radioactivity accumulation from extended areas of bone marrow infiltration to more restricted involvement concentrating around the knees (Figure 16.11). In contrast, several patients not receiving ERT or being treated with inappropriately low ERT doses exhibited worsening of the scintigraphic score at follow-up examination (Figure 16.12).

In those patients who had more than one follow-up evaluation while on ERT, the scintigraphic score continued to improve over time, although generally with the rate decreased at each subsequent evaluation. Despite equal monthly and cumulative ERT doses, significantly more patients reached a normal or near-normal scintigraphic pattern (^{99m}Tc-Sestamibi score between 0 and 2 at the end of follow-up), if ERT was initiated when the duration of symptoms that had led to a diagnosis of Gaucher disease was less than 10 years (33.3% vs. 5.6%, $P = 0.049$).

In this preliminary series of patients, the best correlation of improvement in the ^{99m}Tc-Sestamibi score was found with ERT dose/month ($P = 0.049$; $P<0.001$), ERT duration ($P = 0.043$) and ERT cumulative dose ($P = 0.008$). Correlation with dose/month was highest in patients with prior splenectomy ($P = 0.006$ vs. $P = 0.016$ in nonsplenectomised patients), in whom the presence of bone pain also became significant ($P = 0.038$). Improved ^{99m}Tc-Sestamibi score was highly correlated with improvements in platelet count ($P = 0.022$), and with severity of splenomegaly ($P = 0.028$ to 0.071). Response of the ^{99m}Tc-Sestamibi score was also highly correlated with the SSI response ($P<0.001$), which however did not discriminate patients according to prior splenectomy. The mean ERT dose inducing improvement in the ^{99m}Tc-Sestamibi score was 37 ± 3 U/kg per month, significantly higher than the ERT dose received by those patients in whom the scintigraphic score did not improve (24 ± 5 U/kg per month, $P = 0.004$). Similarly, the average cumulative dose of ERT received by patients in whom the scintigraphic score improved (879 ± 169 U/kg) was significantly higher than the cumulative dose received by those patients who

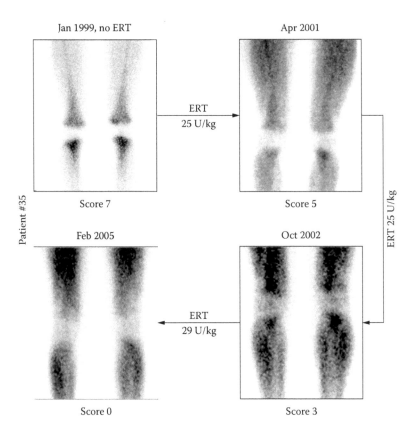

Figure 16.11 Evolution over a 6-year follow-up of the scintigraphic pattern with ^{99m}Tc-Sesta-mibi in a patient with Gaucher's disease under enzyme replacement therapy. There is obvious progressive improvement of the score from a starting level of considerable severity (score 7, January 1999) to complete normalization (score 0, February 2005).

did not improve (496 ± 155 U/kg, P = 0.0112). Statistical analysis also permitted calculation of a dose-effect relationship between ERT and expected improvement of the scintigraphic score (improvement of one unit in the scintigraphic score expected with 24 U/kg per month), as well as the threshold value for the ERT dose above which improvement of the ^{99m}Tc-Sestamibi scintigraphic score is to be expected (in the range between 30–35 U/kg per month).

These results demonstrate that the scintigraphic score derived from the ^{99m}Tc-Sestamibi scan provides a valuable means of monitoring the efficacy of ERT in patients with Gaucher's disease, suitable to identify the range of dose optimal for each single patient. The procedure is easy to perform, reproducible, and widely available. From all the above data accumulated on the use of scintigraphy with ^{99m}Tc-Sestamibi in patients with Gaucher's disease, it can reasonably be stated that:

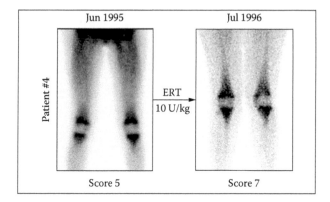

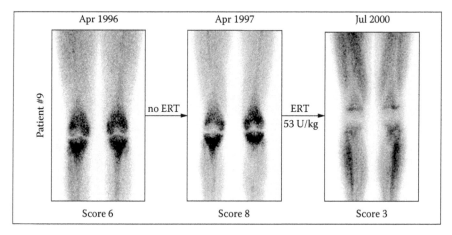

Figure 16.12 Evolution over follow-up of the scintigraphic pattern with 99mTc-Sestamibi in two different patients with Gaucher's disease. Upper panel: 16-year-old patient with type-3 disease treated with very low average doses of enzyme replacement therapy because of a combination of poor compliance and of difficult supply of the enzyme; over the 1-year follow-up there is obvious impairment of the score, from 5 to 7. Lower panel: 12-year-old patient with type 1 disease. There is obvious progression of bone marrow involvement at the 1-year follow-up scan, a period during which the patient did not receive enzyme replacement therapy. In contrast, the 4-year follow-up scan (after about 3 years of enzyme replacement therapy) shows marked improvement of the score, from 7 to 3.

1. Although based on localized evaluation, the 99mTc-Sestamibi score is an excellent predictor of the overall severity of Gaucher disease
2. Changes in the scintigraphic score following ERT are more closely correlated with the main determinants of therapy than other clinical scores (either separately or combined in the Zimran SSI)
3. Better results of ERT on the scintigraphic score are expected in patients with shorter duration of manifestations of disease

4. A dose-response relationship can be identified between ERT and scintigraphic score, and

5. A minimum threshold can be defined for the ERT dose expected to yield positive response of the scintigraphic score to treatment

Therefore, based on extensive experience in large groups of patients, scintigraphy with ^{99m}Tc-Sestamibi can be recommended in patients with Gaucher's disease both as a baseline assessment of the severity of disease and to evaluate bone marrow response to ERT. Major advantages of this procedure are the ability to identify the quantity of enzyme required for skeletal benefit and to correlate this with predictions of optimal dose of exogenous glucocerebrosidase for successful treatment of patients with Gaucher's disease.

REFERENCES

1. Dokal, I.S., Deenmamode, M., and Lewis, S.M., Radioisotope studies in monitoring of Gaucher's disease and its treatment. *Clin. Lab. Haematol.,* 11, 91, 1989.

2. International Committee for Standardization in Haematology, Recommended methods for measurement of red-cell and plasma volume. *J. Nucl. Med.,* 21, 793, 1980.

3. Petit, J.E. et al., Quantitative studies of splenic erythropoiesis in polycythaemia vera and myelofibrosis. *Br. J. Haemat.,* 34, 465, 1976.

4. Das, K.C., Mohanty, D., and Garewal, G., Nutritional megaloblastosis: from morphology to molecular biology, in *Medical Research Monographs*, Sapru, R.P., Ed., Malhotra Publishing House, New Delhi, 1989, 1.

5. Lee, R.E., Balcerzak, S.P., and Westerman, M.P., Gaucher's disease. A morphologic study and measurements of iron metabolism. *Am. J. Med.,* 42, 891, 1967.

6. Lorber, M., Adult-type Gaucher's disease: a secondary disorder of iron metabolism. *Mt. Sinai J. Med.,* 37, 404, 1970.

7. Lorber, M., Iron distribution in adult-type Gaucher's disease [letter]. *N. Engl. J. Med.,* 292, 110, 1975.

8. Silverstein, M.N. and Kelley, P.J., Osteoarticular manifestation of Gaucher's disease. *Am. J. Med. Sci.,* 253, 569, 1967.

9. Bannester, L.H. Haemolymphoid system, in *Gray's Anatomy*, 38th ed., Bannester, L.H, et al., Eds., Churchill Livingstone, New York, 1995, 1399.

10. Chapman, W.C. and Newman, M., Disorders of the spleen, in *Wintrobe's Clinical Hematology*, 10th ed., Lee, G.R., Ed., Williams and Wilkins, Baltimore, 1999, 1969.

11. Dacie, J.W. and Lewis, S.M., Erythrokinetics, in *Practical Haematology*, 8th ed., Dacie, JW. and Lewis, S.M., Eds., Churchill Linvingstone, Edinburgh, 1995, 397.

12. Kim, C.K., Reske, S.N., and Alavi, A., Bone marrow scintigraphy, in *Nuclear Medicine*, Honkin, R.E., et al., Eds., Mosby, St. Louis, 1996, 1223.

13. Ohta, H. et al., Liver scintigraphy in a patients with Gaucher disease. *Ann. Nucl. Med.,* 7, 115, 1993.

14. James, S.P. et al. Liver abnormalities in patients with Gaucher disease. *Gastroenterology*, 80, 126, 1981.

15. Cheng, T.H. and Holman, B.L., Radionuclide assessment in Gaucher's disease. *J. Nucl. Med.,* 19, 1333, 1978.

16. Israel, O., Jerushalami, J., and Front, D., Scintigraphic findings in Gaucher's disease. *J. Nucl. Med.,* 27, 1557, 1986.

17. Noyes, F.S. and Smith, W.S., Bone crisis and chronic osteomyelitis in Gaucher's disease. *Clin. Orthop. Relat. Res.*, 79, 132, 1971.
18. Katz, K. et al. Bone scans in the diagnosis of bone crisis in patients who have Gaucher disease. *J. Bone Joint Surg.*, 73, 513, 1991.
19. Gelfand, M.J. and Hancke, H.T. Jr., Bone scintigraphy in congenital, inherited and idiopatic proliferative bone disease of childhood, in *Bone Scintigraphy*, Silberstein, E., Ed., Futura, Mount Kisco–New York, Futura, 1984, 279.
20. Mariani, G. et al., Scintigraphic findings on ^{99m}Tc-MDP, ^{99m}Tc-Sestamibi and ^{99m}Tc-HMPAO images in Gaucher's disease. *Eur. J. Nucl. Med.*, 23, 466, 1996.
21. Mc Afee, J.G., Reba, R.C., and Majid, M., The muscoloskeletal system, in *Principles of Nuclear Medicine*, 2nd ed., Wagner, H.N. Jr., Ed., Saunders, Philadelphia, 1995, 986.
22. Davidson, A., Kalft, V., and Ryan, P., Bone crisis of Gaucher's disease due to bone ischemia: a case report. *Arthritis. Rheum.*, 28, 218, 1985.
23. Zanzi, I. et al., Scintigraphic and magnetic resonance studies in patients with Gaucher's disease. *Clin. Nucl. Med.*, 13, 491, 1988.
24. Yossipovitch, Z.H., Herman, G., and Makin, M., Aseptic osteomyelitis in Gaucher's disease. *Israel J. Med. Sci.*, 1, 531, 1965.
25. Bell, R.S., Mankin, H.J., and Doppelt, S.H., Osteomyelitis in Gaucher's disease. *J. Bone Joint Surg. Am.*, 68, 1380, 1986.
26. Waldvogel, F.A., Medoff, G., and Swartz, M.N., Osteomyelitis: a review of clinical features, therapeutic consideration and unusual aspects, part I. *N. Engl. J. Med.*, 282, 198, 1970.
27. Handmaker, H. and Leonards, R., The bone scan in inflammatory osseous disease. *Semin. Nucl. Med.*, 6, 95, 1976.
28. Trueta, J., The three types of acute hematogenous osteomyelitis: a clinical and vascular study. *J. Bone Joint Surg.*, 41B, 671, 1959.
29. Elgazzar, A.H. and Abdel-Dayem, H.M., Imaging skeletal infections: evolving considerations, in *Nuclear Medicine Annual*, Feeman L.M., Ed., Lippincott/Williams and Wilkins, Philadelphia, 1999, 157.
30. Mandell, G.A., Imaging in the diagnosis of muscoloskeletal infections in children. *Curr. Probl. Pediatr.*, 26, 218, 1996.
31. Amundsen, T.R, Siegel, M.J., and Siegel, B.A., Osteomyelitis and infarction in sickle cell hemoglobinopathies: differentiation by combining technetium and gallium scintigraphy. *Radiology*, 153, 807, 1984.
32. Miller, J.H., Ortega, J.A., and Heisel, M.A., Juvenile Gaucher disease simulating osteomyelitis. *Am. J. Roentgenol.*, 137, 880, 1981.
33. Schubiner, H., Letourneau, M., Murray, D.L. Pyogenic osteomyelitis versus pseudo-osteomyelitis in Gaucher's disease. Report of a case and review of the literature. *Clin. Pediat.*, 20, 667, 1981.
34. Lachewicz, P.F., Gaucher's disease. *Orthop, Clin, North America*, 15, 765, 1984.
35. Trueta, J. and Caladias, A.X., A study of the blood supply of the long bones. *Surg. Gynec. Obstet.*, 118, 485, 1964.
36. Majd, M. and Frankel, R.S., Radionuclide imaging in skeletal inflammatory and ischemic disease in children. *Am. J. Roentgenol.*, 126, 832, 1976.
37. Szkilas, J.J. et al., Diagnosing osteomyelitis in Gaucher's disease: observation of two cases. *Clin. Nucl. Med.*, 16, 487, 1991.
38. Howie, D.W. et al., The technetium phosphate bone scan in the diagnosis of osteomyelitis in childhood. *J. Bone Joint Surg. Am.*, 65, 431, 1983.

39. Lutzker, L.G. and Alavi, A., Bone and marrow imaging in sickle cell disease: diagnosis of infarction. *Semin. Nucl. Med.*, 6, 83, 1976.
40. Sain, A., Sham, R., and Silver, L., Bone scan in sickle cell crises. *Clin. Nucl. Med.*, 3, 85, 1978.
41. Kim, H.C. et al., Differentiation of bone marrow infarcts from osteomyelitis in sickle cell disorder. *Clin. Nucl. Med.*, 14, 249, 1989.
42. Armas, R.R. and Goldsmith, S.J., Gallium scintigraphy in bone infarction, correlation with bone imaging. *Clin. Nucl. Med.*, 9, 1, 1984.
43. Edwards, C.L. et al., Gallium-68 citrate: a clinically useful skeletal imaging agent. *J. Nucl. Med.*, 7, 363, 1966.
44. Schauwecker, D.S et al., Evaluation of complicating osteomyelitis with Tc-99m MDP, In-111 granulocytes and Ga-67 citrate. *J. Nucl. Med.*, 25, 849, 1984.
45. Seabold, J.E. et al., Procedure guideline for indium-111-leukocyte scintigraphy for suspected infection/inflammation. Society of Nuclear Medicine, *J. Nucl. Med.*, 38, 997, 1997.
46. Datz, F.L. et al., Procedure guideline for technetium-99m-HMPAO-labeled leukocyte scintigraphy for suspected infection/inflammation. Society of Nuclear Medicine, *J. Nucl. Med.*, 38, 987, 1997.
47. McCarthy, K. et al., Indium-111-labeled white blood cells in the detection of osteomyelitis complicated by a pre-existing condition. *J. Nucl. Med.*, 29, 1015, 1988.
48. Schauwecker, DS., Osteomyelitis: diagnosis with In-111-labeled leukocytes. *Radiology*, 171, 141, 1989.
49. Elgazzar, A.H. et al., Multimodality imaging of osteomyelitis, *Eur. J. Nucl. Med.*, 22, 1043, 1995.
50. Johnson, J.E. et al., Prospective study of bone — In-111 labeled white blood cell and gallium scanning for the evaluation of osteomyelitis in the diabetic foot. *Foot Ankle Int.*, 17, 10, 1996.
51. Seabold, J.E. et al., Postoperative bone marrow alterations: potential pitfalls in the diagnosis of osteomyelitis with In-111-labeled leukocyte scintigraphy. *Radiology*, 180, 741, 1991.
52. Palestro, C.J. et al., Infected knee prosthesis: diagnosis with in-111 leukocyte, Tc-99m sulfur colloid and Tc-99m MDP imaging. *Radiology*, 1179, 645, 1991.
53. Bihl, H., Rossler, B., and Borr, U., Assessment of infectious condition in the muscoloskeletal system: experience with Tc-99m HIG in 120 patients, Abstract. *J. Nucl. Med.*, 33, 839, 1992.
54. Love, C. et al., Role of nuclear medicine in diagnosis of the infected joint replacement. *Radiographics*, 21, 1229, 2001.
55. Strand, S.E. and Bergqvist, L., Radiolabeled colloids and macromolecules in the lymphatic system. *Crit. Rev. Ther. Drug. Carrier. Syst.*, 6, 211, 1989.
56. Ikomi, F., Hanna, G.K., and Schmidt-Schonbein, G.W., Mechanism of colloid uptake into the lymphatic system: basic study with percutaneous lymphography. *Radiology*, 196, 107, 1995.
57. Kricon, M.E., Red-yellow marrow conversion: its effect on the location of some solitary bone lesions. *Skel. Radiol.*, 14, 10, 1985.
58. Hermann, G. et al., Gaucher's disease type I: assessment of bone involvement by CT and scintigraphy. *Am. J. Roentgenol.*, 147, 943, 1986.
59. Lorberboym, M. et al., Scintigraphic monitoring of reticuloendothelial system in patients with type 1 Gaucher disease on enzyme replacement therapy. *J. Nucl. Med.*, 38, 890, 1997.

60. Starer, F., Sargent, J.D., and Hobbs, J.R., Regression of the radiological changes of Gaucher's disease following bone marrow transplantation. *Br. J. Radiol.*, 60, 1189, 1987.

61. Stadalnik, R.C., Diffuse lung uptake of Tc-99m-sulfur colloid. *Semin. Nucl. Med.*, 10, 106, 1980.

62. McKusick, K.A. et al., Inhaled-Xenon-133 bone dynamics, Abstract. *J. Nucl. Med.*, 31, 781, 1990.

63. Castronovo, F.P, McKusick, K.A, and Dopplet, S.H., Radiopharmacology of inhaled Xe-133 in skeletal sites containing deposits of Gaucher cells, Abstract, in *Proc. Seventh Int. Symp. Radiopharm.*, Boston, 1991, 22.

64. Matthews, C.M. and Dollery, C.T., Interpretation of 133-Xe lung wash-in and wash-out curves using an analogue computer. *Clin. Sci.*, 28, 573, 1965.

65. Wagner, H.N. Jr., The use of radioisotope techniques for the evaluation of patients with pulmonary disease. *Am. Rev. Respir. Dis.*, 113, 203, 1976.

66. Carey, J.E., Purdy, J.M., and Moses, D.C., Localization of Xe-133 in liver during ventilation studies. *J. Nucl. Med.*, 15, 1179, 1974.

67. Phelps, P., Steele, A., and McCarty, D.J., Significance of xenon-133 clearance rate from canine and human joints. *Arth. Rheum.*, 15, 360, 1972.

68. Lahtinen, T., Karjalainen, P., and Alhava, E.M., Mesurement of bone blood flow with a 133-Xe washout method. *Eur. J. Nucl. Med.*, 4, 435, 1979.

69. Amstutz, H.C. and Carey, E.J., Skeletal manifestations and treatment of Gaucher's disease. Review of twenty cases. *J. Bone Joint Surg.*, 48, 670, 1966.

70. Rosenthal, D.I. et al., Quantitative imaging for Gaucher disease. *Radiology*, 185, 841, 1992.

71. Rosenthal, D.I. et al., Enzyme replacement therapy for Gaucher disease: skeletal response to macrophage-targeted glucocerebrosidase. *Pediatrics,* 96, 629, 1995.

72. Kitani, K. and Winkler, K., *In vitro* determination of solubility of 133 Xenon and 85 Krypton in human liver tissue with varying triglyceride content. *Scand. J. Clin. Lab. Invest.*, 29, 173, 1972.

73. Mariani, G., Unexpected keys in cell biochemistry imaging: some lessons from technetium-99m sestamibi. *J. Nucl. Med.*, 37, 536, 1996.

74. Mariani, G. et al., Bone marrow scintigraphy with ^{99m}Tc-Sestamibi to assess the efficacy of substitutive enzyme therapy in patients with Gaucher's disease, Abstract, in *Proc. Third Meeting Eur. Working Group Gaucher Disease* (EWGGD), Lemnos, 1999.

75. Mariani, G. et al., Bone marrow scintigraphy with ^{99m}Tc-Sestamibi to assess the efficacy of substitutive enzyme therapy in patients with Gaucher's disease. Abstract, *Eur. J. Nucl. Med.,* 26, 1034, 1999.

76. Mariani, G. et al., Assessment of the efficacy of enzyme replacement therapy by bone marrow scintigraphy with ^{99m}Tc-Sestamibi in patients with Gaucher disease. Abstract, in *Proc. Fourth Meeting Eur. Working Group on Gaucher Disease* (EWGGD), Jerusalem, 2000, 34.

77. Mariani, G. et al., Bone marrow scintigraphy with ^{99m}Tc-Sestamibi in patients with Gaucher's disease. Abstract, in *Proc. Fourth Meeting Eur. Working Group Gaucher Disease* (EWGGD), Jerusalem, 2000, 35.

78. Mariani, G. et al., Correlation of serum chitotriosidase levels with semiquantitative scores of clinical severity and of scintigraphic involvement of the bone marrow in patients with Gaucher disease. Abstract, in *Proc. Fifth Workshop Eur. Working Group Gaucher Disease* (EWGGD), Prague, 2002, A33.

79. Aharoni, D. et al., Tc-99m Sestamibi bone marrow scintigraphy in Gaucher disease. *Clin. Nucl. Med.*, 27, 503, 2002.

80. Mariani, G. et al., Severity of bone marrow involvement in patients with Gaucher disease evaluated by scintigraphy with ^{99m}Tc-Sestamibi. *J. Nucl. Med.*, 44, 1253, 2003.

81. Erba, P. et al., Assessment of the long-term efficacy of enzyme replacement therapy by ^{99m}Tc-Sestamibi bone marrow scintigraphy in patients with Gaucher disease. Abstract, *J. Nucl. Med.*, 45, 79P, 2004.

82. Mariani, G. et al., ^{99m}Tc-Sestamibi bone marrow scintigraphy demonstrates the efficacy of long-term enzyme replacement therapy in patients with Gaucher disease. Abstract, in *Proc. Sixth Workshop Eur. Working Group on Gaucher Disease* (EWGGD), Barcelona, 2004, 36.

83. Balleari, E. et al., Technetium-99m-sestamibi scintigraphy in multiple myeloma and related gammopathies: a useful tool for the identification and follow-up of myeloma bone disease. *Haematologica*, 86, 78, 2001.

84. Poll, L.W. et al., Magnetic resonance imaging of bone marrow changes in Gaucher disease during enzyme replacement therapy: first German long-term results. *Skeletal Radiol.*, 30, 496, 2001.

85. Allen, M.J. et al., Pro-inflammatory cytokines and the pathogenesis of Gaucher's disease: increased release of interleukin-6 and interleukin-10. *Q.J.M.*, 90, 19, 1997.

86. Mistry, P.K., Wraight, E.P., and Cox, T.M., Therapeutic delivery of proteins to macrophages: implications for treatment of Gaucher's disease. *Lancet*, 348, 1555, 1996.

87. Lorberboym, M. et al., Scintigraphic evaluation of Tc-99m-low-density lipoprotein (LDL) distribution in patients with Gaucher disease. *Clin. Genet.*, 52, 7, 1997.

88. Zimran, A. et al., The natural history of adult type 1 Gaucher disease: clinical, laboratory, radiologic and genetic features of 53 patients. *Medicine (Baltimore)*, 71, 337, 1992.

89. Hollak, C.E.M. et al., Marked elevation of plasma chitotriosidase activity: a novel hallmark of Gaucher disease. *J. Clin. Invest.*, 93, 1288, 1994.

90. Barton, N.W. et al., Resting energy expenditure in Gaucher's disease type 1: effect of Gaucher cell burden on energy requirements. *Metabolism*, 38, 1238, 1989.

91. Aerts, J.M.F. and Hollak, C.M.E., Plasma and metabolic abnormalities in Gaucher's disease. *Ballières Clin. Haematol.*, 10, 691, 1997.

CHAPTER 17

Epidemiology and Screening Policy

Peter J. Meikle, Maria Fuller, and John J. Hopwood

CONTENTS

INTRODUCTION

Gaucher disease is one of many Mendelian disorders, and although monogenic, the variability in clinical presentation presents a complex picture. Further to this complexity, very little is known about the pathogenesis of the disease. In this respect Gaucher disease is not alone; it belongs to a group of over 50 single gene disorders in which the mechanisms of pathology are poorly understood. These are collectively known as lysosomal storage disorders (LSD), a group of inherited metabolic disorders resulting from a deficiency of any one of a number of lysosomal enzymes or in a few instances from a deficiency of other proteins affecting the lysosome biogenesis or function.

LSD are chronic and progressive diseases and present with a range of clinical pathologies. Nevertheless, they all have the common biochemical feature that the enzyme deficiency/lysosomal dysfunction lead to the accumulation of un-degraded substrates within the lysosomes of affected cells. Gaucher disease belongs to the subgroup of LSD known as the sphingolipidoses based primarily on the accumulation of the sphingolipid, glucosylceramide. In the absence of a family history, diagnosis of Gaucher disease is typically made on the basis of reduced or no enzyme activity. Nonetheless, the value of using the substrate (glucosylceramide) as a marker of disease activity for the diagnosis, prognosis, screening, and monitoring of patients with Gaucher disease should not be understated. This will be exemplified in this chapter describing screening strategies to identify patients with Gaucher disease.

As one of the most common LSD, studies and discussion of Gaucher disease can provide information and insight that can be applied to other LSD. Therefore, in the context of this chapter describing the prevalence and distribution of Gaucher disease, it would seem unjust not to include mention of other LSD. This becomes even more apparent when we consider issues relating to screening policies and technology, as it would seem most likely that a worldwide screening program for any individual LSD, given their relatively low prevalence, could not be justified. Moreover, it would seem that screening for multiple LSD with a combined prevalence of 1 in 5000 births would be feasible.

In this chapter we will review the epidemiology and screening policy of Gaucher disease while at the same time drawing on other LSD to compare and contrast.

EPIDEMIOLOGY

Incidence of Gaucher Disease

Gaucher disease is often reported as the most prevalent of the LSD. In particular, Gaucher disease has a very high prevalence in the Ashkenazi Jewish population. A number of reports estimating the incidence of Gaucher disease in this population have yielded variable results. Early estimates, based on diagnoses divided by birth rate, gave values of 1:7750 [1] and 1:10,000 [2]. These were likely to be underestimates as many patients with the attenuated form of the disease go undiagnosed and therefore were not included in these figures. Later estimates, which were based on determining enzyme activity to identify heterozygotes, gave figures ranging from 1:640 [3] through 1:2003 [4] to 1:3969 [5]. While these estimates produced lower figures than the early estimates, the disparity of these three figures raised some doubt about the accuracy of any particular value. It is possible that these values may result from what is now recognized to be overlap between β-glucosidase activity levels in controls and heterozygotes. Beutler and Grabowski [6] combined the results of two studies in which the carrier frequency of the mutations 1226G and 84GG from 2121 unaffected Ashkenazi Jews were determined [7,8]. These were used to predict a frequency of 1:971 for the 1226G homozygotes and 1:7143 for the 1226G/84GG heterozygotes to give a combined incidence of 1:855 Ashkenazi Jewish births. This figure is likely to represent a more accurate estimate of Gaucher disease in the Ashkenazi Jewish population than earlier estimates. It should be noted that the 1226G mutation and 84GG mutations are reported to represent only 70% and 12% of the Ashkenazi mutations, respectively [9], and so the final incidence figure might be expected to be higher. However, allelic frequencies are calculated from patient alleles and many of the 1226G homozygotes do not present clinically, so the frequency of this allele may be higher than reported. Other common mutations in the Ashkenazi Jewish population include IVS2+1 (2.6% of total mutations), 1448C (3.9%), and 1297T (2.6%) [9].

The reason for the high incidence of Gaucher disease in the Ashkenazi Jewish population has been a matter for debate. Beutler and Grabowski [6] propose that the existence of three common mutations in this population make it clear that the high gene frequency is not the result of a single mutational event in an isolated population but rather the result of selection. They propose a selective advantage for heterozygotes although it is uncertain as to the exact nature of such an advantage; resistance to tuberculosis has been suggested [10]. In contrast to this idea, Slatkin [11] reported that available genetic data was consistent with the isolation of a specific population (founder effect) resulting from a severe bottleneck in population size between 1100 A.D. and 1400 A.D. and an earlier bottleneck in 75 A.D., at the beginning of the Jewish Diaspora. He concluded that a founder effect can account for the relatively high frequency of alleles causing Gaucher disease and three additional lysosomal storage disorders, also prevalent in the Ashkenazi population, if the disease-associated alleles are recessive in their effects on reproductive fitness.

Incidence data of Gaucher disease in other populations have not been comprehensive. A report on the frequency of the 1226G mutation in the Portuguese population suggested an incidence of Gaucher disease of approximately 1:21,500 births, based on the 1226G mutation representing 63% of mutant alleles, which was also determined in the same study [12]. This is significantly higher than the reported prevalence of Gaucher disease in northern Portugal, based upon diagnoses, which was calculated to be 1:74,000 [13]. This suggests that many patients homozygous for the 1226G mutation are at the attenuated end of the clinical spectrum and remain undiagnosed. As with the Ashkenazi population, the disparity in these two estimates also suggests that the 1226G mutation represents more than 63% of mutant alleles in the Portuguese population. Other studies, based on diagnoses, have reported similar incidence figures, with a report of 1:57,000 births in the Australian population [14], 1:40,247 in Italy [15] and 1:86,000 in The Netherlands [16]. In the Australian study [14], the population is predominantly of British decent, with a large contribution from other European countries and to a lesser extent, Asian countries. As such, this population would be comparable to that of most Anglo-Celtic countries and the results could be extrapolated to the white, non-Hispanic populations in the

U.S., Canada, and the United Kingdom, where there have been few studies on the incidence of Gaucher disease.

One of the more common alleles causing Gaucher disease in non-Ashkenazi Jewish populations is the 1448C mutation. This mutation was first characterized in Sweden as the Norrbottnian form of Gaucher disease [17]. The Swedish families of the Norrbottnian type are found in two geographically distinct clusters in Sweden. Each cluster was traced back to a single ancestral couple who were not known to be related to each other. However, the molecular studies were considered compatible with a single founder who arrived in northern Sweden in or before the 16th century [18]. The 1448C mutation is also common in other non-Ashkenazi populations, representing 16% of alleles in the Spanish population [19] and 39% in the non-Ashkenazi Jewish population in Israel [9].

Very few studies have been performed on Asian or African populations to determine the incidence of Gaucher disease. However, of note is the observation that the common mutation 1226G, present at high frequencies in the Ashkenazi Jewish and other Western populations, was not seen at all in the Japanese population after examination of over 100 alleles [20]. Similarly the 1226G mutation was not present in eight black South African patients [21]. The common alleles in the Japanese population were the 1448C and 754A mutations representing 41% and 14% of alleles, respectively [20]. Similarly, the common mutations in Chinese Gaucher patients do not seem to include 1226G or 84GG while the 1448C mutation represented 5 of 12 alleles studied in Chinese patients [22].

Epidemiology of Gaucher Disease Subtypes

Gaucher disease has been classified into types 1, 2, and 3, representing the non-neuronopathic, severe infantile-neuronopathic and the late-onset, intermediate-neuronopathic forms, respectively. In most Caucasian populations, including the Ashkenazi Jews, type 1 is clearly the most prevalent. In the study of LSD prevalence in The Netherlands, a total prevalence of Gaucher disease was calculated to be 1:86,000 births, of these 22% were either type 2 or 3. In the Portuguese study, 3 of 36 diagnoses (8%) made over a 20-year period were type 2 or 3 [13]. The absence of the non-neuronopathic mutation 1226G in the Japanese population potentially leads to a higher proportion of the neuronopathic subtype in this population [20]. In addition, the phenotype of Gaucher disease type 1 in this population has been described as severe and progressive compared to type 1 patients in other populations [23]. Data from the Gaucher Registry on 1698 patients with Gaucher disease, collected worldwide, report fewer than 1% type 2 and only 5% type 3 [24]. However, the Gaucher Registry is a voluntary database and is likely to under-represent the more severe forms of the disease as these patients are not usually on therapy.

Incidence of Other LSD

There have been a limited number of studies on the incidence of LSD. One of the main problems associated with accurate epidemiological studies of these rare disorders is that, in most countries, there are numerous diagnostic centers for LSD, which compounds the problem of collecting and correlating diagnoses. In Australia there are only two such centers, which cover the entire country, New Zealand, and a portion of South East Asia. The National Referral Laboratory for Lysosomal, Peroxisomal, and Related Disorders at the Women's and Children's Hospital, Adelaide, diagnoses approximately 80% of LSD patients in Australia, while the Division of Chemical Pathology, Royal Brisbane Hospitals, Brisbane covers the remainder. In a retrospective case study, Meikle et al. reported on the incidence of LSD in Australia [14]. Over the period January 1980 to December 1996, 470 LSD patient diagnoses were made, representing 27 different disorders. Based on these figures the incidence of LSD range between 1:57,000 for Gaucher disease to as low as 1:4.2 million for sialidosis. The prevalence of LSD as a group was calculated to be 1:7,700 [14], and if prenatal diagnoses were not considered, then the incidence for LSD was 1:9,000 births. However, the neuronal ceroid lipofuscinoses were not included in this study. Estimates of the global incidence of neuronal ceroid lipofuscinoses as high as 1:12,500 have been reported [25]. Based on Australian diagnoses over the period 1998–2003, the combined incidence of the three most prevalent forms (infantile, late-infantile, and juvenile) has been estimated at approximately 1:60,000 (personal communication, M.J. Fietz, Women's and Children's Hospital Adelaide, Australia). In addition, there are likely to be other LSD that are yet to be described clinically and biochemically, suggesting that the combined incidence of LSD is likely to be 1:5000 or less. Based on current figures, Gaucher disease can be calculated to represent approximately 11% of total LSD in the Australian population. The relative incidence of Gaucher disease compared to other LSD in Australia is shown in Figure 17.1.

The incidence of LSD in Australia is not dissimilar to other countries. Poorthuis et al. [16] reported on the frequency of LSD in The Netherlands based on 963 diagnosed cases from 1970–1996. The combined prevalence for all LSD was calculated to be 1:7,100 live births, with glycogen storage disease type II being the most prevalent at 1:50,000. Gaucher disease was calculated to be 1:86,000, somewhat lower than the Australian figure and representing 8% of total LSD. A report from Italy on inborn errors of metabolism stated a combined incidence of LSD as 1:8,275 with Gaucher disease as 1:40,247 (21% of total LSD) [15], while a recent report from Portugal estimated the prevalence of all LSD combined to be 1:4,000 births with Gaucher at 1:86,000 births (5% of total LSD) [13]. Ozkara and Topcu reported the combined incidence of the sphingolipidoses in Turkey as 1:21,500 with Gaucher disease as 1:185,000 [26]. There have also been a number of reports on the prevalence of particular disorders in specific populations. Values as high as 1:18,500 for aspartylglucosaminuria in the Finnish population [27] and 1:3,900 for Tay-Sachs diseases in the Ashkenazi Jewish population [28] have been reported.

For diseases that are very rare, the nonascertainment of even a single case can make a large difference to the calculated birth prevalence. In addition, the methods used for the calculation of prevalence figures must also be considered when comparing data from different studies. In the Australian study, the period used to determine the number of births was the period during which the diagnoses were made; the assumption was made that the late onset patients (born before the study period) were representative of late onset patients born within the study period that had not

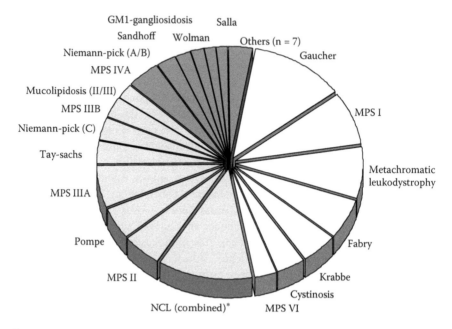

Figure 17.1   Relative incidence[†] and therapy options for lysosomal storage disorders: Pie slices represent the relative prevalence of each disorder. Disorders for which enzyme replacement therapy, drug therapy, or hematopoietic stem-cell transplantation currently exist (unshaded), disorders that are undergoing clinical trials for either enzyme replacement therapy or drug therapy (light shade) and other disorders (dark shade). *NCL (combined) includes infantile, late infantile, and juvenile forms of neuronal ceroid lipofuscinosis. Drug therapy (cystagon) is currently being trialed for the infantile form of the disease. ([†]Incidence values taken from Meikle et al., *JAMA*, 281, 249, 1999, and personal communication, M. J. Fietz, Women's and Children's Hospital Adelaide, Australia. With permission.)

yet presented clinically [14]. In the Dutch and Portuguese studies, the period used to calculate the birth rate was based on the ages of the patients; thus here the assumption is made that all patients born prior to the last patient in the subgroup have been identified [13,16]. There are limitations with both methods and it is likely that ascertainment of the late onset LSD patients was incomplete, leading to underestimation of the prevalence. As stated earlier for Gaucher disease, late-onset patients with other LSD may also go undiagnosed. In Australia, we estimated the incidence of Pompe disease to be approximately 1:200,000. However, studies based on carrier detection in normal populations, have indicated that the incidence may be up to 1 in 40,000 births in both the U.S. [29] and The Netherlands [30]. On the basis of their findings Martiniuk et al. [29] predict that many adult cases at the attenuated end of the clinical spectrum remain undiagnosed.

The burden of LSD on the individuals and families involved is undoubtedly enormous, although there is relatively little documentation in this area. The cost to the community in terms of medical treatment and support is also poorly defined. Therapy of one form or another is likely to be available for the majority of patients in the near future. In addition, a number of technologies for newborn screening have been proposed. If we are to accurately assess the current cost of these disorders to public health systems, and thereby justify the cost of therapy and/or newborn screening, accurate values for prevalence of the disorders will be required.

SCREENING POLICY

Screening can be defined as "the application of a test to a population who are as yet asymptomatic for the purpose of classifying them with respect to their likelihood of having a particular disease." Screening is not a diagnosis but rather a risk assessment. Those who are identified as "high risk" are further counseled and evaluated with diagnostic tests to confirm that they are indeed affected with the disease in question. Newborn screening can be described as a population-based public health program applied regionally with the aim of reducing the morbidity, severity, or mortality of specific metabolic/genetic disorders. Newborn screening programs currently exist in most Western countries, and tests for disorders including phenylketonuria, congenital adrenal hyperplasia, and cystic fibrosis, are commonly employed. The use of electrospray ionization-tandem mass spectrometry (ESI-MS/MS) is becoming more widely accepted as a means to screen for up to 30 different metabolic disorders in the newborn period [31]. Currently there are no established newborn screening programs for Gaucher disease or any other LSD; however a number of groups are working toward the development of such programs [32–37]. The issues relating to the introduction of new tests into existing newborn screening programs are multiple and complex. In 1968 Wilson and Jungner proposed 10 criteria that should be satisfied to justify population-based screening for specific diseases [38]. More recently, Wilcken [39] proposed that the questions to be explored in setting up a screening program can be simplified to: 1) Should we do it? 2) Can we do it? and 3) Can we afford to do it? We will discuss each of these issues, focusing on Gaucher disease and using other LSD where relevant.

Carrier detection for genetic disorders is a form of screening, performed to identify couples at risk of having affected children. Following identification of carrier status (individuals carrying a mutation in a specific gene), couples who are both carriers for the same disorder are offered genetic counseling and usually prenatal diagnosis in subsequent pregnancies. The criteria for selection of diseases for inclusion in carrier detection programs can also be expressed in similar terms to those for newborn screening programs: 1) should we do it? 2) can we do it? and 3) can we afford to do it? Here also, Gaucher disease offers a useful model to discuss the issues relating to LSD in general.

Carrier Testing

Carrier testing for genetic diseases has only been adopted where the incidence of the disease is sufficiently high to warrant testing of a particular population. The Ashkenazi Jewish population is at high risk for a number of genetic diseases including Gaucher disease, as well as Tay-Sachs disease, cystic fibrosis, Canavan disease, familial dysautonomia, Niemann-Pick disease, Fanconi anemia and Bloom syndrome

[40]. Carrier screening for Tay-Sachs disease has been available for over 30 years and over 1 million individuals have been tested. These programs have been effective in reducing the incidence of this disease in the Ashkenazi Jewish population [41]. The early introduction of carrier testing for Tay-Sachs disease was possible as a result of the clear discrimination possible between control and carriers using β-hexosaminidase activity assays. Since the beginning of carrier testing for Tay-Sachs disease, more that 1.4 million individuals have been tested. As a result of subsequent counseling and prenatal testing, the incidence in the Ashkenazi Jewish population has been reduced by more than 90% [42]. More recently, using molecular analysis, the Tay-Sachs program has been expanded in some centers to include other disorders with high prevalence in the Ashkenazi Jewish population. In New York, these include both cystic fibrosis and Gaucher disease and these tests are offered primarily to the Ashkenazi Jewish population. However, carrier testing for cystic fibrosis in the general population, which has a heterozygote incidence of 1:25, is justifiable and becoming more widely adopted in many Western countries.

There are many issues associated with carrier testing including informed consent, test accuracy, pre- and posttesting counseling, including counseling for subsequent prenatal diagnosis, as well as privacy and confidentiality of test results. These are ongoing issues that change as society's values and views on genetic issues evolve. For example, as with most LSD, Tay-Sachs disease can present with a range of clinical phenotypes ranging from the severe infantile form to a more attenuated juvenile and adult onset form. However, all forms involve neuropathology with the infantile form being characterized by developmental retardation, followed by paralysis, dementia and blindness, with death in the second or third year of life. The adult onset form often presents with psychiatric problems, while brain imaging studies reveal marked cerebellar atrophy [43]. In contrast, Gaucher disease presents a new challenge, as the attenuated form of the disease represented by the 1226G mutation may present from early childhood to late in life, or not at all.

Genetic counseling of carriers of Gaucher disease will be a complex process. How will couples decide on a course of action for an affected fetus when a high degree of uncertainty exists as to the resulting phenotype? This is also true for other LSD, most of which can present with a range of clinical phenotypes. A better understanding of the pathological mechanisms of LSD and the ability to predict phenotype in presymptomatic individuals would be advantageous for these disorders.

Newborn Screening

Should We Screen?

The issues relating to newborn screening for Gaucher disease and other LSD are multifactorial and relate to ethical, religious, and cultural differences. It is unlikely that a uniform policy will be defined that is applicable and acceptable in all communities. At best we can identify and discuss the issues relating to newborn screening for these diseases. Individual communities will need to make decisions based on their needs and resources with consideration to the issues associated with these programs.

Consideration of whether we should screen for a particular disorder such as Gaucher disease raises the question of potential good vs. potential harm that may result from such a screening program. Enzyme replacement therapy (ERT) has been available for the non-neuronopathic form of Gaucher disease for over 15 years [44]. The results from this type of therapy have been very positive with substantial improvements in most pathology. However, enzyme does not efficiently cross the blood brain barrier [45], and so ERT has limited application in the treatment of patients with neurological symptoms. Stem cell transplantation has been used in Gaucher disease, type 1; however, the high morbidity and mortality associated with this therapy make ERT a more attractive option for these individuals. Stem cell transplantation has also been performed in type 3 patients with favorable outcomes and may be able to arrest the progression of neurological disease [46]. As ERT appears to be of limited value for type 3 patients, stem cell transplantation may be the therapy of choice when a suitable donor is available. Substrate deprivation therapy has been developed for Gaucher disease [47,48] and other therapies such as gene therapy are under development [49]. These therapies also have the potential for the treatment of neuropathology; however, the efficacy of these therapies will rely heavily upon the early diagnosis and treatment of the disorder before the onset of what is likely to be irreversible pathology.

In addition to improving the outcome of current and proposed therapies, early detection of Gaucher disease has other potential advantages. It will enable genetic counseling of the parents and provide them with reproductive choices; it is not uncommon for families to have two or more affected children before the first one is diagnosed. Early diagnosis as provided by a newborn screening program would also avoid the prolonged and stressful process of diagnosis that is the current situation for many patients and families. The time between a parents' recognition of a problem and a definitive diagnosis can be months or years in some instances. These benefits are primarily directed to the family rather than the newborn but are nonetheless significant. The critical question is whether the benefits flow on to the newborn from a family, which is better informed and prepared for the disease.

On the other hand, the potential harm that could be caused from the diagnosis of late onset Gaucher disease in a newborn must also be considered. Given the relatively high proportion of patients homozygous for the 1226G mutation that are asymptomatic, the potential for false positives in such a newborn screening program is substantial. Even for those patients that will ultimately develop the non-neurono-pathic form of Gaucher disease in adulthood, the benefits to early diagnosis and possible early treatment must be weighed against the disadvantages of living with a disease from an early age that might not manifest until late in adulthood or not at all. These are difficult issues to define and quantify but must be addressed by the community before screening can commence.

Many of the same arguments relevant to screening for Gaucher disease also apply to other LSD. There are multiple treatment strategies for LSD; cystinosis is treated with cysteamine [50], and a number of LSD including MPS I, MPS VI [51–53] and Wolman disease [54] have been responsive to bone marrow (or hemato-poietic stem cell) transplantation. Furthermore, patients with MLD and Krabbe disease, when transplanted before clinical signs were evident, have been reported to develop less central nervous system (CNS) pathology than either untransplanted patients or patients transplanted after clinical signs were present [55]. In addition to ERT for Gaucher disease, ERT for Fabry disease [56,57] and MPS I [58] have recently become available and clinical trials of ERT for MPS II [59], MPS VI [60], and Pompe disease [61] are in progress. While these therapies and clinical trials have shown promising results, as with Gaucher disease, intravenous infusion of enzyme is unlikely to benefit those LSD patients with CNS pathology. Combination ERT and bone marrow or umbilical-cord blood (hematopoietic stem cell) transplantations are under evaluation in MPS I patients with CNS disease. Here, it is proposed that MPS I patients when subjected to enzyme replacement 6 to 12 weeks before the hematopoietic stem cell procedure will show improved engraftment following stabilization of the clinical disease before transplantation. Other therapies such as gene therapy [62,63] and a range of small molecule therapies including substrate deprivation [47,48], chemical chaperone [64,65], and stop codon read through [66] are under development. In addition, directed delivery of the replacement enzyme to the brain has recently been reported in a dog model for MPS I [67] and a mouse model for MPS IIIA [68] for the treatment of CNS pathology.

It is likely that these and other therapies will be established for the treatment of LSDs within the next few years (Figure 17.1). The potential advantages resulting from early detection and treatment of these disorders are the same as described for Gaucher disease, particularly for those LSD involving CNS and bone pathologies. A further consideration, critical to bone marrow transplant therapy, is that early diagnosis of the LSD will allow clinicians to take advantage of the window of opportunity presented by the naturally suppressed immune system of the neonate to maximize the chances of a successful engraftment. Here also, there are potential benefits resulting from genetic counseling of families.

The potential harm resulting from newborn screening for other LSD is also similar to the Gaucher situation. Many LSD have late onset variants that are likely to be identified at birth, resulting in the same dilemma as described above for Gaucher disease. Similarly, as for the infantile form of Gaucher disease (type 2), therapy options for many LSD with neurological symptoms are limited and the question of screening for diseases where there is no effective therapy is contentious, even allowing for the other potential benefits of early detection. Ultimately, individual communities will decide these issues based on their practicalities and ethical, cultural, and religious values.

Can We Screen?

The question "Can we can do it?" addresses the issue of the technology required to achieve acceptable sensitivity and specificity in the screening assay. The potential harm from false positives is considerable, and this must clearly be kept to a minimum and recalls handled with care to minimize the anxiety and stress to the family. The strategies and technology required to achieve a newborn screening program for LSD are under development and will be discussed in the next section.

Can We Afford to Screen?

This question combines both ethical and technical issues and is particularly relevant for rare disorders such as Gaucher disease and other LSD. Clearly a screening program for Gaucher disease alone with an incidence of 1:57,000 is unlikely to be acceptable on the basis of cost for screening per identified case. However, screening for multiple LSD in a single program will increase the number of disorders screened without substantially increasing the cost of the screening program. This is the case with current newborn screening programs utilizing ESI-MS/MS, where many disorders are screened for simultaneously. This type of approach to newborn screening for LSD will address the issue of low incidence of individual disorders and the cost of screening per identified case. Consequently, a program that encompasses the simultaneous screening for multiple LSD could be economically justified and is more likely to be widely adopted than screening assays for individual disorders. Therefore, strategies that enable the simultaneous detection of multiple disorders are required.

Technology to Screen for LSD

A number of strategies and technologies could be used to develop a newborn screening program for Gaucher disease and other LSD. Screening for LSD can involve analyzing the particular enzyme activity or protein deficiency, or alternatively, by the detection of the accumulated substrate(s).

Enzyme Analysis

Following clinical suspicion, diagnosis for Gaucher disease is typically performed by enzymatic analysis of leukocytes using a fluorescent substrate such as 4-methylumbeliferyl-β-D-glucoside. In general, a panel of individual enzyme assays will be performed on a combination of leukocytes and plasma and predominantly include enzymes involved in the digestion of glycosphingolipids and oligosaccharides. Diseases tested for in these panels include Gaucher disease, Niemann-Pick disease types A and B, acid lipase deficiency, G_{M1} and G_{M2} gangliosidoses, Krabbe disease, metachromatic leukodystrophy, mucolipidosis types II and III, fucosidosis, α-mannosidosis, MPS-VII and Schindler's Disease [69]. Chamoles et al. has described enzyme activity assays, performed on dried blood spots, for β-glucosidase and acid sphingomyelinase [36]. Clear discrimination was obtained for all Gaucher disease and Niemann-Pick disease patients from the control population although there was overlap between the carriers of both diseases and the control group [36]. A range of other lysosomal enzyme activities has been determined by direct assay from dried blood spots [35,37]. More specific assays, using immune-capture with monoclonal or polyclonal antibodies, have also been developed for a number of enzymes including; α-glucosidase (deficient in Pompe disease) [70], α-galactosidase (deficient in Fabry disease) [71], and N-acetylgalactosamine-4-sulphatase (deficient in MPS VI) [72]. All these assays are applicable to dried blood spot samples. While both the direct and immune-capture assays are amenable to high throughput screening in 96-well microtitre plates, such a screening program incorporating multiple separate assays would be severely limited by the sample size available on a Guthrie card. In addition, the cost of reagents and personnel to perform such individual assays is likely to be prohibitive in most countries.

Technology is available to miniaturize and multiplex functional enzyme assays. Direct measurement of enzyme activity has been performed by flat protein microarray [73]. Zhu et al. [74] characterized 119 protein kinases for 17 different substrates. Functional protein arrays such as these are primarily intended for use as a high throughput screening tool for activity and substrate specificity, but it is conceivable that they could be applied to quantify multiple enzyme activities in a newborn screening program. Alternatively, measuring enzyme products has been proposed by ESI-MS/MS [34,75–78], although to date this has been confined to three to five enzyme products. Advances in this technology may enable it to be scaled-up to include 10–20 analytes.

Metabolite Profiling

ESI-MS/MS is widely used for multi-analyte screening for inborn errors of metabolism from newborn Guthrie cards [31]. In addition, ESI-MS/MS has been successfully used for the determination of stored oligosaccharide and glycosphingolipid substrates in LSD [79–86]. A retrospective analysis of these small molecule biomarkers in newborn Guthrie cards has demonstrated that for a number of LSD, these analytes were useful screening markers [33]. However, markers suitable for all LSD have yet to be identified, which in part reflects incomplete marker sets, but also the sensitivity of current instrumentation, since many of these markers are present in blood, albeit at low concentrations. For some LSD, the determination of multiple analytes, "a metabolic profile," has provided improved discrimination of control and affected populations [33,86–88], and in some instances information on disease severity [86,87]. The suitability of ESI-MS/MS quantification of biomarkers for a high throughput, cost-effective, newborn screening program from dried blood spots remains to be fully evaluated.

Protein Profiling

Multiplexing to quantify specific protein markers, deficient or elevated in individual LSD, may also provide a feasible platform for newborn screening. Each LSD is associated with a decrease in the activity of a specific lysosomal enzyme or in a few cases, a deficiency of a protein involved in lysosomal biogenesis. In most LSD patients, not only is the level of enzyme activity decreased but the level of mutant protein is also substantially reduced [70–72,89,90]. In addition, in many LSD there is altered biogenesis of other lysosomal proteins [32,91–93]. Thus the generation of "a protein profile" of multiple enzymes/proteins, some elevated and others reduced, could improve patient identification. A protein profile could also extend discriminatory power by reference to control markers, which normalize the population for variability in leukocyte and/or lysosomal content in the newborn sample. Lysosomal markers such as LAMP-1 [32] or a leukocyte-specific marker such as CD45 may fill this role.

It can be postulated that abnormal lysosomal protein profiles may discriminate LSD, even where the deficient protein is not included in the marker set. For example mucolipidosis type II/III can be identified by elevated plasma levels of a range of lysosomal enzymes. Discrimination of other LSD may be possible on the basis of the disturbance caused to lysosomal biogenesis and reflected by a specific pattern of protein expression. Development of algorithms to recognize these profiles could provide accurate identification of a LSD, as well as information on disease severity, by virtue of the level of disturbance.

Issues Related to Newborn Screening

In the near future it is likely that newborn screening will identify Gaucher and other LSD patients at birth, facilitating early treatment of patients before the onset of clinical pathology. In some instances this may be a desirable course of action (where neuropathology is imminent), while in others it may be contraindicated (where clinical presentation is not likely until adulthood if at all, as in some 1226G homozygotes). Diagnosis at birth will remove the ability of physicians to assess disease severity on the basis of clinical presentation. Moreover, a range of therapeutic options will be available and an informed decision on which therapeutic strategy is optimal for a specific patient will become an important issue. Presymptomatic diagnosis will be of limited value in the absence of methods to accurately predict the age of onset, as well as the anticipated rate of progression/clinical course of the disease and thus, assess therapeutic efficacy. It is anticipated that the physician will require a prediction of clinical severity soon after birth, ideally by biochemical analysis of the initial patient sample. In addition, strategies will be required to monitor disease progression in patients receiving therapy, to allow informed decisions on factors such as dosage and the need for further therapeutic intervention.

Prediction of Disease Severity

In some cases, genotype/phenotype correlations can be used to predict clinical severity. In Gaucher disease, the 1226G mutation is protective against the neuronopathic form of the disease. However, within this genotype patients can present with the severe early onset form of Gaucher type 1 or can be asymptomatic until late in life. Thus, where individuals are identified at birth it will be essential to have accurate methods to predict not only the subtype of Gaucher disease but also the severity within subtypes, as this will greatly affect decisions on management and treatment of the individual. To this end we have recently developed immune-based assays to quantify both β-glucosidase protein and activity, the determination of both protein and activity has enabled the calculation of specific activity, which enabled the differentiation of the neuronopathic form of the disease from the non-neuronopathic form (Fuller et al. unpublished observations).

For some LSD, quantification of the level of accumulated substrate may provide an additional parameter to predict clinical severity. Proof of this concept has recently been established using ESI-MS/MS to measure stored glycosaminoglycan derived oligosaccharides in skin fibroblasts from MPS I patients and relating the level of these oligosaccharides to the onset of neuropathology [87]. The immediate application of this new method would be to select patients that are suited to ERT based on the expectation that if untreated they would develop an attenuated, nonneuronal form of MPS I disease, compared to those patients with expected neuropathology that may require alternative therapy such as hematopoietic stem cell transplantation [94].

In Gaucher patients, the primary substrate, glucosylceramide, accounts for only 2% of the increased mass of the liver and spleen [95]. Clearly there are other cellular processes involved in the pathogenesis of this disease. This raises the prospect that while the primary markers of the disease in LSD patients reflect the biochemical defect, they may not be the best indicators of the disease process. Whitfield et al. have reported on the correlation between both primary (glucosylceramide) and secondary (lactosylceramide) lipid markers, as well as the protein markers LAMP-1 and saposin C and disease severity in Gaucher patients [86]. Other markers reflecting disease pathogenesis could also be useful for predicting disease severity and for monitoring patients on therapy. For example, altered lipid profiles have been reported for Fabry disease where in addition to elevated levels of the primary stored substrate trihexosylceramide, there were altered levels of a number of other lipid types, including glucosylceramide, lactosylceramide, and sphingomyelin [88].

It is tempting to speculate that common sites of storage and possibly the nature of secondary storage products may directly relate to the appearance of common pathological features in patients with different LSD, including, for example, coarse facial features, bone abnormalities, organomegaly, and CNS dysfunction. However, we are yet to identify the connections between the primary defects and the molecular mechanisms of pathogenesis. The ability to identify this fundamental connection between storage and pathogenesis will most likely reveal markers that accurately reflect the disease process and which specifically reflect particular problem sites of pathology. These markers will likely be suitable for screening and phenotype prediction.

Monitoring of Presymptomatic Patients

Early diagnosis via a newborn screening program will provide the opportunity to commence therapy in presymptomatic patients. However, the question of how to monitor the efficacy of therapy in presymptomatic individuals will need to be addressed. One obvious parameter is the onset of clinically discernable pathology; however, this is not an optimal strategy, particularly where the CNS is involved. Biochemical markers that detect and quantify this pathology before it is clinically manifested in the patient are desirable.

For Gaucher disease there are a number of biochemical markers already in use for monitoring therapy, including chitotriosidase and acid phosphatase (reviewed elsewhere in this book). However, these are markers of macrophage activation rather than direct markers of disease pathology. The challenge will be to identify markers that will reflect the earliest stages of pathology. Those that connect the storage substrate to the onset of clinical pathology are likely to provide the best indicators of disease progression.

Markers for other LSD have also been reported and relate to either the primary storage substrates stored in each disorder or to secondary metabolites that accumulate as a result of the primary storage. Preliminary studies on MPS [79,81,83] and Fabry [96–98] suggest that these will be very powerful tools for monitoring therapy.

A potential strategy that may address both accurate disease prediction and therapeutic monitoring is biochemical profiling. Previous studies have recognized the ability of multiple parameters to give the "most accurate" prediction of clinical severity in LSD patients [79,86–88]. Biochemical monitoring utilizing protein profiling and/or ESI-MS/MS to generate metabolic profiles for patients before and during therapy promises to be a powerful approach. Quantification of the primary storage substrates for each disorder and the secondary storage products, together with an array of lysosomal protein markers that reflect both the primary defect and its impact on lysosomal structure, biogenesis, and function, are expected to provide an informative picture of the disease state and response to therapy.

REFERENCES

1. Fried, K., Gaucher's disease among the Jews of Israel, *The Bulletin of the Research Council of Israel Proceedings of the Fourth Meeting of the Israel Genetics Circle*, 1958, p. 213.

2. Fried, K., Population study of chronic Gaucher's disease, *Isr J Med Sci*, 9, 1396, 1973.

3. Kolodney, E.H. et al., Phenotypic manufestations of Gaucher disease: clinical features in 48 biochemically verified Type I patients and comment of Type II patients, in *Gaucher Disease: a Century of Delineation and Research*, Desnick, R.J., Gatt, S., and Grabowski, G.A., Eds., Alan R. Liss, New York, 1982, p. 33.

4. Grabowski, G.A. et al., Gaucher type I (Ashkenazi) disease: considerations for heterozygote detection and prenatal diagnosis, in *Gaucher Disease: a Century of Delineation and Research*, Desnick, R.J., Gatt, S., and Grabowski, G.A., Eds., Alan R. Liss, New York, 1982, p. 573.

5. Matoth, Y. et al., Frequency of carriers of chronic (type I) Gaucher disease in Ashkenazi Jews, *Am J Med Genet*, 27, 561, 1987.

6. Beutler, E. and Grabowski, G.A., Gaucher Disease, in *The Metabolic & Molecular Basis of Inherited Disease*, Scriver, C.R., Beaudet, A.L., Sly, W.S., and Valle, D., Eds., McGraw-Hill, New York, 2001, Vol. III, p. 3635.

7. Beutler, E. et al., Gaucher disease: gene frequencies in the Ashkenazi Jewish population, *Am J Hum Genet*, 52, 85, 1993.

8. Zimran, A. et al., High frequency of the Gaucher disease mutation at nucleotide 1226 among Ashkenazi Jews, *Am J Hum Genet*, 49, 855, 1991.

9. Horowitz, M. et al., Prevalence of glucocerebrosidase mutations in the Israeli Ashkenazi Jewish population, *Hum Mutat*, 12, 240, 1998.

10. Kannai, R. et al., The selective advantage of Gaucher's disease: TB or not TB? *Isr J Med Sci*, 30, 911, 1994.

11. Slatkin, M., A population-genetic test of founder effects and implications for Ashkenazi Jewish diseases, *Am J Hum Genet*, 75, 282, 2004.

12. Lacerda, L. et al., Gaucher disease: N370S glucocerebrosidase gene frequency in the Portuguese population, *Clin Genet*, 45, 298, 1994.

13. Pinto, R. et al., Prevalence of lysosomal storage diseases in Portugal, *Eur J Hum Genet*, 12, 87, 2004.

14. Meikle, P.J. et al., Prevalence of lysosomal storage disorders, *JAMA*, 281, 249, 1999.

15. Dionisi-Vici, C. et al., Inborn errors of metabolism in the Italian pediatric population: a national retrospective survey, *J Pediatr*, 140, 321, 2002.

16. Poorthuis, B.J. et al., The frequency of lysosomal storage diseases in The Netherlands, *Hum Genet*, 105, 151, 1999.

17. Dahl, N. et al., Gaucher disease type III (Norrbottnian type) is caused by a single mutation in exon 10 of the glucocerebrosidase gene, *Am J Hum Genet*, 47, 275, 1990.

18. Dahl, N., Hillborg, P.O., and Olofsson, A., Gaucher disease (Norrbottnian type III): probable founders identified by genealogical and molecular studies, *Hum Genet*, 92, 513, 1993.

19. Alfonso, P. et al., Mutation prevalence among 51 unrelated Spanish patients with Gaucher disease: identification of 11 novel mutations, *Blood Cells Mol Dis*, 27, 882, 2001.

20. Eto, Y. and Ida, H., Clinical and molecular characteristics of Japanese Gaucher disease, *Neurochem Res*, 24, 207, 1999.

21. Morar, B. and Lane, A.B., The molecular characterization of Gaucher disease in South Africa, *Clin Genet*, 50, 78, 1996.

22. Choy, F.Y., Wong, K., and Shi, H.P., Glucocerebrosidase mutations among Chinese neuronopathic and non-neuronopathic Gaucher disease patients, *Am J Med Genet*, 84, 484, 1999.

23. Ida, H. et al., Type 1 Gaucher disease: phenotypic expression and natural history in Japanese patients, *Blood Cells Mol Dis*, 24, 73, 1998.

24. Charrow, J. et al., The Gaucher registry: demographics and disease characteristics of 1698 patients with Gaucher disease, *Arch Intern Med*, 160, 2835, 2000.

25. Rider, J.A. and Rider, D.L., Batten disease: past, present, and future, *Am J Med Genet*, Suppl., 5, 21, 1988.

26. Ozkara, H.A. and Topcu, M., Sphingolipidoses in Turkey, *Brain Dev*, 26, 363, 2004.

27. Arvio, M., Autio, S., and Louhiala, P., Early clinical symptoms and incidence of aspartylglucosaminuria in Finland, *Acta Paediatr*, 82, 587, 1993.

28. Petersen, G.M. et al., The Tay-Sachs disease gene in North American Jewish populations: geographic variations and origin, *Am J Hum Genet*, 35, 1258, 1983.

29. Martiniuk, F. et al., Carrier frequency for glycogen storage disease type II in New York and estimates of affected individuals born with the disease, *Am J Med Genet*, 79, 69, 1998.

30. Ausems, M.G. et al., Frequency of glycogen storage disease type II in The Netherlands: implications for diagnosis and genetic counseling, *Eur J Hum Genet*, 7, 713, 1999.

31. Chace, D.H., Kalas, T.A., and Naylor, E.W., The application of tandem mass spectrometry to neonatal screening for inherited disorders of intermediary metabolism, *Annu Rev Genomics Hum Genet*, 3, 17, 2002.

32. Meikle, P.J. et al., Diagnosis of lysosomal storage disorders: evaluation of lysosome-associated membrane protein LAMP-1 as a diagnostic marker, *Clin Chem*, 43, 1325, 1997.

33. Meikle, P.J. et al., Newborn screening for lysosomal storage disorders: Clinical evaluation of a two tier strategy, *Pediatrics*, 114, 906, 2004.

34. Li, Y. et al., Tandem mass spectrometry for the direct assay of enzymes in dried blood spots: application to newborn screening for Krabbe disease, *Clin Chem*, 50, 638, 2004.
35. Chamoles, N.A. et al., Tay-Sachs and Sandhoff diseases: enzymatic diagnosis in dried blood spots on filter paper: retrospective diagnoses in newborn-screening cards, *Clin Chim Acta*, 318, 133, 2002.
36. Chamoles, N.A. et al., Gaucher and Niemann-Pick diseases — enzymatic diagnosis in dried blood spots on filter paper: retrospective diagnoses in newborn-screening cards, *Clin Chim Acta*, 317, 191, 2002.
37. Chamoles, N.A. et al., Hurler-like phenotype: enzymatic diagnosis in dried blood spots on filter paper, *Clin Chem*, 47, 2098, 2001.
38. Wilson, J.M.G. and Jungner, G., Principals and practice of screening for disease, World Health Organization, Geneva, 1968.
39. Wilcken, B., Ethical issues in newborn screening and the impact of new technologies, *Eur J Pediatr*, 162, 14, 2003.
40. Vallance, H. and Ford, J., Carrier testing for autosomal-recessive disorders, *Crit Rev Clin Lab Sci*, 40, 473, 2003.
41. Kaback, M. et al., Tay-Sachs disease — carrier screening, prenatal diagnosis, and the molecular era. An international perspective, 1970 to 1993. The International TSD Data Collection Network, *JAMA*, 270, 2307, 1993.
42. Kaback, M.M., Population-based genetic screening for reproductive counseling: the Tay-Sachs disease model, *Eur J Pediatr*, 159 Suppl. 3, S192, 2000.
43. Neudorfer, O. et al., Late-onset Tay-Sachs disease: Phenotypic characterization and genotypic correlations in 21 affected patients, *Genet Med*, 7, 119, 2005.
44. Barton, N.W. et al., Replacement therapy for inherited enzyme deficiency — macrophage-targeted glucocerebrosidase for Gaucher's disease, *N Engl J Med*, 324, 1464, 1991.
45. Xu, Y.H. et al., Turnover and distribution of intravenously administered mannose-terminated human acid beta-glucosidase in murine and human tissues, *Pediatr Res*, 39, 313, 1996.
46. Ringden, O. et al., Ten years' experience of bone marrow transplantation for Gaucher disease, *Transplantation*, 59, 864, 1995.
47. Butters, T.D. et al., Small-molecule therapeutics for the treatment of glycolipid lysosomal storage disorders, *Philos Trans R Soc Lond B Biol Sci*, 358, 927, 2003.
48. Moyses, C., Substrate reduction therapy: clinical evaluation in type 1 Gaucher disease, *Philos Trans R Soc Lond B Biol Sci*, 358, 955, 2003.
49. Hong, Y.B. et al., Feasibility of gene therapy in Gaucher disease using an adeno-associated virus vector, *J Hum Genet*, 49, 536, 2004.
50. Gahl, W.A. et al., Cysteamine therapy for children with nephropathic cystinosis, *N Engl J Med*, 316, 971, 1987.
51. Hoogerbrugge, P.M. et al., Allogeneic bone marrow transplantation for lysosomal storage diseases. The European Group for Bone Marrow Transplantation, *Lancet*, 345, 1398, 1995.
52. Hoogerbrugge, P.M. and Valerio, D., Bone marrow transplantation and gene therapy for lysosomal storage diseases, *Bone Marrow Transplant*, 21, Suppl. 2, S34, 1998.
53. Hopwood, J.J. et al., Long-term clinical progress in bone marrow transplanted mucopolysaccharidosis type I patients with a defined genotype, *J Inherit Metab Dis*, 16, 1024, 1993.
54. Krivit, W. et al., Wolman disease successfully treated by bone marrow transplantation, *Bone Marrow Transplant*, 26, 567, 2000.

55. Krivit, W., Stem cell bone marrow transplantation in patients with metabolic storage diseases, *Adv Pediatr*, 49, 359, 2002.
56. Beck, M., Agalsidase alfa — A preparation for enzyme replacement therapy in Anderson-Fabry disease, *Expert Opin Investig Drugs*, 11, 851, 2002.
57. Mehta, A., Agalsidase alfa: specific treatment for Fabry disease, *Hosp Med*, 63, 347, 2002.
58. Wraith, J.E. et al., Enzyme replacement therapy for mucopolysaccharidosis I: a randomized, double-blinded, placebo-controlled, multinational study of recombinant human alpha-L-iduronidase (Laronidase), *J Pediatr*, 144, 581, 2004.
59. Muenzer, J. et al., Enzyme replacement therapy in mucopolysaccharidosis type II (Hunter syndrome): a preliminary report, *Acta Paediatr*, Suppl., 91, 98, 2002.
60. Harmatz, P. et al., Enzyme replacement therapy in mucopolysaccharidosis VI (Maroteaux-Lamy syndrome), *J Pediatr*, 144, 574, 2004.
61. Van den Hout, J.M. et al., Long-term intravenous treatment of Pompe disease with recombinant human alpha-glucosidase from milk, *Pediatrics*, 113, e448, 2004.
62. Cheng, S.H. and Smith, A.E., Gene therapy progress and prospects: gene therapy of lysosomal storage disorders, *Gene Ther*, 10, 1275, 2003.
63. Ioannou, Y.A., Enriquez, A., and Benjamin, C., Gene therapy for lysosomal storage disorders, *Expert Opin Biol Ther*, 3, 789, 2003.
64. Matsuda, J. et al., Chemical chaperone therapy for brain pathology in G(M1)-gangliosidosis, *Proc Natl Acad Sci U.S.*, 100, 15912, 2003.
65. Sawkar, A.R. et al., Chemical chaperones increase the cellular activity of N370S beta-glucosidase: a therapeutic strategy for Gaucher disease, *Proc Natl Acad Sci U.S.*, 99, 15428, 2002.
66. Hein, L.K. et al., alpha-L-iduronidase premature stop codons and potential readthrough in mucopolysaccharidosis type I patients, *J Mol Biol*, 338, 453, 2004.
67. Kakkis, E. et al., Intrathecal enzyme replacement therapy reduces lysosomal storage in the brain and meninges of the canine model of MPS I, *Mol Genet Metab*, 83, 163, 2004.
68. Hemsley, K., King, B., and Hopwood, J., Gesellschaft fur MPS e.V., in *8th Int Symp on Mucopolysaccharide and Related Dis*, Mainz, Germany, 2004, p. 37.
69. Meikle, P.J., Fietz, M.J., and Hopwood, J.J., Diagnosis of lysosomal storage disorders: current techniques and future directions, *Expert Rev Mol Diagn*, 4, 677, 2004.
70. Umapathysivam, K., Hopwood, J.J., and Meikle, P.J., Determination of acid alpha-glucosidase activity in blood spots as a diagnostic test for Pompe disease, *Clin Chem*, 47, 1378, 2001.
71. Fuller, M. et al., Immunoquantification of alpha-galactosidase: evaluation for the diagnosis of Fabry disease, *Clin Chem*, 50, 1979, 2004.
72. Hein, L.K. et al., Development of an assay for the detection of mucopolysaccharidosis type VI patients using dried blood-spots, *Clin Chim Acta*, 353, 67, 2005.
73. MacBeath, G. and Schreiber, S.L., Printing proteins as microarrays for high-throughput function determination, *Science*, 289, 1760, 2000.
74. Zhu, H. et al., Analysis of yeast protein kinases using protein chips, *Nat Genet*, 26, 283, 2000.
75. Gerber, S.A., Turecek, F., and Gelb, M.H., Design and synthesis of substrate and internal standard conjugates for profiling enzyme activity in the Sanfilippo syndrome by affinity chromatography/electrospray ionization mass spectrometry, *Bioconjug Chem*, 12, 603, 2001.

76. Gerber, S.A. et al., Direct profiling of multiple enzyme activities in human cell lysates by affinity chromatography/electrospray ionization mass spectrometry: application to clinical enzymology, *Anal Chem*, 73, 1651, 2001.

77. Wang, D. et al., Tandem mass spectrometric analysis of dried blood spots for screening of mucopolysaccharidosis I in newborns, *Clin Chem*, 3, 3, 2005.

78. Li, Y. et al., Direct multiplex assay of lysosomal enzymes in dried blood spots for newborn screening, *Clin Chem*, 50, 1785, 2004.

79. Crawley, A. et al., Monitoring dose response of enzyme replacement therapy in feline mucopolysaccharidosis type VI by tandem mass spectrometry, *Pediatr Res*, 55, 585, 2004.

80. Fuller, M., Meikle, P.J., and Hopwood, J.J., Glycosaminoglycan degradation fragments in mucopolysaccharidosis I, *Glycobiology*, 14, 443, 2004.

81. Fuller, M. et al., Disease-specific markers for the mucopolysaccharidoses, *Pediatr Res*, 56, 733, 2004.

82. Rozaklis, T. et al., Determination of oligosaccharides in Pompe disease by electrospray ionization tandem mass spectrometry, *Clin Chem*, 48, 131, 2002.

83. Ramsay, S.L., Meikle, P.J., and Hopwood, J.J., Determination of monosaccharides and disaccharides in mucopolysaccharidoses patients by electrospray ionisation mass spectrometry, *Mol Genet Metab*, 78, 193, 2003.

84. Whitfield, P.D. et al., Characterization of urinary sulfatides in metachromatic leukodystrophy using electrospray ionization-tandem mass spectrometry, *Mol Genet Metab*, 73, 30, 2001.

85. Whitfield, P.D. et al., Quantification of galactosylsphingosine in the twitcher mouse using electrospray ionization-tandem mass spectrometry, *J Lipid Res*, 42, 2092, 2001.

86. Whitfield, P.D. et al., Correlation among genotype, phenotype, and biochemical markers in Gaucher disease: implications for the prediction of disease severity, *Mol Genet Metab*, 75, 46, 2002.

87. Fuller, M. et al., Prediction of neuropathology in mucopolysaccharidosis I patients, *Mol Genet Metab*, 84, 18, 2005.

88. Fuller, M. et al., Urinary Lipid Profiling for the Identification of Fabry Hemizygotes and Heterozygotes, *Clin Chem*, 51, 688, 2005.

89. Brooks, D.A. et al., Immunoquantification of the low abundance lysosomal enzyme N-acetylgalactosamine 4-sulphatase, *J Inherit Metab Dis*, 13, 108, 1990.

90. Brooks, D.A. et al., Analysis of N-acetylgalactosamine-4-sulfatase protein and kinetics in mucopolysaccharidosis type VI patients, *Am J Hum Genet*, 48, 710, 1991.

91. Chang, M.H. et al., Saposins A, B, C, and D in plasma of patients with lysosomal storage disorders, *Clin Chem*, 46, 167, 2000.

92. Hua, C.T. et al., Evaluation of the lysosome-associated membrane protein LAMP-2 as a marker for lysosomal storage disorders, *Clin Chem*, 44, 2094, 1998.

93. Karageorgos, L.E. et al., Lysosomal biogenesis in lysosomal storage disorders, *Exp Cell Res*, 234, 85, 1997.

94. Staba, S.L. et al., Cord-blood transplants from unrelated donors in patients with Hurler's syndrome, *N Engl J Med*, 350, 1960, 2004.

95. Suzuki, K., Glucosylceramide and related compounds in normal tissues and in Gaucher disease, *Prog Clin Biol Res*, 95, 219, 1982.

96. Boscaro, F. et al., Rapid quantitation of globotriaosylceramide in human plasma and urine: a potential application for monitoring enzyme replacement therapy in Anderson-Fabry disease, *Rapid Commun Mass Spectrom*, 16, 1507, 2002.

97. Mills, K., Johnson, A., and Winchester, B., Synthesis of novel internal standards for the quantitative determination of plasma ceramide trihexoside in Fabry disease by tandem mass spectrometry, *FEBS Lett*, 515, 171, 2002.

98. Nelson, B.C. et al., Globotriaosylceramide isoform profiles in human plasma by liquid chromatography-tandem mass spectrometry, *J Chromatogr B Analyt Technol Biomed Life Sci*, 805, 127, 2004.

CHAPTER 18

Enzyme Replacement Therapy for
Type I Gaucher Disease

Ari Zimran, Bruno Bembi, and Gregory M. Pastores

CONTENTS

Experience with enzyme replacement therapy (ERT) for Gaucher disease has established this approach as the standard of care based on its remarkable safety and effectiveness profile. The first commercially available enzyme formulation was alglucerase (Ceredase™, Genzyme Corp., MA), a mannose terminated placental-derived glucocerebrosidase, which was approved by the FDA as one of the first orphan drugs in 1991 following the successful seminal trial conducted at the NIH by Brady's group in a 9-month study of 12 patients (the majority of whom were children under the age of 18 years) (1). Three years later in 1994, imiglucerase (Cerezyme™, also made by Genzyme), the human recombinant enzyme preparation, was approved following a clinical trial in two centers (at the NIH and MSSM); this study involved 30 patients and was conducted for 12 months (2). The latter study design incorporated a head-to-head comparison of the two enzymatic preparations. Currently, imiglucerase is given to more than 4000 patients worldwide with annual sales expected to reach 1 billion U.S. dollars by the end of 2005. The remarkable improvements in the treated patients' physical and functional well-being, and the corresponding financial windfall associated with ERT in type I Gaucher disease have made it a model for the treatment of other lysosomal storage disorders (LSDs) (with additional enzymes currently approved [for Fabry disease and MPSI] or in advanced clinical trials [for MPS II and Pompe disease] (3)). The history of the development of ERT is recounted in Chapter 1 by The Man — Prof. Roscoe O. Brady — whose vision and determination enabled its realization. Other topics are covered in separate chapters: the biochemical basis of ERT (in Chapter 9), ERT for patients with neuronopathic Gaucher disease (Chapter 12), and the cost of care implications for the patients, physicians and society at large are presented in great detail in the third section of this book (Chapter 23, Chapter 24, Chapter 25, Chapter 26, Chapter 27 and Chapter 28). In this chapter we review the published data on ERT, as well as our collective personal experience with the management of patients with type I Gaucher disease. In the concluding portion, we will briefly mention new enzymatic preparations that are currently in clinical trials or anticipated to be introduced into trials in the near future.

SAFETY AND TOLERANCE

One of the earliest indicators of the projected positive outcome subsequently observed with the use of ERT in Gaucher disease (GD) (and this is true for both alglucerase and imiglucerase) was its amazing safety and acceptance by the patients (despite the need for regular intravenous infusions). Compared with almost any drug prescribed for any medical indication, the safety profile of ERT in GD is remarkable. Most of the side effects listed during clinical trials and noted accordingly in the package insert are rarely observed and most reactions are typically very mild and transient. This has allowed the wide use of home therapy in many countries (4,5), and its administration even during pregnancy, as indicated (6,7) despite the warning in the package insert. The development of antiglucocerebrosidase antibodies has been reported among 5 to 15% of patients (8); mainly consisting of nonneutralizing IgG antibodies, which tend to disappear after 2 years. Only a handful of cases have been associated with IgE antibodies and an anaphylactic-type adverse event. In addition, the majority of patients who have developed IgG antibodies do not show any untoward reactions, and few patients have developed infusion-related reactions such as pruritus, urticaria, and chest discomfort. It has been our impression that the more severe allergic responses tend to occur in patients receiving a higher dosage administered at a rapid rate, while mild side effects (such as dizziness or nausea) have been successfully managed by slowing the rate of the infusion. For patients who have developed urticaria or bronchospasm during the infusion, premedication with antihistamines (with or without steroids) has usually helped to control these problems.

However, there are lingering issues that have not been fully clarified such as weight gain (body weight increase of >5% from baseline), which in some of our patients, especially young women, has been so upsetting that it has led to treatment interruption. In other patients, who are either diabetic or prediabetic, worsening of the control of hyperglycemia has also been noted, and in others, there has been an increase of triglycerides and/or cholesterol levels. Furthermore, small joint bone pains resembling rheumatoid arthritis, sometimes with swelling of the metacarpophalangeal and interphalangeal joints, and morning stiffness have been reported by some patients; these events may be drug related as they usually disappear following a brief period of drug stoppage and have recurred upon re-challenge. Since in most cases these are subjective symptoms, not always reported and typically unaccompanied by any objective parameter (such as abnormal laboratory finding), these findings have not been published. Yet, one of us has observed similar complaints in patients receiving a different enzymatic preparation, further suggesting that this phenomenon is indeed drug related. Finally, the most worrisome, albeit controversial, adverse effect may be the development of primary-like pulmonary hypertension in a small number of patients while on ERT (9,10), usually seen as a potential life-threatening event in splenectomized adult female patients with severe GD at baseline. Regardless of the varying opinions on this matter (11), it has become routine practice to monitor for pulmonary hypertension in patients with Gaucher disease (12), and findings of increased pulmonary artery pressure should prompt appropriate referral and potential consideration of either an interruption of enzyme therapy or the addition of therapy for pulmonary artery hypertension.

There have been severe adverse effects (such as infection and thromboembolic phenomenon) associated with the use of subcutaneous intravenous access device; however, these devices are rarely required as the high frequency regimen is no longer in practice (13).

In summary, ERT is very safe and well tolerated, and the occurrence of drug side effects is rare and usually mild, and only a handful of patients have had to stop ERT because of adverse effects. It has been our general impression that children tolerate the infusions even better than adults and also cope relatively well with the actual procedures involved with intravenous administration.

EFFECTS ON SPLEEN AND LIVER — VOLUME AND FUNCTIONS

Splenomegaly is the most common presenting feature along with its closely linked anemia and thrombocytopenia as a consequence of hypersplenism, as well as other mechanical factors (e.g., abdominal discomfort, splenic infarcts). In the pre-ERT era, some patients, especially those whose treating physicians were concerned about progressive deterioration in the liver and bone involvement (14), developed massive enlargement of the spleen, intruding into the true pelvis and causing abdominal discomfort, urinary frequency, and even dyspareunia. In children, marked splenomegaly among untreated patients was commonly associated with growth (height) retardation. However, no correlation has been shown between the size of the spleen, the related symptoms or even the degree of hypersplenism (15).

Almost all patients with splenomegaly respond to ERT with significant volume reduction, typically noted between 3 and 9 months after starting treatment, regardless of the dosage used. The reduction in splenomegaly is typically accompanied by an increase in hemoglobin level, amelioration of early satiety, growth spurt in children, lessening of the bleeding tendency, but with a less dramatic improvement in platelet counts (particularly among those with marked splenomegaly >25 times normal). In rare cases, such as patients who have suffered from repeated episodes of splenic infarct and in whom a large portion of the splenic mass has been transformed into fibrotic nodules — probably less than 1% of the treated patients — total splenectomy had to be performed due to poor response to ERT alone. Children typically respond faster and more dramatically than adults, in whom sufficient time has transpired for the disease to evolve into secondary complications associated with lipid storage such as fibrosis or infarction.

In contrast, most patients do not suffer from abnormal liver function despite the development of hepatomegaly, which in general, is less dramatic than the splenomegaly (16). The reduction in liver volume is almost universal (which led Beutler to suggest that changes in liver volume should be used as the primary end point for clinical trials, as well as for comparing the response to different ERT regimen (17)); as with the spleen, the response is greatest during the first few months of therapy with a tendency for the response curve to plateau after about 2 years. It has been documented that the greater the degree of organomegaly the more dramatic is the reduction in liver volume, and also that splenectomized patients tend to respond with a very fast and dramatic increase in platelet counts (if thrombocytopenic at baseline — an interesting phenomenon) (18,19). Mildly abnormal liver function tests (also, typically found in splenectomized patients) usually show a normalization within 5 to 12 months. However, patients with the most severe liver involvement (cirrhosis) do not benefit as much from ERT, and there have been reports of pulmonary hypertension (related to closure of intrapulmonary shunts, typical for the hepato-pulmonary syndrome) (20). There have been three cases where hepatocellular carcinoma developed in patients on ERT (this does not indicate any causal relationship, but it is still alarming information and a potential problem that should be considered in the follow up monitoring of treated patients). Therefore, particular caution is indicated when ERT is administered to a patient with cirrhosis.

Although organ volume is a key feature of the disease and an accurate measurement is important for the assessment of the response to therapy, and while both MRI and CT can provide a rather accurate measurement of the organs' volumes, an ultrasound examination to support the physical findings should be acceptable outside the context of clinical trials. There is no justification, for example, for sedation of a 2-year old patient to obtain a more accurate volume of the spleen by MRI, or for the potential exposure to radiation involved with CT (12). In the recent publication by Pastores et al., 30 to 50% reduction in spleen volume and 20–30% reduction in liver volumes are to be expected within the first 24 months of ERT (21).

HEMATOLOGICAL RESPONSE

Most of the patients with an intact spleen who require ERT will have thrombocytopenia and anemia at baseline; if a patient has anemia but no thrombocytopenia, one should look carefully for another cause (16); in general, the patient who receives ERT must maintain normal values of iron as well as vitamin B_{12} (both deficiencies can be found in patients with GD) in order to allow the expected improvement in the hematological parameters.

The anemia usually responds faster to ERT than the thrombocytopenia and will, in most patients (with an intact spleen), be normalized by ERT. In contrast, patients with an intact spleen and severe thrombocytopenia may never normalize even after many years of ERT even when on high dosage (21). In some patients an associated ITP has been diagnosed (22). A few splenectomized patients with thrombocytopenia have shown a dramatic improvement in platelet count, with normalization even with as little as 5 units/kg of enzyme per month (23).

Abnormal platelet functions (both aggregation and adhesion) have recently been found to be another cause of bleeding complications in patients with GD. These abnormalities may be reversed by ERT (24,25).

As a general rule, whenever the therapeutic goals have not been achieved, it is good practice to reassess for another potential cause of persistent or newly developing cytopenia in the treated patient. A particular example may be the development of multiple myeloma, which is more common among patients with GD relative to the general population (26,27), myelodysplastic syndrome (25), or another medical problem leading to blood loss and/or low platelet counts.

A few patients have been reported to benefit from subcutaneous injections of erythropoietin given in the period prior to the availability of ERT (28).

It should be noted from the authors' experience that although many patients have reported improvement in general well-being and indeed have normalized their hemoglobin levels, there remains a significant proportion of patients who despite normal blood counts, and even after correction of cobalamine deficiency, continue to complain of fatigue. Whether or not this symptom is due intrinsically to the metabolic disorder is not certain.

A less common observation has been a progressive thrombocytosis following ERT, usually seen in the splenectomized patient, but it has also been reported in patients with intact spleen (29); this phenomenon remains unexplained.

While most of the attention has been given (with good reason) to changes in hemoglobin levels and platelet counts, many patients also have low white blood cells counts that improve with ERT. Similarly, patients with abnormal chemotaxis typically show normalization of neutrophil function with ERT, which in some cases leads to the resolution of the problem resulting from susceptibility to repeated bacterial infections (30). Recently we have noted that splenectomized patients may develop relative or absolute lymphocytosis with ERT; study of lymphocytes' subpopulations has not revealed any pattern of abnormality (31).

BONE CHANGES ON ERT

Unlike the rather dramatic reversal of the organomegaly and hematological abnormalities seen in patients with GD, the skeletal abnormalities respond more slowly or not at all to enzyme therapy (21). This was the case in the first cohort of patients who took part in the seminal trial of alglucerase, who required 42 months to demonstrate radiological improvement, in contrast to the 6 months it took for the response in visceral and hematological parameters (32,33). It has also been clear from the early days of ERT that pathological damage such as osteonecrosis, bone infarcts and fractures, once it has occurred, is irreversible, and that the maximal benefit derived from ERT in terms of eliciting a response in the skeletal system may occur with early enzyme administration and preventive approaches (16,34). A more recent study of the use of bisphosphonates as an adjuvant therapy has clearly demonstrated the lack of effect of either ERT or alendronate on the lytic lesions associated with GD (35). However, bone pains of all kinds, chronic as well as acute bone crises, are typically much better shortly after institution of ERT (36). There have been conflicting reports related to whether or not higher dosage of ERT is truly required to treat bone-related complications. At a recent meeting of the EWGGD in Barcelona, the study presented by Hollak and von-Dahl retrospectively analyzed the effect of enzyme dosage (very low in the Dutch group and very high in the German group), and demonstrated a statistically significant difference in MRI signals for the high dosage group. However, there are no reports on the relationship between the observed MRI findings and incidence or severity of skeletal complications (37).

The group of experts who have recently defined the goals of treatment with ERT has been well aware of these observations and has suggested the following achievable targets with ERT: lessening and elimination of bone pain and the occurrence of bone crises as well as the prevention of osteonecrosis and osteoporosis, within 1 to 2 years following treatment initiation. In children, the aim would be to achieve normal growth and peak bone mass with increased bone density (also within a period of 2 years) by the time of skeletal maturation; whereas in the adult patient the period given for an increase in cortical and trabecular bone density is 3 to 5 years (21). While avascular necrosis (AVN) of large joints, one of the most debilitating skeletal complications of type I Gaucher disease, has been prevented in most of the children on ERT (even when initiated at a time when significant visceral disease has already developed, and also with the use of low treatment doses) (34), in a limited number of cases episodes of AVN have been reported, regardless of treatment dose. The lack of understanding of the basic pathological process leading to this complication — why it occurs in the first place, why some patients develop a single joint osteonecrosis and are free of any other complications for the rest of their lives (which we have seen in some of our older patients who lived in the "pre-ERT era"), whereas others develop AVN at multiple sites with vertebral fractures as well, and still many more patients (even family members sharing the same genotype) never develop any bony complications — makes it very difficult to assess the actual effects of ERT in these patients. A good management program should definitely include balanced nutrition, regular physical activity, avoidance of exposure to drugs or activities associated with increased risk for osteoporosis and AVN (such as smoking, heavy use of caffeine, steroids, etc.). We also usually prescribe calcium plus vitamin D supplementation to our patients, despite a lack of evidence for the effect of supplemental therapy in its long-term ability to reduce bone-related complications. The addition of bispos-phonates to ERT has been shown to improve bone mineral density (35), but we are not certain what is the best time to start with bisphosphonates in a patient on ERT (especially among our young female patients who have not started a family, given the concerns for the prolonged potential effect on the fetus), how long bisphospho-nates should be given, what are the preferred agents, what may be the role of teriparatide and other stimulators of bone formation, and there are many more open questions relating to some of the most common sources of morbidity in GD.

Not knowing how best to monitor the effects of ERT on the bones (besides determining the incidence and severity of clinical complications) is another example of our limited ability to deal with the bone disease. Frequent MRI bone studies have been recommended by the ICGG coordinators, but this recommendation is based on total lack of systematically collected evidence; moreover, Vellodi and his colleagues have recently reported the lack of any benefit in conducting this level of monitoring in children (probably the most important group of patients where pre-vention may be feasible) (38). MRI with quantitative chemical shift imaging (QCSI) has proven to be a very sensitive and elegant technique to demonstrate the effect of ERT on the bone marrow, with specific advantages in the setting of clinical trials (see Chapter 15). Yet, can we make any evidence-based medical decision according to the numerical results of the QCSI (not to mention its availability in only a limited number of centers at the present time)? There are conflicting data regarding the utility of "bone markers" in the monitoring of patients with GD, and there is uncertainty about the correlation between DEXA studies (degree of osteopenia and osteoporosis) and future outcome, specifically, fracture risk and the development of AVN, which could be key indications for the use of ERT in otherwise asymptomatic patients. However, unlike other modalities, DEXA is readily available, relatively cheap and an easy test for patients, with several recent studies describing its use in GD (39–41).

In summary, ERT has changed the natural history of GD bone disease, largely through its ability to prevent bone crises and AVN when administered early in the course of the disease. On the other hand, it has not been very beneficial for patients already affected by several bone lesions at multiple sites, nor has it shown any significant effect on the lytic lesions of the patients. This is likely the case because of limitations in delivery of ERT to the bony tissue. Optimal monitoring and delin-eating the use of adjuvant medications remain major challenges in the management of our patients.

OTHER CLINICAL EFFECTS

While the liver, spleen, bones, and hematological parameters are the key clinical features of Gaucher disease, patients with severe disease may suffer from other organ involvement such as the lungs. Phenotypic heterogeneity is a hallmark of Gaucher disease, and this applies also to pulmonary involvement, which varies from abnormal pulmonary functions through alveolar, interstitial, or vascular pathologies. ERT has been shown to improve features of lung disease in some patients, but in others, ERT has not been able to affect the lung involvement, and at times, progressive deterioration in pulmonary pathology has even led to the death of patients, primarily those with neuronopathic forms. Pulmonary hypertension (PH) can result from hypoxemia, from intra-pulmonary shunts or it can be related to ERT itself (9–11). There are few patients with severe PH but many different opinions, yet again, the role of ERT whether as an inducing or therapeutic agent awaits further studies. We and others have also seen no benefit from ERT in patients with type IIIc Gaucher disease as far as the cardiac calcifications are concerned.

Renal involvement is rare (42); there is a single report about amelioration of nephritic syndrome by ERT (43).

BIOMARKERS AND OTHER LABORATORY FINDINGS

The availability of several biomarkers, and particularly the two recently identified "surrogate markers," chitotriosidase and CCL18 (both discovered and studied by Aerts and his colleagues), have proven useful in the monitoring of the response to ERT, not just in clinical trials (where their roles will become more important in the future, as new therapeutic modalities are developed and fewer patients with prominent disease features are recruited as subjects for studies). These surrogate markers are described in greater detail in Chapter 14. However, it is noteworthy that there is ambiguity with regard to several issues such as the correlation between biomarker levels and disease severity, the rate and magnitude of reduction in the level of these markers in response to therapy, and their utility for predicting the likelihood of future complications. It is particularly important to refer again to the Dutch-German study where high dose therapy was associated with a statistically significant decrease in chitotriosidase activity on ERT (37).

The earlier generation of biomarkers such as acid phosphatase, angiotensin-converting enzyme (ACE), ferritin and others also show responsiveness to ERT and have been used in the seminal clinical trials of ERT (1,2). However, since chitotriosidase and CCL18 better reflect the actual body "Gaucher cells" burden, they are considered to be more specific, and their reduction with therapy is far more robust. The traditional markers may continue to be of value in the research arena, along with other biomarkers that have been shown to be elevated/abnormal in patients with GD (44).

Liver enzymes in GD tend to remain within the normal range, even in the presence of massive hepatomegaly; yet, if they are elevated — usually in the most severely affected (splenectomized) patients — an improvement will usually occur; if not, one should consider an intercurrent cause (such as viral hepatitis).

Whenever a new laboratory abnormality is found in patients with GD, its correction by ERT provides an indication for the (direct) relation between the abnormality and the metabolic disorder. This has been the case with the defective chemotaxis mentioned above, as well as with several pro-inflammatory cytokines, such as

IL6 and IL10 (45); yet, the lack of response cannot exclude the possibility that the specific abnormality may be a GD manifestation that is unresponsive to therapy.

For example, monoclonal gammopathies, which are rather common among patients with GD, do not respond to ERT, unlike the polyclonal hypergammaglobulinemia, which improves with treatment (46).

Finally, we have observed changes in inflammatory markers (ESR, hsCRP, fibrinogen, and others) among patients with GD, yet improvement in these markers is not always seen, nor do we see a good correlation with the general disease severity (47).

ON DOSE AND INDIVIDUALIZATION

When the first clinical trial was designed, prior to the demonstration of the proof of concept, the decision to use what was proven to be a very high dosage to treat most patients was understandable. Yet, with a lack of formal dosage studies, and based on theoretical considerations for the functionality of a housekeeping enzyme, Beutler recommended the administration of much lower doses given at a higher frequency (48). While early results have confirmed the value of this approach, the dosage controversy has been a sticking point for several years, during which period certain opinion leaders kept glorifying the use of higher dose in the absence of any disclosure of their relationship with the industry. This topic is discussed in more detail in Chapters 23–28, but it is mentioned here to highlight the lack of dose-response studies and the paucity of discussions regarding the associated very high cost of the enzyme. The authors have good experience with the use of different dosing regimens and have shown that a successful outcome can be obtained with this therapeutic modality, with doses as low as 15 to more than 120 units/kg/month.

The first demonstration for the lack of inferiority of the 30 units/kg/infusion vs. the original high dose of 60 units/kg/infusion was presented as early as 1993 (49). Among the first attempts to explore dose-response relationships is a study involving non-Jewish patients with Gaucher disease, albeit a retrospective examination, which showed no statistically significant difference between either low dose (15 units/kg body weight/2 weeks; Amsterdam, The Netherlands) or very high dose (60 units/kg/bi-weekly infusion; Düsseldorf, Germany) regimen; as far as the key clinical features are concerned (liver and spleen volume, hemoglobin and platelets). However, a statistically significant advantage of high dose treatment was noted for two surrogate markers (chitotriosidase and bone marrow MRI score) (37). Since the full study has not yet been published, we do not know whether or not there were any clinical implications in terms of reduction in bone-related complications or other clinical events to justify the huge difference in the cost of care, using the high vs. low dose enzyme schedule (50).

INDICATIONS FOR ERT AND TIMING OF ITS BEGINNING

Fifteen years have gone by since the introduction of ERT, and yet there are no universally accepted standards for treatment indications or a consensus regarding the time to begin therapy. While it is generally accepted that all symptomatic patients with significantly abnormal physical findings and laboratory results should be put on ERT (on this aspect the issue of dosing is secondary, and in most countries will depend on the availability of health care resources), the decision to treat is far more difficult in the milder, asymptomatic or presymptomatic cases, particularly those identified prenatally or as a result of population or family screening studies. Here one must practice the art of medicine when there are no evidence-based parameters to make the decision: while we definitely wish to begin ERT early in the patient at risk, prior to the development of disabling disease features, we are reluctant to commit certain patients to life-long intravenous therapy, especially someone who could live to a ripe old age without any Gaucher-related complications (and all of us have such patients who have remained asymptomatic for most of their lives and have not received any specific treatment). Unlike some of our colleagues who will prescribe ERT to any patient who has the suitable medical insurance to cover the cost of treatment, we tend to follow our patients, and if they belong to a high-risk group (for example, those with a "deleterious" genotype or those with an older sibling with Gaucher-related complications), we would begin ERT when we note significant changes toward deterioration in physical signs (including flattening of growth curves in children (51)) or laboratory parameters (such as persistent and significant drop in platelet counts, accompanied by progressive increases in chitot-riosidase/CCL18 biomarkers). Of course, what would be defined as significant may be viewed differently by various physicians, and this is another issue that resides in the area of unresolved ERT-related topics. The decision to begin ERT in young children who are otherwise well also raises certain dilemmas, with uncertainty about whether getting an intravenous infusion regularly has an impact on their psychosocial development, an aspect of their development that cannot be measured readily like height, weight, or organ volume. At the same time, there are children who may be at risk and in whom early administration of ERT in the presymptomatic phase of their disease may lead to the most rewarding outcome from therapy.

Most patients with GD are given ERT when they present with marked orga-nomegaly with or without significant hypersplenism, if they present with acute bone crisis or develop skeletal complication (such as avascular necrosis of a large joint). Many will also start ERT upon discovery of abnormal laboratory findings. Less common indications include the wish to cope better with bone marrow sup-pression in a patient with hematological or other malignancy but with an otherwise very mild GD, short stature in a mildly affected adolescent, planned pregnancy in a patient who has had a poor obstetric history (such as repeated abortions or massive postpartum bleeding), chronic fatigue when no other cause has been found, etc. These cases are typically treated in different centers as unusual cases, and unfor-tunately they are not reported regularly, to allow better understanding of the appro-priate use of enzyme therapy. It may be the right time to make an effort (possibly via the ICGG) to capture all these cases and summarize what has been achieved with ERT for atypical indications.

MAINTENANCE REGIMENS AND THE EFFECTS OF WITHDRAWAL OF THERAPY

While most of the treating physicians will know exactly what to prescribe for their patients once ERT has been decided (based on each country's policy, physician bias, or personal experience), there are practically no studies of maintenance regimens. Does it make sense to prescribe the same dose of enzyme started in a patient with significant disease burden, 10 years later, when he or she is practically healthy and without any noted abnormalities or signs of an active disease process? Many talk about dose reductions, but very few guidelines are available.

While most think about dose reduction or of a reduction in the frequency of the intravenous infusions as appropriate maintenance approaches, Beutler has suggested (unpublished) the concept of drug "holidays," that is, stoppage and re-instatement of ERT periodically. This approach has not been tested systematically, but several years ago Elstein et al. published the outcome of ERT stoppage in a group of 15 patients with GD who elected to withdraw from treatment for various reasons (including financial constraints, personal reasons, and other concerns such as pulmonary hypertension and "adjuvant-like" arthritis (52). The favorable overall outcome seen in the majority of the patients in this study, despite treatment interruptions, suggest the need to explore this approach in a study design that will combine attention to patients' safety, convenience, and cost effectiveness.

NEW ENZYMATIC PREPARATIONS

It should not come as a surprise to most that the enormous success of ERT in type I Gaucher disease has provided an incentive for more biotech companies to produce their own enzyme and enter the market, once the patent of the current enzymatic formulation and the period of 7 years, exclusivity provided by the FDA's orphan drug law expires. At the time this chapter was written, a gene-activated glucocerebrosidase, which has the authentic wild-type sequence (in contrast to Cerezyme™, which has a single amino acid different from the natural sequence) and is made using human fibroblasts (manufactured by Shire) was being given to patients with type I GD in a clinical trial (phase I/II just completed and phase III trials pending). At the same time, another formulation, a plant-derived human glucocerebrosidase (manufactured by Protalix, Israel) was being given to healthy volunteers in a phase I clinical trial. The new enzymatic preparations, which can be regarded as "bio-similar" agents (despite the fact that both are being developed as novel medications), may lead to price reduction (based on competition for reimbursement from third-party payers) and offers the prospect of an alternative therapeutic option to the few patients who are not tolerating the current enzyme preparation very well or do not achieve reasonable therapeutic goals. A greater expectation by physicians and patients alike is that the new companies will make an effort to modify the enzymatic preparation in a way that would confer an added benefit. An ultimate advantage for a new ERT would be the ability to cross the blood brain barrier and/or to be administered as an oral formulation.

CONCLUSIONS

In 1995, the NIH held a special consensus conference to discuss issues related to the diagnosis and treatment of Gaucher disease. In the final statement, the members of the panel summarized that "The success of enzyme replacement therapy for Gaucher disease is a credit to the investigators, the National Institutes of Health, the pharmaceutical manufacturer, and the many patients and their families" (53). Indeed, the great success of ERT (both for patients and manufacturers) has opened the door for new enzymatic therapies for other rare diseases, as well as for new therapeutic modalities. However, despite this success, there are still many unresolved issues, including optimal dosing, with regard to tissues and organs that do not respond so well or at all to ERT (such as lungs, bones, and brain). These factors underscore the need for novel therapeutic approaches, among which are substrate reduction therapy, pharmacological chaperones, and possibly gene therapy, which alone or in combination may further improve the management of Gaucher disease and hopefully at an affordable cost for all.

REFERENCES

1. Barton, N.W. et al. Replacement therapy for inherited enzyme deficiency — macrophage-targeted glucocerebrosidase for Gaucher's disease. *N. Engl. J. Med.*, 324, 1464, 1991.
2. Grabowski, G.A. et al. Enzyme therapy in type 1 Gaucher disease: comparative efficacy of mannose-terminated glucocerebrosidase from natural and recombinant sources. *Ann. Intern. Med.*, 122, 33, 1995.
3. Desnick, R.J. Enzyme replacement and enhancement therapies for lysosomal diseases. *J. Inherit. Metab. Dis.*, 27, 385, 2004.
4. Zimran, A. et al. Home treatment with intravenous enzyme replacement therapy for Gaucher disease: an international collaborative study of 33 patients. *Blood*, 82, 1107, 1993.
5. Perez Calvo, J.I. and Giraldo Castellano P. Home therapy for type 1 Gaucher disease in Spain. *Med. Clin.* (Barc),119, 756, 2002.
6. Elstein, D. et al. Use of enzyme replacement therapy for Gaucher disease during pregnancy. *Am. J. Obstet. Gynecol.*, 177, 1509, 1997.
7. Elstein, Y. et al. Pregnancies in Gaucher disease: a 5-year study. *Am. J. Obstet. Gynecol.* 190, 435, 2004.
8. Rosenberg, M. et al. Immunosurveillance of alglucerase enzyme therapy for Gaucher patients: induction of humoral tolerance in seroconverted patients after repeat administration. *Blood*, 93, 2081, 1999.
9. Elstein, D. et al. Echocardiographic assessment of pulmonary hypertension in Gaucher's disease. *Lancet*, 351, 1544, 1998.
10. Goitein, O. et al. Lung involvement and enzyme replacement therapy in Gaucher's disease. *QJM*, 94, 407, 2001.
11. Mistry, P.K. et al. Pulmonary hypertension in type 1 Gaucher's disease: genetic and epigenetic determinants of phenotype and response to therapy. *Mol. Genet. Metab.*, 77, 91, 2002.

12. Weinreb, N.J. et al. Gaucher disease type 1: revised recommendations on evaluations and monitoring for adult patients. *Semin. Hematol.*, 41 (Suppl. 5), 15, 2004.

13. Elstein, D. et al. Low-dose low-frequency imiglucerase as a starting regimen of enzyme replacement therapy for patients with type I Gaucher disease. *QJM*, 91, 483, 1998.

14. Ashkenazi, A., Zaizov, R., and Matoth, Y. Effect of splenectomy on destructive bone changes in children with chronic (Type I) Gaucher disease. *Eur. J. Pediatr.*, 145, 138, 1986.

15. Gielchinsky, Y. et al. Is there a correlation between degree of splenomegaly, symptoms and hypersplenism? A study of 218 patients with Gaucher disease. *Br. J. Haematol.*, 106, 812, 1999.

16. Elstein, D. et al. Gaucher's disease. *Lancet*, 358, 324, 2001.

17. Beutler, E. Gaucher Disease: Multiple Lessons from a Single Gene Disease, presented as the Donnall Thomas Lecture at the annual Meeting of the American Society of Hematology, San Diego, Dec. 5–9, 2003.

18. Zimran, A. et al. Low-dose enzyme replacement therapy for Gaucher's disease: effects of age, sex, genotype, and clinical features on response to treatment. *Am. J. Med.*, 97, 3, 1994.

19. Weinreb, N.J. et al. Effectiveness of enzyme replacement therapy in 1028 patients with type 1 Gaucher disease after 2 to 5 years of treatment: a report from the Gaucher Registry. *Am. J. Med.*, 113, 112, 2002.

20. Dawson, A. et al. Pulmonary hypertension developing after alglucerase therapy in two patients with type 1 Gaucher disease complicated by the hepatopulmonary syndrome. *Ann. Intern. Med.*, 125, 901, 1996.

21. Pastores, G.M. et al. Therapeutic goals in the treatment of Gaucher disease. *Semin. Hematol.*, 41 (Suppl. 5), 4, 2004.

22. Lester, T.J. et al. Immune thrombocytopenia and Gaucher's disease. *Am. J. Med.*, 77, 569, 1984.

23. Mistry, P.K. et al. Successful treatment of bone marrow failure in Gaucher's disease with low-dose modified glucocerebrosidase. *QJM*, 83, 541, 1992.

24. Gillis, S. et al. Platelet function abnormalities in Gaucher disease patients. *Am. J. Hematol.*, 61, 103, 1999.

25. Zimran, A. et al. Survey of hematological aspects of Gaucher disease. *Hematology*, 10, 151, 2005.

26. Rosenbloom, B.E. et al. Gaucher disease and cancer incidence: a study from the Gaucher registry. *Blood*, 105, 4569, 2005.

27. Zimran, A. et al. Incidence of malignancies among patients with type I Gaucher disease from a single referral clinic. *Blood Cells Mol. Dis.*, 34, 197, 2005.

28. Rodgers, G.P. and Lessin, L.S. Recombinant erythropoietin improves the anemia associated with Gaucher's disease. *Blood*, 73, 2228, 1989.

29. Dweck. A. et al. Thrombocytosis associated with enzyme replacement therapy in Gaucher disease. *Acta Haematol.*, 108, 94, 2002.

30. Zimran, A. et al. Correction of neutrophil chemotaxis defect in patients with Gaucher disease by low-dose enzyme replacement therapy. *Am. J. Hematol.*, 43, 69, 1993.

31. Attias, D. and Ruchlemer, R. unpublished, 2005.

32. Rosental, D.I. et al. Enzyme replacement therapy for Gaucher disease: skeletal responses to macrophage-targeted glucocerebrosidase. *Pediatrics*, 96, 629, 1995.

33. Poll, L.W. et al. Response of Gaucher bone disease to enzyme replacement therapy. *Br. J. Radiol.* 75, A25, 2002.

34. Zimran, A., Abrahamov, A., and Elstein, D. Children with type I Gaucher disease: growing into adulthood with and without enzyme therapy. *Isr. Med. Assoc. J.*, 2, 80, 2000.

35. Wenstrup, R.J. et al. Gaucher disease: alendronate disodium improves bone mineral density in adults receiving enzyme therapy. *Blood*, 104, 1253, 2004.

36. Weinreb, N.J., et al. Effectiveness of enzyme replacement therapy in 1028 patients with type 1 Gaucher disease after 2 to 5 years of treatment: a report from the Gaucher Registry. *Am. J. Med.*, 113, 112, 2002.

37. De Fost, M. et al. Comparison of long term dose-response data using different dosing regimens for a large group of adult type 1 Gaucher disease patients in the Netherlands and Germany, presented at the EWGGD meeting, Barcelona, Sept. 14–17.

38. Vellodi, A. Routine magnetic resonance imaging of the spine in children with Gaucher disease: does it help therapeutic management? *Pediatr. Radiol.*, 33, 782, 2003.

39. Pastores, G.M. et al. Bone density in Type 1 Gaucher disease. *J. Bone Miner. Res.*, 11, 1801, 1996.

40. Bembi, B. et al. Bone complications in children with Gaucher disease. *Br. J. Radiol.*, 75, A37, 2002.

41. Lebel, U. et al. Bone density changes with enzyme therapy for Gaucher disease. *J. Bone Miner. Metab.*, 22, 597, 2004.

42. Becker-Cohen, R.A. et al. A comprehensive assessment of renal function in patients with Gaucher disease. *Am. J. Kidney Dis.*, 46, 837, 2005.

43. Santoro, D., Rosenbloom, B.E., and Cohen, AH. Gaucher disease with nephrotic syndrome: response to enzyme replacement therapy. *Am. J. Kidney Dis.*, 40, E4, 2002.

44. Aerts, J.M. and Hollak, C.E. Plasma and metabolic abnormalities in Gaucher's disease. *Baillieres Clin. Haematol.*, 10, 691, 1997.

45. Allen, M.J. et al. Pro-inflammatory cytokines and the pathogenesis of Gaucher's disease: increased release of interleukin-6 and interleukin-10. *QJM*, 90, 19, 1997.

46. Brautbar, A. et al. Effect of enzyme replacement therapy on gammopathies in Gaucher disease. *Blood Cells Mol. Dis.*, 32, 214, 2004.

47. Zimran, A. et al. Rheological determinants in patients with Gaucher disease and internal inflammation. *Am. J. Hematol.*, 75, 190, 2004.

48. Figueroa, M.L. et al. A less costly regimen of alglucerase to treat Gaucher's disease. *N. Engl. J. Med.*, 327, 1632, 1992.

49. Pastores, G.M., Sibille, A.R., and Grabowski, G.A. Enzyme therapy in Gaucher disease type 1: dosage efficacy and adverse effects in 33 patients treated for 6 to 24 months. *Blood*, 82, 408, 1993.

50. Anand, G. Uncertain miracle: a biotech drug extends a life, but at what price? *The Wall Street Journal*, November 16, 2005; Page A1.

51. Charrow, J. et al. Enzyme replacement therapy and monitoring for children with type 1 Gaucher disease: consensus recommendations. *J. Pediatr.*, 144, 114, 2004.

52. Elstein, D. et al. Withdrawal of enzyme replacement therapy in Gaucher's disease. *Br. J. Haematol.*, 110, 488, 2000.

53. NIH Technology Assessment Panel on Gaucher Disease. Current issues in diagnosis and treatment. NIH Technology Assessment Panel on Gaucher Disease. *JAMA*, 27, 548, 1996.

CHAPTER 19

Substrate Reduction Therapy

Frances M. Platt and Timothy M. Cox

CONTENTS

The use of substrate reduction therapy for glycosphingolipid storage diseases was first proposed by Dr. Norman Radin.[1] It is noteworthy that Gaucher disease was the "test case" adopted for the experimental development of this postulate. Radin reasoned that, in the classical glycosphingolipid storage diseases (deficiencies of lysosomal hydrolases), if the rate of synthesis of glycosphingolipids were slowed, the burden of new incoming glycosphingolipids for breakdown in the lysosomal compartment would similarly be diminished. Any residual enzymatic activity present in the organelle would be sufficient to break down the stored glycosphingolipids in the lysosome, thus ameliorating the pathology. The goal of this therapeutic strategem was to balance the rate of biosynthesis of glycosphingolipids to their (impaired) rate of catabolism. Even in the absence of any residual enzyme activity in the lysosome, the rate at which storage material accumulated would be slowed, thereby reducing the rate at which the disease evolves. Clearly, in the absence of enzymatic activity, any prospect of arresting the progression of disease would also require a means to augment the function of the enzyme. It is critical to emphasize that the aim of this substrate reduction therapy is only partially to inhibit biosynthesis of glycosphingolipids and not block their formation completely; complete shut-off of glycosphingolipid formation would be predicted to have cytotoxic effects.[2]

Despite its obvious therapeutic potential in the glycosphingolipidoses, only recently has substrate reduction therapy been experimentally tested for salutary effects in cell culture models of storage diseases or in models of human glycosphingolipidoses in intact animals. At the same time, the principle of substrate reduction has been regarded with considerable scepticism because it was unclear as to what extent mammalian glycosphingolipid biosynthesis could be manipulated without inducing deleterious effects. In particular, the abundant expression of glycosphingolipids in the nervous system led to speculation that neurological consequences of glycosphingolipid depletion would inevitably occur in patients receiving such treatment. There is now strong evidence from cell culture models, experimental animals, clinical trials, and postmarketing experience that provides the so-called "proof of principle" of the mode of action of substrate reduction therapy in the glycosphingolipid disorders. Before discussing studies that led to the therapeutic development of substrate reduction therapy, we discuss here the enzyme target for substrate reduction therapy and the principal molecules that are known to inhibit this primary step in the biosynthesis of glycosphingolipids by mammalian cells.

THE TARGET ENZYME FOR SUBSTRATE REDUCTION THERAPY

There are two potential therapeutic approaches for reducing the supply of substrate. The first is to devise a transferase inhibitor specific for each unique step in glycosphingolipid biosynthesis that is responsible for generating the stored glycolipids characteristic of each storage disorder. At present this option is not possible because a full range of "designer" transferase inhibitors is unavailable; moreover, the approach demands the use of carefully adjusted disease-specific therapies, which would probably not be attractive for investment by pharmaceutical companies and for realistic clinical development. The second option is to target the first committed step in the biosynthesis of glycosphingolipids, so that flux through the whole pathway is reduced. This stratagem has the advantage of using a single agent to treat several disorders. Most glycosphingolipidoses are the consequence, direct or indirect, of the storage of glycosphingolipids that are derived from glucosylceramide. Indeed it is the formation of glucosylceramide from ceramide that constitutes the first committed step in the biosynthetic pathway of glycosphingolipids. It is a reaction catalyzed by a glucosyl transferase that transfers the sugar nucleotide, UDP-glucose, to ceramide.[3] This membrane-bound transferase appears to be orientated with its active site on the cytosolic leaflet of an early Golgi compartment: UDP-glucosylceramide transferase is the specific biochemical target for substrate reduction therapy.[4]

POTENTIAL DRUGS FOR SUBSTRATE REDUCTION THERAPY IN THE GLYCOSPHINGOLIPIDOSES

Two classes of compound have been reported to inhibit ceramide glucosyltransferase: the PDMP series and the iminosugars (see Figure 19.1). The PDMP series were the first class of inhibitory molecules to be identified through the pioneering work of Radin and colleagues.[5,6] The second class, the iminosugars, were previously believed to function exclusively as glycohydrolase inhibitors[7]; the iminosugars were

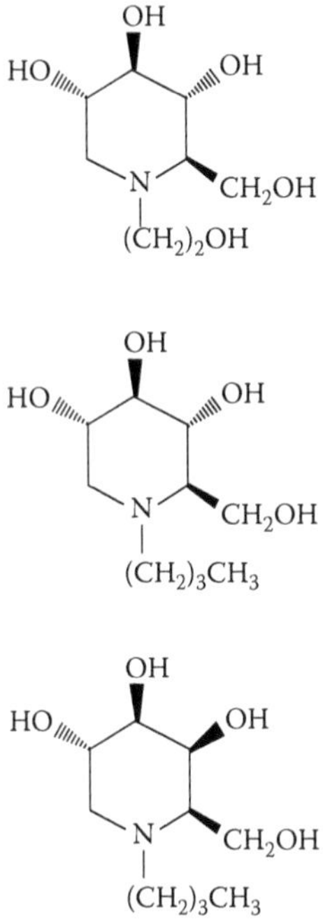

N-hydroxyethyl-DNJ
Miglitol

N-butyl-DNJ
Miglustat

N-butyl-DGJ
(OGT 923)

Figure 19.1 The structures of the iminosugars miglitol, miglustat, and N-butyldeoxygalactonojirimycin.

shown in Oxford more recently to have an action on the biosynthesis of glucosyl-ceramide.[8,9]

1-Phenyl-2-Decanoylamino-3-Morpholino-Propanol (PDMP) Series of Inhibitors

Morpholino- and pyrrolidino-derivatives of the PDMP compounds are mimetics of ceramide and are potent inhibitors of the ceramide glucosyltransferase.[1,5,10] These compounds are competitive and reversible inhibitors and probably act as transition-state analogues. The prototypic compound, D-*threo*-PDMP, is cytotoxic, probably as a result of its hydrophobic properties and because it causes concentrations of ceramide to rise in cells.[10] It appears that the cellular toxicity observed with the PDMP series of compounds is unrelated to the inhibition of glycosphingolipid biosynthesis.[11] Improved compounds in the series, such as the D-*threo*-ethylene-dioxy-P4 compound (D-*threo*-EtDOP4), are less cytotoxic.[10] The P4 compound is better tolerated in animal studies[12] relative to PDMP, although this compound is hydrophobic and so it has a narrow therapeutic window[11]; on drug withdrawal, the compound is cleared only slowly. The preclinical studies with the PDMP series of compounds have recently been reviewed extensively.[11]

Iminosugar Inhibitors

The iminosugar family of compounds are monosaccharide mimetics, many of which occur naturally in certain plants and micro-organisms; for many years they have been used as glycohydrolase inhibitors.[7] Iminosugars have a nitrogen atom in place of the ring oxygen present in monosaccharides such as glucose, and this prevents their metabolism by mammalian cells. The primary use of iminosugars has been in blocking enzymes involved in N-glycan processing and indeed the use of these compounds has assisted in the elucidation of the key steps in this biochemical pathway.

Iminosugars have not solely been used as biochemical reagents, but have also been developed as potential antiviral drugs. One of these compounds, N-butyldeox-ynojirimycin (NB-DNJ), progressed to clinical evaluation as a potential anti-HIV drug in the 1980s.[13,14] NB-DNJ was first reported by Nippon Shinyaku in a patent for use as an antidiabetic agent based on its ability to inhibit α-glucosidases. Its congener, N-hydroxylethyl-1-deoxynojirimycin (6 amino — sorbose) was developed by Bayer and licensed as miglitol for type 2 diabetes. The action of this drug is principally as an inhibitor of intestinal brush-border α-1-4 glucosidases; the drug has inhibitory effects also on the α-1,6 glucosidase activity of glycogen debranching enzyme.

The biochemical targets for the antiviral action of NB-DNJ are the N-glycan-processing enzymes, α-glucosidases I and II,[7] which are resident in the endoplasmic reticulum. Inhibition of these enzymes prevents nascent polypeptides binding the lectin, calnexin, which is resident in the endoplasmic reticulum.[15] The interaction with calnexin appears to be important for the correct folding of some proteins.[16] One protein that is dependent on this interaction to achieve its correct confirmation is the envelope glycoprotein of HIV, gp120.[17–19] In tissue culture infection models, NB-DNJ completely blocks the lifecycle of the virus by preventing post-CD4 binding events that lead to fusion of the viral membrane with the plasma membrane of the host cell.[20] However, in clinical trials of patients with HIV/AIDS, it proved impossible to deliver sufficient concentrations of NB-DNJ to inhibit maturation of gp120 in the endoplasmic reticulum, and NB-DNJ was not further developed for this particular clinical application.[14] Nonetheless, the clinical trial provided a source of comprehensive information on tolerability and unwanted effects of this compound given at very high doses to human subjects. These clinical data in humans moreover proved invaluable for the more recent administration of this drug at 10-fold lower doses in the treatment of human glycosphingolipidoses (see below).

The nitrogen atom of the iminosugars serves as a site for chemical modification and alkylation of glucose (deoxynojirimycin DNJ) or galactose (galactonojirimycin, DGJ) analogues has generated molecules that inhibit the ceramide-specific glucosyl transferase.[21] This was a totally unexpected activity of the deoxynojirimycins. An alkyl chain length of at least three carbon atoms was required for this activity; the methyl- and ethyl-derivatives were inactive and N-propyl had only weak inhibitory activity against the transferase.[22] N-butyl derivatives were chosen for study because they were found to have an optimal alkyl chain length to inhibit glycosphingolipid biosynthesis. NB-DNJ and NB-DGJ are competitive for ceramide and noncompetitive for UDP-glucose.[23] Molecular modeling experiments revealed unexpected homology between these compounds and ceramide.[23] The discovery of these inhibitory properties of alkyl-substituted deoxynojirimycins rendered them suitable for evaluation as potential agents in substrate reduction therapy.

It was found that the potency of this group of iminosugar compounds could be increased further (10-fold) simply by increasing the alkyl chain length; indeed even greater potency (1000-fold) has been achieved using a more lipophilic adamantane compound.[24] The potential disadvantage is that increased potency may go hand-in-hand with cytotoxic effects. With nonyl-alkyl chain deoxynojirimycin compounds, the cytotoxicity could be minimized with only slight reduction in the inhibitory potency, by the use of an ether chain at the 7-carbon position.[25,26]

Inhibitory Constants

NB-DNJ is a potent inhibitor of α-glucosidase I (K_i 0.2 μM) but in tissue culture complete inhibition of this enzyme occurs at 0.5–2 mM NB-DNJ, indicating poor endoplasmic reticulum penetrance. The K_i for UDP glucosylceramide transferase is reported to be 7μM but in cultured cells the operational concentrations required for 50% inhibition of this membrane-bound enzyme are 5–50 μM.

Evaluation of Substrate Reduction Therapy in an *In Vitro* Model of Gaucher Disease

Radin had suggested that the inhibition of glycosphingolipid biosynthesis would have potential application for the treatment of Gaucher disease.[27] However, definitive proof of the value of this application, the so-called "proof of principle," was lacking.

After the discovery at the University of Oxford that certain iminosugars could inhibit the biosynthesis of glycosphingolipids, an *in vitro* model of Gaucher disease was generated and the effects of *N*B-DNJ were tested.[8]

In Gaucher disease, inheritable defects in the gene that encodes lysosomal glucocerebrosidase prevent the complete breakdown of glucosylceramide. Glucosylceramide accumulates over time, leading to storage principally in cells of the monocyte-macrophage system. The murine macrophage cell line WEHI-3B was therefore cultured in the presence of an irreversible inhibitor of glucocerebrosidase, conduritol-β-epoxide (CBE), to induce a Gaucher cytological phenotype. WEHI-3B cells were cultured in the presence or absence of *N*B-DNJ and glucosylceramide concentrations were determined by thin-layer chromatography. After CBE treatment, the WEHI-3B cells accumulated glucosylceramide compared with untreated cells; however, in cultures containing 50 to 500 μM *N*B-DNJ, this accumulation was prevented. At the lower concentration (50 μM), cultures contained glucosylceramide concentrations that were comparable to cells not treated with CBE, whereas at the highest dose (500 μM), glucosylceramide was almost undetectable in the cells. Cells treated with 5 μM *N*B-DNJ were identical to CBE-treated cells alone, demonstrating *in vitro* that a concentration of 50 μM *N*B-DNJ was sufficient to prevent the accumulation of glucosylceramide. There was evidence of storage material in the lysosomes of CBE-treated WEHI-3B cells when examined by electron microscopy, but this was not observed in cultures receiving dual treatment with CBE and *N*B-DNJ, thus indicating that *N*B-DNJ prevented the accumulation of glucosylceramide in lysosomes as a result of partial inhibition of the biosynthesis of glycosphingolipids.

EVALUATION OF SUBSTRATE REDUCTION THERAPY IN MOUSE MODELS OF GLYCOSPHINGOLIPID STORAGE DISEASES

When considering which disorders might benefit from the introduction of substrate reduction, two categories emerge. First, some diseases result directly from a deficiency in breakdown of glycosphingolipids (Gaucher, Fabry, Tay-Sachs, Sandhoff, and GM1 gangliosidosis). Secondly, there are diseases in which glycosphingolipids are stored by mechanisms that are secondary to the underlying genetic defect, e.g., Niemann-Pick disease type C and mucopolysaccharidosis type III. Most of these diseases are associated with neurodegenerative effects associated with neuronal dysfunction and ultimately neuronal loss and the storage of diverse compounds within abnormal lysosomes. Evidence that substrate reduction therapy may benefit these principally neurodegenerative disorders has emerged from the evaluation of the prototypic iminosugar compound, *N*B-DNJ, in numerous experimental models of these conditions in the laboratory mouse.[28]

Substrate Reduction Therapy in a Murine Model of Tay-Sachs Disease

The Tay-Sachs mouse generated by targeted disruption of the gene encoding the α-subunit of β-hexosaminidase A in embryonic stem cells was the first knockout model of a glycosphingolipid storage disease to be generated.[29] Unlike the human

Tay-Sachs disease counterpart, *hexa*[-/-] mice store only modest amounts of GM2 (and GA2) ganglioside in the brain, and very few develop manifestations of neurodegenerative disease within their normal lifespan.[30] Although the indolent phenotype of this animal model precludes evaluation of the clinical efficacy of substrate reduction, the effects of inhibiting glycosphingolipid biosynthesis could readily be evaluated in long-term biochemical and cellular studies.

A major advantage of the iminosugar drugs is their availability after oral administration; a further advantage is their lack of metabolism in mammalian systems, much of the drug being excreted unchanged in the urine. Mice actively clear iminosugars in the urine, but in humans, excretion occurs passively by glomerular filtration. In humans almost constant serum concentrations of the drug can be achieved and higher concentrations are obtained with less intense dosing regimens than are required in mice. Maximum plasma concentrations occur 2.5 hours after oral dosing of *N*B-DNJ in humans — the half-life being 6–7 hours; steady-state concentrations are achieved after 4–6 weeks of treatment.[31]

In preclinical studies, *N*B-DNJ was administered simply as an oral supplement to routine mouse food, thus allowing noninvasive long-term administration of the drug to be sustained at 600–2400 mg/kg/day. In the original experiments, the mice were monitored for reduction of the burden of glycosphingolipid storage in the tissues, and electron microscopy was carried out to evaluate the effects of the intervention on storage bodies within neurons. It was found that substrate reduction therapy slowed the rate of storage of glycosphingolipids in the brain and visceral organs; it was moreover shown that the ultra-structural morphology of storage material in the neurons was significantly improved in those mice that were treated with *N*B-DNJ compared with their untreated litter mates.[32] The serum concentrations of the drug were found to be 18–57 μM and the animals showed up to 70% depletion of tissue glycosphingolipid concentrations in several peripheral tissues. The conclusions from this study were that administration of *N*B-DNJ was effective in reducing the biochemical abnormalities in Tay-Sachs mice and that sufficient drug was able to enter the brain to reduce the rate of glycosphingolipid storage at that site.

A notable feature of these studies was the reduction (15%) in mean body weight of the mice. This is due to a central appetite suppressant effect of the compound in the mouse (Priestman and Platt, unpublished observation). These animals had received 600 mg/kg/d for 50 days; 1200 mg/kg/d for 20 days from the age of 6 weeks but otherwise appeared normal. Dosing the Tay-Sachs mice with 4800 mg/kg/d produced serum concentrations of *N*B-DNJ of ~50 μM, and an estimated concentration of 5 uM in the cerebrospinal fluid; this latter concentration was sufficient to ameliorate GM2 storage in the brain. Since most of the glycosphingolipid diseases affect the brain, these observations were of particular significance when considering a broadening therapeutic application of iminosugars to diseases other than nonneuronopathic Gaucher disease.

Substrate Reduction Therapy in a Mouse Model of Sandhoff Disease

Although the pharmaceutical proof of principle was demonstrated by studies in the Tay-Sachs mouse, the lack of clinical manifestations in this mammalian model precluded evaluation of the impact of NB-DNJ on clinical parameters of disease, including its severity and/or any effect on survival of the experimental animals. For this reason, a second study was carried out using the newly available murine model of Sandhoff disease.[33] Tay-Sachs disease results from defects in the gene encoding the α-subunit of β-hexosaminidase leading to deficiency in β-hexosaminidase activity; in Sandhoff disease, deficient activity of the common β-subunit leads to deficiency of β-hexosaminidases A and B. When the gene encoding the murine β-subunit of hexosaminidases A is disrupted, the resulting mouse model closely resembles human infantile and juvenile onset variants of Tay-Sachs and Sandhoff disease (GM2 gangliosidoses). The reason for the lack of phenotype in the Tay-Sachs mouse (*hexa[-/-]*) is because in the mouse, unlike the human, there is sufficient sialidase activity to convert the primary stored ganglioside GM2, to its asialo-derivative, GA2. GA2 serves as a substrate for the intact β-hexosaminidase B activity. In the mouse, when both hexosaminidases A and B are deficient, a neurodegenerative phenotype occurs as a consequence of extreme accumulation of GM2 and GA2 gangliosides in the nervous system. Sandhoff mice develop tremor, hind limb muscle wasting and rigidity with a stereotypic progression of disease. The mice are moribund by 4 months after birth and thus serve as a useful model to evaluate whether partial substrate inhibition using NB-DNJ can delay symptom onset or even improve life expectancy. The drug was administered orally using the same protocol as for the Tay-Sachs study (2.4–4.8 g/kg/day). It was found that the onset of disease manifestations was significantly delayed by administration of NB-DNJ, including the rate at which motor function declined; life expectancy of the treated animals was extended by approximately 40%.[34] Storage of GM2 and GA2 glycolipids in peripheral organs and the brain was reduced significantly.

Combination Therapy

A major factor that limits the efficacy of substrate reduction in the glycosphingolipidoses is the level of residual enzyme activity. Clearly, knockout mice models of glycosphingolipidoses and patients with the rapid onset of disease in infancy have negligible or near-negligible residual activities of the cognate glycosphingolipid degrading enzyme. Administration of substrate reduction therapy under these circumstances will only slow the rate at which storage develops, and there is no available prospect of balancing the rate of biosynthesis with an impaired rate of glycosphingolipid breakdown. Therefore, to achieve significant clinical benefit in patients with established disease that has already presented in infancy, substrate reduction therapy would require an intervention to augment the intrinsic capacity of the tissues to break down glycosphingolipids.

At the time of writing, the only approved method to enhance enzymatic activity in the brain of patients with these diseases is bone marrow transplantation.[35,36] In the future, however, techniques such as gene therapy and neural stem-cell therapy may prove to be beneficial, provided safe and effective clinical protocols can be worked out.[36–38] It may be possible in the short-term to promote the passage of therapeutic enzymes administered parenterally into the circulation to traverse the blood-brain barrier but no methods for this application are currently approved for clinical use. For example, in Gaucher disease where there is a commercial source of highly effective enzyme replacement therapy available, delivery of active enzyme molecules into the brain might be engineered to benefit the so-called neuronopathic forms of this disease (types 2 and 3). Furthermore, the possibility of therapeutic synergy exists in which substrate reduction is combined with procedures that augment enzymatic activity. The concept of therapeutic synergy was investigated in the Sandhoff mouse using a combination of substrate reducing therapy (NB-DNJ) and marrow transplantation. Combination therapy was found to be more effective than either of the two mono-therapies alone, and in those mice where marrow transplantation induced the highest augmentation of hexosaminidase activity in the brain, significant therapeutic cooperativity (indicating better than additive effects) was demonstrated.[39]

CLINICAL EVALUATION OF NB-DNJ IN GAUCHER DISEASE

At the time of the studies carried out in mouse models of the GM2 gangliosidoses, no viable animal model of Gaucher disease was available. However, the preclinical studies demonstrating proof of principle conducted in a murine model of Sandhoff disease[34] and extensive preclinical toxicity studies combined with trials conducted over a 6-month period (where doses of NB-DNJ up to 3 grams daily were given to human subjects with late-stage HIV infection and AIDS[40–42]) led to the design and approval of a clinical trial in patients with type 1 Gaucher disease.[31] Twenty-eight adult patients with mild-to-moderate type 1 (nonneuronopathic) Gaucher disease who were unable or unwilling to take enzyme replacement therapy were enrolled in this pivotal study, which was carried out in the U.K., The Netherlands, the Czech Republic, and Israel. These patients either had mild hypersplenism or, if they had already had a splenectomy, greatly increased liver volume and thrombocytopenia (less than $100 \times 10^9/l$). Patients received 100 mg miglustat orally, given three times daily.

In relation to Gaucher disease, the mode of action of substrate reduction therapy from the outset was identified as being indirect. The iminosugar inhibitor NB-DNJ was predicted to reduce biosynthesis of glucosylceramide and other related complex glycosphingolipids principally on the surface of newly formed blood cells.[31] In this way, their ultimate ingestion by mononuclear phagocytes in the spleen, liver, and bone marrow would involve phagocytosis of cellular membranes substantially depleted in glycosphingolipids. It was predicted that eventually, in the presence of the residual partial activity of β-glucocerebrosidase in the parent Gaucher macrophage, the inbalance between the rate of glucosylceramide degradation and delivery of incoming glycosphingolipid, would be restored — and with it the pathological storage of glucosylceramide in fixed tissue macrophages. Any therapeutic effect would be predicated on the pathogenetic relationship between storage and the manifestations of Gaucher disease.

The initial clinical trial and subsequent follow-up studies have clearly demonstrated a persistent therapeutic effect of NB-DNJ in patients with mild-to-moderate type 1 Gaucher disease[31,43] (See Figure 19.2). Target plasma concentrations of 1–2

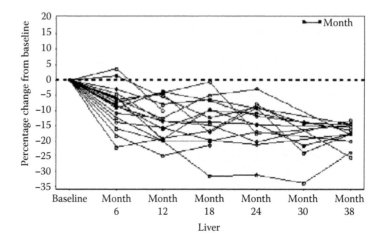

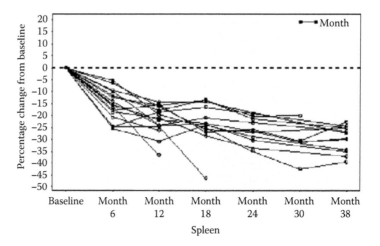

Figure 19.2 Percentage change from baseline in liver and spleen volume over 3 years of treatment with miglustat.

g/ml (5µM) *N*B-DNJ were achieved in the treated patients. At the 1-year time point, treatment with miglustat resulted in a decrease in liver and spleen volumes, (12% and 19%, respectively), a progressive decrease in the activity of the surrogate biomarker of Gaucher disease activity, plasma chitotriosidase activity (decreased by more than 16% over 1 year), and a slow but modest increase in absolute platelet count. In patients whose hemoglobin concentrations were in the anemic range at the start of therapy, there was also a salutary improvement. In extension studies, 13 patients completed treatment with *N*B-DNJ in an open-label continuation of the trial. A dose of 100–300 mg three times daily of *N*B-DNJ was given and a sustained and increasing therapeutic effect on the key clinical, radiological, and laboratory markers of disease activity was observed over a 3-year period[43] (Figure 19.3).

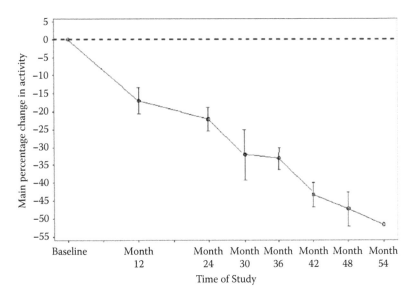

Figure 19.3   Mean percentage change (±SEM) in chitotriosidase activity from baseline to month 54. Only one patient had a visit at 54 months so this point is a single data point.

In a subsequent trial where the dose was reduced to 50 mg three times daily, a weaker therapeutic effect was observed — liver volumes fell by 5.9% and the spleen by 4.5%, respectively, after 6 months of treatment, with little added benefit as a result of attenuated side effects.[44] These results point to a dose-dependent response for the compound. The trials are summarized in Table 19.1.

*N*B-DNJ has thus been vindicated as an effective therapy for Gaucher disease over time, as predicted, and has been generally acceptable to patients, several of whom have continued treatment for many years. At the time of writing, two patients have received treatment for more than 6 years and show continuing regression of disease activity.

As shown in the organ response and hematological parameters, the action of *N*B-DNJ is of slow onset. The pathological manifestations of type 1 Gaucher disease result principally from the accumulation of undegraded glucosylceramide and its congeners, including other glycolipids, within the lysosomes of cells of the macrophage system. These pathological cells are distributed in the spleen, liver (Kupffer cells), bone marrow, and the lungs, where occasional pathology is observed in Gaucher disease. There is good evidence that glucosylceramide is derived mainly from the breakdown of membranes produced by the phagocytosis of effete white and red blood cells. The membranes of these cells are rich in glycolipids derived from glucosylceramide, and it is this material that serves as the source of the accumulating storage macromolecules when glucocerebrosidase activity is reduced as a result of inherited defects.

In clinical trials, early responses to interventions for Gaucher disease can be evaluated by determination of spleen and liver volumes as well as hematological

Table 19.1 Summary of the Current Status of Clinical Studies Aimed at Either Better Understanding Gaucher Disease Type 1 (018) or the Response of Patients with this Disorder to Miglustat

Study	N Main	N Ext	Study design	Patients	Study objectives	Duration (months)
Gaucher Disease Type 1						
Study 001/ 001 extension	28	18	Open label, non-comparative	Naïve to ERT or discontinued ≥ 6 months	Long term efficacy and safety of miglustat 100 mg tid in mild-to-moderate GD type 1 patients	12 months + 24 months extension
Study 003/ 003 extension	18	18	Open label, non-comparative	Naïve to ERT or discontinued ≥ 6 months	Efficacy and safety of miglustat 50 mg tid in mild-to-moderate GD type 1 patients	12 months + 12 months extension
Study 004/ 004 extension	36	10	Randomized, open-label, combination/switch	Patients on stable ERT treatment (≥ 2 years)	Efficacy and safety of miglustat as maintenance therapy in mild-to-moderate GD type 1 patients switching from ERT	6 months randomized, 18 months one arm treatment, open-label
Study 005/ 005 extension (manuscript has been submitted)	12	7	Open label, non-comparative	Patients naïve to ERT or discontinued ≥ 3 months	Efficacy and safety of miglustat 100 mg tid in mild-to-moderate GD type 1 patients, with focus on neurological safety profile	12 months + 12 months extension
Study 016 (reporting Q3 2005)	23		Open label, non-comparative	Patients who participated in Studies-001, -003, and -004	Long-term efficacy and safety of treatment with miglustat	24 month data
Study 018 (Ongoing)	100		Observational study	Patients naïve to or receiving ERT	Baseline prevalence and incidence over 2 years of peripheral neuropathy in patients with GD type 1	24 months
Study 011 (Start 2005)	40		Open label, non-comparative	Patients treated with ERT for at least 5 years	Long-term efficacy and safety of treatment with miglustat as maintenance therapy for mild-to-moderate GD type 1 patients	24 months

parameters, but once stable during maintenance, these clinical parameters are of little value in generating information that relates to disease outcome. In The Netherlands, a means to image the composition of the bone marrow has been developed using magnetic resonance studies to determine the quantitative chemical shift relating to the proportion of triglyceride and water content (fat fraction) of the bone marrow (see Chapter 15). Patients receiving *N*B-DNJ over several years exhibited a gratifying increase in the bone marrow fat fraction in response to *N*B-DNJ, suggesting that this problematic focus of disease also shows salutary and continuous effects as a result of substrate reduction therapy maintained over at least 3 years.[43]

Thus although the speed of response of Gaucher disease to substrate reduction therapy with the iminosugar *N*B-DNJ is less than that expected for all criteria of full dose enzyme replacement therapy at 60 u/kg body weight twice weekly,[46] it is clear that improvements in visceromegaly, hematological abnormalities as well as key indicators of disease activity have been observed. These therapeutic effects are clearly related to the mode of action of the drug. In the pivotal trial, the concentration of GM1 ganglioside measured in leucocytes attained after treatment with *N*B-DNJ for 1 year in 5 patients decreased by approximately 40%.[31] Further evidence that depletion of substrate was achieved during the trial was obtained from assays of plasma glucosylceramide. At baseline, plasma glucosylceramide was 14.2 ± 6.6 μM in 8 subjects; after 12 months of miglustat therapy, this was reduced to 11.4 ± 4.0 μM (data on file, Actelion Pharmaceuticals AG).

Adverse Effects of Miglustat

At the initiation of treatment with miglustat, several predicted unwanted effects occurred. Abdominal bloating, flatulence, and diarrhea were prominent, and it seems likely that these are related to the ability of the agent to inhibit disaccharidases and/or their biosynthesis, on the intestinal brush border. Diarrhea, though common, appears to be transitory and most patients adjust well to the drug with minor dietary alterations.

About one quarter of patients also developed a transient tremor, which proved to be an exaggeration of the physiological or sympathomimetic tremor; in most cases this also resolved spontaneously, although its cause remains unexplained. In patients where this did not spontaneously resolve, either dose reduction or withdrawal of the drug resulted in loss of the tremor. At the same time, several patients developed marked weight loss, partially aggravated by their self-imposed dietary adjustments based on advice to minimize diarrhea. A few patients, unambiguously two in the initial trial, developed a painful axonal neuropathy that has improved slowly on withdrawal of the drug over a prolonged period. It is notable that these patients were among those who had lost the most weight during initiation of the therapeutic trial and peripheral neuropathy does not appear to be a common effect of the administration of *N*B-DNJ. Other patients who had developed peripheral neuropathy have had significant co-morbidities, in particular deficiency of vitamin B_{12}.

A subsidiary trial, which was designed to confirm the previously observed therapeutic effects of miglustat and to evaluate further the neuropsychological and neurological aspects was conducted in 10 adult patients with mild-to-moderate Gaucher disease.

These patients received miglustat 100 mg tid for 12 months. Seven patients continued into the extension phase of the study for another 12 months. At the 24-month time point, the mean percentage change from baseline in liver volume was –6% (p = 0.084) and the change in spleen volume was determined at –15% (p = 0.094). No evidence of significant adverse effects on neurological and neuropsychological function was found during the period of observation, using serial neurological examinations and tremor assessments with accelerometry.[47]

Overall the drug appears to have been well tolerated and represents an acceptable avenue for therapeutic exploration for those patients with Gaucher disease who are unable or unwilling to receive enzyme replacement therapy.

Early toxicity studies in mice using very high doses of NB-DNJ showed smaller spleens and thymus glands with decreased cellularity and disturbed lymphocyte populations. These changes were reversible on withdrawing NB-DNJ and the immune system appeared not to be compromised. NB-DNJ is also associated with aberrant morphology and mobility of spermatozoa in experimental mice.[48] In a study whereby NB-DNJ was administered to mice in the short term, it was shown that the genetic potential of the spermatozoa was not affected and resulted in live offspring that grew normally and with normal fertility.[49] However, appropriate measures for contraception are advised for patients taking miglustat.

Current Status

NB-DNJ is registered as miglustat (Zavesca®) and licensed originally from the Searle company; it has been evaluated by the U.S. Food and Drug Administration (FDA) and by the European Agency for the Evaluation of Medicinal Products under the Orphan Drug Regulatory Procedures. Efficacy data from 80 patients with type 1 Gaucher disease and safety data were examined and in November 2002 European Medicines Evaluation Agency (EMEA) granted approval for marketing; the drug was launched by Actelion in the U.K. in March 2003. The drug was awarded Orphan status by the FDA for treatment of Gaucher disease in October 2000 and was approved by the FDA in July 2003. The Israeli company Teva filed for approval in January 2003.

On the basis of the safety and efficacy data, it was considered that there was a positive balance in favor of therapeutic effects for miglustat in mild-to-moderate Gaucher disease and thus marketing authorization had been granted under the exceptional circumstances of Orphan Drug regulations. Approval was conditional on comprehensive postmarketing surveillance programs, and those who prescribe the drug are enjoined to conduct surveillance and collect safety data for those patients who are receiving it for its licensed (and unlicensed) indications. Since licensing, safety data on 220 patients have been made available to the regulatory authorities by October 2004 and no recommendations were made to alter the evaluation of risk/harm effects of the miglustat. The intensive safety surveillance program has been established by Actelion Pharmaceuticals at the time of marketing authorization.

More than 80% of patients in this program for registration and documentation of their therapeutic responses are within the European Union; about two-thirds have type I Gaucher disease with others representing patients with Niemann-Pick disease type C, GM2 gangliosidosis, type 3 Gaucher disease and including four severely disabled patients suffering from GM1 gangliosidosis. By March 2005, the average exposure of Gaucher disease patients to the drug under the conditions of the surveillance program was almost 1 year. The drug has been discontinued in 13 patients with lysosomal diseases: 5 patients had diarrhea; 3 tremor; 5 with ineluctable progression of GM2 or GM1 gangliosidoses, in three cases, as expected, with death due to their primary neurodegenerative disease. Overall, no new cases of peripheral neuropathy were noted but there were four cases of tremor. Cognitive function was unimpaired in all patients examined although three individuals reported slow mentation or memory. A single elderly patient is recorded in whom Parkinson's disease was considered to be unrelated to his Gaucher disease or its therapy.

An Advisory Council, comprising international experts based in the European Working Group for the Study of Gaucher Disease and Patient Organisations, met to submit a position statement on the indications for the use of miglustat. The emerging consensus document arising from these discussions has now been reported.[50] NB-DNJ (miglustat) is indicated for the oral treatment of mild-to-moderate Gaucher disease, which is defined as a hemoglobin concentration greater than 9 g/dl and a platelet count greater than 50×10^9/l; patients who have rapidly evolving osseous complications of Gaucher disease should not receive miglustat.

Miglustat is at present used for the treatment of Gaucher patients as a second-line agent for those who for one reason or another are unsuitable for enzyme replacement treatment (e.g., venous access or needle phobia), and those who are unwilling or unable to receive the treatment. The drug is not recommended for adolescents and children with Gaucher disease, and contraception is advised for all persons with Gaucher disease who are in the reproductive age group.

A discussion here as to why patients with Gaucher disease may be unsuited for enzyme replacement therapy is warranted. In some circumstances they may have religious or other objections to receiving infusions of enzyme by vein. It would be very unusual for patients to experience adverse reactions to enzyme replacement therapy, and immune reactions are exceptionally rare. However, in our view, a significant number of patients do develop needle phobia or have persistent difficulties with infusion either as a result of poor technique, physical disability, or incapacity to carry out home infusions or visit infusion centers; the group also includes those whose peripheral venous access is severely impaired. A few patients, notably young individuals who have responded to debulking of their disease with enzyme therapy and who are itinerant for reasons of education or employment, for example, prefer to be free of the need for regular infusions and thus would be more suited to the use of an oral treatment.

Future Clinical Trials of Miglustat

At the time of writing, Actelion is sponsoring several important trials in Gaucher disease (Table 19.2). An observational study is underway to establish the prevalence

Table 19.2 Summary of Clinical Trials of Miglustat in Lysosomal Storage Diseases with CNS Pathology and in Fabry Disease

Gaucher Disease Type 3			
Study 006 (Ongoing/ results Q4/ 2005)	30	Phase II, controlled (treatment and no treatment arms), randomized study for the first 12 months; 12 months extension with all patients on active treatment	Adult and pediatric patients	Efficacy and safety of miglustat 200 mg tid. in Gaucher disease type 3 (dose adjustment for pediatric patients)	12 months + 12 months extension
Niemann-Pick C					
Study 007 (Ongoing/ 12 M results Q4 2005)	30	Phase II, controlled (treatment and no treatment arms), randomized study for the first 12 months; 12 months extension with all patients on active treatment	Adults, juveniles and pediatric patients	Efficacy and safety of miglustat 200 mg tid. in Niemann-Pick C patients (dose adjustment for pediatric patients)	12 months + 12 months extension
Lay Onset Tay-Sachs					
Study-009 (Results Q4 2005)	30	Phase II, controlled (treatment and no treatment arms), randomized study for the first 12 months; 12 months extension with all patients on active treatment	Patients aged 18 and over	Efficacy and safety of miglustat 200 mg tid. in late onset Tay-Sachs	12 months + 12 months extension
Fabry Disease					
Study 002	16	Phase I/II, open-label, non-comparative pilot study	Adult male patients	Efficacy and safety of miglustat 100 mg tid. with dose adjustment	12 months

of peripheral neuropathy in patients with Gaucher disease type 1 who have never been exposed to miglustat. In addition to prevalence, the incidence of peripheral neuropathy over the 2-year observation period will be established. Neurological examinations are standardized and evaluated by a central assessor. This study will provide insights into the degree of neurological involvement secondary to Gaucher disease type 1. Another investigation, which has been approved as a multi-centre study by EMEA, will determine the outcome of substrate reduction for therapeutic maintenance in adult patients whose disease has been successfully controlled by enzyme replacement therapy. Over a 2-year period, the parameters of disease activity will be monitored intensively.

Neuronopathic Forms of Gaucher Disease

Recently enzyme replacement therapy in the form of Cerezyme (imiglucerase) has been approved by the Orphan Drug legislation for the treatment of the systemic disease that accompanies the so-called neuronopathic variant of Gaucher disease. Clearly, since Gaucher disease represents a diverse range of severity, there will be other indications for the use of substrate reduction therapy with miglustat. Extensive clinical studies have now shown that neurological manifestations occur in some patients formally categorized as type 1 but who ultimately develop neurosensorial hearing loss, disorders of eye movement, and even Parkinsonism.[51,52] Clearly enzyme therapy is unlikely to offer a prospect of benefit for these manifestations beyond the correction of crippling skeletal disease and the improvement of disabling systemic and visceral manifestations due to its inability to penetrate the blood-brain barrier. There may thus be a justification for the additional use of miglustat in such patients with progressive manifestations of Gaucher disease affecting the brain and where any alternative method of treatment is lacking.

A clinical trial is underway using miglustat in children and adolescents with so-called type 3 neuronopathic Gaucher disease in which preliminary results indicate that the drug penetrates the blood-brain barrier and where cerebrospinal fluid concentrations reach 20–40% of those achieved therapeutically in the plasma (unpublished data); similar concentrations, indicating penetrance of the blood-brain barrier, have been reported in a patient receiving miglustat for Niemann-Pick disease type C.[53]

OTHER NEURODEGENERATIVE DISORDERS

Clearly many sphingolipidoses have neurological components such as Niemann-Pick disease type C in which accumulation of complex gangliosides, including GM2, GM3, and GM1 occurs in neuronal cells. Similarly, glycosphingolipids accumulate in neurons of patients with mucopolysaccharidosis type III. Given the potential therapeutic uses of inhibitors of glycosphingolipid biosynthesis, a number of trials are underway to test the therapeutic effect of miglustat (NB-DNJ). These include late-onset Tay-Sachs disease, juvenile and late-onset forms of Sandhoff disease (both diseases are classified under the category of GM2 gangliosidoses), Niemann-Pick disease type C, as well as neuronopathic type 3 Gaucher disease.

The original clinical trial in Fabry disease was aborted early as a result of untoward neurological effects. Of 16 patients with Fabry disease who received miglustat, 13 developed tremor and 2 complained of parasthesiae. This contributed to the withdrawal of three patients from the study. It seems likely, however, that preexisting renal failure may have led to high plasma concentrations of miglustat in these patients since the main route of excretion of iminosugars is through the kidney. Clearly monitoring the outcomes in patients with chronic neurodegenerative disorders that are slowly progressive represents a major challenge for the design of therapeutic trials. Rigorous clinical, neuropsychological, and radiological evaluations are necessary. Over time there is a need to demonstrate where possible that administration of iminosugars indeed has had a biological effect of relevance to the target condition under study. We have demonstrated that depletion of glycosphingolipids by an iminosugar in the living human reduced pathological lipid storage and corrected the lipid trafficking defect that has been observed in peripheral blood lymphocytes, for example, in a patient with Niemann-Pick disease type C.[53] However, the use of such surrogate biomarkers cannot provide a complete picture of the outcome of substrate reduction therapy at the putative site of action within the brain.

Ultimately, there will be no substitute for the conduct of well-designed trials in the neuroglycosphingolipidoses that will provide clear-cut information as to whether substrate reduction therapy with iminosugars offers the hope of meaningful clinical benefit. At the present time, miglustat, NB-DNJ is licensed only for the treatment of mild-to-moderate Gaucher disease in adults and its use in other human glycosphingolipidoses must be regarded, though theoretically promising, as experimental.

FUTURE COMPOUNDS FOR SUBSTRATE REDUCTION THERAPY

After the demonstration that the alkyl-substituted deoxynojirimycins, NB-DNJ and N-butyldeoxygalactonojirimycin (NB-DGJ), serve as inhibitors of glucosylceramide synthase at low micromolar concentrations, other animal studies have been undertaken at the University of Oxford. These showed that the galactose analogue NB-DGJ appears to have the same therapeutic efficacy as its glucose congener but substantially lacks the undesirable gastrointestinal side effects.[54] This analogue does not inhibit the α-glucosidases I and II present in the endoplasmic reticulum[22] and does not inhibit sucrase or maltase in the brush border membrane and does not have an unexplained effect on lymphoid organ reduction, which has been observed in experimental animals receiving high doses of NB-DNJ.

Administration of NB-DGJ appears to be unassociated with weight loss in experimental animals.[54] It should be noted that neither this sugar nor NB-DNJ inhibits the galactosylceramide transferase that is important for the biosynthesis of galactosylceramide and sulphatide. Thus the NB-DGJ, which appears to be a more selective inhibitor of glycosphingolipid biosynthesis than its licensed analogue miglustat, may ultimately prove to be the favored compound for patients with glycosphingolipidoses where substantial inhibition of glycosphingolipid biosynthesis is needed in the nervous system.[21,23,26,54]

The search for other inhibitors of glucosidases by chance identified a more potent inhibitor of glucosylceramide synthase than *N*B-DNJ. Adamantane-pentyl-deoxynojirimycin was found to inhibit glycolipid bioynthesis at concentrations at least 1000-fold lower in the plasma than those achieved with *N*B-DNJ. Preliminary reports indicate that this agent was also able to prevent the accumulation of globotriasylceramide in the Fabry knockout mouse.[24,55] With the prototype now encouragingly licensed for use, the whole field of substrate reduction therapy and modulation of sphingolipid biosynthesis for disorders of metabolism, infection, and immunity is brimming with innovative clinical and laboratory research activity.

REFERENCES

1. Vunnan, R.R. and Radin, N.S. Analogs of ceramide that inhibit glucocerebroside synthetase in mouse brain. *Chem-Phys-Lipids*, 26, 265, 1980.
2. Yamashita, T. et al. A vital role for glycosphingolipid synthesis during development and differentiation. *Proc. Natl. Acad. Sci. U.S.A.*, 96, 9142, 1999.
3. Sandhoff, K. and Kolter T. Biosynthesis and degradation of mammalian glycosphingolipids. *Philos. Trans. R. Soc. Lond B. Biol. Sci*, 358, 847, 2003.
4. Trinchera, M., Fabbri, M. and Ghidoni, R. Topography of glycosyltransferases involved in the initial glycosylations of gangliosides. *J. Biol. Chem.*, 266, 20907, 1991.
5. Inokuchi, J. and Radin, N.S. Preparation of the active isomer of 1-phenyl-2-decanoylamino-3-morpholino-1-propanol, inhibitor of murine glucocerebroside synthetase. *J. Lipid Res.*, 28, 565, 1987.
6. Radin, N.S., Shayman, J.A. and Inokuchi J. Metabolic effects of inhibiting glucosylceramide synthesis with PDMP and other substances. *Adv. Lipid Res.* 26, 183, 1993.
7. Winchester, B. and Fleet, G.W. Amino-sugar glycosidase inhibitors: versatile tools for glycobiologists. *Glycobiology*, 2, 199, 1992.
8. Platt, F.M. et al. N-butyldeoxynojirimycin is a novel inhibitor of glycolipid biosynthesis. *J. Biol. Chem.*, 269, 8362, 1994.
9. Platt, F.M. et al. Inhibition of substrate synthesis as a strategy for glycolipid lysosomal storage disease therapy. *J. Inherit. Metab. Dis.* 24, 275, 2001.
10. Lee, L., Abe, A. and Shayman, J.A. Improved inhibitors of glucosylceramide synthase. *J. Biol. Chem.* 274, 14662, 1999.
11. Platt, F.M. and Butters, T.D. Inhibition of substrate synthesis: a pharmacological approach for glycosphingolipid storage disease therapy. In *Lysosomal Disorders of the Brain*, Platt, F.M. and Walkley, S.U. Eds., Oxford University Press, 2004.
12. Abe, A. et al. Reduction of globotriaosylceramide in Fabry disease mice by substrate deprivation. *J. Clin. Invest.*, 105, 1563, 2000.
13. Dwek, R.A. et al. Targetting glycosylation as a therapeutic approach. *Nature Drug Discovery*, 1, 65, 2002.
14. Fischl, M.A. et al. The safety and efficacy of combination N-butyl-deoxynojirimycin (SC-48334) and zidovudine in patients with HIV-a infection and 200-500 CD4 cells.mm³. *J. Acquir. Immune Defic. Syndro.* 7, 139, 1994.
15. Bergeron, J.J. et al. Calnexin: a membrane-bound chaperone of the endoplasmic reticulum. *Trends Biochem. Sci.* 19, 124, 1994.

16. Helenius, A. et al. Calnexin, calreticulin and the folding of glycoproteins. *Tr. Cell Biol.* 7, 193, 1997.
17. Karlsson, G.B. et al. Effects of the imino sugar N-butyldeoxynojirimycin on the N-glycosylation of recombinant gp120. *J. Biol. Chem.* 268, 570, 1993.
18. Fischer, P.B. et al. N-butyldeoxynojirimycin-mediated inhibition of human immun-odeficiency virus entry correlates with impaired gp120 shedding and gp41 exposure. *J. Virol.*, 70, 7153, 1996.
19. Fischer, P.B. et al. N-butyldeoxynojirimycin-mediated inhibition of human immun-odeficiency virus entry correlates with changes in antibody recognition of the V1/V2 region of gp120. *J. Virol.* 70, 7143, 1996.
20. Fischer, P.B. et al. The alpha-glucosidase inhibitor N-butyldeoxynojirimycin inhibi-tors human immunodeficiency virus entry at the level of post-CD4 binding. *J. Virol.* 69, 5791, 1995.
21. Butters, T.D. et al. Molecular requirements of imino sugars for the selective control of N-linked glycosylation and glycosphingolipid biosynthesis. *Tetrahedron Assymetry* 11, 113, 2000.
22. Platt, F.M. et al. N-butyldeoxygalactonojirimycin inhibits glycolipid biosynthesis but does not affect N-linked oligosaccharide processing. *J. Biol. Chem.* 269, 27108, 1994.
23. Butters, T.D. et al. Small-molecule therapeutics for the treatment of glycolipid lyso-somal storage disorders. *Philos. Trans. R. Soc. London B. Biol. Sci.* 358, 927, 2003.
24. Overkleeft, H.S. et al. Generation of specific deoxynojirimycin-type inhibitors of the non-lysosomal glucosylceramidase. *J. Biol. Chem.* 273, 26522, 1998.
25. van den Broek, L.A.G.M. et al. Synthesis of oxygen-substituted N-alkyl 1-deoxynojir-imycin derivatives: aza sugar α-glucosidase inhibitors showing antiviral (HIV-1) and immunosuppressive activity. *Recueil des Travaux Chimiques des Pays-Bas.* 113, 507, 1994.
26. Butters, T.D. et al. Molecular requirements of imino sugars for the selective control of N-linked glycosylation and glycosphingolipid biosynthesis. *Revue Roumaine de Biochimie.* 35, 75, 1998.
27. Radin, N.S. Treatment of Gaucher disease with an enzyme inhibitor. *Glycoconj. J.* 13, 153, 1996.
28. Platt, F.M. et al. Substrate reduction therapy in mouse models of the glycosphingolip-idoses. *Phil. Trans. R. Soc. Lond. B.* 358, 947, 2003.
29. Yamanaka, S. et al. Targeted disruption of the Hexa gene results in mice with biochemical and pathologic features of Tay-Sachs disease. *Proc. Natl. Acad. Sci. U.S.A.,* 91, 9975, 1994.
30. Jeyakumar, M. et al. An inducible mouse model of late onset Tay-Sachs disease. *Neurobiology of Disease.* 2002.
31. Cox, T. et al. Novel oral treatment of Gaucher's disease with N-butyldeoxynojirimycin (OGT 918) to decrease substrate biosynthesis. *Lancet.* 355, 1481, 2000.
32. Platt, F.M. et al. Prevention of lysosomal storage in Tay-Sachs mice treated with N-butyldeoxynojirimycin. *Science.* 276, 428, 1997.
33. Sango, K. et al. Mouse models of Tay-Sachs and Sandhoff diseases differ in neuro-logic phenotype and ganglioside metabolism. *Nat. Genet.* 11, 170, 1995.
34. Jeyakumar, M. et al. Delayed symptom onset and increased life expectancy in Sand-hoff disease mice treated with N-butyldeoxynojirimycin. *Proc. Natl. Acad. Sci. U.S.A.,* 96, 6388, 1999.
35. Ringden, O. et al. Ten years' experience of bone marrow transplantation for Gaucher disease. *Transplantation.* 59, 864, 1995.

36. Dobrenis, K. Cell-mediated delivery systems, in *Lysosomal Disorders of the Brain.* Platt, F.M. and Walkley, S.U. Eds. Oxford University Press, 339, 2004.
37. Sands, M.S. Gene therapy, in *Lysosomal Disorders of the Brain.* Platt, F.M. and Walkley, S.U., Eds. Oxford University Press, 409, 2004.
38. Snyder, E.Y., Daley, G.Q. and Goodell, M. Taking stock and planning for the next decade: realistic prospects for stem cell therapies for the nervous system. *J. Neurosci. Res.* 76, 157, 2004.
39. Jeyakumar, M. et al. Enhanced survival in Sandhoff disease mice receiving a combination of substrate deprivation therapy and bone marrow transplantation. *Blood.* 97, 327, 2001.
40. Fenouillet, E. and Gluckman, J.C. Effect of a glucosidase inhibitor on the bioactivity and immunoreactivity of human immunodeficiency virus type 1 envelope glycoprotein. *J. Gen. Virol,* 72, 1919, 1991.
41. Fischl, M.A. et al. The safety and efficacy of combination N-butyldeoxynojirimycin (SC-48334) and zidovudine in patients with HIV-1 infection and 200-500 CD4 cells.mm³. *J. Acquir. Immune, Defic. Syndr.* 7, 139, 1994.
42. Tierney, M. et al. The tolerability and pharmacokinetics of N-butyldeoxynojirimycin in patients with advanced HIV disease (ACTG100). The AIDS Clinical Trials Group (ACTG) of the National Institute of Allergy and Infectious Diseases. *J. Acquir. Immune Defic. Synd. Hum. Retrovirol* 10, 549, 1995.
43. Elstein, D. et al. Sustained therapeutic effects of oral miglustat (Zavesca, N-butyldeoxynojirimycin, OGT 918) in type 1 Gaucher disease. *J. Inherit. Metab. Dis.* 27, 757, 2004.
44. Heitner, R. et al. Low-dose N-butyldeoxynojirimycin (OGT 918) for type 1 Gaucher disease. *Blood Cells Mol. Dis.* 28, 127, 2002.
45. Hollak, C.E.M. et al. Dixon quantitive chemical shift imaging is a sensitive tool for the evaluation of bone marrow response to enzyme supplementation in type 1 Gaucher disease. *Blood Cell Mol. Dis.* 27, 1005, 2002.
46. Grabowski, G.A. et al. Enzyme therapy in type 1 Gaucher disease: comparative efficacy of mannose-terminated glucocerebrosidase from natural and recombinant sources. *Ann. Intern. Med.* 122, 33, 1995.
47. Pastores, G.M., Barnett, N.L. and Kolodny, E.H. An open-label, noncomparative study of miglustat in type 1 Gaucher disease: efficacy and tolerability over 24 months of treatment. *Clin. Ther.* 27, 1215, 2005.
48. van der Spoel, A.C. et al. Reversible infertility in male mice following oral administration of alkylated imino sugars: a non-hormonal approach to male contraception. *P.N.A.S.* 99, 17173, 2002.
49. Suganuma, R. et al. Alkylated imino sugars, reversible male infertility-inducing agents, do not affect genetic integrity of male mouse germ cells despite induction of sperm deformities. *Biology of Reproduction.* 72, 805, 2005.
50. Cox, T.M. et al. The role of the iminosugar N-butyldeoxynojirimycin (miglustat) in the management of type 1 (non-neuronopathic) Gaucher disease: a position statement. *J. Inherit. Metab. Dis.* 26, 513, 2003.
51. Wong, K. et al. Neuropathology provides clues to the pathophysiology of Gaucher disease. *Mol. Genet. Metab.* 82, 192, 2004.
52. Neudorfer, O. et al. Occurrence of Parkinson's syndrome in type 1 Gaucher disease. *Quart. J. Med.* 89, 691, 1996.
53. Lachmann, R.H. et al. Treatment with miglustat reverses the lipid-trafficking defect in Niemann-Pick disease type C. *Neurobiol. Dis.* 16, 654, 2004.

54. Andersson, U. et al. N-butyldeoxygalactonojirimycin: a more selective inhibitor of glycolipid biosynthesis than N-butyldeoxygalactonojirimycin, *in vitro* and *in vivo*. *Biochem. Pharmacol.* 59, 821, 2000.

55. Aerts, J.M. et al. Biochemistry of glycosphingolipid storage disorders: implications for therapeutic intervention. *Philos. Trans. R. Soc. Lon. B. Biol. Sci.,* 358, 905, 2003.

CHAPTER 20

Pharmacologic Chaperone Therapy for Lysosomal Diseases

Robert J. Desnick and Jian-Qiang Fan

CONTENTS

INTRODUCTION

Remarkable progress has been made in our ability to treat lysosomal storage diseases. Enzyme replacement therapy (ERT), bone marrow transplantation, and substrate deprivation are currently being used to treat these disorders, and the potential of gene and stem cell therapies are being investigated.[1–4] Forty years ago, Christian de Duve first suggested that lysosomal diseases could be treated by replacing the defective enzyme with its normal counterpart.[5] However, it was not until the early 1990s that ERT became a reality with the demonstration of its safety and effectiveness in patients with type 1 Gaucher disease (Chapter 18). Subsequently, ERT was approved for Fabry disease and mucopolysaccharidoses I and VI, and approval of ERT for Pompe disease and mucopolysaccharidosis II is anticipated in the near future. Bone marrow transplantation in lysosomal storage diseases was first reported in a patient with mucopolysaccharidosis I (Hurler's disease) in 1981[6] and has been performed for mucopolysaccharidoses II, VI, Gaucher disease types 1–3, Krabbe disease, metachromatic dystrophy and several others with varying clinical benefit in successfully engrafted recipients.[2,7] Substrate deprivation has been approved and evaluated for Gaucher type 1 disease.[8,9] The effectiveness and limitations of bone marrow replacement and substrate deprivation for Gaucher disease are described in Chapter 19 and Chapter 22, respectively.

Pharmacologic chaperone therapy is a novel and attractive approach for the treatment of protein misfolding/mistrafficking disorders by rescuing misfolded/mistrafficked mutant proteins (for recent reviews, see[10–12]). For lysosomal storage diseases, this therapeutic strategy uses small molecules that are reversible, enzyme inhibitors that can rescue misfolded/mistrafficked enzymes that would otherwise be transported by the endoplasmic reticulum-associated degradation (ERAD) machinery to the proteosome for degradation (for reviews, see[12–14]). Pharmacologic chaperones are attractive therapeutic agents as they can be orally administered and presumably can gain access to most or all cell types.

The first studies of pharmacologic chaperone therapy for lysosomal disorders, carried out in Fabry disease (deficient α-galactosidase A [α-Gal A] activity), demonstrated that mutant α-Gal A enzymes encoded by various missense mutations could be effectively rescued by various α-Gal A reversible inhibitors, including 1-deoxygalactonojirmycin (DGJ).[15,16] Subsequently, specific pharmacologic chaperones have been identified for Gaucher disease, Tay-Sachs and Sandhoff diseases, and GM1 gangliosidosis.[17–21] In a murine model of GM1 gangliosidosis, studies have shown that a pharmacologic chaperone can cross the blood-brain barrier and decrease the neural lysosomal substrate accumulation.[18] Recognizing the potential therapeutic effectiveness of pharmacologic chaperone therapy, investigators are evaluating potential compounds as chaperones for other lysosomal storage diseases. In addition, clinical proof-of-concept has been demonstrated in a patient with later-onset Fabry disease.[22] These findings have been translated into the first FDA-approved clinical trial of pharmacologic chaperone therapy for a lysosomal disease. A Phase 1 clinical trial of pharmacologic chaperone therapy for Fabry disease has been successfully completed (Amicus Therapeutics; http://www.amicustherapeutics.com) and a Phase

2 clinical trial is underway (http://www.clinicaltrials.gov/ct/show/NCT00214500? order=1).

Here, we provide an overview of the molecular basis of pharmacologic chaperone therapy for lysosomal diseases, and describe recent *in vitro* and *in vivo* studies that highlight the potential of this therapeutic strategy in which the mutant lysosomal enzyme can be rescued and its stability, delivery to the lysosome, and function increased. This therapeutic strategy offers the potential to prevent or effectively reverse the disease manifestations in lysosomal diseases resulting from protein misfolding/mistrafficking. It is anticipated that even a small increment in enzymatic activity can markedly attenuate a severe phenotype, as illustrated by the fact that the less severe subtypes of Gaucher disease and other lysosomal disorders are due to the presence of mutant enzymes that have very low levels of residual activity (even <1% of wildtype).

LYSOSOMAL ENZYME BIOSYNTHESIS AND DEGRADATION

Lysosomal proteins are synthesized in the cytoplasm and then secreted into the lumen of the endoplasmic reticulum (ER) in a largely unfolded state. In general, protein folding is governed by the principle of self-assembly.[23] Newly synthesized polypeptides and glycopolypeptides fold into their native (active) conformation based on their amino acid sequences in a thermodynamic fashion with help from the resident molecular chaperones (e.g., BiP, calnexin, HSPs, etc.). In order to monitor the *in vivo* folding process, the ER has evolved a "quality control" mechanism, termed ERAD, which uses molecular chaperones to bind and rescue unstable misfolded conformers to facilitate their proper folding and assembly and to prevent the aggregation of nonnative forms through binding and release cycles.[12,24–26] This machinery ensures that only properly folded and assembled proteins are transported to the Golgi complex for further maturation. Improperly folded proteins are retained in the ER by molecular chaperones and then transported to the cytosol for uniquitination and degradation within the cytosolic proteosomes.[24] In this way, misfolded or unstable (normal or mutant) proteins are eliminated from the cell.[24,27–30] However, even after interaction with molecular chaperones, it is estimated that up to 30% of normal proteins do not achieve their functional state, misfold and/or aggregate, and are rapidly degraded within minutes of their synthesis by the cell's quality control machinery.[31]

MUTATIONS CAUSING LYSOSOMAL DISEASES

Most lysosomal disorders result from a variety of mutations in their respective disease gene that render the encoded enzyme or protein nonfunctional. Nonsense and frameshift mutations encode truncated and markedly altered polypeptide sequences that cannot function and are rapidly transported and degraded, presumably by the ERAD machinery, to the cytosol for proteosome degradation. Alternatively, transcripts containing premature termination codons caused by nonsense and frameshift mutations may be degraded through nonsense-mediated decay. Splicing defects either result in no enzyme protein or in an insufficient amount of normal enzyme, which results in substrate accumulation. Certain missense mutations and some small inframe triplet deletions may not (or may minimally) impair the functional domains of the mutant protein (i.e., the active site, receptor-binding site, etc.), but may cause polypeptide misfolding, aggregation, instability, and/or altered trafficking to the lysosome. Presumably, mutations that cause misfolding slow the normal folding process, thereby resulting in higher concentrations of "folding intermediates," which can self-aggregate.[12,32] Such mutant lysosomal proteins are retained in the ER where they become associated with molecular chaperones (e.g., calnexin, BiP, etc.) that attempt to restore the native conformation or, having failed to form a functional state, undergo rapid degradation by the quality control system.

In many lysosomal disorders, certain missense mutations produce mutant enzymes that retain a small amount of residual enzymatic activity (even 1%). These missense mutations are associated with the less severe and/or later-onset disease phenotypes. The presence of residual activity presumably results from the small amount of the mutant glycopeptide that was properly folded, assembled, posttranslationally modified, and trafficked to the lysosome. Such mutations are excellent candidates for pharmacologic chaperone therapy (see "Criteria for Selection of Pharmacologic Chaperones" below). Indeed, even mutations that totally misfold are potentially rescuable by pharmacologic chaperones.

PHARMACOLOGIC CHAPERONES CAN STABILIZE MISFOLDED PROTEINS

"Pharmacologic chaperones" are specific, low molecular weight, ligands that bind to and rescue misfolded and/or mistargeted proteins that can augment the cell's molecular chaperones to enhance correct folding and organelle targeting of misfolded mutant conformers. By stabilizing the corrected conformation of the mutant protein, they prevent its degradation. Pharmacologic chaperones include substrate analogues, active-site inhibitors, cofactors, effector molecules, receptor agonists and antagonists, or other modulators and epitope-directed ligands. The mechanisms by which these chaperones function also may include stabilizing a specific conformation of the misfolded protein, reducing aggregation, and/or preventing nonproductive interactions with other resident proteins.

As proposed for enzymes by Fan et al.,[15] such compounds can diffuse into the cell and bind site-specifically to folding intermediates of the mutant enzyme protein, thereby stabilizing the intermediate that is rate-limiting for the folding and/or trafficking of the mutant enzyme, and rescuing a portion of it from degradation. Enzyme inhibitors can be effective pharmacologic chaperones because of their high affinity to the catalytic domain. Such inhibitors serve as a folding template for those mutant proteins with fragile conformational structures during the protein folding process and induce proper folding. In this way, the pharmacologic chaperone prevents the excessive degradation of the misfolded mutant enzyme that may retain full or partial activity if it can be properly folded, processed, and transported to its normal site of action. Once the mutant enzyme/inhibitor complex is secreted out of the ER, the inhibitor at sub-inhibitory concentrations can be replaced by higher concentrations of substrate for metabolic action.[33]

Ulloa-Aguirre et al.[12] have proposed general guidelines to identify the nature of rescuable and nonrescuable missense mutations for rescuable enzymes:

> The missense mutation should not alter critical residues essential for ligand binding or substrate/cofactor binding in the case of enzymes which are, predictably, non-functional.
>
> Loss or gain of a cysteine residue may potentially disrupt required sulphuryl bridges or form inappropriate bridges that may be so significantly disruptive to the protein's structure that rescue cannot occur.
>
> Loss or gain of a proline residue may limit or even impede pharmacological rescue since this amino acid typically causes forced turns in the protein sequence, which may dramatically alter the structure, rendering it nonrescuable. In some proteins, an abrupt turn is likely a requisite for correct structure and cannot be corrected by pharmacologic chaperones.
>
> Substitutions that impede or promote hydrogen bond formation may reduce the ability of the pharmacologic chaperone to rescue due to the inability of the mutant protein to establish correct interactions between its different domains.
>
> Substitutions by larger amino acid residues (valine, tryptophan, threonine, and cysteine) may or may not be destabilizing, depending on their position, while replacements with smaller residues (glycine or alanine) may allow for more steric freedom, may be accommodated in the folding process and, therefore, are potentially strong candidates for pharmacologic chaperone rescue.
>
> If the enzyme or protein has been crystallized, molecular modeling studies can be used to predict the effect of the amino acid substitution on the conformation of the mutant enzyme and the impact on its active site.

For lysosomal enzymes, these pharmacologic chaperones must bind reversibly, so that when the rescued protein arrives in the lysosomes, the chaperone will be displaced by the high concentration and greater affinity of the already accumulated natural substrate(s). In contrast to ERT with recombinant lysosomal enzymes small hydrophobic molecules given orally may cross the blood-brain barrier, diffuse through connective tissue matrices, and reach target sites of pathology that infused lysosomal enzymes cannot, or only can when administered at very high doses. Thus, pharmacologic chaperone therapy is particularly attractive for the treatment of neurodegenerative lysosomal diseases.

EXPERIMENTAL STUDIES OF PHARMACOLOGIC CHAPERONE THERAPY IN LYSOSOMAL DISORDERS

The lysosomal storage diseases are excellent candidates for pharmacologic chaperone therapy. As noted above, most lysosomal storage diseases have less severe and/or later onset subtypes resulting from mutations encoding mutant enzymes with a small amount of residual activity. Such mutant enzymes are targets for pharmacologic chaperone therapy. To identify potential pharmacologic chaperones for lysosomal disorders, investigators have evaluated known or chemically modified substrate inhibitors and analogues, or have performed high throughput screens of chemical libraries.[34] These compounds can be evaluated in tissue culture systems or transgenic mice to determine their *in vitro* and *in vivo* ability to rescue certain mutant enzymes (see examples below). Pharmacologic chaperone therapy may be particularly useful in disorders resulting from common missense mutations that are rescuable, such as β-glucocerebrosidase N370S in type 1 Gaucher disease,[17] β-hexosaminidase α-chain G269S in chronic Tay-Sachs disease,[20] and α-Gal A N215S in Fabry disease. However, the use of a specific enzyme inhibitor that binds to the active site has the potential to rescue a variety of missense mutations whose mutant proteins retain residual activity, even <1% of wildtype. Experimental studies of pharmacologic chaperone therapy in lysosomal diseases are listed in Table 20.1 and described below. Extension of these studies to mutant enzymes encoded by other missense mutations and to missense mutations encoding mutant proteins in other lysosomal diseases is anticipated.

Fabry Disease (Deficient α-Gal A Activity)

This X-linked recessive disorder results from the deficient activity of the lysosomal exogalactosidase, α-Gal A.[35,36] There are two major subtypes, the severe childhood-onset "classical phenotype" and the "later-onset phenotypes," including atypical, cardiac, or renal variants.[36–40] To date, over 400 α-Gal A mutations have been identified, including missense, nonsense, splicing, and frameshift mutations due to small and large deletions and insertions (Human Gene Mutation Database).[41] Most

Table 20.1 Experimental and Clinical Studies of Pharmacologic Chaperones in Lysosomal Diseases

Disease	Deficient Enzyme Activity	Pharmacologic Chaperone	Model System	Ref.
Fabry	α-Galactosidase A	1-Deoxygalactonojirimycin (DGJ)	Cultured Cells	15
			Transgenic Mice	15
			Phases 1 and 2 Clinical Trials[a]	
Gaucher	β-Glucocerebrosidase	N-(n-Nonyl)deoxynojirimycin (NN-DNJ)	Cultured Cells	17
		N-Octyl-β-valienamine (NOV)	Cultured Cells	19
		N-Butyl-deoxynojirimycin (NB-DNJ)	Expression in COS Cells	21
GM1-Gangliosidosis	β-Galactosidase	N-Octyl-4-epi-β-valienamine (NOEV)	Cultured Cells	18
			Transgenic Mice	18
Tay-Sachs Sandhoff	β-Hexosaminidase A β-Hexosaminidase A & B	N-Acetylglucosamine-thiazoline (NGT)	Cultured Cells	20

[a] Amicus Therapeutics; http://www.amicustherapeutics.com

mutations are "private," occurring in a single or a few families, except for those that occur at CpG dinucleotides which are known mutational hotspots.[42,43]

The crystal structure of α-Gal A has been determined and the location and predicted effect of various missense mutations have been correlated with the observed clinical phenotype.[44,45] Molecular modeling[46] and expression studies[47,48] have been used to characterize the missense mutations and to predict or assess their residual activities. In affected males with the classic phenotype, all types of mutations have been reported.[36] However, most patients with the later-onset phenotype had missense mutations that encoded mutant enzymes with low levels of residual α-Gal A activity,[38–40] the exception being a splicing mutation that also results in a very low activity level of the normally spliced enzyme.[49]

Previously, galactose, a weak α-Gal A inhibitor, was shown to stabilize the residual activity of certain α-Gal A missense mutations, but not others when these mutant alleles were overexpressed in COS-1 cells.[50] Fan and colleagues evaluated a series of α-Gal A substrate analogues and identified the iminosugar, DGJ, a compound that bound to several mutant α-Gal A enzymes at sub-inhibitory concentrations and most effectively increased their residual activities.[15,16,33] For example, the residual α-Gal A activities in cultured lymphoblasts from patients with the Q279E and R301Q mutations were enhanced seven- to eight-fold after incubation with 20 μM DGJ for 4 days (Figure 20.2). In addition, these investigators generated transgenic mice carrying the rescuable R301Q mutation and demonstrated that oral administration of DGJ caused a dose-dependent increase in α-Gal A activity in the tissues of the mice (Figure 20.3). No toxic effects were observed in transgenic mice treated with DGJ for 140 days.[15] Independent studies in R301Q transgenic mouse fibroblasts demonstrated that DGJ treatment released the rescuable R301Q mutant enzyme from the molecular chaperone BiP in the ER. The rescued mutant enzyme was then trafficked to the lysosome via the mannose-6-phosphate receptor-mediated pathway where it cleared the accumulated glycosphingolipid substrate, globotriaosylceramide (GL-3).[51] Thus, these "*in vitro*" findings indicated that DGJ stabilized the mutant α-Gal A glycoprotein such that more of the enzyme was transported to the lysosome, where it is functional.

Gaucher Disease (Deficient β-Glucocerebrosidase Activity)

This autosomal recessive disorder results from the deficient activity of the lysosomal exoglycosidase, β-glucocerebrosidase.[52,53] There are three major subtypes: type 1, a nonneurologic disorder with clinical onset in childhood to late adult life, type 2, a severe neurodegenerative disorder of infancy, and type 3, a neurologic form with juvenile or late-juvenile onset.[53] Genotype/phenotype studies have revealed that certain mutations (e.g., N370S) have residual activity, are neuro-protective, and are responsible for the nonneurologic type 1 phenotype. To date, over 190 β-glucocerebrosidase mutations have been identified, including missense, nonsense, splicing, and small deletions and insertions (Human Gene Mutation Database).[41] Of interest, the functional β-glucocerebrosidase gene is 16 kb upstream from a pseudogene, and crossovers and/or gene conversions have resulted in a variety of complex mutations.[53] Several of the β-glucocerebrosidase mutations are common in various populations.

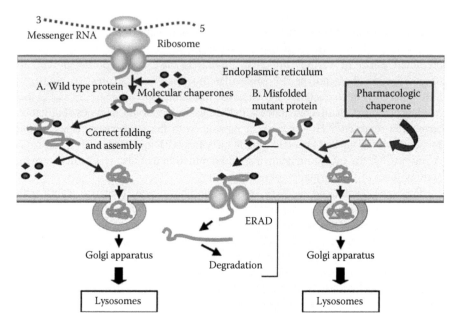

Figure 20.1 Mechanism of pharmacologic chaperone rescue for mutant lysosomal enzymes. A. Schematic of the processing, correct folding, and assembly of newly synthesized lysosomal enzymes. The newly synthesized enzyme is translocated into the endoplasmic reticulum (ER), where molecular chaperones facilitate its proper folding and subunit assembly. The molecular chaperones then dissociate from the folded, assembled enzyme, which moves to the Golgi apparatus and then to lysosomes, where the enzyme is stable and active in the acidic environment of these organelles. B. Misfolding and mis-assembly of mutant lysosomal enzymes. The mutant proteins are retained in the ER and are eventually transported to the cytosol by the ERAD machinery where they are degraded. Pharmacologic chaperones (yellow triangles) bind to the active site of the enzyme, promote folding, and stabilize the mutant enzyme for proper subunit assembly. This promotes transport of the mutant enzymes to the Golgi apparatus and the lysosomes, thereby increasing the concentration of the mutant enzyme and its residual enzyme activity. (From Fan, J.Q., *Trends Pharmacol Sci*, 24, 355, 2003. With permission.)

For example, the "neuro-protective" N370S mutation occurs in about 70% of alleles in the Ashkenazi Jewish type 1 patients and also is common in European type 1 patients. In addition, the common L444P missense mutation is pan-ethnic and when homoallelic causes type 3 disease. The crystal structure of β-glucocerebrosidase recently has been determined,[54] and studies have attempted to correlate various genotypes with the disease phenotypes.[53,55,56]

Recent studies in cultured cells from patients with various β-glucocerebrosidase genotypes have shown that certain mutant enzymes encoded by missense mutations, such as K157Q, D409H, P415R, and L444P, are retained in the ER where they become bound to calnexin and are then transported by the ERAD pathway to the cytosol for proteosomal degradation.[57] Several of these misfolded enzymes have been the subject of pharmacologic chaperone rescue by active site-directed substrate analogues. Lin and colleagues demonstrated that N-octyl-β-valienamine (NOV)

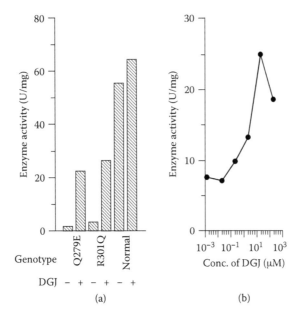

Figure 20.2 Enhancement of α-Gal A in lymphoblasts from patients with Fabry disease. A. α-Gal A activity in lymphoblasts (Q279E, R301Q and normal) cultured in the presence (+) or absence (-) of 20 μM DGJ. B. Effect of DGJ concentration on α-Gal A activity. Lymphoblasts (R301Q) were cultured with DGJ at various concentrations before being collected for enzyme assay. (From Fan, J.Q. et al., *Nat Med*, 5, 112, 1999. With permission.)

rescued the residual F213I β-glucocerebrosidase activity in cultured cells from type 1 Gaucher patients.[19] However, NOV did not rescue other β-glucocerebrosidase mutant enzymes encoded by N370S, L444P, and 84GG. In contrast, Kelley and colleagues evaluated several substrate analogues of β-glucocerebrosidase and found that *N*-(*n*-nonyl)deoxynojirimycin (NN-DNJ) at sub-inhibitory concentrations (10 M) led to a twofold increase in the residual activity of the common N370S mutant enzyme.[17] Of note, NN-DNJ did not rescue the mutant enzyme encoded by the more severe and common L444P missense mutation. That the L444P glycoprotein was not rescuable with NN-DNJ was either due to the intrinsic nature of the mutant enzyme (e.g., less stable, more rapidly degraded, etc.) or its active site might be altered such that the substrate analogue binds poorly. More recently, Kelly and colleagues evaluated 34 potential enzyme inhibitors to determine if they could rescue β-glucocerebrosidase mutant enzymes N370S, L444P, and G202R in cultured skin fibroblasts. Of these, several compounds increased the activity of N370S and G202R, but not L444P.[84] Thus, certain mutant proteins for a given enzyme may require different pharmacologic chaperones. However, it is more likely — and attractive — that a particular active-site directed, small molecule chaperone will bind most, if not all, the mutant proteins with residual activity encoded by the same gene, as has been shown for Fabry disease (Fan, J.Q., personal communication, 2005). This logic should hold for other mutant proteins with residual receptor, effector, or epitope function.

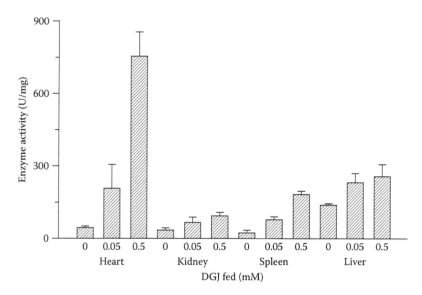

Figure 20.3 *In vivo* enhancement by DGJ of α-Gal A missense mutation R301Q in transgenic Mice. DGJ was administrated orally to R301Q transgenic mice at the dosages indicated (horizontal axis). Data were obtained by subtracting the endogenous α-Gal A activity in nontransgenic mice fed DGJ from the activity in transgenic mice for the organs indicated (horizontal axis). Mean values ± 1 standard deviation are shown. (From Fan, J.Q. et al., *Nat Med*, 5, 112, 1999. With permission.)

It is of interest to note that miglustat (*N*-butyl-deoxynojirimycin, *N*B-DNJ), an inhibitor of ceramide glucosyltransferase used in type 1 Gaucher patients for substrate deprivation, also may be a pharmacologic chaperone for β-glucocerebrosidase.[21] Studies of transfected COS cells with constructs expressing several mutant β-glucocerebrosidase enzymes, including N370S, S364R, V15M, and M123T had 1.5- to 9.9-fold increases in activity when incubated with 10 μM *N*B-DNJ, whereas the iminosugar did not rescue the activities of the mutant L444P, L336P, and S465del β-glucocerebrosidases, suggesting that *N*B-DNJ was both a substrate deprivation molecule for glycolipid synthesis and a pharmacologic chaperone for β-glucocerebrosidase. Thus, several substrate analogues for β-glucocerebrosidase have been shown to serve as pharmacologic chaperones for various mutant enzymes. The prevalence of the type 1 subtype, and particularly the frequency of the N370S mutation, makes Gaucher disease an excellent candidate disease. Moreover, if a pharmacologic chaperone can be identified to rescue the L444P mutant protein, this therapeutic approach might stabilize or reverse the neurologic manifestations in patients with type 3 disease.

GM1 Gangliosidosis (Deficient Acid β-galactosidase Activity)

This autosomal recessive disorder results from the deficient activity of the lysosomal exogalactosidase, acid β-galactosidase. There are at least three major subtypes, all of which have neurologic involvement due to the accumulation of GM1 and related ganglioside substrates.[58] The infantile form is a rapidly progressive neuro-degenerative disease with demise in the first years of life. The juvenile- and adult-onset forms have residual β-galactosidase activity but at such low levels that the substrate(s) accumulates, although to a lesser degree in the adult-onset form. To date, over 50 mutations have been identified in the β-galactosidase gene (Human Gene Mutation Database),[41] and genotype/phenotype correlations have been proposed.[58] The crystal structure of the human enzyme has not been resolved.

Suzuki and colleagues reported the rescue of mutant β-galactosidase activity in deficient human cultured fibroblasts and in a transgenic mouse model of GM1 Gangliosidosis.[18] Using the galactose derivative, N-octyl-4-epi-β-valienamine (NOEV), they demonstrated increased β-galactosidase activity in cultured fibroblasts from unrelated patients with GM1 Gangliosidosis. Based on these in vitro studies, they introduced a rescuable mutation, R201C, into β-galactosidase null mice. These transgenic mice were given NOEV ad libitum in their drinking water for 1 week (~1.4 mg/day) and then sacrificed. Increased β-galactosidase activity was observed in all tissues, including the brain, as determined by activity assays and histochemical staining. In addition, the accumulated ganglioside substrates, GM1 and GA1 were decreased in the brain as shown by immunostaining. These studies indicated that the orally administered pharmacologic chaperone gained access to the brain, rescued the newly synthesized β-galactosidase mutant enzyme, and decreased the already accumulated substrates in brain lysosomes.

These studies demonstrate that small molecular ligands can rescue a variety of missense mutations and that pharmacologic chaperones can be administered orally and may readily cross the blood-brain barrier, making this approach particularly attractive.

Tay-Sachs and Sandhoff Diseases (Deficient β-hexosaminidase A and β-hexosaminidase AB Activities, Respectively)

Tay-Sachs disease results from the deficient activity of β-hexosaminidase A, and Sandhoff disease results from the deficient activities of both β-hexosaminidase A and B isozymes.[59] Both diseases are autosomal recessive disorders. β-Hexosaminidase A is a heterodimer with α and β subunits, each encoded by a different gene. β-hexosaminidase B is a homodimer of β subunits. Mutations in the α-subunit gene cause β-hexosaminidase A deficiency in Tay-Sachs disease, whereas mutations in the common β-subunit gene result in the deficient activity of both isozymes.[59] There are three major subtypes of Tay-Sachs disease, the infantile form being a severe neurodegenerative disorder with demise typically by 5 years, a juvenile-onset neurodegenerative disease, and a chronic form characterized primarily by progressive ataxia and muscle weakness beginning in childhood or adolescence and later-onset mild to moderate intellectual impairment. The infantile-onset subtype is prevalent in the Ashkenazi Jewish population with a carrier frequency of about 1 in 25. The less-frequent later-onset form also occurs primarily in individuals of Ashkenazi Jewish descent. To date, over 100 α-chain mutations have been identified that cause Tay-Sachs disease (Human Gene Mutation Database).[41] Two common Jewish mutations cause the infantile form, a four base exonic insertion and a splicing defect, both of which result in no enzyme protein and the infantile phenotype. In contrast, most later-onset forms result from a common missense mutation, G269S.[59,60]

Sandhoff disease also has three major subtypes, the infantile-, juvenile- and adult-onset phenotypes.[59] The disease is rare and is not common to any particular population. To date, over 30 β-chain mutations have been identified, and like the respective Tay-Sachs phenotypes, infantile-onset Sandhoff disease is caused by mutations that result in no enzyme protein or activity. In contrast, the juvenile- and adult-onset phenotypes are caused by mutations that have residual activity of <5% of wildtype.[59]

Tropak and colleagues evaluated various known β-hexosaminidase inhibitors to identify a pharmacologic chaperone that could increase β-hexosaminidase α-chain mutant G269S residual enzymatic activity.[20] Of the seven inhibitors tested, N-acetylglucosamine-thiazoline (NGT), specifically and most effectively increased the amount of mature α-chain glycopolypeptide and β-hexosaminidase activity in the lysosomal subcellular fraction of cultured fibroblasts from a later-onset patient who was homoallelic for the G269S α-chain mutation. Increasing concentrations of NGT in the growth media resulted in higher activities of residual activity with days in culture and was not toxic to the cells. Similarly, NGT was able to enhance the β-hexosaminidase β-chain P504S mutant enzymatic activity in fibroblasts from a patient with adult-onset Sandhoff disease. Thus, these studies demonstrated that NGT bound to and stabilized the mutant enzymes in the ER, resulting in increased amounts of active mutant enzymes that exited the ER and were transported to the lysosomes, where they could function at their preferred acidic pH.

CLINICAL PROOF OF CONCEPT: PHARMACOLOGIC CHAPERONE THERAPY FOR FABRY DISEASE

The clinical efficacy of pharmacologic chaperones for lysosomal disorders has been investigated in the "cardiac variant" of Fabry disease. These patients have residual α-Gal A activity and a later-onset phenotype.[36,38,39] As proof of the concept, a cardiac variant for Fabry disease who had severe heart disease and was a candidate for cardiac transplantation was treated with galactose, a reversible competitive inhibitor of α-Gal A. Galactose is not as effective an inhibitor as DGJ, so 1 gm/kg was administered intravenously three times weekly and the infusions were well tolerated. The biochemical, histologic, and clinical effects of the infusions were monitored at 3 months and 2 years of therapy.[22] After 3 months, there was evidence of improvement, and after 2 years of continuous treatment, there was marked improvement in cardiac contractility (an increase in the left ventricular ejection fraction), a moderate reduction in ventricular-wall thickness, and a reduction in cardiac mass (Table 20.2). These improvements, which persisted for more than 3 years, were confirmed by the findings of independent observers, by two-dimensional echocardiography, and by cardiac MRI studies. Cardiac transplantation was no longer required in this patient, because of the clinical improvement (from New York Heart Association functional

Table 20.2 Clinical Effect of Pharmacologic Chaperone Therapy in a
Patient with the Later-Onset Cardiac Variant of Fabry Disease

	Treatment		
	Before	3 months	2 years
Echocardiographic data:			
LV end-diastolic diameter (mm)	68	59	58
LV end-systolic diameter (mm)	53	45	44
Interventricular septum thickness (mm)	20	16	16
LV posterior wall thickness (mm)	16	14	14
Shortening fraction (%)	22	24	24
LV ejection fraction (%)	33	55	55
Hemodynamic data:			
LV end-diastolic pressure (mm Hg)	25	14	—
Cardiac index (liters/min/m^2)	1.7	2.9	—
Cardiac MRI data:			
Mean LV wall thickness (mm)	18	15	14
Mean RV wall thickness (mm)	10	7	6
LV mass (g)	293	235	228
LV ejection fraction (%)	32	51	55

Source: Frustaci, A. et al., N Engl J Med, 345, 25, 2001. With permission.

class IV to class I). The galactose served as a competitive inhibitor that could bind
to the active site and rescue the mutant enzyme (Figure 20.1), thereby promoting
the proper folding, dimerization, and processing of the mutant enzyme, and preventing the proteasomal degradation of misfolded, mutant enzyme glycopeptides. Thus,
for patients with the later-onset cardiac variant of Fabry disease whose residual α-
Gal A activity can be enhanced *in vitro*, active site-specific pharmacologic chaperone
therapy may prove safe and therapeutically effective. As noted below, a Phase 2
clinical trial is underway sponsored by Amicus Therapeutics, Inc.

CRITERIA FOR SELECTION OF PHARMACOLOGIC CHAPERONES

Implicit in the selection of a pharmacologic chaperone for the treatment of
patients with a lysosomal or other disease is the demonstration that the compound
is nontoxic, readily eliminated in the urine so it or its metabolites do not accumulate,
and optimally rescues a variety of disease-causing mutations. In addition, molecules
that could function as pharmacologic chaperones should be cell permeable and have
the ability to reach the ER and remain undegraded/unmetabolized long enough to
stabilize the target mutant protein and transport it to the lysosome.[12] For neurodegenerative disorders, the compound must be able to reach sufficient concentration
in the brain. Finally, clinical studies must demonstrate long-term safety, biochemical
effectiveness (increased enzymatic activity and/or substrate clearance) and clinical
benefit without adverse reactions.

Table 20.3 Some Experimental Studies of Pharmacological Chaperones in Nonlysosomal Disorders

Disease	Defective Protein	Reference
Enzymes:		
Retinitis pigmentosa 17	Carbonic anhydrase IV	71
Channel Proteins:		
Cystic fibrosis	CFTR	34,72
Long QT syndrome	HERG K+ channel	73,74
Receptor Defects:		
Nephrogenic diabetes insipidus	V2 vassopressin receptor	63
Retinitis pigmenosa (auto dom)	Rhodopsin	62,75,76
Familial hypercholesterolemia	LDL receptor	77
Hypogonatropic hypodonadism	Gonadotropin releasing hormone receptor	78,79
Autosomal dominant nocturnal frontal lobe epilepsy	{alpha}4{beta}2 AChRs	80
	Delta-Opioid receptor	79
DNA transcriptional protein:		
Cancer	p53	81
Conformational diseases:		
ATT deficiency	α1-antitrypsin	82
Amyloidoses	Transthyetin	83

Source: Based on Ulloa-Aguirre, A. et al., *Traffic*, 5, 821, 2004; Fan, J.Q., *Trends Pharmacol Sci*, 24, 355, 2003; and Desnick, R.J. and Schuchman, E.H., *Nat Rev Genet*, 3, 954, 2002. With permission.

PHARMACOLOGIC CHAPERONE THERAPY FOR NONLYSOSOMAL DISEASES

Pharmacologic chaperones have been sought and/or evaluated for a variety of nonlysosomal diseases (Table 20.3). These include disorders such as Alzheimer,s disease[61] and retinitis pigmentosa,[62] where there is mutant protein misfolding and aggregation and in other diseases such as cystic fibrosis,[62] and nephrogenic diabetes insipidus[63] where there are also trafficking defects as well as misfolding of mutant proteins. Pharmacologic chaperones have been the subject of recent reviews.[10,12,33,64–69]

CLINICAL TRIALS OF PHARMACOLOGIC CHAPERONE THERAPY

Efforts are underway to develop pharmacologic chaperone therapy for Fabry disease based on the preclinical studies in Fabry cultured fibroblasts and lympho-blasts,[15] the safety and effectiveness in the transgenic R301Q mice,[70] and the clinical proof of concept in the later-onset cardiac variant of Fabry disease.[22] Under the sponsorship of Amicus Therapeutics, Inc., a Phase 1 safety study was conduced in normal individuals that established the pharmacologic chaperone AT1001's safety in humans and also demonstrated that the rescue of normal α-Gal A activity was dose-responsive (Amicus Therapeutics; http://www.amicustherapeutics.com). Based on these results, a Phase 2 clinical trial is currently under way (http://www.clinical-trials.gov/ct/show /NCT00214500?order=1). This clinical trial will assess the extent to which an active-site directed pharmacologic chaperone will increase the residual activity of various mutant enzymes and determine their dose-response relationships. Such studies may also establish a threshold of activity that must be achieved to stabilize and/or reverse the substrate accumulation in different cell types, and ameliorate or modify various clinical manifestations. If pharmacologic chaperone therapy proves clinically effective for patients with Fabry disease who have residual α-Gal A activity, investigators will be stimulated to carry out clinical trials of pharmacologic chaperone therapy in other lysosomal diseases, particularly those with neurologic manifestations.

CONCLUSIONS AND FUTURE DIRECTIONS

The past 5 years have seen remarkable changes for scientists and clinicians working on lysosomal diseases. The "era" of lysosomal disease treatment has led to a new awareness of these disorders and renewed excitement among researchers and patients. Enzyme replacement therapy is available for four disorders (Gaucher and Fabry diseases, and MPS I and VI), and under development for several others. Experimental and clinical experience with other approaches including bone marrow transplantation, substrate deprivation, and gene and stem cell therapies have identified their useful applications and limitations.[1-4] However, despite these remarkable advances, significant hurdles still remain before the goal of treating all or most lysosomal diseases can be realized. Primary among them is the need to treat the neurodegenerative lysosomal diseases. Pharmacologic chaperone therapy represents an attractive new strategy that uses small molecules that can be administered orally to rescue misfolded/mistargeted lysosomal enzymes (or other proteins), thereby resulting in their increased activity and therapeutic benefit. It is likely that pharmacologic chaperone therapy strategies will have a significant therapeutic effect, as recently shown for the cardiac variant of Fabry disease.[22] In addition, pharmacologic chaperone therapy offers the possibility of treating disorders with neurologic involvement, since small molecular weight pharmacologic chaperones may be designed to cross the blood-brain barrier. Of note, they may have markedly different biodistributions than replaced recombinant human enzymes whose tissue delivery is primarily dependent on their respective receptor-mediated delivery and do not cross the blood-brain barrier. Thus, for certain diseases, combined pharmacologic chaperone and enzyme replacement therapies may be beneficial, particularly for the treatment of neurodegenerative lysosomal and other diseases of the central nervous system. In summary, it is anticipated that the early years of the 21st century will witness the development of safe and effective pharmacologic chaperone therapies for patients suffering from lysosomal and other inborn errors of metabolism.

ACKNOWLEDGMENTS

The authors thank Dr. Kenneth H. Astrin for assistance with the manuscript. This work was supported in part by grants from the National Institutes of Health, including a research grant (5 R37 DK34045 Merit Award) and a grant (5 M01 RR00071) for the Mount Sinai General Clinical Research Center from the National Center for Research Resources. Dr. Fan is an inventor of patents licensed to Amicus Therapeutics, Inc., and Drs. Desnick and Fan serve as consultants and hold stock options in the company. Dr. Desnick is a consultant and receives a research grant from the Genzyme Corporation and is an inventor of patents licensed to the company.

REFERENCES

1. Treacy, E.P., Valle, D., and Scriver, C.R., Treatment of Genetic Disease, in *The Metabolic and Molecular Bases of Inherited Diseases*, Scriver CR et al., Eds., McGraw-Hill, New York, 2001, 175.
2. Schiffmann, R. and Brady, R.O., New prospects for the treatment of lysosomal storage diseases, *Drugs*, 62, 733, 2002.
3. Ellinwood, N.M., Vite, C.H., and Haskins, M.E., Gene therapy for lysosomal storage diseases: the lessons and promise of animal models, *J Gene Med*, 6, 481, 2004.
4. Eto, Y. et al., Treatment of lysosomal storage disorders: cell therapy and gene therapy, *J Inherit Metab Dis*, 27, 411, 2004.
5. de Duve, C., From cytases to lysosomes, *Fed. Proc*, 23, 1045, 1964.
6. Hobbs, J.R. et al., Reversal of clinical features of Hurler's disease and biochemical improvement after treatment by bone-marrow transplantation, *Lancet*, 2, 709, 1981.
7. Krivit, W., Peters, C., and Shapiro, E.G., Bone marrow transplantation as effective treatment of central nervous system disease in globoid cell leukodystrophy, metachromatic leukodystrophy, adrenoleukodystrophy, mannosidosis, fucosidosis, aspartylglucosaminuria, Hurler, Maroteaux-Lamy, and Sly syndromes, and Gaucher disease type III, *Curr Opin Neurol*, 12, 167, 1999.
8. Platt, F.M. and Butters, T.D., New therapeutic prospects for the glycosphingolipid lysosomal storage diseases, *Biochem Pharmacol*, 56, 421, 1998.
9. Platt, F.M. et al., Inhibition of substrate synthesis as a strategy for glycolipid lysosomal storage disease therapy, *J Inherit Metab Dis*, 24, 275, 2001.
10. Perlmutter, D.H., Chemical chaperones: a pharmacological strategy for disorders of protein folding and trafficking, *Pediatr Res*, 52, 832, 2002.
11. Cohen, F.E. and Kelly, J.W., Therapeutic approaches to protein-misfolding diseases, *Nature*, 426, 905, 2003.
12. Ulloa-Aguirre, A. et al., Pharmacologic rescue of conformationally-defective proteins: implications for the treatment of human disease, *Traffic*, 5, 821, 2004.
13. Hampton, R., ER-associated degradation in protein quality control and cellular regulation, *Curr Opin Cell Biol*, 14, 476, 2002.
14. Meusser, B. et al., ERAD: the long road to destruction, *Nat Cell Biol*, 7, 766, 2005.
15. Fan, J.Q. et al., Accelerated transport and maturation of lysosomal alpha-galactosidase A in Fabry lymphoblasts by an enzyme inhibitor, *Nat Med*, 5, 112, 1999.
16. Asano, N. et al., *In vitro* inhibition and intracellular enhancement of lysosomal α-galactosidase A activity in Fabry lymphoblasts by 1-deoxygalactonojirimycin and its derivatives, *Eur J Biochem*, 267, 4179, 2000.

17. Sawkar, A.R. et al., Chemical chaperones increase the cellular activity of N370S beta-glucosidase: a therapeutic strategy for Gaucher disease, *Proc Natl Acad Sci U.S.A.*, 99, 15428, 2002.

18. Matsuda, J. et al., Chemical chaperone therapy for brain pathology in GM1-gangliosidosis, *Proc Natl Acad Sci U.S.A.*, 100, 15912, 2003.

19. Lin, H. et al., N-octyl-beta-valienamine up-regulates activity of F213I mutant beta-glucosidase in cultured cells: a potential chemical chaperone therapy for Gaucher disease, *Biochim Biophys Acta*, 1689, 219, 2004.

20. Tropak, M.B. et al., Pharmacological enhancement of beta-hexosaminidase activity in fibroblasts from adult Tay-Sachs and Sandhoff Patients, *J Biol Chem*, 279, 13478, 2004.

21. Alfonso, P. et al., Miglustat (NB-DNJ) works as a chaperone for mutated acid beta-glucosidase in cells transfected with several Gaucher disease mutations, *Blood Cells Mol Dis*, 35, 268, 2005.

22. Frustaci, A. et al., Improvement in cardiac function in the cardiac variant of Fabry's disease with galactose-infusion therapy, *N Engl J Med*, 345, 25, 2001.

23. Anfinsen, C.B. and Scheraga, H.A., Experimental and theoretical aspects of protein folding, *Adv Protein Chem*, 29, 205, 1975.

24. Ellgaard, L., Molinari, M., and Helenius, A., Setting the standards: quality control in the secretory pathway, *Science*, 286, 1882, 1999.

25. Sabatini, D.D. and Adesnik, M.B., The biogenesis of membranes and organelles, in *The Metabolic and Molecular Bases of Inherited Diseases*, Scriver CR et al., Eds., McGraw-Hill, New York, 2001, 475.

26. Hartl, F.U. and Hayer-Hartl, M., Molecular chaperones in the cytosol: from nascent chain to folded protein, *Science*, 295, 1852, 2002.

27. Fewell, S.W. et al., The action of molecular chaperones in the early secretory pathway, *Annu Rev Genet*, 35, 149, 2001.

28. Schubert, U. et al., Rapid degradation of a large fraction of newly synthesized proteins by proteasomes, *Nature*, 404, 770, 2000.

29. Princiotta, M.F. et al., Quantitating protein synthesis, degradation, and endogenous antigen processing, *Immunity*, 18, 343, 2003.

30. Oda, Y. et al., EDEM as an acceptor of terminally misfolded glycoproteins released from calnexin, *Science*, 299, 1394, 2003.

31. Yewdell, J.W., Not such a dismal science: the economics of protein synthesis, folding, degradation and antigen processing, *Trends Cell Biol*, 11, 294, 2001.

32. Horwich, A., Protein aggregation in disease: a role for folding intermediates forming specific multimeric interactions, *J Clin Invest*, 110, 1221, 2002.

33. Fan, J.Q., A contradictory treatment for lysosomal storage disorders: inhibitors enhance mutant enzyme activity, *Trends Pharmacol Sci*, 24, 355, 2003.

34. Pedemonte, N. et al., Small-molecule correctors of defective DeltaF508-CFTR cellular processing identified by high-throughput screening, *J Clin Invest*, 115, 2564, 2005.

35. Brady, R.O. et al., Enzymatic defect in Fabry's disease. Ceramidetrihexosidase deficiency, *N Engl J Med*, 276, 1163, 1967.

36. Desnick, R.J., Ioannou, Y.A., and Eng, C.M., α-Galactosidase A deficiency: Fabry disease, in *The Metabolic and Molecular Bases of Inherited Diseases*, 8th ed., Scriver CR et al., Eds., McGraw-Hill, New York, 2001, 3733.

37. Elleder, M. et al., Cardiocyte storage and hypertrophy as a sole manifestation of Fabry's disease. Report on a case simulating hypertrophic non-obstructive cardiomyopathy, *Virchows Arch A Pathol Anat Histopathol*, 417, 449, 1990.

38. von Scheidt, W. et al., An atypical variant of Fabry's disease with manifestations confined to the myocardium, *N Engl J Med*, 324, 395, 1991.
39. Nakao, S. et al., An atypical variant of Fabry's disease in men with left ventricular hypertrophy, *N Engl J Med*, 333, 288, 1995.
40. Nakao, S. et al., Fabry disease: Detection of undiagnosed hemodialysis patients and identification of a "renal variant" phenotype, *Kidney Int*, 64, 801, 2003.
41. Stenson, P.D. et al., Human Gene Mutation Database (HGMD): 2003 update, *Hum Mutat*, 21, 577, 2003.
42. Cooper, D.N. and Youssoufian, H., The CpG dinucleotide and human genetic disease, *Hum Genet*, 78, 151, 1988.
43. Kondrashov, A.S., Direct estimates of human per nucleotide mutation rates at 20 loci causing Mendelian diseases, *Hum Mutat*, 21, 12, 2003.
44. Garman, S. and Garboczi, D., Structural basis of Fabry disease, *Mol Genet Metab*, 77, 3, 2002.
45. Garman, S.C. and Garboczi, D.N., The molecular defect leading to Fabry disease: structure of human alpha-galactosidase, *J Mol Biol*, 337, 319, 2004.
46. Matsuzawa, F. et al., Fabry disease: correlation between structural changes in alpha-galactosidase, and clinical and biochemical phenotypes, *Hum Genet*, 117, 317, 2005.
47. Yasuda, M. et al., Fabry disease: characterization of alpha-galactosidase A double mutations and the D313Y plasma enzyme pseudodeficiency allele, *Hum Mutat*, 22, 486, 2003.
48. Dominissini, S. et al., Comparative *in vitro* expression study of four Fabry disease causing mutations at glutamine 279 of the alpha-galactosidase A protein, *Hum Hered*, 57, 138, 2004.
49. Ishii, S. et al., Alternative splicing in the alpha-galactosidase A gene: increased exon inclusion results in the Fabry cardiac phenotype, *Am J Hum Genet*, 70, 994, 2002.
50. Okumiya, T. et al., Galactose stabilizes various missense mutants of alpha-galactosidase in Fabry disease, *Biochem Biophys Res Commun*, 214, 1219, 1995.
51. Yam, G.H., Zuber, C., and Roth, J., A synthetic chaperone corrects the trafficking defect and disease phenotype in a protein misfolding disorder, *FASEB J*, 19, 12, 2005.
52. Brady, R.O., Kanfer, J.N., and Shapiro, D., Metabolism of Glucocerebrosides. Ii. Evidence of an Enzymatic Deficiency in Gaucher's Disease, *Biochem Biophys Res Commun*, 18, 221, 1965.
53. Beutler, E. and Grabowski, G., Gaucher Disease, in *The Metabolic and Molecular Bases of Inherited Disease*, Scriver CR et al., Eds., McGraw-Hill, New York, 2001, 3635.
54. Dvir, H. et al., X-ray structure of human acid beta-glucosidase, the defective enzyme in Gaucher disease, *EMBO Rep*, 4, 704, 2003.
55. Whitfield, P.D. et al., Correlation among genotype, phenotype, and biochemical markers in Gaucher disease: implications for the prediction of disease severity, *Mol Genet Metab*, 75, 46, 2002.
56. Brautbar, A. et al., The 1604A (R496H) mutation in Gaucher disease: genotype/phenotype correlation, *Blood Cells Mol Dis*, 31, 187, 2003.
57. Ron, I. and Horowitz, M., ER retention and degradation as the molecular basis underlying Gaucher disease heterogeneity, *Hum Mol Genet*, 14, 2387, 2005.
58. Suzuki, Y., Oshima, A., and Nanba, E., β-Galactosidase deficiency (β-galactosidosis): GM1 gangliosidosis and Morquio B disease, in *The Metabolic and Molecular Bases of Inherited Diseases*, Scriver CR et al., Eds., McGraw-Hill, New York, 2001, 3775.
59. Gravel, R.A. et al., The GM2 Gangliosidoses, in *The Metabolic and Molecular Bases of Inherited Diseases*, Scriver CR et al., Eds., McGraw-Hill, New York, 2001, 3827.

60. Kaback, M. et al., Tay-Sachs disease — carrier screening, prenatal diagnosis, and the molecular era. An international perspective, 1970 to 1993. The International TSD Data Collection Network, *JAMA*, 270, 2307, 1993.

61. Sacchettini, J.C. and Kelly, J.W., Therapeutic strategies for human amyloid diseases, *Nat Rev Drug Discov*, 1, 267, 2002.

62. Noorwez, S.M. et al., Retinoids assist the cellular folding of the autosomal dominant retinitis pigmentosa opsin mutant P23H, *J Biol Chem*, 279, 16278, 2004.

63. Morello, J.P. et al., Pharmacological chaperones rescue cell-surface expression and function of misfolded V2 vasopressin receptor mutants, *J Clin Invest*, 105, 887, 2000.

64. Welch, W.J. and Brown, C.R., Influence of molecular and chemical chaperones on protein folding, *Cell Stress Chaperones*, 1, 109, 1996.

65. Kuznetsov, G. and Nigam, S.K., Folding of secretory and membrane proteins, *N Engl J Med*, 339, 1688, 1998.

66. Morello, J.P. et al., Pharmacological chaperones: a new twist on receptor folding, *Trends Pharmacol Sci*, 21, 466, 2000.

67. Gregersen, N. et al., Defective folding and rapid degradation of mutant proteins is a common disease mechanism in genetic disorders, *J Inherit Metab Dis*, 23, 441, 2000.

68. Carrell, R.W. and Lomas, D.A., Alpha1-antitrypsin deficiency — a model for conformational diseases, *N Engl J Med*, 346, 45, 2002.

69. Desnick, R.J. and Schuchman, E.H., Enzyme replacement and enhancement therapies: lessons from lysosomal disorders, *Nat Rev Genet*, 3, 954, 2002.

70. Ishii, S. et al., Transgenic mouse expressing human mutant alpha-galactosidase A in an endogenous enzyme deficient background: a biochemical animal model for studying active-site specific chaperone therapy for Fabry disease, *Biochim Biophys Acta*, 1690, 250, 2004.

71. Bonapace, G. et al., Chemical chaperones protect from effects of apoptosis-inducing mutation in carbonic anhydrase IV identified in retinitis pigmentosa 17, *Proc Natl Acad Sci U.S.A.*, 101, 12300, 2004.

72. Dormer, R.L. et al., Correction of delF508-CFTR activity with benzo(c)quinolizinium compounds through facilitation of its processing in cystic fibrosis airway cells, *J Cell Sci*, 114, 4073, 2001.

73. Zhou, Z., Gong, Q. and January, C.T., Correction of defective protein trafficking of a mutant HERG potassium channel in human long QT syndrome. Pharmacological and temperature effects, *J Biol Chem*, 274, 31123, 1999.

74. Ficker, E. et al., The binding site for channel blockers that rescue misprocessed human long QT syndrome type 2 ether-a-gogo-related gene (HERG) mutations, *J Biol Chem*, 277, 4989, 2002.

75. Saliba, R.S. et al., The cellular fate of mutant rhodopsin: quality control, degradation and aggresome formation, *J Cell Sci*, 115, 2907, 2002.

76. Noorwez, S.M. et al., Pharmacological chaperone-mediated *in vivo* folding and stabilization of the P23H-opsin mutant associated with autosomal dominant retinitis pigmentosa, *J Biol Chem*, 278, 14442, 2003.

77. Li, Y. et al., Receptor-associated protein facilitates proper folding and maturation of the low-density lipoprotein receptor and its class 2 mutants, *Biochemistry*, 41, 4921, 2002.

78. Janovick, J.A., Maya-Nunez, G. and Conn, P.M., Rescue of hypogonadotropic hypogonadism-causing and manufactured GnRH receptor mutants by a specific protein-folding template: misrouted proteins as a novel disease etiology and therapeutic target, *J Clin Endocrinol Metab*, 87, 3255, 2002.

79. Petaja-Repo, U.E. et al., Ligands act as pharmacological chaperones and increase the efficiency of delta opioid receptor maturation, *EMBO J*, 21, 1628, 2002.
80. Kuryatov, A. et al., Nicotine acts as a pharmacological chaperone to upregulate human {alpha}4{beta}2 AChRs, *Mol Pharmacol*, epub, 2005.
81. Foster, B.A. et al., Pharmacological rescue of mutant p53 conformation and function, *Science*, 286, 2507, 1999.
82. Mahadeva, R. et al., 6-mer peptide selectively anneals to a pathogenic serpin conformation and blocks polymerization. Implications for the prevention of Z alpha(1)-antitrypsin-related cirrhosis, *J Biol Chem*, 277, 2002.
83. Hammarstrom, P. et al., Prevention of transthyretin amyloid disease by changing protein misfolding energetics, *Science*, 299, 713, 2003.
84. Sawkar, A.R. et al., Gaucher disease-associated glucocerebrosidases show mutation-dependent chemical chaperoning profiles, *Chem Biol*, 12, 1235, 2005.

CHAPTER 21

The Significance of the Blood-Brain Barrier for Gaucher Disease and Other Lysosomal Storage Diseases

David J. Begley

CONTENTS

INTRODUCTION

Of the 50 or more lysosomal storage diseases so far described in the literature, the majority are neuronopathic and involve neurological changes, degeneration, and damage, the severity of which is variable. Of the lysosomal storage diseases, Gaucher disease is the most common and may be divided into types I, II, and III. Both types II and III are neuronopathic with involvement of the central nervous system (CNS). In type II Gaucher disease neurodegeneration is rapid and severe and survival times are generally short (<2 years), whereas in type III the progression is much slower with patients surviving into adulthood [1]. It is intriguing that, although the defective enzyme β-glucosidase (glucocerebrosidase/glucosylceramidase) is the same, albeit with different mutation in the same gene, in Gaucher types 1, 2, and 3, only types 2 and 3 have neuronopathic involvement [1,2]. This suggests that other factors may be involved with the development of neuronopathy in these cases, but these remain obscure [2].

With the introduction of enzyme-replacement therapy (ERT) in the early 1990s and the intravenous infusion of mannose-terminated glucocerebrosidase both the quality of life and its expectancy have improved for many patients. Unfortunately the replacement enzyme with a molecular weight of circa 60 kDa does not cross the blood-brain barrier (BBB), and therefore enzyme-replacement for the brain has been difficult to achieve. An increase in life expectancy, coupled with the failure of ERT to reach the brain [1–4], may possibly allow a further progression in the severity of neurodegeneration and unmask additional neuronopathic pathology, especially in older type 1 patients. An understanding of the BBB and its role in the etiology and treatment of Gaucher and other lysosomal storage diseases is imperative. The structure and function of the BBB has been recently reviewed in detail [5,6].

THE BLOOD-BRAIN BARRIER

All organisms with a well-developed central nervous system (CNS) have a blood-brain barrier (BBB). In mammals and man this barrier is created by the endothelial cells of the brain and spinal cord microvasculature, which constitute by far the largest surface area for exchange (at least 12 m^2 in the human adult) between blood and brain. In addition, the epithelial cells of the choroid plexus facing the cerebrospinal fluid constitute the blood-cerebrospinal fluid barrier, (BCSFB). Finally the avascular arachnoid membrane, underlying the dura, which completely encloses the CNS, completes the seal between the extracellular fluid of the CNS and the rest of the body. The relationship of these barriers within the brain is shown in Figure 21.1 [7] and the detailed structure of the cell association of the blood-brain barrier is shown in Figure 21.2 [8].

The blood-brain barrier is necessary as the central nervous system needs to maintain an extremely stable internal fluid environment, which is an absolute requirement for reliable synaptic communication between nerve cells. The quantal chemical communication that takes place across synapses and the complex neural spatial and temporal summation and integration of signals that occurs between neurons could not occur efficiently if brain extracellular fluid were allowed to fluctuate as markedly as does the somatic extracellular fluid. The blood plasma for instance contains high levels of the neuroexcitatory amino acids glutamate and glycine, which fluctuate significantly after the ingestion of food. If these excitatory neurotransmitters are released into the brain in an uncontrolled manner, as for example from neurons during ischemic stroke, considerable permanent neurotoxic/neuroexcitatory damage can occur.

In this sense, the blood-brain barrier is a protective barrier that shields the CNS from neurotoxic substances that circulate in the blood. These neurotoxins may be naturally occurring metabolites or xenobiotic substances that are ingested in the diet or otherwise acquired from the environment. Fully differentiated neurons are not

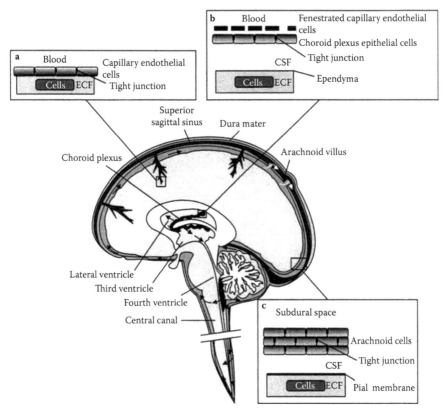

Figure 21.1 Barriers of the brain: There are three principal barrier sites between blood and brain. (a) Firstly there is the BBB proper, which is created at the level of the cerebral capillary endothelial cells by tight junction formation. It is by far the largest surface area for exchange and in the adult human is between 12 and 18 m^2 in surface area. No brain cell is further than about 25 μm from a capillary, so once the BBB is crossed, diffusion distances to the target for drugs are short. Targeting a drug across the BBB is therefore the favored route for global delivery of drug to all brain cells. (b) The blood-CSF barrier (BCSFB) lies at the choroid plexi in the lateral and fourth ventricles of the brain and tight junctions are formed between the epithelial cells at the CSF-facing surface of the plexi. In the human, because of the villous surface of the choroid plexi, the surface area of the BCSFB is approximately 10% of that of the BBB. Some drugs and solutes enter the brain principally across the choroid plexi into CSF; others enter via the BBB and BCSFB. (c) The arachnoid barrier. The brain is enclosed by the arachnoid membrane lying under the dura. The arachnoid membrane is avascular but lies close to the superior sagital sinus and is separated from it by the dura. The cells of the arachnoid membrane have tight junctions forming an effective seal. Arachnoid villi project into the sagital sinus through the dura and a significant amount of CSF drains into the sinus through these valve-like villi, which only allow CSF movement out of the brain to blood. Transport across the arachnoid membrane is not an important route for the entry of solutes into brain. (Adapted from Kandel, E.R., Schwartz, J.H., and Jessel, T.M., *Principles of Neural Science*, 4th ed. New York, McGraw-Hill, 2000, 1294. With permission.)

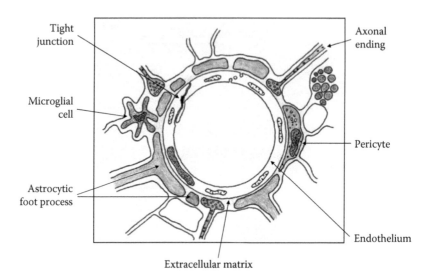

Tight
junction

Axonal
ending

Microglial
cell

Pericyte

Astrocytic
foot process

Endothelium

Extracellular matrix

Figure 21.2 The neurovascular unit (cell association) forming the BBB: The cerebral endothelial cells form tight junctions at their margins, which completely seal the
aqueous paracellular diffusional pathway between the cells. Pericytes are distributed discontinuously along the length of the cerebral capillaries and partially
surround the endothelium. Both the cerebral endothelial cells and the pericytes
are surrounded by and contribute to their local extracellular matrix, which forms
a perivascualar matrix, distinct in its composition from the extracellular matrix of
the brain parenchyma. Foot processes from astrocytes form a network completely
surrounding the capillaries, and it is this close association that is thought to
maintain barrier properties. Axon projections from neurons are also applied
closely to the capillary endothelial cells and contain vasoactive neurotransmitters
and peptides. These axonal endings may play a part in modulating BBB permeability on a short-term basis. Microglia (perivascular macrophages) are the resident immunocompetant cells of the brain and are derived from circulating
monocytes and macrophages. The movement of solutes across the BBB is either
passive, driven by a concentration gradient from plasma to brain with more lipid
soluble substances entering most easily, or may be facilitated by passive or active
transporters in the endothelial cell membranes. Efflux transporters in the endothelium may limit the CNS penetration of a wide variety of solutes. (From Begley
2004, with permission.)

able to divide and replace themselves under normal circumstances, and there is a
continuous steady rate of neuronal cell death from birth throughout life in the healthy
human brain. Any acceleration in this rate of cell death resulting from an increased
ingress of neurotoxins into the brain would become prematurely debilitating.

Thus the blood-brain barriers have a dual function; they provide the especially
stable fluid environment that is necessary for complex neural function and protect
the CNS from chemical insult. Typical concentration differences between plasma
and cerebrospinal fluid for some selected solutes are listed in Table 21.1.

Table 21.1 Typical Plasma and Cerebrospinal Fluid Concentrations for Some Selected Solutes[a]

Solute	Units	Plasma	CSF	Ratio
Na^+	mM	140	141	~1
K^+	mM	4.6	2.9	0.63
Ca^{++}	mM	5.0	2.5	0.5
Mg^{++}	mM	1.7	2.4	1.4
Cl^-	mM	101	124	1.23
HCO_3^-	mM	23	21	0.91
Osmolarity	mOsmol	305.2	298.5	~1
pH		7.4	7.3	
Glucose	mM	5.0	3.0	0.6
Total Amino Acid	μM	2890	890	0.31
Leucine	μM	109	10.1–14.9	0.10–0.14
Arginine	μM	80	14.2–21.6	0.18–0.27
Glycine	μM	249	4.7–8.5	0.012–0.034
Alanine	μM	330	23.2–32.7	0.07–0.1
Serine	μM	149	23.5–37.8	0.16–0.25
Glutamic acid	μM	83	1.79–14.7	0.02–0.18
Taurine	μM	78	5.3–6.8	0.07–0.09
Total Protein	mg/ml	70	0.433	0.006
Albumin	mg/ml	42	0.192	0.005
Immunoglobulin G (IgG)	mg/ml	9.87	0.012	0.001
Transferrin	mg/ml	2.6	0.014	0.005
Plasminogen	mg/ml	0.7	0.000025	0.00004
Fibrinogen	mg/ml	325	0.00275	0.000008
α 2-macroglobulin	mg/ml	3	0.0046	0.0015
Cystatin C	mg/ml	0.001	0.004	4.0

[a] Compiled from various sources, values mostly human.

STRUCTURE AND FUNCTION OF THE BLOOD-BRAIN BARRIER

The BBB to macromolecules and polar solutes is created by the formation of tight junctions between the cerebral endothelial cells, the choroid plexus epithelial cells and the cells of the arachnoid membrane. These tight junctions effectively abolish any aqueous paracellular diffusional pathways between these cells from the blood plasma, or somatic extracellular fluid, to the brain extracellular fluid, which effectively prevents the free diffusion of polar solutes from blood to brain [5]. This barrier to paracellular diffusion would potentially seal the brain off from essential polar nutrients such as glucose and amino acids, and, therefore, the BBB must contain a number of specific transporters to supply the CNS with its requirements for these substances. Examples of BBB solute carriers (SLC transporters) and ATP-binding cassette (ABC) transporters are listed and described in Table 21.2. The formation of tight junctions essentially gives the BBB the properties of a cell membrane, both in terms of the diffusional characteristics imposed by the lipid

Table 21.2 Examples of Transporters Present in the Blood-Brain Barrier

Transporter (SLC/ABC)[a]	Abbreviation and/or Transporter Subtype	BBB Location	Orientation[d]	Example of Endogenous Substrates/ Mechanism
Glucose (SLC)	$GLUT_1$	Luminal Abluminal	Blood to brain	Glucose (Facilitative, bi-directional)
Sodium-dependent glucose transporter (SLC)	$SGLT_1$	Abluminal	Brain to endothelium	glucose
Sodium myoinisitol cotransporter (SLC)	SMIT	Luminal	Blood to endothelium	Myoinisitol (Sodium dependent)
Aminoacid (SLC)	L_1/LNNA	Luminal Abluminal	Blood to brain	glutamine, histidine, isoleucine, leucine, methionine, phenylalanine, tryptophan, threonine, tyrosine, valine. (Facilitative, bi-directional)
Amino acid (SLC)	y^+	Luminal Abluminal	Blood to brain	arginine, lysine, ornithine (Facilitative, bi-directional)
Amino acid (SLC)	A	Abluminal	Brain to endothelium	alaninine, glutamine, glycine, proline (Sodium dependent)
Amino acid (SLC)	ASC	Abluminal	Brain to endothelium	alanine, leucine (Sodium dependent)
Amino acid (SLC)	$B^{o,+}$	Abluminal	Brain to endothelium	alanine, serine, cysteine (Sodium dependent)
Amino acid (SLC)	X^-_{AG}	Luminal Abluminal	Blood to endothelium Brain to endothelium	glutamate, aspartate (Sodium dependent)
Amino acid (SLC)	β	Luminal Abluminal		taurine, -alanine (Sodium dependent)
Nucleoside, Nucleotide, nucleobase (SLC)	ENT_1/ENT_2	Luminal	Blood to endothelium	Nucleosides, nucleotides, Nucleobases (Facilitative, equilibrative)
Nucleoside, Nucleotide, nucleobase (SLC)	CNT_1/CNT_2/ CNT_3	Abluminal	Endothelium to brain	Nucleosides, nucleotides, Nucleobases (Sodium dependent exchange)
Monocarboxylic acids (SLC)	MCT_1	Luminal Abluminal	Blood to brain	Ketone bodies
Monocarboxylic acids (SLC)	MCT_2	Abluminal	Brain to endothelium	Lactate (Proton exchanger)

Table 21.2 Examples of Transporters Present in the Blood-Brain Barrier (Continued)

Transporter (SLC/ABC)[a]	Abbreviation and/or Transporter Subtype	BBB Location	Orientation[d]	Example of Endogenous Substrates/ Mechanism
Organic anion transporters (SLC)	OAT$_3$	Luminal	Blood to brain	Dicarboxylate exchange with α-ketoglutarate, bicarbonate, Cl-
Organic anion transporting polypeptide (SLC)°	OATP-B	Luminal Abluminal	Blood to endothelium Endothelium to brain	Organic anion/bicarbonate exchangers
Organic cation transporters (SLC)°	OCT	Luminal	Blood to endothelium	Organic cation/proton exchange
Novel organic cation transporter (SLC)°	OCTN$_1$ OCTN$_2$	Luminal Abluminal[c]	Blood to endothelium Endothelium to brain	Organic cation/proton exchange
P-glycoprotein (ABC)	PgP	Luminal	Endothelium to blood	Many endogenous and therapeutic lipid soluble compounds (ATP driven)
Multidrug resistance-associated proteins (ABC)˙	MRP1 MRP2 MRP3 MRP5	Luminal Abluminal˙	Endothelium to blood and to brain	Organic anions and drugs conjugated to glutathione, sulphate, glucuronate (ATP driven)
Breast cancer related protein (ABC)	BCRP	Luminal	Endothelium to blood	Many endogenous and therapeutic lipid soluble compounds (ATP driven)

SLC — solute carrier transport protein; ABC — ATP-binding cassette transporter.
° Reversible depending on substrate gradients.
˙ Possible species variation in membrane insertion and expression.

In order to transport a substrate across the blood-brain barrier from blood to brain, the transporter must be expressed in both cell membranes and be bidirectional. Alternatively one transporter may carry substrate into the BBB endothelial cells and another out of the cells. (For example, an organic anion may enter the endothelium by OAT become conjugated and leave by means of one of the MRPs). If a transporter is inserted into one membrane only it will transport out of the endothelium or accumulate substrate within the endothelium. Exchangers are driven by the concentration gradient of substrate and exchange ion/molecule and can reverse if the concentration gradient is changed. ATP consuming ABC transporters can transport against a concentration gradient.

bilayer, and the directionality and properties of the specific transport proteins (SLC) that are present in the cell membrane.

In terms of the cerebral endothelial cells, the expression of transporters in the cell membrane is often polarized in the luminal membrane (facing the blood) or the abluminal membrane (facing the brain) so that transport can be directional, either predominantly into, or out of, the CNS. A number of ABC efflux transporters are present in the blood-brain barrier and remove many potentially harmful lipid soluble neurotoxins and reduce their CNS penetration. In addition the activity of ABC transporters substantially reduces the brain penetration and, thus, the efficacy, of a

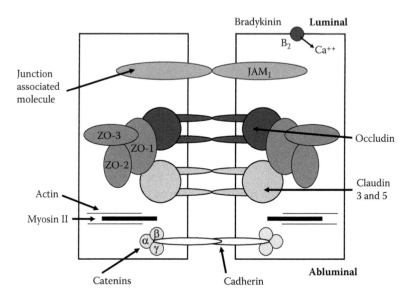

Figure 21.3 Structure of BBB tight junctions: The tight junctional complexes comprise the junction associated molecules (JAM), occludin, claudins 3 and 5 and the cadherin molecules. The cadherins provide structural integrity and attachment between the cells but do not contribute to barrier function. The barrier to diffusion and the high electrical resistance of the BBB appears to be largely due to the properties of claudins 3 and 5 (see text). The claudins and occludin are linked to the scaffolding proteins ZO-1, ZO-2, and ZO-3. These are in turn linked to the actin/myosin cytoskeletal system within the cell. Activation of the actin/myosin system by a rise in free intracellular calcium, for example, arising from ligand binding to the B_2 bradykinin receptor, can withdraw the claudins and occludin from the paracellular space and modify the tight junctional properties. The JAM molecules are members of the immunoglobulin superfamily and are thought to act as cell-adhesion molecules for leukocytes.

considerable number of potentially useful pharmaceutical compounds within the CNS [8]. A summary of the routes across the BBB for cells, solutes and proteins is shown in Figure 21.4 [5].

The cerebral endothelial cells are joined completely around their margins by the tight junctions. These junctional complexes consist of adherens junctions (AJ) where cadherin proteins span the extracellular space and are linked into the cell cytoplasm by the scaffolding proteins alpha, beta, and gamma catenin. These junctional complexes are present between cells and hold the cells together, giving the tissue structural support. They do not appear to participate in the formation of tight junctions (TJ) between the cells. The tight junctions consist of a further complex of proteins spanning the paracellular space and consist of junction associated molecules (JAM) and the proteins occludin and claudin [9] (see Figure 21.3). The occludins and claudins are linked to the cytoplasmic regulatory and scaffolding proteins ZO1, ZO2, ZO3, and cingulin. There are some 20 isoforms of claudin (claudins1–20) [10], and it has been shown that in experimental allergic encephalomyelitis (EAE) and glioblastoma multiforme (GBM) there is a selective loss of claudin-3, but not claudin-5 or occludin, from BBB tight junctions, and this disappearance is associated with

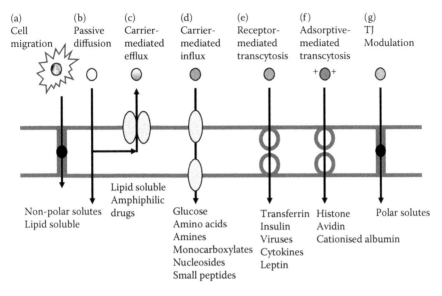

(a)	(b)	(c)	(d)	(e)	(f)	(g)
Cell migration	Passive diffusion	Carrier-mediated efflux	Carrier-mediated influx	Receptor-mediated transcytosis	Adsorptive-mediated transcytosis	TJ Modulation

Non-polar solutes
Lipid soluble

Lipid soluble
Amphiphilic
drugs

Glucose
Amino acids
Amines
Monocarboxylates
Nucleosides
Small peptides

Transferrin
Insulin
Viruses
Cytokines
Leptin

Histone
Avidin
Cationised albumin

Polar solutes

Figure 21.4 Routes of transport across the BBB: (a) Leukocytes cross the BBB by a process of dipedesis through the endothelial cells. They penetrate close to the tight junctional regions, but not through them. The junction-associated molecules (JAM) and the cell surface protein CD99 are thought to interact with the leukocytes to initiate dipedesis. (b) Solutes may passively diffuse through the cell membrane and cross the endothelium. Greater lipid solubility favors this process. (c) Active efflux carriers (ABC transporters) may intercept some of these passively penetrating solutes and pump them out of the endothelial cell. (c) Carrier-mediated influx, which may be passive or secondarily active, can transport many essential polar molecules such as glucose, amino acids and nucleosides into the CNS. (e) RMT can transport macromolecules such as peptides and proteins across the cerebral endothelium. (f) AMT appears to be induced nonspecifically by positively charged macromolecules and can also transport across the endothelium. (g) Tight junction modulation can occur or be induced pharmacologically, which "relaxes" the junctions and wholly or partially opens the paracellular aqueous diffusional pathway. (Modified from Begley, D.J. and Brightman, M.W., *Prog. Drug Res.*, 61, 39, 2003. With permission.)

a loss of BBB integrity. In the case of EAE this is associated with inflammatory events at the BBB. Also mice that have claudin-5 genetically "knocked out" have a severely compromised and leaky blood-brain barrier and die shortly after birth [12]. Therefore it appears clear that the removal of either claudin-3 or claudin-5 from the tight junctional complexes can result in a compromised BBB. It appears to be a unique feature of the tight junctions of the BBB that they are extremely sensitive to locally produced CNS and circulating factors, which can, on a minute-to-minute basis, modulate the properties and the function of the paracellular pathways [11].

It is the presence of these tight junctions that seal the aqueous paracellular diffusional pathway between the endothelial cells and block the free movement of macromolecules, polar solutes, and ions from blood plasma into the extracellular fluid of brain. It is the impediment to the movement of ions that results in the high electrical resistance of the blood-brain barrier *in vivo*, of circa 1800 $\Omega \cdot cm^2$ [13]. The function of the tight junctions appears to be regulated via the intracellular regulating scaffolding proteins, which link the junctional molecules claudin and occludin to intracellular actin and the cytoskeleton [9]. Alterations in both intracellular and extracellular calcium can modulate tight junction assembly [14,15] and alter the electrical resistance across the cell layer and the effectiveness of the tight junctions as a barrier in the paracellular diffusive pathway.

Induction of many blood-brain barrier properties, including formation of tight junctions, and the polarity of transporters in the luminal and abluminal membranes of the cerebral endothelial cells, is believed to result from a close association between astrocytes and the endothelial cells [16]. Tight junctions can be induced to form between cerebral endothelial cells in culture with the use of astrocyte conditioned media, and, therefore, some of the BBB inducing factors are thought to be soluble in nature [17]. It is probable that these factors are multiple and complex-involving both a two-way exchange of signalling molecules and cell-cell contact [8,18].

DEVELOPMENT AND ONTOLOGY OF THE BLOOD-BRAIN BARRIER

The BBB develops during fetal life and is well formed, especially to proteins and macromolecules, by birth [6,19–25]. In the mouse the BBB begins to form between E11 and 17 [6] by which time identifiable tight junctions are present. The presence of tight junction will be highly restrictive to the trans-endothelial movement of polar molecules and macromolecules. In mammals that are born in a relatively immature state, such as the rat and mouse, many of the characteristic BBB transport mechanisms may continue to mature and only become fully expressed and functional in the perinatal or postnatal period.

In the mouse an RT-PCR signal for the ABC transporter mdr2 is present in the brain by E13 and all three isoforms mdr1a, mdr1b, and mdr2 show strong signals by E18 [26]. In the rodent it is the mdr1a and mdr1b isoforms that are the principal drug transporting isoforms; in the human, only a single P-glycoprotein gene product MDR1 transports drugs [27]. MDR1 in the human and mdr1a in the rodent are both expressed in the luminal membrane of the BBB endothelial cells, and mdr1b is expressed in glia and elsewhere in the CNS of rodents [27]. Expression of P-glycoprotein in the luminal endothelial cell membranes of the rat BBB was detected by immunoblotting by P7 and reaches a plateau of expression by P28 [28].

The high electrical resistance characteristic of the BBB is exhibited in the BBB of rats by E21 [29], indicating that the tight junctions are formed prenatally, and a low conductivity demonstrates that they already form an effective barrier to the movement of ions. Keep et al. [23] have demonstrated that the BBB permeability to the ion ^{86}Rb is significant at E21 with an influx rate constant of $42.5\pm4.3uL.g^{-1}/min^{-1}$ but within 2 days postnatally this has declined to $12.2\pm0.6\ uL.g^{-1}/min^{-1}$, with a further slow decline to $7.0\pm0.3\ uL.g^{-1}/min^{-1}$ by 50 days postnatally. Tight junction formation seems to be a very early feature of BBB development, and a barrier to the free movement of protein is formed at the very early stages of brain development [25]. Tight junctions are formed as blood vessels invade the brain at E10 in the mouse and E11 in the rat. As gliogenensis does not initiate until E17 and continues postnatally, tight junction formation per se would appear to be more induced by neural signals than glial signals in the first instance [25].

Preston et al. [24] have shown that BBB permeability to the nonmetabolizable, but slowly BBB penetrant, tracer mannitol (182 Da) is between 0.19–0.22 uL.g^{-1}/min^{-1} in the brain of rats of 1 week of age and that this permeability is identical to that of adult rats. The vascular space occupied by the tracer mannitol (the initial volume of distribution V_i) falls from 1.23 at 1week of age to 0.75 ml.100g^{-1} brain in the adult rat [24], indicating either a greater vascular volume, (greater capillary density or capillary diameter) in the neonatal rat or a significant degree of internalization of the mannitol by the endothelium, possibly by endocytosis into the cerebral capillary endothelial cells compared to the adult. There is nothing to suggest that the human BBB is not at least equally well formed as that of the rat at birth. Occludin and claudin-5 expression is present in the capillary endothelium of the brain of the 14-week human fetus and has the same cellular distribution as seen in the adult [30]. Pioneering studies by Grontoft [31] in stillborn human fetuses from approximately 12 weeks, gestation and in perinatal deaths have demonstrated a BBB to trypan blue present from at least the start of the second trimester, which is comparable to that of the adult.

TRANSPORT ACROSS THE BLOOD-BRAIN BARRIER

The endothelial cells forming the BBB express a number of transport systems in their cell membranes (Figure 21.4 and Table 21.2 and Table 21.3). Some of these transport proteins are highly polarized and are inserted into the luminal or abluminal membrane only; others are inserted into both membranes [5,8,32–34]. The particular orientation of these transporters may therefore preferentially transport substrate into the endothelial cell or across the endothelial cell, and the orientation of the transport may be from blood-to-brain or in the opposite direction. The presence of the tight junctions probably also has the function of segregating transport proteins and lipid rafts to either the luminal or abluminal membrane domain and to prevent their free movement from one side of the BBB to the other, thus preserving polarity of the barrier.

Lipid soluble molecules can diffuse though the BBB and enter the brain passively. There is a general correlation between lipid solubility, usually determined as the logD octanol/buffer partition co-efficient at pH7.4, and the rate at which most drugs will enter the CNS. Factors that reduce the entry of compounds into the CNS are a high polar surface area (PSA) especially in excess of 80Å^2, a tendency to form hydrogen bonds (especially in excess of 6) which will increase the free energy requirements of moving from an aqueous phase into the lipid of the cell membrane, a large number of rotatable bonds in the molecule, and a molecular weight in excess of 450Da. A high affinity of binding to plasma proteins can also significantly reduce CNS penetration. However, these molecular and physico-chemical factors are not always an absolute indication for CNS penetration and activity, and there are a number of examples of CNS active drugs that do not comply with these general rules for BBB penetration [35].

Table 21.3 Strategies for Enhancing Drug Penetration of the BBB

Approach	Strategy	Advantages/Disadvantages
Optimizing physico-chemical properties	Increase lipid solubility Reduce polar surface area Increase plasma AUC Reduce hydrogen bonding potential Reduce molecular weight	Drug potency may be reduced Lipidization increases Pgp and toxicological liability and reduces aqueous solubility
Pro-drugs and chemical delivery systems	Administer drug in an inactive form, which is converted to active drug in brain	Plasma AUC can be increased Drug stability in circulation improved Only active at desired site Drug can be locked into brain by conversion to more polar active drug
Intracerebral injection	Direct stereotaxic injection into desired brain region Introduction of slow-release implant or colony of stem cells Transfection	Damage may result from injection or implantation Movement of brain extracellualr fluid may carry drug away from intended site Infection/low repeat rate Stem cells or transfection irreversible
Olfactory route	Small molecular weight lipid-soluble drugs can be transferred from olfactory mucosa to subarachnoid CSF	Patients can "self-titrate" drug if effect is apparent Noninvasive Large molecular weight cut-off (c40kDa) Peptidases present in mucosa
Modulating/ opening the BBB	Use of an osmotic agent or drug to open/relax tight junctions Use of heating effect of electromagnetic radiation	Nonselective and allows other possibly damaging molecules to enter brain Rapidly reverses giving controlled window of delivery Radiation can be focused to specific region
Reducing the influence of efflux transporters	Co-administration of an inhibitor of one or more of the ABC efflux transporters Design of drugs that have reduced affinity for ABC efflux transporters	Long-term inhibition of ABC transporters may have side-effects resulting from an increased entry of substrates into the CNS The SAR of substrate ABC transporter interaction is poorly understood making rational drug design difficult
Delivery using endogenous transporters	The design of drugs as pseudo-substrates for transport systems, which normally carry essential metabolites into the CNS	Apart from the glucose and large neutral amino acid transport system the capacity (V_{max}) of these transporters is not high Both drug and natural substrates will compete for transport producing possible on/off effects

Table 21.3 Strategies for Enhancing Drug Penetration of the BBB (Continued)

Approach	Strategy	Advantages/Disadvantages
Cell-penetrating peptides	A drug, peptide or gene can be attached to a cell-penetrating peptide, which will carry it across the cell membrane	As this delivery system is not tissue specific it could be used to carry ERT into both somatic cells and the CNS
Liposomes and nanoparticles	Large constructs that can act as vectors to carry drug, enzyme, or genes across the BBB	Can be targeted to the CNS These systems are capable of carrying large therapeutic payloads

Source: These strategies have been recently fully reviewed in Begley, D.J., *Pharmacol. Ther.*, 104, 29, 2004.

With respect to lipid solubility a very large number of drugs have a much lower CNS penetrance than might be suggested by their logD value. These substances and many of their metabolites are actively effluxed from the brain and the capillary endothelium forming the BBB by members of the ABC transporter cassette [27]. The strategy of increasing the lipid solubility of a drug to make it more brain penetrant may sometimes be counterproductive as it may increase the likelihood of the molecule becoming a substrate for ABC transporters [27]. In the BBB the ABC transporters of greatest significance are P-glycoprotein (Pgp, Multidrug Resistance Protein, ABCB-1), Breast Cancer Related Protein (BRCP, ABCG-2) and the Multi-drug Resistance-associated Proteins (MRPs, ABCC-1,2,4,5), [27].

Transcytosis of macromolecules across the BBB via endocytic mechanisms is a significant route of transport for a number of larger molecules. Although the majority of large blood-borne molecules are physically prevented from entering the brain by the presence of the blood-brain barrier, transcytotic mechanisms exist to transport a variety of large molecules and complexes across the BBB. A summary of these transcytotic mechanisms is given in Table 21.3 [36–51]. These mechanisms involve either receptor-mediated transcytosis (RMT) or absorptive-mediated transcytosis (AMT). In RMT the binding of a ligand to a specific receptor on the cell surface triggers an endocytic event, and the receptor(s) and the bound ligand are internalized into the endothelial cell and routed across the cytoplasm to be exocytosed at the opposite pole of the cell. Dissociation of the ligand and receptor presumably can occur during cellular transit or during the exocytic event. In AMT an excess positive charge on the molecule appears to induce endocytosis with subsequent transcytosis. In both case to achieve transcytosis of an intact protein or peptide, the lysosomal compartment within the cell must be avoided by the routing of the endocytic vesicles and their contents away from this compartment. This does not occur in many endothelia and may be a specialized phenomenon in the BBB, where transcytosis of some ligands becomes a necessity [52]. In many tissues an avoidance of the lysosomal compartment does not occur, and this is exactly what is required for effective ERT and delivery of enzyme in these tissues to the lysosome.

Electron microscopic studies of BBB endothelial cells suggest the presence of few endocytic profiles compared to other endothelia. For example, the BBB contains only a fifth to a sixth of the observable endocytic profiles of muscle capillary endothelia [53]. However, when a comparison is made of the ability of endothelia in different tissues to trancytose protein, there is a very poor correlation between protein permeability of a microvessel and the number of observable endocytic profiles [53].

Smaller peptides may employ fluid-phase endocytosis or RMT mechanisms to cross the BBB, but it may also be possible for them to use a transporter protein, inserted into the cell membrane, in a similar manner to metabolites [54,55]. In some cases the receptor that transduces the signal at the cell membrane can also act as the transporter for the peptide and is co-opted to initiate RMT (or other transport system); in other cases the transporter protein for a peptide is distinct from the receptor peptide [46].

Mononuclear cells are continually penetrating the BBB and taking up residence in the CNS as microglia. These microglia are also able to return to the general circulation via the CSF drainage, and thus there is a slow continuous turnover and exchange of the microglial cell population with the cells of the peripheral immune system. The CNS microglia are the immune competent cells of the CNS and become activated in inflammation and pathological states. Circulating mononuclear cells and neutrophils are attracted to sites of BBB inflammation, penetrate the barrier, and form cuffs around the capillaries in the perivascular space. In both the normal BBB and inflammatory states (EAE) the mononuclear cells appear to penetrate by a process of diapedesis directly though the cytoplasm of the endothelial cells and not by the paracellular route through the tight junctional complexes as has been previously suggested [56,57]. This enables the cells to cross the BBB without temporarily opening it. In inflammatory states involving the BBB, the tight junctions may be relaxed and cells may then enter by both transcellular and paracellular routes.

THE BLOOD-BRAIN BARRIER AND THE DELIVERY OF CENTRALLY ACTIVE DRUGS

As previously mentioned the BBB forms a formidable obstacle to the delivery of drugs to the CNS [5,8,35,40]. Many drugs are polar in nature and cannot diffuse easily across the barrier. In addition, those that have significant lipid solubility tend to be substrates for ABC transporters and are actively effluxed from the endothelial cells of the cerebral capillaries and thus only have a limited access to the brain [27].

A number of strategies exist for enhancing drug uptake across the blood-brain barrier. These center around three approaches (i) either pharmacologically opening the barrier by modifying tight junctions and thus allowing a free entry of small molecules, or (ii) altering the physico-chemical properties of drugs to enhance their passive penetration of the barrier or (iii) by exploiting or manipulating the transport systems present in the BBB. These approaches have recently been fully reviewed elsewhere [8] and are summarized in Table 21.4.

The need in Gaucher disease and other lysosomal storage diseases with neuronopathic sequelae is to deliver replacement enzyme across the BBB to the brain tissue. As replacement enzymes are large molecular weight proteins of several

Table 21.4 Examples of Transcytosis/Transport of Large Molecules and Complexes Across the BBB

Transport System	Abbreviation (Receptor)	Example Ligands	Type	BBB Direction	Reference
Transferrin	Tf$_R$	Fe-transferrin	RMT	Blood to brain	Visser et al. 2004[36]
Melanotransferrin	MTf$_R$	Melanotransferrin (p97)	RMT	Blood to brain	Demeule et al. 2002[37]
Lactoferrin	Lf$_R$	Lactoferrin	RMT	Blood to CSF	Talkuder et al. 2003[38]
Apolioprotein E receptor 2	ApoER2	Lipoproteins and molecules bound to ApoE	RMT	Blood to Brain	Hertz and Marchang 2003[39]
LDL-receptor related protein 1 and 2	LRP1 LRP2	Lipoproteins, Amyloid-, lactoferrin, α 2-macroglobulin, melanotarnspferrin (p97), ApoE	RMT	Bi-directional	Hertz and Marchang 2003[39]; Gaillard et al. 2005[40]
Receptor for advanced glycosylation end-products	RAGE	Glycosylated proteins, Amyloid-, S-100, amphoterin	RMT	Blood to brain	Stern et al. 2002[41]; Deane et al. 2004[42]
Immunoglobulin G	Fc-R	IgG	RMT	Blood to brain	Zlokovic et al. 1990[43]
Insulin	—	Insulin	RMT	Blood to brain	Banks 2004[44]
P-glycoprotein	Pgp	Amyloid-	RMT	Brain to blood	Lam et al. 2001[45]
Leptin	—	Leptin	RMT	Blood to brain	Banks et al. 2002[46]
Tumor necrosis factor	—	TNFα	RMT	Blood to brain	Pan and Kastin 2002[47]
Epidermal growth factor	—	EGF	RMT	Blood to brain	Pan and Kastin 1999[48]
Heparin-binding epidermal growth factor-like growth factor (diptheria toxin receptor)	HB-EGF (DTR)	Diphtheria toxin and CRM197 (protein)	RMT	Blood to brain	Gaillard et al. 2005[40]
Leukemia inhibitory factor	LIFRa (gp190)	LIF	RMT	Blood to brain/spinal cord	Pan et al. 2000[49]
Cationized proteins	+	Cationized albumin	AMT	Blood to brain	Pardridge 1990[50]
Cell-penetrating peptides	+	SynB5/pAnt-(43–58)	AMT	Blood to brain	Drin et al. 2003[51]

Note: Many of the receptors involved in RMT are poorly defined and are multifunctional and multiligand in nature. Thus some ligands may be transported by more than one system and some receptors may with time turn out to be one and the same (- receptor uncharacterized; + not receptor-mediated, nonspecific).

kilo-Daltons, the only available strategies available will be the use of a transcytotic route involving an endocytic mechanism or by direct injection into the brain.

In the case of β-glucosidase (glucocerebrosidase) the enzyme can be internalised by neurons *in vitro* [58], and mannose-terminated glucocerebrosidase, when injected directly into the white matter of the striatum of rats, is convectively delivered to the cortex, where it is taken up by neurons [59]. Similar work with a mouse model of MPS-IIIA, where sulphamidase was injected into the brains of mice from 6 weeks of age, also delays the development of neuropathology [4]. However, repeated injection of replacement enzyme intracerebrally in humans is probably not a viable strategy for long-term treatment.

As the enzymes used in current replacement therapy (ERT) for lysosomal storage diseases do not cross the BBB [1,2,4,58], alternative methods for achieving transcytosis into the CNS need to be explored. Terminating the enzyme with a ligand for one of the RMT mechanisms for large molecules may provide a mechanism. For example, glycation of many proteins results in an enhanced penetration of the BBB [60], possibly by rendering the molecule a substrate for the receptor for advanced glycosylation end products (RAGE), although the mechanism involved is uncertain. Monoclonal antibodies to the transferrin receptor (OX26) have been used as a vector to transport peptides such as vasoactive intestinal polypeptide (VIP), nerve growth factor (NGF), glial derived neurotrophic factor (GDNF), and brain derived neurotrophic factor (BDNF) into the CNS [61]. The theory behind using a monoclonal antibody to the transferrin receptor is that this avoids competition with endogenous circulating transferrin and interaction of antibody with the receptor appears to initiate RMT. β-galactosidase enzyme has been successfully delivered to the mouse brain by directly coupling it to an 8D3 monoclonal antibody to the transferrin receptor [62]. Cationizing the enzyme might induce AMT but may also result in a loss of enzymic activity. Attaching a cell penetrating peptide may also offer an opportunity. Cell penetrating peptides have been used to carry small molecules into the CNS such as doxorubicin [63] but can also carry large proteins into cells [64] and across the BBB [65].

Liposomes and nanoparticles are large constructs and can theoretically carry significant amounts of adsorbed or encapsulated enzyme, drug, or other content across the BBB. Peglylated immunoliposomes targeted with an antitransferrin receptor monoclonal (8D3) have been used to transfect the β-galactosidase and luciferase gene into rat brain [66,67]. Nanoparticles have been largely used to target smaller peptides and drugs to the brain [68], but large proteins could be adsorbed onto or incorporated into their structure.

Some recent studies suggest that there may be a transport system in the BBB present in early life for the enzymes involved in two different mucopolysaccharidoses. In MPS IIIA the damaged enzyme is sulfamidase, and Gliddon and Hopwood [69] have shown in a mouse model of the disease that if replacement enzyme is given intravenously from birth, there is a significant delay in the development of neuronopathy. Also Vogler [70] has shown in young MPS VII mice treated intravenously with β-glucuronidase there was a marked delay in the development of CNS pathology. Both Gliddon and Hopwood [69] and Vogler [70] attribute this to the BBB being incomplete at birth and the tight junctions not fully sealing and closing the barrier until 10–14 days of age, citing Stewart and Hayakawa [71]. As the earlier discussion in this chapter has pointed out, tight junction formation is an early feature of BBB formation (E11 in the mouse) and would be an effective barrier to these high molecular weight replacement enzymes in the rodent from birth if they were the sole factor involved in BBB permeability.

A further study by Urayama et al. [72] shows that in the normal 2-day-old mouse there is a saturable BBB transcytotic mechanism mediated by mannose 6-phosphate (M6P) for transporting β-glucuronidase (GUS), whose deficiency produces MPS VII, into the CNS. This M6P/IGF2R receptor-mediated mechanism disappears in adult life [72]. This study also demonstrates that the BBB is tight to the smaller protein albumin at 2 days postnatally showing the presence of an effective barrier to protein at this age. This transient BBB transport is probably the underlying explanation for CNS enzyme entry in the Gliddon and Hopwood [69], Vogler et al. [70] and Urayama et al. [72] studies and may apply to a significant number of large molecules in demand by the rapidly growing neonatal brain. These transport mechanisms in the BBB may become down-regulated as the animal ages but retained in the endothelia of some peripheral tissues. In the case of some lysosomal enzymes, the CNS requirement in the neonatal period may be greater than can be supplied within the brain and the need is made up by the transport of circulating enzyme. There is probably nothing "immature" in the sense of an underdevelopment of the perinatal BBB. The specific transport mechanisms that it expresses, however, may just be different to those in the adult and adapted to the needs of the animal at the time. A good example is probably the monocarboxylic acid transporter MCT1, which can transport both ketone bodies and lactate, and which is highly expressed at the BBB in the suckling animal and then down-regulates somewhat in the adult but is always expressed [73]. Peripheral endothelial cells express the receptor for albumin, (albondin), which can initiate RMT, whereas the cerebral endothelial cells do not [74]. It can be argued that a specialization of the adult BBB is that the luminal membrane of the endothelial cells expresses fewer and probably different receptors that can initiate RMT than the neonatal BBB or the endothelial cells of microvessels in peripheral tissues.

These studies raise the interesting suggestion that a mechanism(s) that might carry lysosomal enzymes across the BBB may be present in early life and then become down-regulated as the individual ages. To what extent this occurs in the human neonate is unknown. A down-regulated mechanism may be capable of reactivation and of transporting ERT into the CNS. This type of mechanism is not unique, certainly in the rodent, and applies to the transport of immunoglobulin-G (IgG) across the neonatal gut for the first 21 days of life [75], which then becomes down-regulated postweaning.

An alternative strategy applied to Gaucher and other lysosomal storage diseases is to prevent the over-accumulation of storage product by limiting its generation. N-butyl-deoxynojirimycin (*N*B-DNJ) is an iminosugar derivative and an inhibitor of the enzyme glucosylceramide synthetase (glucosyltransferase), which is the initial enzyme in the pathway of glycosphingolipid synthesis. This approach is termed *substrate reduction therapy* (SRT) and prevents an over-accumulation of storage product in sphingolipidoses in the degradation pathway at a point prior to the defective enzyme. *N*B-DNJ is effective at preventing the development of symptoms in a Tay-Sachs mouse model and also in Gaucher disease patients [76]. *N*B-DNJ is a small moderately lipophilic molecule and apparently crosses the blood-brain barrier. However the mechanism by which it does this has not been explored and its brain penetration has not been quantified. Its lipophilicity may give it sufficient passive permeability to enter the CNS by purely diffusive mechanisms, but it also contains a sugar residue, which may be reactive with a BBB transporter.

The alkyl chain present in *N*B-DNJ and other similar substrate-reducing drugs resembles that present in alkyl glycerols, which can be employed to open the BBB [77], with a similar protocol to that used in hyperosmotic opening of the barrier. Co-administration of alkyl glycerol and methotrexate greatly increases the brain content of the cytotoxic drug [78]. It is proposed that the mechanism of alkyl glycerol is by disrupting tight junctions and thus allowing other co-administered solutes into the CNS [79]. However for the CNS delivery of methotrexate the drug is applied at 200 mM intra-arterially to open the BBB. If injected systemically, the volume of distribution of the agent is too large to produce a BBB opening effect. Although *N*B-DNJ is present in blood in far smaller quantities, its interactions with cells and cell membranes may have a similar basis. As the alkyl chain of these iminosugars is lengthened, the compound becomes more lipophilic and thus may cross the BBB more readily. The hydrophobic sugar residue appears to be the enzyme inhibitory domain of the molecules and at least a three carbon alkyl chain is required for cell entry [76]. If the alkyl chain is too long, the compounds become cytotoxic probably by disturbing the structure of the cell membrane [76]. The mechanism by which *N*B-DNJ enters cells and crosses the BBB needs to be fully elucidated so that the necessary physico-chemical properties for BBB penetration can be preserved in second generation substrate-depleting drugs possibly aimed at different steps in lysosomal degradation pathways intended for application in a number of different lysosomal storage diseases. Combination therapies might become optimal with one form of the drug designed for peripheral substrate reduction and another designed for BBB penetration.

PATHOLOGY OF THE BLOOD-BRAIN BARRIER IN LYSOSOMAL STORAGE DISEASE

Some studies have suggested that in some lysosomal storage diseases the BBB itself may be altered. A study by Jeyakumar et al. [80] with mouse models of Sandhoff disease and GM1 gangliosidosis have shown an increased extravasation of Evans blue-albumin and endogenous immunoglobulin G in both of these mouse models. As Evans blue binds to albumin in the circulation, this represents an increased BBB permeability to a tracer of circa 67kDa. This permeabilization of the BBB could be linked to inflammatory processes occurring in the CNS. CNS microglia in this Sandhoff mouse model are activated as shown by an increase in MHC class II expression by the cells and an increased production of the pro-inflammatory cytokines TNFα and interleukin-1 in the brains of both mouse models [80]. Interleukin-1 is able to induce changes in the permeability of brain blood vessels when injected directly into brain, with juvenile rats younger than 21 days being much more sensitive to the cytokine and showing a marked inflammatory BBB opening [81,82]. These studies suggest that the BBB is still differentiating postnatally in this respect and its sensitivity to pro-inflammatory cytokines is declining.

TNFα has also been shown to induce synthesis and release of matrix metalloproteinase 9 (MMP-9), which can re-model the basal lamina of the cerebral endothelial cells [83] by degrading laminin and cell adhesion molecules. A combination of pro-inflammatory factors may further recruit and facilitate the entry of mononuclear leucocytes into the CNS, increasing cytokine release and reinforcing a loss of BBB integrity. Microglial activation has also been noted in mouse models of mucopolysaccharidosis I and IIIB Sanfilippo disease [84]. Extravazation of immunoglobulin has also been observed in a mouse model of Juvenile Batten Disease (NCL3) [85].

Thus it seems clear that the BBB may be damaged and altered in a number of lysosomal storage diseases and consequently become more permeable to many blood-borne solutes. Any nonspecific increase in permeability of the BBB would freely admit the low molecular weight neuroexcitatory amino acids glutamic acid and glycine into the CNS, as the concentration gradient from plasma to brain is significant (see Table 21.1). An increase in the ECF concentration of these neurotransmitters may induce neuroexcitatory damage, cell death, and seizures. The significant plasma proteins albumin, prothrombin, and plasminogen are normally almost completely excluded from brain by the BBB and, if they are introduced experimentally into the brain, are intensely irritant to the brain cells and can trigger a number of processes that can initiate glial scarring, oedema, seizures, and neuronal cell death [86]. For example glia, microglia and neurons all express the thrombin receptor PAR$_1$. Activation of PAR receptors influences a number of second messengers, inducing an inhibition of adenyl cyclase activity, stimulation of phosphoinositide turnover, activation of PKC, and activation of mitogen activated kinase (MAP-kinase) and phosphoinositide 3-kinase (PI 3 kinase) [86]. PAR$_1$ activation also rapidly increases intracellular calcium in primary rat hippocampal neuron cultures [87,88] and PAR$_1$ activation also potentiates NMDA receptor stimulation allowing cytotoxic levels of calcium to enter neurons [87]. The protease inhibitors protease nexin (PN-1) and neuroserpin are produced by neurons, astroglia (whose end-feet abut the cerebral endothelium), and the smooth muscle cells of the cerebral arterioles [85]. The cysteine protease inhibitor cystatin C is found in significant concentrations in CSF and is unusually one of the few proteins in CSF found in higher concentration than plasma (see Table 21.1). Cystatin C is synthesised and released locally in the CNS by the choroid plexus [89]. These protease inhibitors are therefore well placed to inactivate a limited and transient extravazation of prothrombin and other serine proteases which may occur in normal physiology but may be overwhelmed by a significant pathological increase in BBB permeability. It is interesting to note in this context that the administration of anti-inflammatory drugs such as indomethecin and aspirin together with NB-DNJ greatly improves the survival of Sandhoff mice [90], possibly by limiting inflammatory damage to the BBB and elsewhere.

It has been shown that agonist-induced calcium release in neurons derived from the brains of acute neuronopathic Gaucher patients is also greatly enhanced, especially in response to glutamic acid stimulation [3,58,91]. This increase in calcium release may directly result from the accumulation of storage product in these neurons or may be induced secondarily to BBB pathology. A rise in free intracellular calcium in cerebral endothelial cells would also relax the tight junctions of the BBB and nonspecifically increase its permeability to plasma solutes [15], including glycine, glutamic acid, and serine proteases. It is noteworthy that glutamic acid has been identified as a substrate for P-glycoprotein [92] and this, coupled with System A and X^-_{AG} for amino acids, which transport both glycine and glutamic acid from brain extracellular fluid into the cerebral endothelial cells at the abluminal membrane, will provide a direct transport from brain to blood, certainly for glutamic acid. This transport combined with excitatory amino acid re-uptake into neurons and glia will combine under normal circumstances to keep the brain extracellular levels of excitatory amino acids low. A compromised BBB would disturb these homeostatic mechanisms and put the neural and glial re-uptake mechanisms under constant and severe load.

A rise in free calcium within the endothelial cells is thought to activate the actin/myosin system attached to the scaffolding ZO proteins and initiate withdrawal of claudin and occludin from the tight junctional regions [13,14]. Putting this information together it becomes probable that directly, or indirectly, pathological changes in BBB permeability may be directly involved in some of the neuronopathic consequences of a significant number of the lysosomal storage diseases. It is interesting to speculate as to whether the BBB may be more robustly maintained in Type 1 Gaucher disease compared to Types 2 and 3. At present no adequate animal models exist for the Gaucher sub-types.

CONCLUSIONS

The physiological role of the BBB creates a very significant challenge to the delivery of many therapeutic drugs to the CNS, especially those that are polar and/or of high molecular weight.

The introduction of substrate reduction therapy has opened the door to new approaches to treating a number of lysosomal storage diseases. A major attraction is that as the enzyme inhibitors act at an early point in glycosphingolipid metabolism, a single drug may be used to treat a number of diseases. The development of new second-generation substrate-reducing drugs requires a full understanding of how they may reach the CNS so that new compounds can be designed in a physicochemically informed manner in order to retain a significant BBB penetrance and not create further problems for CNS drug delivery.

A considerable pressure exists for developing various vectors (liposomes/nanoparticles/targeting molecules), which can carry replacement enzyme across the BBB. The attraction of these types of vector is that they are potentially capable of delivering a significant payload of enzyme/drug across the BBB. A second attractive feature of these systems is that they are not drug specific, so they could be used to carry a wide variety of nonpenetrant drugs into the CNS and have considerable use in a large number of CNS disease states.

Bone marrow transplantation (BMT) is now gaining establishment as a treatment for a number of lysosomal storage diseases. If some of the transplanted bone marrow cells capable of synthesizing and releasing some free lysosomal enzyme are progenitors for circulating mononuclear cells, which are capable of crossing the BBB and becoming microglia synthesising enzyme, this might become, with time, a means of introducing enzyme synthesis into the CNS. Alternatively it may be possible to develop stem cells producing the required enzyme, which can be injected into the CNS, taking up residence, and thus providing a local source of enzyme.

Early diagnosis, together with effective CNS delivery of enzyme into the CNS, coupled with early BMT, and possibly combined with an appropriate substrate depleting drug, may possibly form the most effective initial strategies for treating neuronopathic disease. With time the CNS may then acquire sufficient enzyme producing microglia to protect the brain, without the need for further specific CNS targeting.

In addition the involvement of the BBB in Gaucher and other neuronopathic lysosomal diseases needs to be better understood. Damage to the barrier may contribute significantly to CNS damage, and it is important to assess this role. Is the barrier always damaged in neuronopathic disease? Does treatment of the disease with either enzyme replacement or substrate reduction restore the barrier, and is there scope for reinforcing the barrier if damage is suspected? Clearly the BBB is relevant to both the pathology of the lysosomal storage diseases and their effective treatment.

REFERENCES

1. Barranger, J.A. and O'Rourke, E., Lessons learnt from the development of therapy for Gaucher disease, *J. Inherit. Metab. Dis.,* 24 (Suppl. 2), 89, 2001.
2. Futerman, A.H. and van Meer, G., The cell biology of lysosomal storage disorders, *Nature Rev. Mol. Cell. Biol.,* 5, 554, 2004.
3. Pelled, D. et al., Enhanced calcium release in the acute neuronopathic form of Gaucher disease, *Neurobiol. Dis.,* 18, 83, 2005.
4. Savas, P.S., Hemsley, K.M., and Hopwood J.J., Intracerebral injection of sulfamidase delays neuropathology in murine MPS-IIIA, *Mol. Gen. Metab.,* 82, 273, 2004.
5. Begley, D.J. and Brightman, M.W., Structural and functional aspects of the blood-brain barrier, *Prog. Drug Res.,* 61, 39, 2003.
6. Ballabh, P., Braun, A., and Nedergaard, M., The blood-brain barrier: an overview. Structure, regulation, and clinical implications, *Neurobiol. Dis.,* 16, 1, 2004.
7. Kandel, E.R., Schwartz, J.H., and Jessel, T.M., *Principles of Neural Science,* 4th ed., McGraw-Hill, New York, 2000, 1294.
8. Begley, D.J., Delivery of therapeutic agents to the central nervous system: the problems and the possibilities, *Pharmacol. Ther.,* 104, 29, 2004.
9. Wolburg, H. and Lippoldt, A., Tight junctions of the blood-brain barrier: development, composition and regulation, *Vasc. Pharm.,* 38, 323, 2002.
10. Mitic, L.L., van Itallie, C.M., and Anderson, J.M., Molecular physiology and pathophysiology of tight junctions. I. Tight junction structure and function: lessons from mutant animals and proteins, *Am. J. Physiol.,* 297, G 250, 2000.

11. Wolburg, H. et al., Localisation of claudin-3 in tight junctions of the blood-brain barrier is selectively lost during experimental autoimmune encephalomyelitis and human glioblastoma multiforme, *Acta Neuropathol.*, 105, 586, 2003.

12. Nitta, T. et al., Size-selective loosening of the blood-brain barrier in claudin-5 deficient mice, *J. Cell Biol.*, 161, 653, 2003.

13. Butt, AM., Effect of inflammatory agents on electrical resistance across the blood-brain barrier in pial microvessels of anaesthetised rats, *Brain Res.*, 696, 145, 1995.

14. Balda, M.S. et al., Assembly and sealing of tight junctions: possible participation of G-proteins, phospholipase C, protein kinase C and calmodulin, *J. Membr. Biol.*, 122, 193, 1991.

15. Abbott, N.J., Role of intracellular calcium in regulation of brain endothelial permeability, in *Introduction to the Blood-Brain Barrier: Methodology and Biology*, W.M. Pardridge, Ed., Cambridge University Press, Cambridge, 1998, 345.

16. Rubin, L.L. et al., Differentiation of brain endothelial cells in cell culture, *Ann. NY Acad. Sci.*, 633, 420, 1991.

17. Neuhaus, J., Risau, W., and Wolburg, H., Induction of blood-brain barrier characteristics in bovine brain endothelial cells by rat astroglial cells in transfilter coculture, *Ann. NY Acad. Sci.*, 633, 578, 1991.

18. Abbott, N.J., Astrocyte-endothelial cell interactions and blood-brain barrier permeability, *J. Anat.* 200, 629, 2002.

19. Olsson, Y. et al., Blood-brain barrier to albumin in embryonic, new born and adult rats, *Acta Neuropathol.*, 10, 117, 1968.

20. Tauc, M., Vignon, X., and Bouchaud, C., Evidence for the effectiveness of the blood-CSF barrier in the fetal rat choroid plexus. A free-fracture and peroxidase diffusion study, *Tiss. Cell*, 16, 65, 1984.

21. Saunders, N.R., Development of the blood-brain barrier to macromolecules, in *Barriers and Fluids of the Eye and Brain*, M.B. Segal, Ed., Macmillan Press, London, 1992, 128.

22. Moos, T. and Mølgård, K., Cerebrovascular permeability to azo dyes and plasma proteins in rodents of different ages, *Neuropath. Appl. Neurobiol.*, 19, 120, 1993.

23. Keep, R.F. et al., Developmental changes in blood-brain barrier potassium permeability in the rat: relation to brain growth, *J. Physiol.* 488, 439, 1995.

24. Preston, J.E., Al-Saraff, H., and Segal, M.B., Permeability of the developing blood-brain barrier to ^{14}C-mannitol using the rat *in situ* brain perfusion technique, *Dev. Brain Res.*, 87, 69, 1995.

25. Saunders, N.R., Knott, G.W., and Dziegielewska, K.M., Barriers in the immature brain, *Cell Mol. Neurobiol.*, 20, 29, 2000.

26. Scheingold, M. et al., Multidrug resistance gene expression during the murine ontogeny, *Mech. Aging Devel.*, 122, 255, 2001.

27. Begley. D.J., ABC transporters and the blood-brain barrier, *Curr. Pharm. Des.*, 10, 1295, 2004.

28. Matsuoka, Y. et al., Developmental expression of P-glycoprotein (multidrug resistance gene product) in rat brain, *J. Neurobiol.*, 39, 383, 1999.

29. Butt, A.M., Jones, H.C., and Abbott, N.J., Electrical resistance across the blood — brain barrier in anaethetised rats: a developmental study, *J. Physiol.*, 429, 47, 1990.

30. Virgintino D. et al., Immunolocalisation of tight junction proteins in the adult and developing human brain, *Histochem. Cell Biol.*, 122, 51, 2004.

31. Grontoft O., Intracranial haemorrhage and blood-brain barrier problems in the newborn, *Acta Pathol. Microbiol. Scand.* (Suppl. C), 1, 1964.

32. Betz, A.L., Firth, J.A., and Goldstein, G.W., Polarity of the blood-brain barrier: distribution of enzymes between the luminal and abluminal membranes of brain capillary endothelial cells, *Brain Res.*, 192, 1728, 1980.

33. Begley, D.J., The blood-brain barrier: principles for targeting peptides and drugs to the central nervous system, *J. Pharm. Pharmacol.*, 48, 136, 1996.

34. Mertsch, K. and Maas, J., Blood-brain barrier penetration and drug development from an industrial point of view, *Curr. Med. Chem. — Central Nervous System Agents* 2, 187, 2002.

35. Bodor, N. and Buchwald, P., Brain-targeted drug delivery: experiences to date, *Am. J. Drug Targ.*, 1, 13, 2003.

36. Visser, C.C. et al., Characterisation of the transferrin receptor on brain capillary endothelial cells, *Pharm. Res.*, 21, 2004.

37. Demeule M. et al., High transcytosis of melanotransferrin (P97) across the blood-brain barrier, *J. Neurochem.* 83, 924, 2002.

38. Talkuder, M.J.R., Takeuchi, T., and Harada, E., Receptor-mediated transport of lacto-ferrin into the cerebrospinal fluid via plasma in young calves, *J. Vet. Med. Sci.*, 65, 957, 2003.

39. Hertz J. and Marchang P., Coaxing the LDL receptor into the fold, *Cell*, 112, 289, 2003.

40. Gaillard, P., Visser, C.C., and de Boer, A.G., Targeted delivery across the blood-brain brain barrier, *Expert Opin. Drug. Del.*, 2, 299, 2005.

41. Stern, D. et al., Receptor for advanced glycation endproducts: a multiligand receptor magnifying cell stress in diverse pathologic settings, *Adv. Drug Del. Rev.*, 54, 1615, 2002.

42. Deane, R., Wu, Z., and Zlokovic, B.V., RAGE (Yin) versus LRP (Yang) balance regulates Alzheimer amyloid beta-peptide clearance through transport across the blood-brain barrier, *Stroke*, 35, 2628, 2004.

43. Zlokovic, B.V. et al., A saturable mechanism for transport of immunoglobulin G across the blood-brain barrier of the guinea pig, *Exp. Neurol.*, 107, 263, 1990.

44. Banks, WA., The source of cerebral insulin, *Eur. J. Pharmacol.*, 490, 5, 2004.

45. Lam FC. et al., Beta-amyloid efflux mediated by P-glycoprotein, *J. Neurochem.*, 76, 1121, 2001.

46. Banks W.A. et al., Leptin transport across the blood-brain barrier of the Koletsky rat is not mediated by a product of the leptin receptor gene, *Brain Res.*, 950, 130, 2002.

47. Pan, W. and Kastin, A., TNFα transport across the blood-brain barrier is abolished in receptor knockout mice, *Exp. Neurol.*, 174, 193, 2002.

48. Pan, W. and Kastin, A., Entry of EGF into brain is rapid and saturable, *Peptides*, 20, 1091, 1999.

49. Pan, W., Kastin, A.J., and Brennan, J.M., Saturable entry of leukaemia inhibitory factor from blood to the central nervous system, *J. Neuroimmunol.*, 106, 172, 2000.

50. Pardridge, W.M. et al., Evaluation of cationised albumin as a potential blood-brain barrier drug transport vector, *Exp. Neurol.*, 255, 893, 1990.

51. Drin G. et al., Studies on the internalisation mechanism of cationic cell-penetrating peptides, *J. Biol. Chem.*, 278, 31192, 2003.

52. Nag, S. and Begley, D.J., Blood-brain barrier, exchange of metabolites and gases, in *Pathology and Genetics. Cerebrovascular Diseases*, Kalimo, H. Ed., ISN Neuropath Press, Basel, 2005, 22.

53. Stewart, P.A., Endothelial vesicles in the blood-brain barrier: are they related to permeability? *Cell Mol. Neurobiol.*, 20, 149, 2000.

54. Banks, W.A., Role of the blood-brain barrier in communication between the central nervous system and the peripheral tissue, in *Blood Spinal Cord and Brain Barriers in Health and Disease,* Sharma H.S and Westman J., Eds. Elsevier/Academic Press, San Diego, 2004, 73.

55. Kastin, A. and Pan, W., Peptide transport across the blood-brain barrier, *Prog. Drug Res.*, 61, 79, 2003.

56. Wolburg, H., Wolburg-Bucholz, K., and Engelhardt, B., Diapedisis of mononuclear cells across cerebral vessels during experimental autoimmune encephalomyelitis leaves tight junctions intact, *Acta Neuropathol.*, 109, 181, 2005.

57. Engelhardt, B. and Wolburg, H., Transendothelial migration of leucocytes: through the front door or around the side of the house, *Eur. J. Immunol.*, 34, 2955, 2004.

58. Pelled, D., Shogomori, H., and Futerman A.H., The increased sensitivity of neurons with elevated glucocerebroside to neurotoxic agents can be reversed by imiglucerase, *J. Inherit. Metab. Dis.*, 23, 175, 2000.

59. Zirzow, G.C. et al., Delivery, distribution and neuronal uptake of exogenous mannose-terminal glucocerobrosidase in the intact brain, *Neurochem. Res.*, 24, 310, 1999.

60. Poduslo, J.F. and Curran G.L., Glycation increases the permeability of proteins across the blood-nerve and blood-brain barriers, *Mol. Brain Res.*, 23, 157, 1994.

61. Bickel, U. et al., *In vivo* demonstration of subcellular localization of antitransferrin receptor monoclonal antibody-colloidal gold conjugate in brain capillary endothelium, *J. Histochem. Cytochem*, 42, 1493, 1994.

62. Zhang, Y. and Pardridge, W.M., Delivery of beta-galactosidase to mouse brain via the blood-brain barrier transferrin receptor, *J. Pharm. Exp. Ther.*, 313, 1075, 2005.

63. Rousselle C. et al., New advances in the transport of doxorubicin by a peptide vector-mediated strategy, *Mol. Pharmacol.*, 57, 679, 2000.

64. Fawell, S. et al., Tat-mediated delivery of heterologous proteins into cells, *Proc. Natl. Acad. Sci. U.S.A.*, 91, 664, 1994.

65. Schwarze, S.R. et al., *In vivo* protein transduction: delivery of a biologically active protein into the mouse, *Science*, 285, 1569, 1999.

66. Shi, N. et al., Brain-specific expression of an exogenous gene after iv administration, *Proc. Natl. Acad. Sci. U.S.A.*, 98, 12754, 2001.

67. Pardridge, W.M., Drug and gene targeting to the brain with molecular Trojan horses, *Nat. Rev. Drug Discov.*, 1, 131, 2002.

68. Kreuter, J., Nanoparticle systems for brain delivery of drugs, *Adv. Drug Deliv. Revs.*, 47, 65, 2001.

69. Gliddon, B.L. and Hopwood, J.J., Enzyme replacement therapy from birth delays the development of behaviour and learning problems in mucopolysaccharidosis type III A mice, *Pediatr. Res.*, 56, 65, 2004.

70. Vogler, C. et al., Enzyme replacement in murine mucopolysaccharidosis type VII: neuronal and glial response to beta-glucouronidase requires early initiation of enzyme replacement therapy, *Pediatr. Res.*, 45, 838, 1999.

71. Stewart, P.A. and Hayakawa, E.M., Interendothelial junctional changes underlie the developmental tightening of the blood-brain barrier, *Dev. Brain Res.*, 32, 271, 1987.

72. Urayama A. et al., Developmentally regulated mannose 6 phosphate receptor-mediated transport of a lysosome enzyme across the blood-brain barrier, *Proc. Natl. Acad. Sci. U.S.A.*, 101, 12658, 2004.

73. Gerhart, D.Z. et al., Expression of mono-carboxylate transporter MCT1 by brain endothelium and glia in adult and suckling rats, *Am. J. Physiol.*, 273, E207, 1997.

74. Pardridge, W.M., Eisenberg, J., and Cefalu, W.T., Absence of albumin receptor on brain endothelial cells *in vivo* or *in vitro*, *Am. J. Physiol.*, 249, E264, 1985.

75. Morris, B. and Begley, D.J., The transmission of ^{125}I labelled globulins to the circulation in young rats, *J. Zool.*, 169, 101, 1973.
76. Butters, T.D., Dwek, R.A., and Platt, F.M., New therapeutics for the treatment of glycosphingolipid lysosomal storage diseases, *Adv. Exp. Med. Biol.*, 535, 219, 2003.
77. Lee, H.J., Zhang, Y., and Pardridge, W.M., Blood-brain barrier disruption following the internal carotid arterial perfusion of alkylglycerols, *J. Drug. Targ.*, 10, 463, 2002.
78. Erdlenbruch, B. et al., Intracarotid administration of short chain alkylglycerols for increased delivery of methotrexate to the rat brain, *Br. J. Pharmacol.*, 139, 685, 2003.
79. Erdlenbruch, B. et al., Alkyglycerol opening of the blood-brain barrier to small and large fluorescent markers in normal and C6 glioma-bearing rats and isolated brain capillaries, *Br. J. Pharmacol.*, 140, 1201, 2003.
80. Jeyakumar, M. et al., Central nervous system inflammation is a hallmark of pathogenesis in mouse models of GM1 and GM2 gangliosidosis, *Brain*, 126, 974, 2003.
81. Anthony, D.C. et al., Age-related effects of interleukin-1 beta on polmorphonuclear neutrophil-dependent increases in blood-brain barrier permeability in rats, *Brain*, 120, 435, 1997.
82. Bolton, S.J. Anthony, D.C., and Perry, V.H., Loss of tight junction proteins occludin and zonula-occludens-1 from cerebral vascular endothelium during neutrophil-induced blood-brain barrier breakdown *in vivo*, *Neuroscience*, 86, 1245, 1998.
83. Rosenberg, G.A. et al., Tumour necrosis factor-α-induced gelatinase B causes delayed opening of the blood-brain barrier and expanded therapeutic window, *Brain Res.*, 703, 151,1995.
84. Ohmi, K. et al., Activated microglia in cortex of mouse models of mucopolysaccharidoses 1 and IIIB, *Proc. Natl. Acad. Sci. U.S.A.*, 100, 1902, 2003.
85. Guerin, C. et al., Blood-brain barrier defects in mouse models of NCL. Presented at NCL 2003, Chicago, April, 2003.
86. Gingrich, M.B. and Traynelis, S.F., Serine proteases and brain damage — is there a link? *Trends in Neuroscience*, 23, 399, 2000.
87. Yang, Y. et al., Thrombin receptor on rat primary hippocampal neurons: coupled calcium and cAMP responses, *Brain Res.*, 761, 11, 1997.
88. Gingrich, M.B. et al., Potentiation of NMDA receptor function by the serine protease thrombin, *J. Neurosci.*, 20, 4582, 2000.
89. Smith, D., Johanson, C.E., and Keep, R.F., Peptide and peptide analog transport systems at the blood-CSF barrier, *Adv. Drug Del. Revs.*, 56, 1765, 2004.
90. Jeyakumar, M. et al., NSAIDs increase survival in the Sandhoff disease mouse: synergy with N-butyldeoxynojirimycin, *Ann. Neurol.*, 56, 642, 2004.
91. Korkotian E. et al., Elevation of intracellular glucosylceramide levels results in an increase in endoplasmic reticulum density and in functional calcium stores in cultured neurones, *J. Biol. Chem.*, 274, 21673, 1999.
92. Liu, X-D. and Liu, G-Q., P-glycoprotein regulated transport of glutamate at the blood-brain Barrier, *Acta Pharmacol. Sin.*, 22, 111, 2001.

CHAPTER 22

Hematopoietic Stem Cell Transplantation, Stem Cells, and Gene Therapy

Charles Peters and William Krivit*

CONTENTS

* Deceased after submission of manuscript.

INTRODUCTION

Gaucher disease is a lysosomal glycolipid storage disorder characterized by the accumulation of glucosylceramide (glucocerebroside) due to deficient activity of the enzyme acid β-glucosidase. Three types of Gaucher disease have been delineated. Type 1 is the most common and is distinguished from type 2 and type 3 disease by the lack of primary central nervous system (CNS) involvement. Chapter 10 discusses type 1 disease in greater detail. The acute neuronopathic form of Gaucher disease (i.e., type 2) has an early onset with severe CNS involvement and death typically within the first several years of life. The subacute neuronopathic form (i.e., type 3) is characterized by neurologic symptoms that are later in onset with a more chronic course than that observed in type 2 disease. Type 2 and 3 disease are reviewed in greater detail in Chapter 11. Patients with Gaucher disease have hepatosplenomegaly, bone lesions, and in some cases involvement of the lungs and other organs. This chapter will focus on the use of hematopoietic and other stem cells as therapeutic modalities to treat Gaucher disease. A discussion of gene transfer/therapy will also be presented.

HEMATOPOIETIC STEM CELL TRANSPLANTATION (HSCT)

The first successful transplantations of allogeneic hematopoietic stem cells (HSCs) were performed in 1968 in three children with immunodeficiencies [1–4]. For Gaucher disease, the first successful HSCT was performed at Huddinge Hospital in Stockholm, Sweden by Prof. Ringdén and his team [5]. In this section, an overview of hematopoietic stem cell transplantation (HSCT) will be provided. Outcomes observed in patients with Gaucher disease according to disease type following HSCT will be presented later in this chapter.

Overview of Hematopoietic Stem Cell Transplantation

The term hematopoietic stem cell transplantation refers to the intravenous infusion of blood stem cells obtained from bone marrow, peripheral blood, or umbilical cord blood. Significant advances in supportive care made the development of HSCT possible. These advances included the use of central venous catheters, hyperalimentation, specific antiemetics, growth factors, human leukocyte antigen (HLA) typing, virtual banks for performing searches for unrelated donors, and banks for umbilical cord blood. Concomitant with these advances have been the generation of new knowledge in immunology, transfusion medicine, and infectious diseases. Finally, acceptance of the concept that HSCT could serve as a form of therapy for single gene defects characterized by deficiency of a lysosomal hydrolytic enzyme was critical. This overview of HSCT will include brief discussions of histocompatibility, preparative regimens and regimen-related toxicities, graft-vs.-host disease (GVHD) prophylaxis and treatment, sources of HSCs, and complications of HSCT.

Histocompatibility and Donor Selection

Transplantation of tissues between two genetically dissimilar individuals almost always results in an immune reaction that leads to rejection of the graft unless immunosuppressive medications are given. This immunologic reactivity involves activation of T- and B-lymphocytes. It stems from differences between the transplant host and donor for cell surface markers known as histocompatibility antigens. Histocompatibility antigens that prompt the strongest reactions are encoded by a group of genes residing in the major histocompatibility complex (MHC). Antigens encoded by MHC genes are expressed on essentially all nucleated cells. Characterization of the human leukocyte antigen (HLA) system, standardization of HLA-typing methods, and the development of accepted nomenclature have been significant advances over the past four decades. With the advent of DNA-based typing methods, it is now possible to define each class of HLA molecule by its unique sequence [6]. HLA antigens are encoded on chromosome 6 and are clustered in three distinct regions designated class I, class II, and class III. Both class I and class II genes encode polypeptides that are critical in controlling T-lymphocyte recognition and determining histocompatibility in transplantation. A characteristic feature of HLA genes is their profound polymorphism. Three loci in the class I region encoding HLA-A, -B, and -C alloantigens constitute the major class I determinants important for matching in HSCT. Class II region genes include HLA-DR, -DQ, and -DP, which also show striking degrees of polymorphism and hence are critical to matching in the setting of HSCT. There have been dramatic advances in the laboratory methods used to define HLA genes and alloantigens. Historically, HLA class I antigens were defined by serologic methods using a complement-dependent microcytotoxicity assay and panels of alloantisera containing HLA antibodies. A second method of HLA typing has involved testing T-lymphocytes *in vitro* for their ability to recognize certain HLA antigens. The most commonly used cellular assay was the mixed lymphocyte culture (MLC) reaction, a test in which disparity for class II HLA-D region antigens led to lymphocyte activation and proliferation. The development of DNA-based typing methods for analysis of HLA genes has greatly advanced knowledge of MHC diversity, the role of class I and class II molecules in the immune response, and the factors important in the selection of volunteer unrelated donors of bone marrow, peripheral blood, or umbilical cord blood for HSCT. A suitable donor can be identified for approximately 50–80% of patients for whom an unrelated donor search is initiated. The likelihood of identifying a donor is increased if the donor and patient share the same ethnic or racial background. Important work has also been performed on delineating the relevance of the vector of HLA compatibility in assessing the risk for graft failure and acute GVHD. A detailed presentation of this information is beyond the scope of this chapter. However, briefly, the clinical utility of DNA-based typing methods for donor selection is twofold: 1) the identification and prioritization of well-matched donors (e.g., more complete and accurate matching is associated with lower risks of graft failure and GVHD and with improved survival); and 2) identification and avoidance of donors mismatched for multiple alleles (e.g., multilocus disparity is associated with increased risks of graft failure, GVHD, and mortality) [7].

An additional consideration in the selection of a potential donor of HSCs for transplantation into a patient with Gaucher disease is the donor's glucocerebrosidase enzyme activity level in white blood cells. It is recommended that this enzyme activity level be assessed prior to transplanting HSCs from that donor. With full donor-derived engraftment and chimerism after HSCT, recipients demonstrate an enzyme activity level that is equal to that of the donor.

Preparative Regimens and Regimen-Related Toxicities

Generally, treatment regimens used for HSCT must accomplish two goals: (1) tumor cytoreduction and ideally disease eradication; and (2) immunosuppression to overcome host rejection of the graft. In the case of storage disorders such as Gaucher disease, there is no malignant clone to be eradicated; however, the obstacle to achieving successful long-term donor-derived engraftment is significant since the immune system is intact in patients with storage disorders who are undergoing HSCT. In addition, the success of HSCT is limited, in part, by transplant-related mobidity and mortality. A regimen-related toxicity grading system that evaluates nonhematologic toxicity focuses on nine organ systems: heart, bladder, kidneys, lungs, liver, mucosa, CNS, gastrointestinal tract, and skin [8]. This is particularly relevant for storage diseases in which accumulated substrate in organs and tissues such as the lungs, liver, and gastrointestinal tract could contribute to additional toxicity and complications during and after the HSCT preparative regimen.

Graft-vs.-Host Disease, Prophylaxis, and Treatment

HLA disparity between the HSC donor and recipient is the most important factor governing the severity and kinetics of acute GVHD. The incidence of acute GVHD is much increased with HLA-nonidentical related marrow donors when compared with HLA-identical related donors. The importance of class I and class II allele matching in reducing the risk for GVHD has been widely documented in unrelated donor cohorts. Sex mismatching and female donor parity have been associated with an increased risk of acute GVHD. Age of the host is another key factor associated with the development of acute GVHD; increasing age is associated with an increasing risk. Among unrelated HSCT recipients, increasing donor age is also associated with an increased risk of developing GVHD. The source of HSCs may also influence the development of GVHD. In a study of unrelated donor HSCTs, the incidence of acute GVHD was significantly lower in recipients of umbilical cord blood compared to recipients of bone marrow. Peripheral blood HSCT appears to be associated with an increased risk for the development of chronic GVHD [9].

Clinical features of acute GVHD are most often noted in the skin as a maculo-papular exanthem, in the liver with cholestatic jaundice, and in the gastrointestinal tract. In the latter, symptoms of acute GVHD of the distal small bowel and colon include profuse diarrhea, intestinal bleeding, crampy abdominal pain, and ileus. Upper GI tract GVHD typically causes anorexia and dyspepsia [9].

Prevention of acute GVHD is promoted through the identification of the best available, matched donor, immunosuppressive therapy such as methotrexate, cyclosporine, corticosteroids, tacrolimus, mycophenolate mofetil, antibody prophylaxis, and bone marrow T-lymphocyte depletion [9]. Primary therapy for acute GVHD has traditionally been glucocorticoids for the initial management. Secondary therapy is typically less effective in achieving control of acute GVHD. A number of established as well as novel therapies have been used or are under study [9].

Chronic GVHD can affect numerous organs and tissues of the body beginning at 6 months following HSCT. These tissues include: skin, liver, eyes, mouth, lungs, GI tract, neuromuscular system, and others. [9] Therapy is primarily prolonged administration of immunosuppressive agents.

Sources of Hematopoietic Stem Cells: Bone Marrow, Peripheral Blood, and Umbilical Cord Blood

Potential sources of allogeneic hematopoietic stem cells for transplantation include bone marrow, peripheral blood, and umbilical cord blood. The cells may come from a related or unrelated donor. The National Marrow Donor Program (NMDP) was founded in 1987 as a cooperative effort of the American Association of Blood Banks (AABB), the America Red Cross, the Council of Community Blood Centers, and the U.S. Navy to facilitate volunteer, unrelated donor marrow transplantation. Standards established by the NMDP cover cells obtained from bone marrow, peripheral blood, and umbilical cord blood. In addition to the NMDP, there are numerous other registries throughout the world that cooperate in identifying unrelated donors of HSCs. The growth and development of registries of unrelated potential donors of adult HSCs and umbilical cord blood has been successful due to the following reasons: 1) need for a source of hematopoietic stem cells for the 70% of the population that does not have an available HLA-matched donor within the immediate family and 2) the need of the voluntary unrelated donor to help and to be of service. These volunteer donors may be called upon to donate HSCs from bone marrow or peripheral blood. In addition, banking of frozen units of umbilical cord blood has occurred worldwide. Several potential advantages of unrelated donor umbilical cord blood transplantation over unrelated donor bone marrow or peripheral blood are the ready availability of banked umbilical cord blood units, a shorter time to acquisition of umbilical cord blood with a more rapid time to transplantation and the ability to tolerate greater degrees of HLA disparity. Clinical experience suggests that HLA mismatches of up to two antigens out of six can result in acceptable clinical outcomes [10]. Ongoing studies are investigating the optimal dose of umbilical cord blood cells as well as the possible concomitant use of two units of umbilical cord blood for transplantation, particularly in adult recipients [11,12].

Complications of HSCT and Their Management

A number of complications can occur following allogeneic transplantation, including acute and chronic graft-vs.-host disease (GVHD); infections such as bacterial, fungal, and viral; pulmonary; gastrointestinal and hepatic; neurological; hemolytic anemias; delayed immune reconstitution; pain; oral; impaired growth and development; as well as delayed complications. While it is beyond the scope of this chapter to review these in detail, it may be reasonably concluded that significant progress has been made in recent years to prevent many of these complications, reduce their severity, or more effectively treat them. This has significant ramifications for whether HSCT is chosen as a potential therapy for various inherited metabolic disorders including Gaucher disease.

HEMATOPOIETIC STEM CELL TRANSPLANTATION FOR GAUCHER DISEASE

The field of HSCT for Gaucher disease dates back over two decades. In this section of the chapter, attention will focus on the clinical experience with HSCT for all types of Gaucher disease. Conclusions will be drawn about the effectiveness of HSCT for each type of Gaucher disease.

Overview

The principle underlying the use of HSCT for selected inherited metabolic storage diseases is that of cross-correction. In 1968, Fratantoni and co-workers established the basis of our understanding of transferable lysosomal enzymes by demonstrating metabolic cross-correction of defects in co-cultures of fibroblasts from Hurler and Hunter syndrome patients [13]. This observation and the later demonstration of correction by lymphocyte extracts or serum [14,15] led Prof. John Hobbs to employ HSCT as a permanent source of enzyme in a Hurler patient [16]. Dramatic improvement in disease phenotype paved the way for HSCT in many inherited metabolic diseases. In the case of Gaucher disease, the phenotype of type 1 patients is expressed as a result of changes in macrophages, which are the progeny of HSCs; hence, the logic behind using HSCT as a potential therapy in this disorder. The first patient with Gaucher disease (type 3) to be transplanted showed normalization of glucocerebrosidase enzyme activity in the peripheral blood early after the transplant and clearance of marrow Gaucher cells within approximately 6 months [17]. This patient had an advanced stage of type 3 Gaucher at the time of transplant; during the post-HSCT course, there were no significant changes in his clinical status. This observation is consistent with subsequent clinical experience when HSCT is performed in lysosomal storage disease patients whose disease is far advanced. The patient died of an infectious complication prior to 1 year. Subsequently a number of patients with Gaucher disease type 1 have been treated by HSCT [18–20]. Responses in surviving patients have been favorable and will be discussed (see below).

Type 1 Gaucher Disease

The early experience with HSCT for type 1 patients dates to the mid-1980s at the Westminster Children's Hospital in London, England. Prof. Hobbs and co-workers reported on the beneficial effect of pre-HSCT splenectomy in patients with type 1 Gaucher disease and their rapid and remarkable improvement [19]. A biochemical marker of Gaucher disease, plasma chitotriosidase, declined to normal levels in patients with Gaucher disease by 5 to 12 years after their HSCTs [21]. Reports have focused on the marked improvement in Gaucher bone disease following HSCT, including resolution of bone crises and resumption of normal growth [19,22,23].

Type 2 Gaucher Disease

Due to the rapidly progressive nature of the acute neuronopathic form (i.e., type 2) of Gaucher disease, attempts to perform HSCT in these patients have been limited and without clinical benefit. Alternative approaches to neurologic disease in Gaucher disease are being investigated. They include substrate depletion and chaperone therapy, which are discussed in greater detail in Chapter 18 and Chapter 19, respectively.

Type 3 Gaucher Disease

A study undertaken by clinicians in Sweden at the Huddinge Hospital led to the characterization of the Norrbottnian type of Gaucher disease. This study focused on CNS symptomatology and function, correlation of clinical signs with laboratory, neuropathological and biochemical findings, the effects of splenectomy on the course and severity of the disease, and the effect of HSCT on this form of type 3 Gaucher disease [24]. The clinical pattern observed was that of a usually alert child with normal intelligence, short stature, splenomegaly, a bleeding tendency, and ocular manifestations. The course was slowly progressive in these children whose median age at diagnosis was approximately 2 years. There was considerable variability in the clinical course with a median life expectancy of 12 years. Early motor development was delayed in the lower extremities but generally normal in the upper limbs. Intelligence declined with age. Ocular changes included abducens nerve (cranial nerve VI) weakness and abnormalities in horizontal gaze. Splenectomized patients tended to demonstrate more marked EKG changes and retinal infiltrates. Postmortem examinations revealed glucosylceramide in high concentrations in the cerebellum and cerebral subcortical white matter. This characterization of the natural history and clinical course of patients with Norrbottnian Gaucher disease served as an important foundation by which the experience with HSCT could be judged. For patients with Gaucher type 3 disease, the group led by Prof. Olle Ringdén in Stockholm has the largest clinical experience in transplantation [25–29]. These patients had correction of their hepatomegaly, disappearance of Gaucher cells, resolution of bone pain, and experienced growth spurts. Furthermore, encouraging results have been reported in the long-term follow-up (i.e., 10 years) of the neurologic and neuropsychologic function of these patients [30]. Neurocognitive function has stabilized in the normal or above normal range in most patients. This was a critical observation that underscored the value of HSCT for patients with Norrbottnian Gaucher disease. As noted, without HSCT, patients with the Norrbottnian type of Gaucher disease deteriorated mentally. During the Huddinge Hospital HSCT experience for Gaucher disease, mixed donor-recipient chimerism was observed in some patients. A question was raised as to whether the beneficial effects of HSCT would still be realized. There is now clinical evidence to support the notion that patients who demonstrate mixed chimerism after HSCT are still able to derive benefit from the transplant [31,32]. Further underscoring the significant benefits of HSCT has been the inability of long-term enzyme replacement therapy (ERT) to stabilize the neurologic and neuropsychologic status of patients with Norrbottnian (type 3) Gaucher disease [33].

STEM CELLS

Stem cells can be defined as single cells that are capable of both self-renewal and differentiation. Hematopoietic stem cells are therefore blood-forming cells that can duplicate by self-renewal as well as give rise to all types of differentiated blood cells.

Overview

The primary focus of allogeneic transplantation for Gaucher disease and other lysosomal storage disorders has been and continues to be on the use of HSCs. However, there is keen interest and much investigatory work on novel or alternative stem cells for use in transplantation as well as for gene transfer/therapy. This section will focus on only two types of stem cells, mesenchymal and neural.

Mesenchymal Stem Cells

Approximately 40 years ago, Friedenstein described stromal cells in the bone marrow that were spindle-shaped and proliferated to form colonies. These cells attached to plastic and could differentiate *in vitro* into multiple cell types including osteoblasts, chondroblasts, adipocytes, etc. More recently these cells have been isolated from postnatal bone marrow and called mesenchymal stem cells (MSCs) or stromal stem cells. Despite extensive attempts to characterize MSCs, definitive *in vivo* markers have not been identified. Mesenchymal stem cells have been used extensively for transplantation experiments in animals as well as for some therapeutic trials in humans. However, further study is needed to better define the molecular mechanisms of MSC differentiation to evaluate their full potential for tissue regeneration [34].

Two examples of therapeutic trials using MSCs will be presented. First, intravenous MSCs were administered early after acute myocardial infarction (MI) in a swine model. Price and co-workers at Cedars-Sinai Medical Center at UCLA then noted improvement in left ventricular ejection fraction and limited wall thickening in the remote noninfarcted myocardium consistent with a beneficial effect on post-MI ventricular remodeling [35]. Second, after observing that patients with Hurler syndrome and metachromatic leukodystrophy (MLD) develop significant skeletal and peripheral nervous system defects respectively despite successful HSCT, Koc and Krivit hypothesized that some of these defects might be corrected by infusion of allogeneic, multi-potential, bone marrow-derived MSCs [36]. There was no infusion-related toxicity. In most recipients, culture-purified MSCs at 2 days, 30 to 60 days, and 6 to 24 months after MSC infusion were still host in origin. In two patients, bone marrow-derived MSCs contained 0.4 and 2.0% donor MSCs by FISH when examined 60 days after MSC infusion. In four patients with MLD, small improvements in some peripheral nerve conduction velocities were noted. The bone mineral density was either maintained or slightly improved in all patients. However, there was no clinically apparent change in patients' overall health, mental and physical development following MSC infusion. The authors concluded that donor allogeneic MSC infusion was safe and may be associated with reversal of disease pathophysiology in some tissues. Further study into the clinical use of MSCs for transplantation is needed.

Neural Stem Cells

Transplantation of stem cells or their differentiated progeny and mobilization of endogenous stem cells in the postnatal brain have been proposed as potential therapies for various brain disorders such as Parkinson's disease and stroke and genetic neurodegenerative disorders such as Gaucher disease (type 2 and 3) and the leukodystrophies [37–39]. The best source of stem cells remains unknown. Research has focused on embryonic stem cells and stem cells from embryonic or postnatal brain or from other tissues. A current review described the progress to date on neural differentiation of postnatally derived bone marrow and umbilical cord blood cells [40]. While historically cell-based therapies for the CNS have been derived from fetal or embryonic origin, recently the neural potential of readily-available and accessible adult bone marrow and umbilical cord blood stem cells has been elucidated.

GENE THERAPY

Autologous hematopoietic stem cells have been used as targets for gene transfer/therapy, with applications in inherited disorders, cell therapy, and acquired immunodeficiency states. The types of cells include hematopoietic progenitor/stem cells, lymphocytes, and MSCs.

Overview

To date, inherited disorders for which gene transfer/therapy clinical trials have been performed include severe combined immunodeficiency (SCID), common variable gamma-chain immunodeficiency, chronic granulatomous disease and Gaucher disease, type 1. Preclinical gene transfer/therapy efforts are ongoing for thalassemia, sickle cell disease, Wiskott-Aldrich syndrome, and Fanconi anemia. Clinical trials of immune-based therapy with genetically modified lymphocytes are being conducted for malignant disorders. To treat patients infected with HIV, clinical trials using antiviral agents in conjunction with autologous transplantation are active. Gene therapy vectors are being developed to eradicate tumor cells that contaminate autologous stem cell products. However, the risk of insertional mutagenesis and the potential for the development of leukemia was highlighted by the first gene therapy trials for patients with SCID that had conferred therapeutic benefit. Despite the slow progress in the field of gene transfer/therapy, there is great promise for its therapeutic applications in the future [41].

Experience with Gene Therapy for Gaucher Disease

There is extensive laboratory and limited clinical experience with glucocere-brosidase gene transfer by various techniques into human and animal HSCs. In this final section of the chapter, a broad overview of the field of gene transfer/therapy for Gaucher disease will be presented.

In the early 1990s, work by at least three groups showed the following: 1) expression of human glucocerebrosidase following retroviral vector-mediated trans-duction of murine HSCs [42], 2) retroviral-mediated transfer of the human gluco-cerebrosidase gene into cultured Gaucher bone marrow [43], 3) transfer and sus-tained expression of the human glucocerebrosidase gene in mice and their macrophages after transplantation of retrovirally transduced bone marrow [44], and 4) high titer amphotrophic vectors containing the glucocerebrosidase gene capable of transducing human CD34+ HSCs at high efficiency (i.e., 70%) [45]. Subsequently, in the mid-1990s, further laboratory progress was made [46–50] leading to the first clinical trials of gene transfer in patients with type 1 Gaucher disease [51]. While there was great promise for initial efforts at achieving therapeutic gene transfer in patients with Gaucher disease, significant clinical benefit was not observed by any medical groups performing such clinical trials. Consequently, investigators redi-rected their focus to improving methods for gene transfer in the setting of ever-improving success rates with HSCT for various lysosomal storage disorders and leukodystrophies. Much of the attention was directed toward the use of viral vectors for gene transfer, particularly retroviral ones [52–55]. New insights into the selection of transduced cells were also gained through the use of methotrexate and the green fluorescent protein gene, respectively [56,57].

Currently, there are high hopes tempered by modest expectations for gene transfer/therapy for both genetic diseases and cancer. These perspectives are reflected in the numerous reviews and commentaries published recently [58–63]. Recent developments have also been aided by the development of a mouse model of Gaucher disease. Gene transduction of hepatocytes with a plasmid DNA vector encoding human glucocerebrosidase led to high-level expression and secretion of enzyme into the systemic circulation with resultant normalization of levels of glucocerebrosidase in Kupffer cells [64]. Due to limitations experienced with viral vectors, alternative approaches have been evaluated with and without success. An attempt to use a chimeraplast strategy (i.e., correction of single-base mutations by mismatch repair mechanisms using chimeric RNA/DNA oligonucleotides, called chimeraplasts) for a mutation (i.e., the 1448C->T) causing Gaucher disease was inefficient [65]. Other recent approaches have included: 1) knockdown of chimeric glucocerebrosidase by green fluorescent protein-directed small interfering RNA [66], 2) vascular and hepatic delivery of a HIV-1-based lentivirus vector encoding human glucocerebrosidase cDNA leading to the production of therapeutic levels of glucocerebrosidase by cultured fibroblasts derived from patients with Gaucher disease [67], and 3) recombinant adeno-associated viral vectors containing the glucocerebrosidase cDNA driven by the human elongation factor 1-alpha promoter resulting in efficient expression of human glucocerebrosidase in human Gaucher fibroblasts maintained over a period of 20 weeks [68].

QUESTIONS, CONCLUSIONS, AND FUTURE DIRECTIONS

The history of treatment of Gaucher disease started with splenectomy, continued with bone marrow transplantation, and for the past 15 years has focused primarily on ERT. Despite these major therapeutic advances, many questions and clinical management issues remain [69]. These include: 1) how to monitor ERT to determine the optimal dosage and therapeutic effect, 2) how to treat mild disease, 3) whether intermittent treatment is a reasonable option, 4) the cause of neurologic signs and how best to treat the CNS, 5) management of pulmonary hypertension, 6) monitoring and treatment of bone disease, 7) the role of novel therapies such as substrate depletion and chaperone therarpy, 8) who are the most appropriate candidates for HSCT and when in the course of their disease should they undergo transplant. One could conclude that ERT is the safest therapy currently available; however, with advances in the field of HSCT current success rates are surpassing 90% in selected patient populations [70]; thus, perhaps the role of HSCT in patients with type 1 and 3 Gaucher should be re-examined. Perhaps, the time has come for the judicious use of multiple therapeutic agents and modalities tailored to the patient's clinical condition and type of Gaucher disease. While the fields of gene transfer/gene therapy and stem cell transplantation hold great promise, both require further research before they can yield clinical benefits for patients with Gaucher disease. Experts from around the world in the field of Gaucher disease have written extensively and eloquently on these and other issues [71–76].

REFERENCES

1. Bach, F.H. et al., Bone-marrow transplantation in a patient with the Wiskott-Aldrich syndrome. *Lancet*, 2, 1364, 1968.
2. Gatti, R.A. et al., Immunological reconstitution of sex-linked immunological deficiency. *Lancet*, 2, 1366, 1968.
3. Good, R.A. et al., Successful marrow transplantation for correction of immunological deficit in lymphopenic agammaglobulinemia and treatment of immunologically induced pancytopenia. *Exp. Hematol.*, 19, 4, 1969.
4. De Konig, J. et al., Transplantation of bone-marrow cells and fetal thymus in an infant with lymphopenic immunological deficiency. *Lancet*, 1, 1223, 1969.
5. Ringdén, O. et al., Long-term follow-up of the first successful bone marrow transplantation in Gaucher disease. *Transplantation*, 46, 66, 1988.
6. Bodmer, J.G. et al., Nomenclature for factors of the HLA system. *Tissue Antigens*, 57, 236, 2001.

7. Mickelson, E. and Petersdorf, E.W., Histocompatibility, in *Thomas' Hematopoietic Cell Transplantation*, Blume, K.G., Forman, S.J., and Appelbaum, F.R., Eds., 3rd ed., Blackwell Publishing, Malden, MA, 2004, chap. 4.

8. Bensinger, W.I. and Spielberger, R., Preparative regimens and modification of regimen-related toxicities, in *Thomas' Hematopoietic Cell Transplantation*, Blume, K.G., Forman, S.J., and Appelbaum, F.R., Eds., 3rd ed., Blackwell Publishing, Malden, MA, 2004, chap. 13.

9. Sullivan, K.M., Graft-vs.-host disease, in *Thomas' Hematopoietic Cell Transplantation*, Blume, K.G., Forman, S.J., and Appelbaum, F.R., Eds., 3rd ed., Blackwell Publishing, Malden, MA, 2004, chap. 50.

10. Broxmeyer, H.E. and Smith, F.O., Cord blood hematopoietic cell transplantation, in *Thomas' Hematopoietic Cell Transplantation*, Blume, K.G., Forman, S.J., and Appelbaum, F.R., Eds., 3rd ed., Blackwell Publishing, Malden, MA, 2004, chap. 43.

11. Barker, J.N. and Wagner, J.E., Umbilical cord blood transplantation: current practice and future innovations. *Crit. Rev. Oncol. Hematol.*, 48, 35, 2003.

12. Barker, J.N. et al., Transplantation of 2 partially HLA-matched umbilical cord blood units to enhance engraftment in adults with hematologic malignancy. *Blood*, 105, 1343, 2005.

13. Fratantoni, J., Hall, C., and Neufeld, E., The defect in Hurler and Hunter syndromes. II. Deficiency of specific factors involved in mucopolysaccharide degradation, *Proc. Natl. Acad. Sci. U.S.A.*, 64, 360, 1969.

14. Di Ferrante, N., Nichols, B., Donnelly, P., Neri, G., Hrgovcic, R., Berglund, R., Induced degradation of glycosaminoglycans in Hurler's and Hunter's syndromes by plasma infusion, *Proc. Natl. Acad. Sci. U.S.A.*, 68, 303, 1971.

15. Knudson, A., Di Ferrante, N., and Curtis, J., Effect of leukocyte transfusion in a child with type II mucopolysaccharidosis, *Proc. Natl. Acad. Sci. U.S.A.*, 68, 1738, 1971.

16. Hobbs, J. et al., Reversal of clinical features of Hurler's disease and biochemical improvement after treatment by bone-marrow transplantation, *Lancet*, 2, 709, 1981.

17. Rappeport, J.M. and Ginns, E.I., Bone-marrow transplantation in severe Gaucher disease. *N. Engl. J. Med.*, 311, 84, 1984.

18. Hobbs, J.R., Experience with bone marrow transplantation for inborn errors of metabolism. *Enzyme*, 38, 194, 1987.

19. Hobbs, J.R. et al., Beneficial effect of pre-transplant splenectomy on displacement bone marrow transplantation for Gaucher's disease. *Lancet*, 1, 1111, 1987.

20. August, C. et al., Bone marrow transplantation (BMT) in Gaucher's disease. *Pediatr. Res.*, 18, 236a, 1984.

21. Young, E. et al., Plasma chitotriosidase activity in Gaucher disease patients who have been treated either by bone marrow transplantation or by enzyme replacement therapy with alglucerase, *J. Inherit. Metab. Dis.*, 20, 595, 1997.

22. Yabe, H. et al., Secondary G-CSF mobilized blood stem cell transplantation without preconditioning in a patient with Gaucher disease: report of a new approach which resulted in complete reversal of severe skeletal involvement, *Tokai J. Exp. Clin. Med.*, 30, 77, 2005.

23. Jones, S. et al., DBMT for Gaucher's disease, in *Correction of Certain Genetic Diseases by Transplantation*, Hobbs, J.R., Ed., Headstart Printing Co., London, 1989, 23.

24. Erikson, A., Gaucher disease — Norrbottnian type (III). Neuropaediatric and neurobiological aspects of clinical patterns and treatment, *Acta Paediatr. Scand.* Suppl., 326, 1, 1986.

25. Erikson, A. et al., Clinical and biochemical outcome of marrow transplantation for Gaucher disease of the Norrbottnian type. *Acta Paediatr. Scand.,* 79, 680, 1990.

26. Ringdén, O. et al., Long-term follow-up of the first successful bone marrow transplantation in Gaucher disease. *Transplantation,* 46, 66, 1988.

27. Lonnqvist, B. et al., Cytomegalovirus infection associated with and preceding chronic graft-versus-host disease. *Transplantation,* 38, 465, 1984.

28. Tsai, P. et al. Allogeneic bone marrow transplantation in severe Gaucher disease. *Pediatr. Res.,* 31, 503, 1992.

29. Svennerholm, L. et al., Norrbottnian type of Gaucher disease — Clinical, biochemical and molecular biology aspects: successful treatment with bone marrow transplantation. *Dev. Neurosci.* 13, 345, 1991.

30. Ringdén, O. et al., Ten years' experience of bone marrow transplantation for Gaucher disease. *Transplantation,* 59, 864, 1995.

31. Chan, K.W., et al., Bone marrow transplantation in Gaucher's disease: effect of mixed chimeric state. *Bone Marrow Transplant.,* 14, 327, 1994.

32. Yen, C.C. et al., Allogenic bone marrow transplantation for Gaucher disease — a case report, *Zhonghua Yi Xue Za Zhi* (Taipei), 59, 372, 1997.

33. Erikson, A., Astrom, M., and Mansson, J.E., Enzyme infusion therapy of the Norrbottnian (type 3) Gaucher disease, *Neuropediatrics,* 26, 203, 1995.

34. Kalervo Vaananen, H., Mesenchymal stem cells, *Ann. Med.,* 37, 469, 2005.

35. Price, M.J. et al., Intravenous mesenchymal stem cell therapy early after reperfused acute myocardial infarction improves left ventricular function and alters electrophysiologic properties, *Int. J. Cardiol.,* Epub ahead of print, 2005.

36. Koc, O.N. et al., Allogeneic mesenchymal stem cell infusion for treatment of metachromatic leukodystrophy (MLD) and Hurler syndrome (MPS-IH), *Bone Marrow Transplant.,* 30, 215, 2002.

37. Lindvall, O. and Kokaia, Z., Stem cell therapy for human brain disorders. *Kidney Int.,* 68, 1937, 2005.

38. Leker, R.R. and McKay, R.D., Using endogenous neural stem cells to enhance recovery from ischemic brain injury. *Curr. Neurovasc. Res.,* 1, 421, 2004.

39. Mueller, D. et al., Transplanted human embryonic germ cell-derived neural stem cells replace neurons and oligodendrocytes in the forebrain of neonatal mice with excitotoxic brain damage, *J. Neurosci. Res.,* Epub ahead of print, 2005.

40. Ortiz-Conzalez, X.R. et al., Neural induction of adult bone marrow and umbilical cord stem cells, *Curr. Neurovasc. Res.,* 1, 207, 2004.

41. Becker, P.S., The current status of gene therapy in autologous transplantation, *Acta Haematol.,* 114, 188, 2005.

42. Weinthal, J. et al., Expression of human glucocerebrosidase following retroviral vector-mediated transduction of murine hematopoietic stem cells, *Bone Marrow Transplant.,* 8, 403, 1991.

43. Nolta, J.A. et al., Retroviral-mediated transfer of human glucocerebrosidase gene into cultured Gaucher bone marrow, *J. Clin. Invest.,* 90, 342, 1992.

44. Ohashi, T. et al., Efficient transfer and sustained high expression of the human glucocerebrosidase gene in mice and their functional macrophages following transplantation of bone marrow transduced by a retroviral vector, *Proc. Natl. Acad. Sci. U.S.A.,* 89, 11332, 1992.

45. Karlsson, S., Correll, P.H., and Xu, L., Gene transfer and bone marrow transplantation with special reference to Gaucher's disease, *Bone Marrow Transplant.,* 11, Suppl. 1, 124, 1993.

46. Krall, W.J. et al., Cells expressing human glucocerebrosidase from a retroviral vector repopulate macrophages and central nervous system microglia after murine bone marrow transplantation, *Blood,* 83, 2737, 1994.

47. Kiem, H.P. et al., Gene therapy and bone marrow transplantation, *Curr. Opin. Oncol.,* 7, 107, 1995.

48. Wells, S. et al., The presence of an autologous marrow stromal cell layer increases glucocerebrosidase gene transduction of long-term culture initiating cells (LTCICs) from the bone marrow of a patient with Gaucher disease, *Gene Ther.,* 2, 512, 1995.

49. Hallek, M. and Wendtner, C.M., Recombinant adeno-associated virus (rAAV) vectors for somatic gene therapy: recent advances and potential clinical applications, *Cytokines Mol. Ther.,* 2, 69, 1996.

50. Havenga, M. et al., Development of safe and efficient retroviral vectors for Gaucher disease, *Gene Ther.,* 4, 1393, 1997.

51. Barranger, J.A. et al., Gaucher's disease: studies of gene transfer to haematopoietic cells, *Baillieres Clin. Haematol.,* 10, 765, 1997.

52. Takiyama, N. et al., Comparison of methods for retroviral mediated transfer of glucocerebrosidase gene to CD34+ hematopoietic progenitor cells, *Eur. J. Haematol.,* 61, 1, 1998.

53. Robbins, P.D. and Ghivizzani, S.C., Viral vectors for gene therapy, *Pharmacol. Ther.,* 80, 35, 1998.

54. Havenga, M.J. et al., Methotrexate selectable retroviral vectors for Gaucher disease, *Gene Ther.,* 5, 1379, 1998.

55. Dunbar, C.E. et al., Retroviral transfer of the glucocerebrosidase gene into CD34+ cells from patients with Gaucher disease: *in vivo* detection of transduced cells without myeloablation, *Hum. Gene Ther.,* 9, 2629, 1998.

56. Havenga, M. et al., *In vivo* methotrexate selection of murine hemopoietic cells transduced with a retroviral vector for Gaucher disease, *Gene Ther.,* 6, 1661, 1999.

57. Shimizu, T. et al., A simple and efficient purification of transduced cells by using green fluorescent protein gene as a selection marker, *Acta Paediatr. Jpn.,* 40, 586, 1998.

58. McIvor, R.S., Gene therapy of genetic diseases and cancer, *Pediatr Transplant.,* 3, Suppl. 1, 116, 1999.

59. Barranger, J.A., Rice, E.O., and Swaney, W.P., Gene transfer approaches to the lysosomal storage disorders, *Neurochem. Res.,* 24, 601, 1999.

60. Barranger, J.A. and Novelli, E.A., Gene therapy for lysosomal storage disorders, *Expert Opin. Biol. Ther.,* 1, 857, 2001.

61. Cabrera-Salazar, M.A., Novelli, E., and Barranger, J.A., Gene therapy for the lysosomal storage disorders, *Curr. Opin. Mol. Ther.,* 4, 349, 2002.

62. De Fost, M., Aerts, J.M., and Hollak, C.E., Gaucher disease: from fundamental research to effective therapeutic interventions, *Neth. J. Med.,* 61, 3, 2003.

63. Grabowski, G.A., Perspectives on gene therapy for lysosomal storage diseases that affect hematopoiesis, *Curr. Hematol. Rep.,* 2, 356, 2003.

64. Marshall, J. et al., Demonstration of the feasibility of *in vivo* gene therapy for Gaucher disease using a chemically induced mouse model, *Mol. Ther.,* 6, 179, 2002.

65. Diaz-Font, A. et al., Unsuccessful chimeraplast strategy for the correction of a mutation causing Gaucher disease, *Blood Cells Mol. Dis.,* 31, 183, 2003.

66. Campbell, T.N. and Choy, F.Y., Knockdown of chimeric glucocerebrosidase by green fluorescent protein-directed small interfering RNA, *Genet. Mol. Res.,* 3, 282, 2004.

67. Kim, E.Y. et al., Expression and secretion of human glucocerebrosidase mediated by recombinant lentivirus vectors *in vitro* and *in vivo*: implications for gene therapy of Gaucher disease, *Biochem. Biophys. Res. Commun.*, 318, 381, 2004.

68. Hong, Y.B. et al., Feasibility of gene therapy in Gaucher disease using an adeno-associated virus vector, *J. Hum. Genet.*, 49, 536, 2004.

69. Erikson, A., Remaining problems in the management of patients with Gaucher disease, *J. Inherit. Metab. Dis.*, 24, Suppl. 2, 122, 2001.

70. Peters, C. et al., Cerebral X-linked adrenoleukodystrophy: the international hematopoietic cell transplantation experience from 1982 to 1999, *Blood*, 104, 881, 2004.

71. Jmoudiak, M. and Futerman, A.H., Gaucher disease: pathological mechanisms and modern management, *Br. J. Haematol.*, 129, 178, 2005.

72. Germain, D.P., Gaucher's disease: a paradigm for interventional genetics, *Clin. Genet.*, 65, 77, 2004.

73. Grabowski, G.A., Recent clinical progress in Gaucher disease, *Curr. Opin. Pediatr.*, 17, 519, 2005.

74. Elstein, D. et al., Gaucher disease: pediatric concerns, *Paediatr. Drugs*, 4, 417, 2002.

75. Schiffmann, R. and Brady, R.O., New prospects for the treatment of lysosomal storage diseases, *Drugs*, 62, 733, 2002.

76. Balicki, D. and Beutler, E., Gene therapy of human disease, *Medicine* (Baltimore), 81, 69, 2002.

CHAPTER 23

Ethical Concerns in Treating Rare Diseases with Expensive Therapy

Deborah Elstein and Avraham Steinberg

CONTENTS

INTRODUCTION TO THE ETHICAL CONTEXT

Health care professionals are united in concern for their patients' welfare and are committed to certain ethical norms that reflect both the unique relationship of the care giver to the patient as well as that of man to his/her fellow man. Above all, there are the values of the sacredness of and respect for human life and the personal and professional integrity of the health care provider. The sacredness of life has been defined as respect for human life that should be assisted insofar as is feasible and consistent with the ethical values enumerated below. Yet, there is less ethical value in prolonging life by virtue of respect for life if this produces a significant lack of quality of life. Prolongation of life is not always equivalent to the best interests of the patient (if known) under all ethical situations, and may indeed occasionally be in conflict with the moral right of respect for and dignity of life. Of equal importance is appreciation of the role of the doctor as a care giver with the expectation that he/she respects the commonly held ethical norms and can be expected to uphold them when choosing therapeutic alternatives and administering potentially harmful procedures. Thus, there are no simple solutions to these complicated and often intersecting values. Therefore, many societies and religions have created mechanisms that apply, often in a dispassionate way, to what inevitably will be called "tragic choices" since, in the ultimate analysis, the subject of the discourse is a fellow human being who has no recourse but to be dependent on the decisions of others. This is especially true when never-competent patients are the subject of these tragic choices.

In discussing universal values as the context for issues that impact decision making by a doctor, four ethical principles have been advocated, which many moral societies accept as ultimate moral goals. Nonetheless, as will be shown, modifications due to the ethical/religious perspective of the individual or the society are often inherent in the system and possibly even desirable.

1. Autonomy. Each care giver must respect the autonomy of his/her patient who is his/her fellow human, including respecting the patient's decisions regardless of the processes that may have been used to arrive at those decisions, including religious, moral, or societal concerns that conflict with those of the care giver. This implies that the care giver must fully inform the patient of all data concerning the diagnosis and management options, in language that is comprehensive and unambiguous. In the current era there is no place for attitudes of inequality, i.e., the health professional as superior to the patient by virtue of his status. That said, the autonomy of the doctor must also be respected but not in greater measure than that of the patient; however, because of the responsibility of the doctor to provide the best possible service to his/her patient, it is the doctor's responsibility to keep the patient's welfare foremost among his/her concerns. Autonomy of the individual, patient or doctor, is limited by prevailing precepts that are codified in the legal and normative procedures of a particular society. One cannot presuppose that even in the most tragic of circumstances, the needs of an individual could contravene that of another or that of society as a whole.

2. Nonmalfeasance. The commitment to avoid harm is a basic tenet of moral codes. This value is underscored in the Jewish aphorism "avoid harm and do good" wherein the supposition is that both are very important but not necessarily identical. Thus, while it may not be in the nature of men to be kindly to each other, there is an imperative not to induce injury *per se*. The philosophical approach of utilitarianism, which advocates attempting to provide the greatest good to the greatest number even if a few do not benefit or are harmed, also argues that what is morally right is proportional to the benefit derived.

3. Beneficence. Beyond the admonition not to induce harm, there is a parallel value (to do good as opposed to an obverse value of avoiding harm). For the health professional this may be an obvious standard which that elude those not in the caring professions, yet as a corollary to the value of nonmalfeasance, doing good highlights an active elemental requirement to seek the best possible management for a patient, and thereby suggests seeking expert opinions and providing complete

and evidence-based options as the underpinnings of the information and care provided to the patient. An interesting form of beneficence is the process whereby pharmaceuticals are introduced for sale.

4. Justice. Inequality among men/women is morally wrong, and therefore it is to be expected that each would deal with his or her fellow justly and fairly. Nonetheless, no two individuals interact in a vacuum but rather are placed within societal constraints, admittedly with unique and diverse expectations that are dependent on historical, geographic, religious, ethnic, economic, and many other variables. Thus, justice may be proscribed by well-defined criteria in comparing the individual's need for justice relative to that of the organization/system/society of which he is a member. In instances where total equality is impossible, there is a need for adjusted justice based on different modes of considerations. One of the greatest constraints for justice is the economic problem of societies whereby there are scarce resources, insufficient to provide all societal needs, whether medical or other (education, security, culture, etc.). Hence, there is a moral obligation to allocate the scarce resources in the fairest way, and the ethical issues revolve around distributive justice, rather than absolute justice.

The above four values are classic in the sense that they are part of the fabric of most ethical skeins. There are three further values that mark many other systems, especially that of the Jewish or Judeo-Christian perspective.

1. Value of Life: The spark of "human-ness" inherent in every being is not to be belittled and must be part of the equation of interpersonal relationships. Beyond being able to assess what life is worth to any other individual is the nature of the hubris of a man who wants to make that judgment. Thus, as a value, life *per se* is not relative to any other value but is an inherent value and should not be qualified by any other value.

2. Responsibility/Solidarity of Society: Unlike an interpersonal relationship where both sides may have equally strong claims, society is not an individual with parochial concerns. Society as a group has more complex rules that must be applied to maintain the fabric of its existence. Hence, it may be acceptable for a society to invoke systems that are by their nature directives, particularly when the values of autonomy and justice, for instance, are unclear.

3. Compassion for the Weak: Some societies may be unable to accept responsibility for their weak, be they defined as poor and without health insurance, terminally sick, or incapable of ever contributing to the general good. Compassion and by extension, underwriting the welfare of those who cannot care for themselves, is a value of enlightened religions, and it is a value in socialistic democratic nations and as such may be a further source of "tragic choices."

INTRODUCTION TO LYSOSOMAL DISORDERS WITH LETHAL INFANTILE FORMS

Lysosomal storage disorders by virtue of similar enzymatic and biochemical pathophysiology are consequently also united by the signs and symptoms that patients present since these are due to intracellular storage of (the various) glycolipids. The organs of storage are specific to each disorder yet almost all have neuronopathic signs in addition to visceral involvement. Most importantly, the "one gene, one enzyme" thinking has made each disease a potential target for both gene therapy and enzyme therapy. Whereas the latter has been successfully applied for visceral improvement in a few diseases, only gene therapy holds the potential for a curative approach. The recent introduction of substrate inhibition has refined the goals of therapy since it may now be possible to effect neurological improvement as well as visceral change since small molecule therapy may pass the blood-brain barrier. These innovations may induce an ethical distinction between management of neonatal and infantile or juvenile forms of these diseases that evince progressive fatal decline in neurological functioning relative to therapeutic options permissible for the nonfatal adult or chronic forms of these same diseases. Because of the values of respect for and dignity of life, combined with the problem of scarce resources, there have heretofore been good and moral reasons to disallow therapy where treatment merely prolonged life without significantly affecting quality of life for the patients with neurological forms. In the most recent era, with the potential for both prolongation of life and improved quality of life using small molecule therapy, the concern over ethical decisions of this nature might lead to revisions, since the decisions will be made for babies who have no decision-making capacity, but the application of the new scientific developments might have a positive result for them, hence be in their best interest.

BABIES WHO CANNOT MAKE DECISIONS

When evolving ethical principles for intervention on behalf of those who have no ability to make the choice for themselves, there is a difference between those who may never develop that capacity and those who will live long enough to eventually be capable of decision making on their own behalf. In assuming that both categories require surrogate or guardian decision makers, the choice of these may also require ethical adjudication.

The application of the importance of quality of life as a moral principle in the case of a baby that may not live to make comparable decisions, or in equal measure, may not develop the intellectual capacity to make moral decisions, is different than when applying this value to an individual who does have decision-making abilities. What is "good" quality of life or potentially "better" quality of life are concepts that may conflict with issues of societal value or personal worth. The question of improved quality of life is not morally justifiable in these cases if the values aspired to are based on values of the doctor or the surrogate/guardian, or as suggested above, life in itself is a sufficient value.

It may be questioned whether the doctor should be privy, morally, to this process to preclude the possibility that the doctor may overstate the case for prolongation of life as a result of his responsibility to respect life. The doctor, certainly, has a professional responsibility to supply relevant data for the decision-making process, but has much less authority in the actual decisions, which in most instances are based on values and societal norms.

It is well accepted that parents as the natural surrogates of babies make decisions on their behalf. In cases where the parents are inappropriate or incapable of serving as surrogate decision makers, someone else must make a decision, and it will, by its nature, be paternalistic and possibly without an ethical guiding principle. In such cases a hospital Ethics Committee or the courts may be required to decide. Thus, when the patient is incapable and the natural surrogate, the parents, are conflicted, society imposes other agents whose ethical values may be complementary but will invariably be directive in order to be expedient.

CONFLICTING INTERESTS

In balancing between the principles of quality of life and the value of mere prolongation of life, the needs of the patient may not be the only concern. Rather, other interests may hold sway equally since, as suggested above, no interpersonal interaction occurs in a societal vacuum. Therefore the following interests must be included in a moral equation: that of the infant patient, that of the family and/or surrogate, that of the care giver(s), that of the medical care provider (e.g., the medical center, the third-party insurance group, or the legal establishment), and that of society as the ultimate arbitrator of provision of resources.

In many cases where the patient is incompetent, resolution is based on evidence of common interests among several of the above players and creative solutions based on a case-sensitive approach (see above).

When a surrogate or guardian is interposed between the doctor and his/her patient, the moral responsibilities of the doctor to respect the integrity of the patient are transferred to the surrogate/guardian in equal measure regardless of potential conflicts between the interests of the patient and that of the surrogate/guardian. The trust placed in the doctor's ability to deal in an ethically acceptable manner in all cases reflects the multipartite interests that may exist. Therefore, doctors must be open to peer review either informally or within the context of an Ethical Institutional Review Board. In each circumstance there must be source documentation of a plan of action taken with rationale for the course, preferably supported by experience of others, dictated by the principles of good clinical practice and based upon the best interest of the infant, since his/her autonomous wishes cannot be determined.

BABIES WITH LYSOSOMAL STORAGE DISORDERS

In the instance of babies with lysosomal diseases where the sole option for therapy is an investigational drug or regimen, transparency is an ethical imperative. The doctor must first reveal any potential for conflicts of interest that may impinge on his integrity in the specific instance. Although careful consideration of management opportunities may be circumscribed to a handful of choices, the weighing of risks vs. benefits may be skewed if the treating physician has a vested interest in a particular regimen. This in itself does not preclude the doctor from dealing with great integrity and respect in the patient's best interest, but does require identification of the source of the conflict of interest to the patient and/or surrogate/guardian, and documentation (under the best of circumstances) that this has been understood by the latter.

Of note is the realization that the interplay between the moral obligations of the doctor and that of other interests in the case of a baby who does not currently and may never in the future develop decision-making capabilities, may be viewed from diverse vantage points based on religion, national resources, some for whom the decision may be appropriate and so on. The models for some ethical decisions, particularly with regard to prolongation of life, for instance, are in some countries based on scarce resources, in some on the principle of greatest good, and in some as a compromise between societal benefits relative to individual needs. Thus, even in the current era, moral suppositions are *per force* seen in a nationalistic context that may be directive in its most literal sense.

SCARCE RESOURCES AND ETHICAL DECISIONS

As mentioned above, several countries have predicated national policy on availability of resources rather than on ethical decisions about prolongation of life. This indeed is an irrefutable reality of the modern era where there are instances of true scarcity, such as insufficient enzyme for treatment of all patients with an enzyme deficiency as well as relative scarcity such as compelling but competing claims by adults with milder disease who will enjoy the result of the treatment to better their quality of life and who contribute to society relative to claims by babies who may suffer from incurable progressive diseases.

When scarcity is included in the ethical decision making, the interplay among the moral values such as between justice and patient autonomy or justice and beneficence requires societal attention. Allocation of scarce resources is a topic of considerable interest in itself, but when focusing the ethical dilemma on the case of orphan diseases for which the specific therapy is customized and expensive, the task is made more difficult.

Justice as a value makes the supposition that all those requiring a specific treatment are equal; for society to justify unequal allocation of resources because of scarcity, some suppositions are expected: that of the need for the treatment, that of the relative success rate depending on specific case-by-case situations, that of the ability of both individuals and groups to control their possessions (i.e., money), that of the effect of unequal allocation on the mores and morals of the society, and whether the personal merits of the individuals are part of the decision-making process.

THE GREATER NEED

Although difficult to quantify, it is a morally defensible position that the greater the need, the greater the claim for unequal allocation of scarce resources. In this case, infantile forms of lysosomal storage disorders would therefore be candidates for expensive enzyme replacement therapy. Furthermore, however, under conditions of scarcity, the probability of "success" impacts on the choice of candidates for unequal allocation of scarce resources. Thus, providing expensive enzyme replacement for babies with neurological progression would not be an ethically acceptable choice since enzyme therapy would not impact the trajectory of their disease and its ultimate outcome. A further issue is the question of whether a patient with a specific need that cannot be fully met by any resource but partially by a scarce resource has a greater claim than a person with a lesser need whose need can be completely fulfilled by the scarce resource. If so, reduction in organomegaly may be an attainable goal for mildly affected patients and an unattainable goal for severely affected patients with declining cognitive function.

The respect for control over possessions is an outgrowth of the respect for autonomy. However, in the case of scarce resources, it is difficult to sanction treatment for the wealthy only because of respect for their possessions and wealth. This position flies in the face of the standard of equality and the above principle of need. When a sick poor child and a rich older individual require the same resource, condoning the autonomy of the wealthy without regard to the needs of the less wealthy is not ethical. In its most universal application, when medical resources are scarce, allocation of these resources should be based first on benefits of equal value and only secondarily on need or respect for possessions.

However, there may be a more flexible principle if one were to allow for private acquisition of a scarce resource vs. both vying for national revenues to acquire the treatment that is in scarce supply. Conversely, if one considers expert medical services as a scarce resource, then one must question whether doctors who receive private patients are ethically required to provide comparable services to patients who cannot pay privately. In the case of Ethical Boards or national governments predicating therapy on management by experts in the field, then noncompliance with this *a priori* requisite denies the importance of both need and justice. Similarly, diagnosis and screening services (and counseling where appropriate) should be available to all, but dealing with the results of these tests potentially may be resource dependent. Justice would then be compromised if only those with means are capable of follow-up. For example, ethnic screening for lysosomal storage disorders may elicit universal response if provided gratis to an at-risk population (such as Ashkenazi Jews); however, since therapy for some of these diseases is expensive and not curative, while for others there is no therapeutic option at all, there may only be limited ethical justification for large-scale screening. On the other hand, there might be significant benefit for parents who wish to plan their family and avoid the birth of sick children. This, obviously, is of universal importance, unrelated to the economic status of individuals.

RARE GENETIC DISEASES AND SCARCE RESOURCES

Patients with rare genetic diseases that are consequently more difficult to diagnosis at birth, and for which therapy and cure are not always available, suffer from a number of ethical issues that compound each other. Is screening a moral imperative if the disease is not lethal but "merely" reduces quality of life? Is the inability to achieve immediate diagnosis a lack of integrity on the part of the medical establishment? When experts are unavailable because of scarcity, is there a moral right to require equal availability for all patients regardless of ability to pay? If treatment exists, but it is dependent on scarce resources and therefore cannot be equally applied, would it be justified to apply it unequally? And is there an obligation by governmental or private individuals to supply alternative treatment when there is no profit motive?

The concept of "merit" as part of the moral considerations when viewing allocation of scarce resources implies that the need for these very same resources is a result of passive or active disregard for self-preservation on the part of the patient (e.g., noncompliance with other drug regimens or drug abuse). Whether this is an actual principle is unclear, but society reserves the right to deal with such patients in an unequal way by dint of the ethical value of justice.

ADOPTING THE ORPHANS

Dealing with conflicts between and among the ethical values because of the imposition of scarcity may be inimical to systems of social justice. However, in thinking about some guidelines, the foremost recommendation would be presentation of the need for ethical principles as gate-keepers in allocating medical resources. The importance of updating information about efficacy, safety, and adverse events (long-term surveillance) may facilitate appropriate allocation of resources when patient needs are unequal.

In addressing these issues as reflecting the reality of rare diseases, the current decade of expansion of orphan drugs for lysosomal storage disorders may be a hallmark in ethical decision making. Whereas the Congressional Orphan Drug Act of 1983 may be viewed more cynically, the adoption of the orphan diseases in a formal way was a morally righteous societal decision to correct inequalities and provide justice to patients who suffer from uncommon ailments. A decade after this proactive legislation, in the years 1994–2003, an average of 14 orphan drugs per year (range 6–24 drugs) have been approved by the American Food and Drug Administration.

Thus, in returning to the values of autonomy and justice, the obligation to provide equal access to available resources to all citizens is a societal value, rather than an obligation that can be imposed on pharmaceutical companies because of the ethical value of beneficence. Even if one were to subscribe to the argument that the Orphan Drug Act provides public monies for research into rare disorders and hence manufacturers should feel obligated to pursue research into rare diseases, there would be no specific obligation on the part of any one company to do so. The value of beneficence is not equal to a profit motive although the case may be made that there is a moral red line between developing maximally profitable drugs and not developing any orphan drugs at all. Again, the responsibility to provide beneficence is both a societal and individual value, and this is constrained by resource availability, so that even a wealthy society may not be required to provide a disease-free existence to all its citizens regardless of cost. Nonetheless, it seems to be morally correct and praiseworthy on the part of the rich, whether individuals or companies or nations, to help the poor, even in face of reduction in profit.

CONCLUSION

The ethical solution is difficult to achieve; possibly it is an impossible task, and not equally attainable by all societies. The nature of the solution demands a balance between the needs/values of the minority based on compassion, societal responsibility, and the value of life, relative to the expectation of the majority to benefit commensurate with justice and the greater good. Based on beneficence, philanthropy by individuals or the drug companies could be encouraged. Alternatively, based on solidarity of society, drug company profits from all drugs could be required by society to be the source provided to pay for the needs of the weak, i.e., orphan diseases.

Keeping ethical principles in mind, it may prove to be the case that modern society, by acknowledging either compassion or responsibility to the weak, may single out a health category such as rare diseases that primarily affect infants and choose to treat them from a non-health costs source. However, using the same values of society wanting to help the weak, one could recommend finding a means to include dozens of ultra-minorities in a cohesive group that no longer begs the question of greatest good.

RECOMMENDATIONS

An outgrowth of technological advances relevant to rare diseases is that of application of therapies based on genotyping that enhance efficacy and safety. If the case were to be made that all lysosomal storage disorders have more attributes that bind them than symptoms and signs that mark them as diverse, and if consequent to this logic, the therapeutic goal would be to develop a single modality applicable to all (with some potential modifications), then this entire class of diseases would become a single albeit conglomerate entity. The application of this principle would eliminate the issue of scarce resources that currently results in competition among rare disorders for disease-specific orphan drugs that are expensive and not all of equal safety or efficacy. Too, the ethical advantage would be the ability to adjudicate individual claims for justice and autonomy on a societal level since the resources would be more abundant in the sense that the archetypical modality would be allocated to patients on the basis of need.

Another approach to solve the ethical dilemma of orphan diseases is to globalize the issue, namely, to identify all patients with lysosomal storage disorders throughout the world, allocate funds for research and treatment through national budgets, which should be relative to the number of such patients in the countries and to the total national budgets, and distribute the treatment according to individual needs unrelated to individual ability to pay. The coordination of such a noble project could be performed by international agencies.

RECOMMENDED READING

1. Hughes, D.A., Tunnage, B., and Yeo, S.T., Drugs for exceptionally rare diseases: do they deserve special status for funding? *QJM.* 98, 829, 2005.
2. Gericke, C.A., Riesberg, A., and Busse, R., Ethical issues in funding orphan drug research and development. *J Med Ethics.* 31, 164, 2005.
3. Gross, M.L., Ethics, policy, and rare genetic disorders: the case of Gaucher disease in Israel. *Theor Med Bioeth.* 23, 151, 2002.
4. Elstein, D., Abrahamov A., and Zimran A., Ethical considerations for enzyme replacement therapy in neuronopathic Gaucher disease. *Clin Genet.* 54, 179, 1998.
5. Beutler, E., The cost of treating Gaucher disease. *Nat Med.* 2, 523, 1996.
6. Stanley, J.M., Appleton International Conference 1991. The Appleton International Conference: developing guidelines for decisions to forgo life-prolonging medical treatment — Preamble, Parts I, II, III, and IV. *J Med Ethics,* 18 (Suppl.) S3, 1992.

CHAPTER 24

Societal Aspects in Treating Rare Diseases with Expensive Therapy

Deborah Elstein and Avi Israeli

CONTENTS

INTRODUCTION

One of the purposes of a government is to cope with difficult or even tragic choices, i.e., as the ultimate arbitrator of interpersonal conflict, be it at the level of two individuals or members of small groups, or at the level of a majority vs. one or more minorities. In one sense this may imply control of the burden of values and dealing with extraordinarily expensive decisions with high public profiles. In questions of health and prevention of disease, we look to our national government and our health care providers in both the public and private sectors for guidance and support in response to issues that go beyond personal needs and which may be in conflict with those of other individuals or groups.

This chapter is devoted to the thorny question of whether a governmental policy to provide very expensive therapy to a very small segment of the population can be justified when the needs of the many are great and the resources are limited. In addition, a sizeable portion of those who vie for every health dollar (or shekel or pound or lira), are suffering from deadly diseases, infectious diseases, acute eruptions, and sudden relapses: how can a government deal fairly with a very small minority of persons who suffer from rare diseases and who unfortunately require a disproportionately expensive therapy? What guideposts can be employed in developing the policy and should further guidelines direct the government once it enacts a policy that is inherently not egalitarian? Finally, should societal directives be employed *a priori* as part of a democratic approach to health care?

In adopting a governmental health policy that is realistic and to some extent wise, the basic rules of surveillance, response, and national infrastructure are invoked. Thus, one outcome of national policy is to develop a plan for response to requests for unequal shares of scarce resources, not the least trivial of which is financial resources. In delineating such a policy for rare diseases such as Gaucher disease where the cost of specific therapy may prove to be a societal burden, the governmental plan needs to adhere to ethical constraints as well. Nonetheless, primary objectives would include the following cardinal aspects that will guide the discussion in this chapter: limiting the burden of the disease for affected individuals, minimizing the social disruption of treating a nonterminal disease that may in fact be at the expense (literally and figuratively) of more common and fatal disorders, and reducing economic losses on a national scale. In the case of Israel, we are governed by remnants of a socialistic approach to national health care and as such, every resident is protected from the specter of having no health care at all; similarly, while not overtly democratic, this approach ensures a modicum of good clinical care (including medications) for every man, woman, and child regardless of concurrent disorders. In contrast, some wealthier countries with ability to supply even expensive therapies to those whose insurance carriers are committed to all levels of care may unfortunately not have a governmental mandate or ability to pay for citizens without third-party carriers.

GOVERNMENTAL HEALTH POLICY

The basic rules of a governmental health policy are described below for the interface of rare diseases with expensive orphan drugs, specifically, Gaucher disease.

1. Surveillance. Monitoring the effect of response to therapy is the key pivotal point in this policy since therapeutic intervention implies safety and efficacy. The capacity to track the natural history of the disease and measure the impact of specific therapy would at minimum document the benefit to individual patients who receive treatment. While this may not be acceptable as the sole yardstick of a governmental policy, it does provide a structure. In the early years after the advent of enzyme replacement therapy for Gaucher disease, the Israeli Ministry of Health formed a Committee of Experts that determined eligibility for treatment. This Committee devised criteria based on severity, since their combined clinical experience had taught them that not all patients with Gaucher disease will require specific treatment. At the same time, and again based on the clinical acumen of Prof. Ernest Beutler, a low-dose regimen was accepted as the initial starting dose for all patients. In effect, these two policy decisions succeeded in providing care for those patients with Gaucher

disease who would benefit the most, minimized social disruption by not adopting a blanket policy of care for all patients regardless of severity, and also reduced the national economic outlay by recognizing that low-dose was equally effective in the long-run and at a quarter of the cost demanded by the manufacturer for their "standard" care. Unfortunately, governmental control ended once the patient was given approval to receive treatment. Surveillance should include much more than the seminal description of its policy. In the case of Israel's Committee, which in the intervening decade has changed chairmen twice, and has re-formulated the criteria of approval to acknowledge that the current and future candidates for therapy will not be as severely affected as those of previous eras, no rules for monitoring have been enacted. The problematic aspect of this is the dichotomy between outcome and response as monitoring options. Developing systems for detecting plateaus in response, poor compliance, and actual benefit based on organ system involvement should be incorporated into the current scheme. The Israeli Ministry of Health has apparently been remiss in not providing the simple solution of secretarial (human or computer) assistance in tracking the outcome of patients approved for therapy. Too, there are also no means to track patients rejected by the Committee as to deterioration based on clinical data. There is no official collaboration among the physicians most expert in the disease with the governmental agencies that approve therapy except at the behest of the individual physician. There is similarly no national referral clinic with international caliber experts who would serve patients with the highest degree of expertise and interact with either the Committee or the Ministry of Health in tracking patients both untreated and treated. At present, surveillance of outcome of the Committee-based system remains anecdotal and sporadic.

2. Response. Clear definitions of therapy goals, including adjunct therapies such as orthopedic surgery, antibiotic prophylaxis, and so on must be prioritized. For example, if the goal of enzyme replacement therapy for patients with Gaucher disease is reduction of hepatosplenomegaly, the means assessment of organomegaly should be standardized, universally available and noncontroversial: if ultrasound examinations are sufficient for first-pass evaluation, computerized tomography (CT) should not be employed; if annual examinations are adequate for adults, bi-annual examinations should not be reimbursed. In another patient, if the goal of enzyme replacement therapy is to prevent bone involvement, then evidence-based medicine should allow one to assess the effect of treatment on bone density, incidence of fractures, and/or the degree of osteopenia or osteoporosis, or other accepted measures of skeletal disease. However, if these are lacking or equivocal, then alternative and/or adjunct treatments should be encouraged such as bisphosphonates to ameliorate bone density. The inventory of therapeutic options should not be limited to or by the fact that there is specific enzyme replacement. Surgeries for replacement of affected joints may be sufficient onto themselves to improve quality of

life and ensure a relatively pain-free existence; this choice should not be dismissed simply because the patient has been approved for enzyme therapy. Children whose sole symptom of Gaucher disease was painful bone crises, and hence have been given many years of enzyme therapy with no overt signs or symptoms of disease progression, should be re-evaluated in early adulthood as to the necessity of such expensive pro-phylaxis. Women whose sole indication for enzyme therapy is recurrent abortions should first undergo routine evaluations of both common and rare independently sorting problems, such as male factor and cobalamin deficiency (levels of which decrease per year of enzyme therapy), before using enzyme therapy as prophylaxis. And in the final analysis, the gov-ernment should play watchdog in the sense that it should require full disclosure of clinical response on an annual basis for each disease-specific parameter before renewal of its approval for continued drug supplies. Admittedly, when private funding can be recruited, as in third-party insur-ance carriers or in wealthier individuals or countries, this equation may be less of a balance between demands and responses since private sources may skew the need for responses.

3. Infrastructure. This refers to the national resources that can be brought to bear in support of the above requisite responses. One component is improvement in the Committee system, although this in itself is an inno-vative solution that highlights the fact that in Gaucher disease the natural history is not death and not decreased longevity, but the therapeutic options are among the most expensive in the world. Improvements are myriad and easily categorized for Israel, but also for other countries with comparable economies; the trick is in its implementation. Here again the gatekeeper function of government is applied as requisite to the good of the many vs. protection of the few. Screening for eligibility, compliance, and documentation of efficacy and safety are basic data that would be included in the purview of a reorganized committee of peers. In many countries, such committees are not only composed of medical experts but also of lay persons with an eye to economic rather than medical issues: government, as protector of individual rights, should decry the biasing of choosing therapy on any basis other than medical need. However, infra-structure may be limited and as such, the overall budget may be circum-scribed so that medical criteria are more stringently adhered to and patients are carefully monitored. The application of electronic clinical reporting will, in the long-run, expedite the process of approval of candidates for treatment and facilitate long-term surveillance of safety and efficacy. Yet, society would be ill-served by merely enabling tabulation of clinical signs and symptoms. Evidence-based medicine requires understanding of emer-gent trends upon which to alter the course of clinical management. An unparalleled example of this based on a decade of experience with Gau-cher disease is the undeniable effect of enzyme replacement therapy on hematological parameters and hepatosplenomegaly, and the equally undeniable fact that a plateau in clinically significant responsiveness is

achieved, regardless of age, disease severity, splenectomy, immunological status, dosing regimen or frequency of infusions, within 2 to 3 years from advent of treatment. This critically important bit of data has not been incorporated into the equation of societal concerns. Surprisingly, no consensus forum has dealt with the issue of maintenance/tapering-off regimens and/or drug vacations/withdrawal. To ascribe to a life-time of expensive therapy with putative evidence of diminishing returns runs counter to the government's mandate as mediator of resources in an optimum fashion, i.e., prudently and wisely. Thus, when evaluating the infrastructure available to society in allocating scarce resources, first and foremost budgetary resources, the cost of *continuing* to provide enzyme replacement therapy is without precedent and possibly without clinical basis. In rectifying this situation to the benefit of all, long-term monitoring (which is largely available) would establish a new norm for rare diseases that require costly therapies. A few options for "maintenance regimens" are: alternate on and off periods of predetermined duration with clinical evaluation at regular intervals; decreased dosages after 3 years of full-dose therapy (and here again one needs to question what is the minimal effective starting dose as a matter of public knowledge); and setting clinically achievable goals such as near-normalization of hematological parameters and/or organ volumes. Each of the above would serve the interests of the patient by limiting the burden of disease, minimize social disruption by not allowing some individuals the right to limitless resources without societal restraints, and also reduce national economic losses.

ROLE OF THE PRIVATE SECTOR

Currently there are at least three companies with active products for Gaucher disease. Each of these companies has and will provide a therapeutic option for other related rare diseases. Yet apparently only Genzyme Corporation (Cambridge, MA, U.S.) is capable of providing philanthropic assistance to patients in developing countries ("emerging markets"). Despite global marketing of high-dose enzyme therapy for Gaucher disease and other rare diseases, which is certainly lucrative, the high risk of development has possibly restricted potential competition and in equal measure has circumscribed the extent of involvement of the private sector in providing compassionate use of these drugs. Thus, whereas government by its nature would benefit from greater commitments by the private sector to defray exorbitant health costs, besides facilitating clinical trials that enable volunteer patients to accept investigational drugs without payment, the governmental policy cannot depend on outside resources for rare diseases. Nonetheless, it behooves government to encourage the private sector by incentives such as user-friendly applications for clinical trials.

Conversely, consortia of the various health care sectors should include the private sector as well in the hope of maximizing good clinical practice. This would underscore who are the "players" with clear interests and how to analyze those health policy aspects with clear deficiencies. Conventional economic policies of the

government generally establish a framework in which social objectives can be incorporated into the decision-making process. The private sector, on the other hand,
understandably motivated in large measure by a profit margin, may overlook the
social/societal aspirations, accepting that there are "winners" and "losers" even in
the health care equation. The challenge, therefore, is to achieve the support of the
private sector when there is a perceived inequality that the government must be
attentive to, but which the private sector may choose to ignore.

Private nonprofit organizations may also be viewed as potential partners of
government in supplying good health care to its members. The recent proliferation
of such specialized organizations as spokespersons and sources of financial support
for research and treatment may be viewed as privatization of a governmental function
that should be encouraged when governmental policy is inadequate.

CONCLUSION

With the development of more expensive orphan drugs for diseases even more rare
than Gaucher disease, it is to be hoped that efforts towards national reorganization of
priorities will progress at a more rapid pace to accommodate these equally pressing and
long-awaited therapies. Unlike Gaucher disease, which served as a model for designer
drugs for orphan diseases, these other rare metabolic diseases are invariably lethal in
early childhood and hence the societal responsibilities to individuals who have heretofore suffered and died for lack of effective care will inevitably be based on slightly
different criteria. But, with these considerations in mind and with more orphan drugs
on the immediate horizon, governments must prepare adequately so that when national
health budgets are squeezed, those with deadly diseases, infectious diseases, acute
eruptions, and sudden relapses will not suffer the consequences of restricted access to
scarce resources while expensive orphan drugs are distributed without the normal checks
and balances expected by an enlightened society.

RECOMMENDED READING

1. Thamer M., Brennan N., Semansky R., A cross-national comparison of orphan drug
   policies: implications for the U.S. Orphan Drug Act. *J Health Polit Policy Law*, 23,
   265, 1998.
2. Mrsic M., Stavljenic-Rukavina A., Fumic K., Labar B., Bogdanic V., Potocki K.,
   Kardum-Skelin I., Rovers D., Management of Gaucher disease in a post-communist
   transitional health care system: Croatian experience. *Croat Med J*, 44, 606, 2003.
3. Maeder T., The orphan drug backlash. *Sci Am*, 288, 80, 2003.
4. Olauson A., The Agrenska centre: a socioeconomic case study of rare diseases.
   Pharmacoeconomics, 20 Suppl. 3, 73, 2002.
5. Clarke J.T., Amato D., Deber R.B., Managing public payment for high-cost, high-
   benefit treatment: enzyme replacement therapy for Gaucher's disease in Ontario.
   CMAJ, 165, 595, 2001.
6. Cox T.M., Cost-effectiveness controversy. *Pharmacoeconomics*, 8, 82, 1995.

CHAPTER 25

Gaucher Disease as a Model for an Orphan Disease: Medical Aspects

Jill Waalen and Ernest Beutler

CONTENTS

INTRODUCTION

The development of enzyme replacement therapy (ERT) for Gaucher disease marked a true triumph of rational drug design. The discovery of mannose receptors on macrophages and the subsequent modification of glucocerebrosidase to expose mannose to target the enzyme to macrophages led to clinical success in treating this disease for which previously there had been no specific therapy.

Establishment of optimal dosing regimens for enzyme replacement therapy — first with alglucerase (Ceredase), and later the recombinant product, imiglucerase (Cerezyme) — however, has not had such a rational base and has often been driven more by economics rather than sound science. Higher doses than needed for clinical response continue to be widely used despite clear evidence that lower doses are equally effective, leading to confusion among physicians regarding the best approach to treating individual patients. The history of enzyme replacement therapy for Gaucher illustrates how economic forces, many of which are rooted in the Orphan Drug Act of 1983, can impede adequate clinical evaluation of drugs for orphan diseases.

THE ECONOMICS OF ORPHAN DISEASES

Barriers to development of rational drug dosing of enzyme replacement therapy in Gaucher disease are best understood in the context of the Orphan Drug Act of 1983. Until the early 1980s, the development of treatments for orphan diseases such as Gaucher disease was greatly impeded by simple economics. While the cost of developing a drug for a rare disease is as high as that for any other drug, the potential market, and thus potential profit, is much smaller. As a result, in 1983, the year the Act was passed, only a handful of drugs that would have been designated orphan drugs were on the market.

The Orphan Drug Act of 1983 — a group of federal laws applicable to treatments for diseases affecting fewer than 200,000 Americans — quickly changed the economic equation. A key provision is 7-year market exclusivity, allowing approval of the same drug by a prospective competitor only if it is demonstrated to be safer, more effective, or easier to take. More direct financial incentives also included under the Act are a 50% tax credit on all clinical trial costs, exemption from user fees that the Food and Drug Administration (FDA) normally charges drug sponsors (currently more than $500,000), and grants awarded through the FDA to support development.

The Act also promotes expedited approval of orphan drugs. Acknowledging that for orphan diseases there are too few patients to easily demonstrate efficacy, patients are frequently dispersed throughout the country and great efforts (and money) are required to bring them to referral centers, and the small numbers of patients also affect the evaluation of different dosages and dosing frequencies, the FDA is flexible regarding study design and primary study end points for treatments. Thus, unlike other drugs that are required to be tested in double-blinded, controlled, randomized clinical trials, sponsors of orphan drugs have been allowed to utilize open protocol, open label, historical control, or crossover trials and use surrogate endpoints.[1] Examples of orphan designated products with FDA market approval using surrogate endpoints include Fabrazyme for Fabry's disease, Gleevec for gastrointestinal stromal tumors, and Orfadin for hereditary tyrosinemia Type I.

The Act's effect on availability of treatments for rare diseases was almost immediate. By 1985, 90 drugs for orphan diseases were under development. In 2004, 250 products for orphan diseases have received approval, benefiting a potential patient population of approximately 12 million.[1] Many of the approved drugs are for diseases that affect very few patients. Only 12 children in the U.S. had adenosine deaminase (ADA) deficiency, for example, at the time polyethylene glycol (PEG)-ADA therapy was being developed.[2] Approval of sacrosidase, an enzyme for treating sucrose-isomaltase reductase deficiency, was based on trials enrolling a total of 41 patients.[2] Fabrazyme is expected to be another leading product for its manufacturer Genzyme, even though less than 4000 men in the U.S. are known to be affected by Fabry disease.[3] And the number of orphan drugs in development continues to grow for ever rarer diseases. Considering lysosomal storage diseases alone, in addition to Gaucher and Fabry diseases, treatments are currently in the pipeline for Pompe disease, Hurler-Scheie disease, Hunter disease, and Niemann-Pick disease.[4]

The Act has not only stimulated development of drugs for treating diseases that were previously untreatable, but has also driven the development of new products for diseases with available treatments. Hemophilia, a rare disease with a long history of replacement therapy with clotting factors, predating the Orphan Drug Act by decades, is a case in point. The discovery of clotting factors and their subsequent purification and development as treatments progressed over the last century — from treatment with whole blood, then citrated plasma in the first half of the century to the use of cryoprecipitates by the late 1960s. The first recombinant Factor VIII product emerged in the era of the Orphan Drug Act, becoming commercially available in 1987. In the decades since, more recombinant products have been developed and marketed at ever higher prices.[5] The demand for newer products is driven by the legitimate desire for purer, safer products. The superiority of the new products, in this regard, however, is assumed rather than clinically proven. Controversy also exists whether continuous "prophylactic therapy" with clotting factors is superior to "on demand" therapy limited to bleeding episodes and optimal regimens for prophylactic therapy, as well as the appropriate time to start, have not been defined.[6]

As with other orphan diseases, the limited patient pool coupled with exclusivity protections and incentives for manufacturers to maintain existing price structures make it unlikely that these issues will be formally evaluated. Meanwhile, the price tag for treatment continues to escalate. Assuming current prices and treatment modalities, it is estimated that the cost of treating of a baby born with hemophilia in 2001 with recombinant clotting factor concentrates would reach $1 million — a typical lifetime cap for insurance policies — by age 10 if clotting factor were used prophylactically and by age 20 if clotting factor were used only to treat active bleeding episodes.[7]

Largely driven by the incentives provided by the Orphan Drug Act of 1983, treatment of rare diseases is now touted as a particularly lucrative niche for biotechnology companies and other small- to medium-sized pharmaceutical companies.[8] Exclusivity arrangements provided by the Act as well as pressure from advocacy groups on third-party payers to pay for these treatments have led to pricing untethered to market forces. Appeal of the market is further enhanced by the fact that many orphan diseases are single genetic mutations involving single proteins that are distinctive targets and prime products for production by biotechnology. In addition, with ever more precise targeting of therapies to subgroups defined by pharmacogenetics, it is speculated that as a result even drugs for common conditions may end up qualifying as orphan products.

GAUCHER DISEASE

There is no better example of how lucrative the orphan drug market can be than enzyme replacement therapy for Gaucher Disease. The therapy has been hugely profitable for its manufacturer Genzyme based on a pricing structure that did not change when production methods changed from extraction from placentas — to produce Ceredase — to less expensive recombinant methods — to produce Cerezyme.[2] Cerezyme was the company's biggest seller in 2003, with an anuual revenue of $700 million,[3] making it "the world's most expensive medicine."[2] Steady growth in profits made from the drug over the first 12 years of its marketing has been attributed by the company's CEO to "finding all the patients" for whom treatment may be indicated.[3]

Given the apparent saturation of the market and a pricing structure based on high doses, it is not surprisingly that there has been little incentive for evaluating lower doses. Decreasing the usual dose from 60 U/kg every 2 weeks to 7 U/kg every week, for example, would decrease the cost of treating a 70 kg patient from $436,000 to $87,360, translating into significantly reduced profits. However, evidence has been available since shortly after the drug was approved that lower doses are very effective. Physicians treating patients with Gaucher disease should examine the evidence as outlined below in determining the best regimen for their patients.

Enzyme Replacement Therapy for Gaucher Disease: The Original Dosing Regimen

Original approval of mannose-enriched glucocerebrosidase for the treatment of type I Gaucher disease was based on a very small trial of 12 patients with moderate to severe disease.[9] The trial was limited to only four adults and eight children, presumably because the enzyme was in short supply. A very high dose of 60 U/kg was given every 2 weeks to 10 of the patients and every week to 2 patients. The infrequent dosing was likely based on practical issues: many of the patients in the study did not reside in the Washington, D.C., area where the clinical trial was being conducted. The investigators wisely used a very generous dose of enzyme to max- imize the probability that the trial would be successful; it was unlikely that a second study could have been launched if the first had failed.

After treatment for 9 to 12 months, patients had clinically significant increases in hemoglobin and platelet counts, and significant decreases in spleen size and liver size. As a result of these relatively dramatic changes, the drug was approved based on the one small trial and the alglucerase was promptly marketed by Genzyme Corporation, under the trade name of Ceredase. The company built its pricing structure on the high dose regimen used in the trial, setting the cost at about $4 per unit, translating into $16,800 every 2 weeks for a 70 kg patient.

Optimizing the Regimen

From a pharmacologic basis, the large infrequent dose of 60 U/kg administered one to two times a month could hardly be considered optimal. For a preparation with a circulating half-life of only about 12 minutes, a dosing interval of 1 to 2 weeks makes little sense. In addition, specific mannose receptors are sparsely dis- tributed on macrophages and are outnumbered by three or four orders of magnitude by nonspecific, low affinity receptors present on endothelial cells and many other cells. As a consequence, very little of the infused enzyme reaches the macrophage. Theoretically, a much larger proportion of a smaller dose would reach the target.

In fact, evidence that much smaller doses given at more frequent intervals were fully effective began accumulating shortly after drug approval from trials conducted by independent investigators without the manufacturer's support. In a trial published in 1992, only a year after the original trial, 14 patients with moderately severe to severe disease were given a total of 30 U alglucerase/kg/month in fractionated doses given either daily (1 U/kg) or three times a week (2.3 U/kg).[10] The effects of treatment on blood counts and size of liver and spleen were found to be the same as those seen with much larger doses administered every 2 weeks. Increasing the dose given three times a week by fourfold (9.2 U/kg), did not significantly improve the response.

A subsequent trial of 25 patients reported in 1995 showed that most patients had good responses to even lower doses of alglucerase, some as low as 1.15 U/kg three times weekly for a total of 15 U/kg/month.[11] Meta-analyses of data from trials utilizing a number of different low dose/high frequency regimens have clearly shown that patients treated with 15–30 U/kg/month in fractionated doses three times weekly or daily have similar decreases in liver and spleen size to patients treated with high dose/low frequency regimens (86–130 U/kg/month administered at 2-week intervals) (Figure 25.1).[12] Six months of therapy with either regimen results in significant increases in hemoglobin concentration, with the most anemic patients having the largest response (Figure 25.2). Changes in cortical bone thickness take much longer, with significant increases detected patients treated with either regimen for 30–40 months (Figure 25.3). It should be noted that reports on the effects of ERT published since this 2000 review have not included consistent measurements of organ size[13,14] or have not differentiated between various doses and dose frequencies[15,16] and could therefore not be used to expand the earlier meta-analyses.

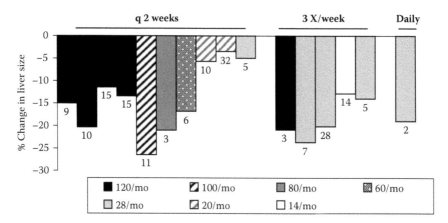

Figure 25.1   Effect of enzyme replacement therapy on liver size by dose and dose frequency at 6 months. Studies included are from references 9–11, 24–28. Numbers below bars indicate number of patients studied. (From Beutler, E., *Blood Cells Mol. Dis.*, 26, 304, 2000. With permission.)

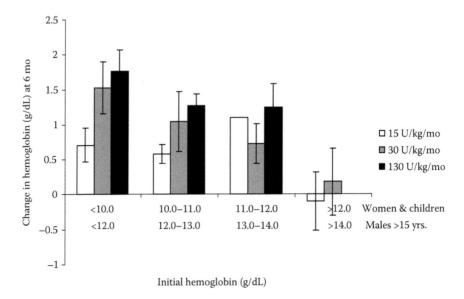

Figure 25.2 Effect of enzyme replacement therapy on hemoglobin by dose at 6 months. Studies included are from references 11, 20, 25, 29. (From Beutler, E., *Baillere's Clin. Haematol.*, 10, 756, 1997. With permission.)

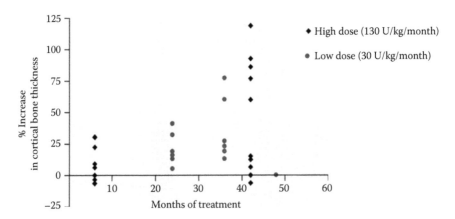

Figure 25.3 Effect of enzyme replacement therapy on cortical bone thickness by dose and duration. Studies included are from reference 30 (low dose) and reference 31 (high dose). (From Beutler, E., *Baillere's Clin. Haematol.*, 10, 759, 1997. With permission.)

Individualizing Therapy

Despite the evidence that good responses are achieved in Gaucher patients with enzyme replacement using much less drug, use of the high dose regimen continues to be common. While the package insert for Cerezyme states that "Initial dosages range from 2.5 U/kg 3 times/week to 60 U/kg once every two weeks," it is also noted that "60 U/kg every 2 weeks is the dosage for which the most data are available." The reason for this self-serving statement is obvious: with one exception, Genzyme sponsored only trials in which only the large dose was given. As a result, in the U.S., most Gaucher patients are started on a regimen of 65 U/kg/month, with fewer than 25% started on doses of 15–30 U/kg/month. [17] Average starting dose, however, has been noted to vary significantly by country. The vast majority of patients (>75%) in the U.K. and Israel, for example, is started on doses of 25–30 U/kg/month.[17] The decreased dosages used in Israel have been estimated to have cut the total cost of treating all Gaucher patients in that country from $80 million to $20 million per year.[18]

Because most physicians treat only a few Gaucher patients, recommendations from consultants, who are often supported by the manufacturer, are particularly influential in determining dosing regimens. In addition, patient advocacy groups often encourage patients to request high dose treatment as standard of care. Controversy surrounding dosing can make it difficult for physicians to determine the best approach to therapy in individual patients.

Based on the totality of the evidence, a general approach can be suggested. First, it must be determined which patients need to be treated. Although genotype can be helpful in predicting disease severity, even patients with the same genotype can differ substantially in regard to clinical manifestations.[19] Severity of symptoms at presentation provides the best indication of need for treatment. It is important to realize that the disease is only slowly progressive in adults and often not progressive at all [20]; more progression occurs in patients who have less favorable genotypes,[21] but even in these, progression is measured over a period of months or years. There is plenty of time to evaluate the course of the patient before committing to a course of therapy, and little justification for initiating therapy to "prevent" problems. Patients with severe bone lesions, however, need to be considered for therapy, even if they are not symptomatic, since skeletal disease, unlike visceral disease, may not be reversible. In children the situation is somewhat different. Progression does occur, particularly in children under the age of 10, and this must be taken into account in making therapeutic decisions.

The data support initiation of therapy at doses of 15–30 U/kg/month in fractionated doses administered at least once a week. Failure to respond is not necessarily an indication for increasing the dose; there is no evidence that increasing dose will improve response (Figure 25.4). Symptomatic treatment is effective and should be provided throughout.

Costs in Perspective

In addition to the concerns regarding inadequate clinical evaluation of orphan drugs as outlined in this chapter, the continued development of expensive therapies for increasingly rare diseases raises the important question of whether, and when, society will decide the price is too high. For enzyme replacement therapy for Gaucher disease alone, the cost to the health care system will approach $6 billion dollars by the end of 2005, 14 years after its market approval, with revenues from Cerezyme currently nearing $800 million per year to treat fewer than 4,000 patients worldwide.[22] By comparison, it is projected that the global polio vaccination program will

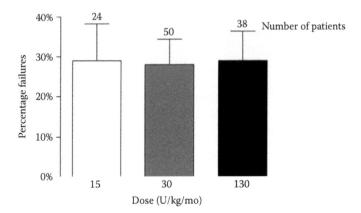

Figure 25.4 Failure rates[a] at different monthly doses of enzyme replacement therapy ([a]using the criteria of Hollak et al. based on a combination of hematological and visceral responses.[11]) The analysis includes all studies of five or more patients with evaluation of responses at 6 months.[11,24,25,29,32] Error bars represent standard error of the mean. (From Beutler, E., *Baillere's Clin. Haematol.*, 10, 757, 1997. With permission.)

have cost $67 billion to prevent 855,000 deaths and 4 million cases of paralysis between 1970 and 2050.[23] To date, questions regarding financial limits have largely been left unaddressed by the parties involved — government regulatory bodies, patients, physicians, manufacturers, and payers. It is clear that in the future, as much rationality must be applied to the postapproval use of treatments for orphan diseases such as Gaucher disease as went into their original development in the laboratory.

REFERENCES

1. Haffner, M.E., Developing treatments for inborn errors: incentives available to the clinician, *Mol. Genet. Metab.*, 81, Suppl. 1, S63–S66, 2004.
2. Maeder, T., The orphan drug backlash, *Sci. Am.*, 288, 80, 2003.
3. Tsao, A., Genzyme: beyond "orphan" diseases. *BusinessWeek Online.* 1–9–2004. The MacGraw-Hill Companies, 10–28–2004. Ref Type: Electronic Citation. URL:www. businessweek.com/technology/content/jan2004/tc2004019_6626_tc006.htm?campaign _id=search
4. Fan, J.Q., A contradictory treatment for lysosomal storage disorders: inhibitors enhance mutant enzyme activity, *Trends in Pharmacol. Sci.*, 24, 355, 2003.
5. Kingdon, H.S. and Lundblad, R.L., An adventure in biotechnology: the development of haemophilia A therapeutics — from whole-blood transfusion to recombinant DNA to gene therapy, *Biotechnol. Appl. Biochem.*, 35, 141, 2002.
6. Aledort, L.M., Can costs of hemophilia products be curtailed? Not as we do business today! *Thromb. Haemost.*, 88, 541, 2002.
7. Rogoff, E.G., et al., The upward spiral of drug costs: a time series analysis of drugs used in the treatment of hemophilia, *Thromb. Haemost.*, 88, 545, 2002.
8. Milne, C.P., Orphan products — pain relief for clinical development headaches, *Nat. Biotechnol.*, 20, 780, 2002.

9. Barton, N.W., et al., Replacement therapy for inherited enzyme deficiency — Macrophage-targeted glucocerebrosidase for Gaucher's disease, *N. Engl. J. Med.*, 324, 1464, 1991.

10. Figueroa, M.L., et al., A less costly regimen of alglucerase to treat Gaucher's disease, *N. Engl. J. Med.*, 327, 1632, 1992.

11. Hollak, C.E.M., et al., Individualised low-dose alglucerase therapy for type 1 Gaucher's disease, *Lancet*, 345, 1474, 1995.

12. Beutler, E., Commentary: Dosage-response in the treatment of Gaucher disease by enzyme replacement therapy, *Blood Cells Mol. Dis.*, 26, 303, 2000.

13. Ida, H., et al., Effects of enzyme replacement therapy in thirteen Japanese paediatric patients with Gaucher disease, *Eur. J. Pediatr.*, 160, 21, 2001.

14. Dweck, A., et al., Type I Gaucher disease in children with and without enzyme therapy, *Pediatr. Hematol. Oncol.*, 19, 389, 2002.

15. Weinreb, N.J., et al., Effectiveness of enzyme replacement therapy in 1028 patients with type 1 Gaucher disease after 2 to 5 years of treatment: a report from the Gaucher Registry, *Am. J. Med.*, 113, 112, 2002.

16. Perez-Calvo, J., et al., Extended interval between enzyme therapy infusions for adult patients with Gaucher's disease type 1, *J. Postgrad. Med.*, 49, 127, 2003.

17. Cox, T., personal communication, 2003.

18. Gross, M.L., Ethics, policy, and rare genetic disorders: the case of Gaucher disease in Israel, *Theor. Med. Bioeth.*, 23, 151, 2002.

19. Beutler, E., Gaucher disease as a paradigm of current issues regarding single gene mutations of humans, *Proc. Natl. Acad. Sci. U.S.A.*, 90, 5384, 1993.

20. Beutler, E., et al., The clinical course of treated and untreated Gaucher disease. A study of 45 patients, *Blood Cells Mol. Dis.*, 21, 86, 1995.

21. Maaswinkel-Mooij, P., et al., The natural course of Gaucher disease in The Netherlands: implications for monitoring of disease manifestations, *J. Inherit. Metab. Dis.*, 23, 77, 2000.

22. Based on published annual earning reports. Anard, G., Why Genzyme can charge so much for Cerezyme, Wall Street Journal, November 16, 2005, A15.

23. Khan, M.M. and Ehreth, J., Costs and benefits of polio eradication: a long-run global perspective, *Vaccine*, 21, 702, 2003.

24. Fallet, S., et al., Enzyme augmentation in moderate to life-threatening Gaucher disease, *Pediatr. Res.*, 31, 496, 1992.

25. Grabowski, G.A., et al., Enzyme therapy in type 1 Gaucher disease: Comparative efficacy of mannose-terminated glucocerebrosidase from natural and recombinant sources, *Ann. Intern. Med.*, 122, 33, 1995.

26. Altarescu, G., et al., Comparative efficacy of dose regimens in enzyme replacement therapy of type 1 Gaucher disease, *Blood Cells Mol. Dis.*, 26, 285, 2000.

27. Zimran, A., et al., Replacement therapy with imiglucerase for type 1 Gaucher's disease, *Lancet*, 345, 1479, 1995.

28. Barton, N.W., Brady, R.O., and Dambrosia, J.M., Treatment of Gaucher's disease, *N. Engl. J. Med.*, 328, 1564, 1993.

29. Zimran, A., et al., Low-dose enzyme replacement therapy for Gaucher's disease: effects of age, sex, genotype, and clinical features on response to treatment, *Am. J. Med.*, 97, 3, 1994.

30. Elstein, D., et al., Effect of low-dose enzyme replacement therapy on bones in Gaucher disease patients with severe skeletal involvement, *Blood Cells Mol. Dis.*, 22, 104, 1996.

31. Rosenthal, D.I., et al., Enzyme replacement therapy for Gaucher disease: skeletal responses to macrophage-targeted glucocerebrosidase, *Pediatrics*, 96, 629, 1995.
32. Beutler, E., Kuhl, W., and Vaughan, L.M., Failure of alglucerase infused into Gaucher disease patients to localize in marrow macrophages, *Mol. Med.*, 1, 320, 1995.

CHAPTER 26

Meeting the Needs of Patients with Gaucher Disease: Pioneering a Sustainable Model for Ultra-Orphan Diseases

David P. Meeker and Henri A. Termeer

CONTENTS

Since its implementation in 1983, the U.S. Orphan Drug Act has been an unquestionable success. The legislation provides incentives for the biotechnology and pharmaceutical industry to develop therapies for diseases with a very low prevalence. Diseases with a patient prevalence of less than 200,000 individuals in the U.S. qualify as an orphan disease, and tax incentives and 7 years of market exclusivity reward companies that successfully develop a therapy.

Since the legislation was enacted, over 1400 product candidates have been designated for orphan diseases, with 265 therapies having been successfully developed with orphan disease designation. Similar legislation was passed in Europe in 2000 with a comparable patient population prevalence (<5/10,000) to define an orphan disease and has been comparably successful, with 220 designations and 21 approved therapies to date.

Although this legislation has achieved the goal of increasing the number of rare diseases with specific therapies, the issue of patient access remains largely unaddressed. Today there are a number of specific barriers to developing therapies for rare diseases and ensuring patients with that disease can access the therapy, which we will explore more fully in this chapter.

Genzyme's experience developing and commercializing enzyme replacement therapy (ERT) for Gaucher disease highlights many of the challenges inherent to meeting the needs of a rare disease patient population. Enzyme replacement therapy was approved in 1991 in the U.S. and in 1994 in Europe for the treatment of Gaucher disease. To date more than 4400 patients are receiving therapy worldwide. Fourteen years after therapy was approved, there are approximately 1400 patients on therapy in the U.S.

The 100-fold difference in magnitude between what is considered an orphan disease and populations such as Gaucher disease, which might be considered "ultra-orphan," creates unique challenges even within the orphan disease group. Genzyme's experience developing therapies for an "ultra-orphan" disease population has led us to accept a responsibility that extends well beyond simply making the therapy available. Regulatory approval is only one step towards creating sustainable health care systems capable of caring for the patient with a rare disease. As the manufacturer of the only life-saving or potentially life-saving therapy for a rare disease, we must work with health care systems around the world to ensure that patients in need can access therapy. That commitment can only be fulfilled through an in-depth understanding of the huge disparities in existing health care systems around the world and a sustainable business model.

THE BEGINNING

Genzyme began its effort to develop enzyme replacement therapy in the early 1980s, building upon the pioneering work of Dr. Roscoe Brady of the National Institutes of Health (NIH), who first theorized 20 years earlier that the symptoms of Gaucher disease could be addressed by replacing the deficient enzyme, glucocerebrosidase. Henry Blair, Genzyme's scientific founder, produced enzyme material for Dr. Brady's early clinical trials at his Tufts University laboratory beginning in 1977. By 1981, Genzyme had been founded and the first patient was infused with Genzyme-supplied glucocerebrosidase at the NIH.

Despite the invaluable pioneering contribution that Dr. Brady provided — and substantial work at Genzyme and the NIH to remodel the enzyme to increase its uptake in affected cells — the potential viability of ERT remained unclear. Even if it were possible to demonstrate clinical efficacy, it would be extremely difficult to supply enzyme for all Gaucher patients given that the only known source of the enzyme was human placentas.

The first evidence of the potential efficacy of enzyme replacement therapy came in 1983, when a 4-year-old boy, Brian Berman, began treatment at the NIH. Brian's mother brought him to the NIH determined to provide one last opportunity to avoid having his spleen removed, a common procedure for Gaucher patients before the introduction of enzyme replacement therapy. Like many untreated patients, Brian's spleen was so enlarged that he appeared to have a basketball in his belly.

Brian responded to treatment almost immediately. Within 3 months his spleen and liver shrunk notably and he began to show the energy of a typical 4-year-old. It was precisely the response that had proven so elusive in earlier experiments. Unfortunately, Brian was the only one to benefit. Seven other patients who enrolled in the NIH trial with Brian failed to show any clinical benefit. Why he alone responded was unclear at that time but Brian's progress was so profound that we had no choice but to press ahead.

In the years that followed, Genzyme bet its future on a series of unlikely occurrences. We bet that we could figure out why Brian was benefiting, and replicate his success with others; that we could find financial backers who would share our vision for helping this small patient population in need; and that we could find a way to produce enough enzyme despite considerable obstacles. When we had successfully developed a therapy, we bet that private and public payers would share our urgency to support the needs of people with Gaucher disease over the long term.

Along with Brian's mother, we visited potential investors throughout the summer and fall of 1987. Investors were not excited about a potential therapy for an unknown disease, with minimal clinical data, sourced from human placentas at the height of the HIV/AIDS crisis. Financial support was obtained in October 1987, just days before the stock market crash might have shut down any possibility of funding.

ENZYME PRODUCTION

We prepared for our first clinical trial while also undertaking the long-term planning necessary to create a sustainable supply of enzyme. We were guided by the knowledge that under no circumstance could we allow a patient's treatment to be interrupted once it began. Again, this was a lesson learned from Brian Berman, who missed several scheduled infusions during the first clinical trial because we were unable to produce glucocerebrosidase reliably. Each time he went off treatment, Brian's condition almost immediately worsened, only to be alleviated after he was treated again. Companies developing treatments like enzyme replacement therapy become a lifeline to patients that cannot be interrupted.

To understand the challenge of producing glucocerebrosidase on a large scale, one must consider that it required approximately 22,000 placentas to extract enough enzyme to treat one patient for one year. Treating the entire population of people with Gaucher disease would require millions of placentas each year. The risk of viral contamination was a particular concern given that Ceredase would be the only placenta-derived medicine on the market at the height of the HIV/AIDS crisis.

Following our successful clinical trial, Ceredase was approved by the FDA in the spring of 1991, exactly 10 years after the company was formed. By the end of that year, 298 patients were on therapy. Within 5 years this number had grown to more than 1000. To supply these patients, more than a million placentas each year were sent to a plant in Lyon, France, where enzyme was extracted and Ceredase was purified for worldwide use. From there the material was sent to Cambridge, MA, for further production and purification, and then to a specialty plant in New Mexico for fill/finish and worldwide distribution.

We understood early on that we would need a more reliable source of enzyme than the placenta-derived program. Genzyme began the work to create a recombinant source of enzyme well before we had gained regulatory approval for Ceredase, and we broke ground in 1994 on what today is our flagship protein manufacturing facility in Boston. We committed to spending more than $200 million on this facility before we knew that the recombinant source would work. It did, and we continue to produce Cerezyme there today.

CLINICAL DEVELOPMENT

With most ultra-orphan diseases, relatively little may be known about a disease state prior to the advent of a therapy. In these cases, there is less urgency to make an accurate diagnosis because only symptomatic therapy is available. Many affected individuals may not know they have the disease. The awareness among physicians is correspondingly low. A recent survey of patients with Fabry disease, a similar rare genetic lysosomal storage disease, revealed that the average age of diagnosis was 29 years of age and the number of physicians seen before a correct diagnosis was made was nine despite the fact that patients develop symptoms in their teens or earlier. Patient, physician, and societal awareness of Gaucher disease was no different in the late 1980s.

The regulatory agencies, similarly, had little experience evaluating therapies for such a rare disease. The pivotal trial design for Ceredase, which involved only 12 patients and did not have an active control patient population, reflected the developmental challenges at that time. Essentially all available enzyme was used to treat the 12 patients in that trial. The dose of enzyme was chosen based on our best understanding of the dose of enzyme that had led to improvement in the small number of previously treated patients. The trial was successful with all 12 patients showing significant improvement in multiple disease parameters. Subsequent experience from around the world has led to the recommendation that the dose of enzyme should be individualized, tailored to the need of that specific patient — a clear recognition of the significant disease heterogeneity present in Gaucher disease patients.

Because therapies for rare diseases are often approved with relatively few patients having been treated, postapproval commitments have become increasingly important in the development of therapies for rare diseases. Preapproval trials are small in comparison to traditional drug development trials such as those for hypertension or diabetes, which involve thousands of patients. Increasingly, regulatory agencies are requiring placebo, controlled trials, which can be particularly challenging given the life-threatening nature of some of these diseases and the absence of alternative therapies. In that design, it is critical that the entry criteria define a homogeneous population (subset) from within the rare disease population that can be compared. Hard clinical endpoints may not be addressable in the context of a relatively short placebo-controlled randomized trial. Long-term follow-up of the treated population is therefore required to understand the full benefits of therapy. Regulatory agencies have recognized this challenge and will approve therapies for life-threatening diseases based on changes in a surrogate marker that is judged reasonably likely to predict clinical benefit. In the case of Gaucher disease, the uniform clinical response among the 12 patients coupled with the important individual histories of patients like Brian Berman who improved on therapy and worsened when enzyme was no longer available strongly supported approval of the therapy.

Genzyme recognized from the beginning the importance of creating a registry to allow long-term follow-up of patients both on therapy and those who did not require enzyme replacement therapy. The Gaucher Registry has been tracking the progress of patients around the world both on and off therapy since the early 1990s. Given the clinical heterogeneity and rarity of the condition, this global effort has resulted in a far better understanding of the disease than has ever been available before.

In late 2002, a leading set of authors published the most extensive data yet on the clinical outcomes of patients who have received Cerezyme therapy for as long as 5 years (Table 26.1) and the success they described is striking.

According to the authors, enzyme replacement therapy effectively reversed and prevented the progression of the major clinical manifestations of type 1 Gaucher disease, leading to complete or partial elimination of anemia and platelet deficiency, reduction of the enlargement of the liver and spleen, and improvement or prevention of bone pain and bone crises.

Approximately 90% of all patients with anemia achieved normal levels of hemoglobin within 2 years of initiation of therapy, and this response was sustained for up to 5 years. Similarly, patients' low platelet counts typically normalized within 1 year if their spleen had been previously removed, a procedure that is now rarely necessary since the availability of enzyme replacement therapy. In those with an intact spleen, platelet counts increased by 50 to 100% within 12 months, with further improvements through 5 years. Liver sizes decreased by 20–30% in 1 to 2 years and up to 40% in 5 years. Improvement in spleen size was even more pronounced, with a decrease by 30% in 1 year and 50% after 2 years. Resolution of bone pain occurred in half of all symptomatic patients within 1 to 2 years, while recurrent or new episodes of bone crises were substantially reduced as well.

SUSTAINABLE HEALTH CARE SYSTEMS

Over time and through the use of a worldwide registry, the clinical benefit of enzyme replacement therapy for Gaucher disease has been thoroughly defined. An ongoing challenge, even 14 years after therapy first became commercially available is to ensure access of this therapy to patients around the world. The creation of a sustainable health care system capable of caring for the patient with a rare disease begins long before the approval of a specific therapy. Arguably, whether one lives in the developed world with the most sophisticated health care system or in the developing world, health care systems specifically designed to take care of the patient with an "ultra orphan" disease do not exist.

The basic components of that system are as follows: Societal awareness should be high, ensuring that patients and physicians have a good understanding of the

Table 26.1 Response to Enzyme Replacement Therapy after 2 to 5 Years in Patients with Abnormal Values at the Inception of Treatment*

	Duration of Treatment (Years from Start)			
	2	3	4	5
Hemoglobin (g/dL increase from baseline				
Number of patients "with spleen"	135	105	76	45
Mean ± SD	2.5 ± 1.7	2.6 ± 1.5	2.7 ± 1.3	3.0 ± 1.4
Number of patients "without spleen"	49	33	31	14
Mean ± SD	2.4 ± 1.6	2.7 ± 1.6	2.3 ± 1.8	2.4 ± 1.6
Platelet count (% increase from baseline)				
Number of patients "with spleen"	222	170	119	61
Mean ± SD	74 ± 77	80 ± 87	82 ± 70	104 ± 82
Number of patients "without spleen"	15	12	8	4
Mean ± SD	197 ± 221	236 ± 230	259 ± 247	316 ± 355
Liver volume (% decrease from baseline)				
Number of patients "with spleen"	94	56	37	17
Mean ± SD	2.9 ± 14	36 ± 15	38 ± 16	41 ± 13
Number of patients "without spleen"	35	21	11	10
Mean ± SD	38 ± 15	41 ± 16	50 ± 17	47 ± 13
Spleen volume (% decrease from baseline)				
Number of patients "with spleen"	96	55	36	16
Mean ± SD	49 ± 17	54 ± 18	57 ± 16	56 ± 18

* Patients included in 2-, 3-, 4-, or 5-year results based on availability of data; results for each year are based on baseline and follow-up data.

Source: Reprinted from Table 3, *The American Journal of Medicine*, V113(2): 112–119, Weinreb, N.J., "Effectiveness of enzyme replacement therapy with Type 1 Gaucher Disease after 2 to 5 Years of Treatment: A Report from the Gaucher Registry © 2002 Excerpta Medica, Inc.

disease in question and the ability to make a rapid, accurate diagnosis. Ideally, in addition to a general level of awareness, medical centers with a lead clinical expert and supporting subspecialties capable of caring for the multi-systemic nature of these diseases should be available within an acceptable proximity. Following an appropriate evaluation, patients who are judged in need of treatment must have access to an infusion site where they can receive the weekly or biweekly infusions required to replace the missing enzyme as is the case with Gaucher disease. Beyond the medical intervention, patients with rare disease benefit from and ideally should have access to the support that comes from being part of a patient organization. Individual countries vary tremendously in their availability of these resources.

From the beginning, Genzyme has committed to making enzyme replacement therapy available to patients in need. A potential barrier to accessing therapy for any orphan disease is the cost of therapy. Three basic factors determine the cost of therapy:

1. The cost of production, which as we have discussed in the case of a biologic product, such as enzyme replacement therapy, is high.
2. The cost of development, which is not necessarily less simply because the disease is rare and the number of patients enrolled in the clinical trials is small. In fact, the cost of that development may actually be increased as a function of trying to identify and enroll the necessary number of patients required to demonstrate safety

and efficacy. Because the number of patients treated prior to approval is relatively small, in most cases regulatory approval is associated with significant postapproval commitments.

3. Finally, the fact that those costs must be recouped over such a small patient population, which dictates that the cost of therapy for an individual patient will be high.

Although the per-patient cost of Cerezyme in developed economies makes it an expensive investment for health officials on a per-patient basis, the total cost to society is finite and relatively small due to the size of the patient populations involved. Most are willing to assume responsibility for improving the health of all patients with Gaucher disease, even in countries where health budgets are tight. Such use of a highly effective and safe treatment for a disease that affects so few patients represents a prudent and compassionate use of resources.

Central to our ability to bring treatment to as many patients as possible around the globe is a two-tier pricing policy whereby countries with the means to do so all pay roughly the same price for enzyme replacement therapy. We recognize that as the sole manufacturer of a potentially life-saving therapy, our commitment extends beyond the developed world. Our model of support is not simply to supply enzyme but to partner with local physicians, patients, and, ultimately, the responsible authorities to create the awareness and build the infrastructure that is required to care for the patient with rare diseases.

In countries where health systems are less developed, patients gain access to care through the Gaucher Initiative, whose goal it is along with Project HOPE and other nonprofit organizations to provide free treatment in countries such as Algeria, Belarus, Chile, China, Cuba, Ecuador, Egypt, Haiti, India, Jamaica, Kenya, Malaysia, Ukraine, Tanzania, Pakistan, Palestine, Philippines, South Africa, Sri Lanka, Yugoslavia, and Vietnam. There have been hundreds of Gaucher patients treated through these initiatives. As a group, these patients form the fastest growing segment of the entire population treated with enzyme replacement therapy

Although Cerezyme is a very safe and effective medicine, it is not a cure. In the fullness of time, gene or stem cell therapy may provide an actual cure for Gaucher disease, but it is not a realistic near-term possibility. Enzyme replacement therapy provides a bridge to the day when a more definitive therapy becomes available. Patients treated with ERT have avoided the devastating irreversible complications of Gaucher disease and have a chance to benefit from the next generation of therapy.

Our work on gene therapy programs for lysosomal storage disorders has progressed over the past decade. We have developed one of the largest and most advanced gene therapy research organizations in the world. Gene therapy-based products are in clinical development for cardiovascular and oncologic indications. Gene therapies for both central nervous system and noncentral nervous system targets are moving towards preclinical development. Progress to date has been incremental and has not developed as quickly as any of us would hope. In fact, we often remind ourselves that in the early days of Genzyme we asked our scientific advisory board whether we should pursue development of an enzyme replacement therapy for Gaucher disease. Their consensus was that gene therapy was "just around the corner" and progressing so rapidly that it would be available as quickly as an effective enzyme replacement therapy. Fortunately, an ERT was developed in parallel with continued research into the elusive gene therapy approach.

The orphan disease legislation had done much to foster the development of therapies for rare diseases. Development of a therapy for Gaucher disease has played a unique role in that success in that it defined a viable business model for treating the patient with rare diseases. Multiple companies in addition to Genzyme are now pursuing possible therapies for other orphan diseases. As a result, there is unprecedented activity on behalf of patients who had little hope even 10 years ago. Genzyme has recently introduced new enzyme replacement therapies for Fabry and MPS-I diseases and has recently gained approval of a similar treatment for Pompe disease. We are planning a clinical trial for Niemann Pick B disease, and we are continuing discussions with partners to introduce enzyme replacement therapies for other rare diseases.

In 2005, we invested nearly 20% of our total revenue on research and development. Our investment in treatments for lysosomal storage disorders is by far the largest commitment of its kind anywhere — in government, academia, or industry. The cost to bring our treatment for Pompe disease to the market will exceed $500 million by the end of 2005 and will grow even higher as we extend our clinical development program further.

Despite the cost of development, there is no other choice but to move forward. Pompe disease is a devastating illness, which for those affected by the infantile form, will typically result in death before a child's first birthday. After years of development, a first generation treatment is now tantalizingly close. Today, more than 270 Pompe patients are receiving enzyme replacement therapy, and we are contacted almost weekly with requests to add more patients on a compassionate use basis.

To some observers, the current level of clinical activity offers evidence that the field of orphan drug development has "matured." From our vantage point, the journey has just begun. Significant barriers persist and will require all of our attention in the years to come if we are truly to do all that we can for patients.

From a financial perspective, orphan drug development remains far too expensive and risky. Sustaining this model given the staggering upfront development costs and the small patient populations over which to spread these costs remains a formidable issue for industry, government, patients, and society at large, especially as budget pressures mount.

Forty years after Roscoe Brady first posited that enzyme replacement therapy could help patients with Gaucher disease, more than 4400 people in 80 countries around the world are benefiting from his vision. Brian Berman, who gave the first glimpse of what an effective therapy could bring, has graduated from college, married, and is raising a family. Thousands more like him have been given the chance to enjoy a better life by enzyme replacement therapy.

When the definitive chapter is written about our involvement with Gaucher disease, our hope is that it will tell how we developed a real cure for the disease while helping patients thrive until that cure became available.

CHAPTER 27

Patients' Perspective

Susan Lewis, Tanya Collin-Histed, Jeremy Manuel, and Greg Macres

CONTENTS

Before the early 1990s, most people suffering from Gaucher disease and their families knew very little about the condition and were seen by doctors who often could not explain the disorder in any detail. This lack of information was common to many rare genetic conditions. Doctors had rarely seen another patient with the disease and even if two or more patients were treated at the same hospital, they were often seen by different doctors and even in different departments. Patients or their parents were told not to look up information in a library, even a medical one, as it was likely to be inaccurate or out of date.

Patients and their families felt alone and isolated. They had never met anyone else with the disease, so they could not compare symptoms or get any prognosis of what would be the likely outcome of the condition. In addition, some patients felt stigmatized by having a rare genetic condition and kept the disease hidden, sometimes even from members of their families. If symptoms such as nose bleeds or the need for bone surgery did come to the notice of relatives and friends, the reasons for them were kept secret.

However, the development of enzyme replacement therapy in the early 1990s was to change the lives of patients for the better, and the Internet, which arrived a couple of years later, provided families with easy access to helpful information.

PATIENTS' SUPPORT GROUPS — PRE AND POST ENZYME REPLACEMENT THERAPY

The first patients' support group was established in 1975 in the northern (Norrbottian) part of Sweden by three families whose children were diagnosed with the juvenile neuronopathic form of the disorder, type 3 Gaucher disease. They set out to gather information regarding care, treatment, rehabilitation, and social security that they could pass on to others. In 1983, two affected sisters, Tineke and Jenny Timmerman, together with Ria Guijt, started a Dutch support group with 15 patients and the same number of family, friends, physicians, and scientists. Meanwhile in the U.S., Dr. Robin Berman, on discovering that her son Brian had Gaucher disease in 1983, searched for anyone carrying out research in the field. She discovered that Dr. Roscoe Brady and his team at the National Institutes of Health (NIH), not far from her own home in Maryland, were working on a possible treatment. Although medically qualified, she offered to work for Dr. Brady without charge to help in his research and no less importantly worked with Henri Termeer, the chief executive of the newly formed Genzyme Corporation, to help raise the necessary capital to develop and manufacture the first enzyme replacement therapy called Ceredase, which was licensed by the U.S. Food and Drug Administration (FDA) in 1991.

In May 1991, eight English patients or their parents met in a north London community hall. The meeting was organized by Jeremy Manuel, who with his family had founded the Helen Manuel Foundation in memory of their mother who had suffered from Gaucher disease. Of the patients who attended, none knew each other (except for two sisters) although they lived within a radius of a few miles. Prof. Victor Hoffbrand, a consultant hematologist at the Royal Free Hospital, London, described the symptoms of Gaucher disease, which several of the patients recognized but had not realized were part of their condition. He spoke about the new enzyme replacement therapy, which was due to be licensed in the U.S. Prof. Mia Horowitz (then from the Weizmann Institute of Science and now of Tel Aviv University) spoke about the genetic basis of the disease and the mutations that had been discovered.

After the meeting, patients and parents exchanged stories. One mother, who had been told that her daughter would not live to the age of 20, met a married woman with Gaucher disease who had children and a career. A recently diagnosed man in his 50s met two sisters both of whom had been diagnosed 30 years earlier while in their teens. The sense of relief of those who were able to exchange their life stories was obvious and it was agreed that more information and emotional support was needed. This meeting led to the foundation of the U.K. Gauchers Association.

Similar groups were forming in other countries in Europe and around the world, and in 1992 an international symposium for clinicians, scientists, and patients' support groups was sponsored by the Genzyme Corporation in Amsterdam. Representatives of patients' support groups from many countries were able to establish a network of contacts which still flourishes. Both the German and Italian patients' support groups were established following the symposium, spurred on by the longer established groups.

When Patients Became Health Care Professionals

In each country, patient support groups for Gaucher disease were started by a few patients and their families seeking information. Many of these founders remain active in their groups to this day. Although a few are from the medical profession, the majority come from diverse backgrounds, from lawyers to writers, entrepreneurs to office workers. A doctor whose son has Gaucher disease also initiated a patient group. However, all involved soon became experts in this narrow field of lysosomal storage diseases, acquiring knowledge and experience from a variety of sources and contacts, which turned them into health information and health care professionals. They have been able to provide valuable information to families and participate in discussions with doctors, scientists, and health authority officials on a local and national level. The knowledge gained and their personal drive led them to negotiate with high-level representatives from the pharmaceutical industry to obtain factual information, devoid of marketing bias, which could be verified by clinical experts.

Patients' Needs

Patients need accurate, easy-to-understand information about their disease, its symptoms and progression. This information needs to be in written form, be easily communicable on the telephone or at meetings, and over the Internet. Patients also need specialist doctors who have experience of the disease and are keen to learn more for their patients' benefit. Not least, patients need access to available treatment.

The exchange of experiences between patients and their families remains an important aspect of a patient's support group's work. This can be done either by telephone or by e-mail. Meetings with a guest speaker and at annual conferences or workshops, where information is presented by physicians and scientists, also create an environment for affected families, especially newly diagnosed families or those having had no contact with other families, which helps them obtain the emotional and practical support they seek. Patients talk to other patients; parents speak to other parents, be they parents of young children or those facing the readjustment of passing on the responsibility for the condition to adult children. Husbands, wives, and caregivers can also share their thoughts with others in a similar position.

Information

The booklet *Living with Gaucher Disease* written by Dr. Norman Barton, Dr. Robin Berman, Dr. Ernest Beutler, Dr. Roscoe Brady, and Dr. Gregory Grabowski and published by the Genzyme Corporation in 1991 proved the perfect layman's guide. In everyday English, it explained how the accumulation of Gaucher cells affected the spleen, liver, and bone marrow, discussed inheritance patterns and how the disease is diagnosed, and went into the emotional and social aspects, including the pain, fatigue, and disability, which many patients experience. The booklet became a bible for many patients who share it with their doctors, teachers, and others. It formed part of an information pack sent out by the patient groups in the U.S. and U.K. and was supplied to many overseas sufferers. It has been translated into many languages and remains as useful today as it did when it was first published.

The U.K. Gauchers Association began to produce a newsletter called *Gauchers News* in 1992 published every six months, this has developed into a magazine of up to 24 pages and is currently distributed to 1300 families, doctors, scientists, and others in the U.K. and worldwide. It includes the latest information on every aspect of the disease, and its content is verified by a variety of specialists. It regularly publishes a personal story from a patient's or a parent's experience of living with the disorder. Often emotional, these stories help others to understand their own condition.

In addition fact sheets and brochures on specific aspects of life for a patient with Gaucher disease are produced in the U.K. Leaflets explaining neuronopathic (type 3) Gaucher disease and patients' special educational needs were written by Dr. Ashok Vellodi, who leads a pediatric Gaucher Centre in London and by Tanya Collin-Histed, mother of a type 3 child. A fact sheet on type 2 Gaucher disease was written by a family who had suffered the tragedy of losing their young child, where they explain in detail their story and the help they needed and obtained. A panel of physicians contributed additional information. On reading this fact sheet, a father who had recently lost his own child to the disease said: "You describe my situation exactly. How could you have written about our experience without knowing us?"

Relevant articles from the newsletters and all other information supplied by the U.K. Gauchers Association including *Living with Gaucher Disease* are available on its website at www.gaucher.org.uk. The Internet has enabled members of the general public to obtain instant information via their personal computers. The U.K. website is translated into Spanish and Russian, and links are provided to all Gaucher patient groups around the world.

A genetic diseases e-mail discussion list was established in the late 1990s and has enabled patients with Gaucher and other rare diseases to compare notes on many issues.

Publicity

The patients and families who started the support groups realized that the help they obtained by speaking to each other and exchanging information could benefit others. Research had shown that there was anticipated to be a significant number of sufferers in each country. They resolved to publicize the availability of information and assistance for those who might feel the same isolation and ignorance of current advances that they had experienced.

Gradually the number of people with Gaucher disease contacting the U.K. patient group increased. Eight people with Gauchers disease attended the inaugural meeting in 1991; by 2005, the Gauchers Association was in contact with nearly 270 patients in the U.K. During the same period, the number of identified patients increased significantly in Europe, North and South America and other countries around the world.

The Internet has played a crucial role in raising awareness of Gaucher disease throughout the world. The U.K. office regularly receives inquiries from many countries including Australia, Malaysia, Pakistan, South Africa, South American countries and even the U.S. Each inquiry is dealt with and where possible directed back to the patient group in the inquirer's own country.

Specialist Centers

Patients need knowledgeable doctors, and in the U.K. it gradually became apparent that there were only a few doctors who were experienced or showed an active interest in Gaucher disease. Many sufferers who contacted the Gauchers Association did not have a doctor who knew much about the disorder and these patients were advised to ask their local doctor to refer them to one of the experienced doctors. Thus centers of excellence gradually evolved.

In 1997, Prof. Timothy Cox, who had developed a clinic caring for a large number of patients with Gaucher disease at Addenbrooke's Hospital, Cambridge, applied for supra regional funding from the U.K. Department of Health through the National Specialist Commissioning Advisory Group. By this method, central funding was obtained for the assessment and management of patients at Addenbrooke's Hospital together with two pediatric clinics at Great Ormond Street Hospital in London and the Royal Manchester Children's Hospital, which already treated a number of children with the disease. A year later, a second adult center at the Royal Free Hospital in London was included. Central funding meant that the centers did not need to seek individual funding for each patient from his or her local health authority for the assessment and management of their disease, which greatly facilitated the care of the patients.

Even after the specialist centers were formed, the patient support role continued. Apart from directing families to the centers, occasionally a patient or family did not fully understand what was said to them by their doctor or at a clinic: "What exactly did he mean?" If the patient representative was at all unsure of how to answer, he or she would contact the doctor for clarification. Patient groups have developed a panel of trusted doctors who will explain and expand on any issue.

Availability of Treatment

When patients and their families first heard that enzyme replacement therapy could improve many of their symptoms or stop them getting worse, they were ecstatic. This excitement was soon tempered when they realized how much the treatment cost. Would their medical insurance companies or national health system be able to afford it? The debate over treatment using a high dose vs. a low dose took on a new dimension, and Dr. Ari Zimran's well chosen words: "It's not a matter of high dose vs. low dose but low dose or no dose" resounded in many countries around the world. Not only was there a debate about dosage levels but if a low dose was to be prescribed, it would have to be divided so instead of a patient receiving 60 units per kilogram of body weight every 2 weeks, it was 15 units per kilogram of body weight every 2 weeks, to be given three times a week, that is 2.3 units per kilogram of bodyweight every 2 days.

In the U.S. and in Europe the argument took place openly and most significantly during patient meetings. Even if low dose and frequent infusions were medically acceptable, the logistics of travelling to hospital three times a week was personally difficult especially by those patients who worked or studied. An hour long transfusion usually necessitated at least 2 hours from the time the nurse or doctor arrived in the clinic to when the solution was mixed, the needle or catheter inserted into the patient's vein and the giving set flushed after the infusion. In addition the patients and perhaps parents had to travel sometimes long distances to reach the hospital, which took time and money.

Dr. Ernest Beutler and Dr. Ari Zimran advocated home infusions, and in several countries this became an acceptable, safe, and preferred method of treatment. In some countries patients preferred a nurse or doctor to visit the home but in several others, notably Holland and the U.K., most patients learned to infuse themselves or parents learned to infuse their children. In some cases, a patient carries out the whole procedure, including siting the needle into a vein, entirely on their own. Many patients have now been carrying out this procedure for over 12 years with only an occasional call for help, usually if the needle was too difficult to site.

Obtaining Treatment

Even low dose treatment is expensive and during the early 1990s in many countries, it often became part of the patients' support groups' role to lobby a health authority to provide treatment to a patient, often working together with the patient's family. Those responsible for meeting the cost of enzyme replacement therapy often said that the treatment was "experimental" as well as expensive. In the U.K., local health authority officials and Members of Parliament were approached to facilitate the provision of therapy, and every case was settled in favor of the patient. During the following years enzyme replacement therapy and subsequently substrate reduction therapy for those who could not use enzyme replacement therapy for medical reasons, have become a recognized expense. However, with new expensive treatments for other rare diseases becoming available, the need to justify genuine need for enzyme replacement therapy remains an important issue for patients with Gaucher disease.

European Gaucher Alliance

Many physicians and scientists have played an active role in supporting and encouraging the work of patient groups. In 1994 the first meeting of the European Working Group on Gaucher Disease (EWGGD) was held in Trieste, Italy, and representatives of the known European patient groups, including the Israeli Gaucher Association, were invited to join with the doctors and scientists. While the patient representatives understood relatively little of the detailed science discussed, their presence was seen to encourage the professionals and much was also learned during the coffee, lunch, and dinner breaks by all parties.

Patient representatives continued to attend EWGGD meetings in Maastricht, Holland, in 1997 and in Lemnos, Greece, in 1999. During these meetings, the patient representatives took time to discuss their own priorities and concerns. It became apparent that the patient groups needed more time for discussion and at the fourth EWGGD meeting in Jerusalem, Israel (2000), a separate day was set aside for a special European Gaucher Alliance (EGA) meeting. At subsequent EWGGD meetings in Prague, Czech Republic, (2002) and Barcelona, Spain, (2004), the EGA held their own separate meetings to discuss issues of common interest.

During the late 1990s patients from former Communist countries in central and eastern Europe began to contact members of the EGA, and in 2001, an EGA delegation from the U.K. and Italy travelled to Bulgaria to hear about the plight of five children with the disease. Participants included Susan Lewis (U.K.), Fern Torquati (Italy), and Dr. Ari Zimran (Israel). They were shocked to see the severity of the children's disease but were pleased that funding had recently been obtained for them. Later in the year, Fern Torquati received an e-mail from the sister of a young Yugoslavian boy needing treatment. She arranged for the child and his mother to travel to Italy and obtained enzyme replacement therapy for him. She then went to Yugoslavia to evaluate the situation there and found many patients in need of treatment.

In the meantime the U.K. invited all EGA delegates to their 10th anniversary conference in London in 2001. Patients' representatives from Russia, Ukraine, Bulgaria, Romania, and Yugoslavia arrived to tell their own stories of patients in desperate need of treatment. The EGA took on the role of negotiating for humanitarian aid with Genzyme Corporation and in 2004, the company initiated the European Cerezyme Access Programme, which provides free treatment for severely affected patients whose health services cannot or will not meet the cost of treatment. Genzyme had previously provided humanitarian aid in Egypt, China, and parts of India and Pakistan through Project Hope and the International Cerezyme Access Programme.

Facilitating Interactions

The patients' support groups' focus on the needs of the individual patient has put them in a unique position to act as an intermediary in many discussions both on a national and international basis.

In individual countries patients' support groups have acted as facilitators between doctors and treatment centers providing the support and structure to focus on future activities and projects. The mere taking on of the responsibility of organizing a meeting (freeing the clinicians of the time-consuming details) has resulted in successful outcomes. In the U.K., the Gauchers Association has commissioned a national research study on bone disease to be a collaborative project between the four national centers.

On an international level, because there are only a relatively small number of clinicians and scientists in the field and due to the initiative to involve patients in scientific and medical meetings, patients' support group leaders have developed personal relationships with doctors and scientists from around the world and have, through their professional approach, earned their respect and confidence. This has enabled individual patients' support groups to play an active role in enhancing collaboration between medical centers and individual patient groups in countries where this approach is still novel.

Patients' support group leaders also appreciate the commercial expectations of the pharmaceutical industry and its drive to develop and market their product. The patients' support groups have a role to do what they can to ensure that the collaboration between clinicians and scientists with industry focuses on the advancement of treatment most beneficial to patients.

NEURONOPATHIC GAUCHER DISEASE (NGD)

Although many of the issues described above cover all types of Gaucher disease, patients with neuronopathic Gaucher disease (type 2 and 3 Gaucher disease) have additional challenges associated with the neurological aspects of the disease.

In 2004, enzyme replacement therapy was licensed for the treatment of type 3 Gaucher disease by the European Commission, although prior to the license, patients were already receiving enzyme therapy off-label due to the dramatic improvement of their visceral disease. Although enzyme replacement therapy has improved the life expectancy and quality of life for many patients, there is no evidence that it crosses the blood-brain barrier, and the neurological aspects of this disease are being increasingly seen. Therefore the challenge to develop a treatment that crosses the blood-brain barrier still remains.

The neurological aspects of the disease remain challenging for patients and their families and present many unknowns regarding quality of life, life expectancy, ability to function in society, and the impact on education and employment.

Many patients with type 3 Gaucher disease receive enzyme replacement therapy in high doses, and one infusion can take many hours. Increasingly a small portable device called an Intermate (manufactured by Baxter) is being used. This has the advantage of operating without electricity and has no moving parts. It has enabled patients to move around freely and go out while receiving their treatment.

In 2003 a clinical trial of substrate reduction therapy with miglustat (Zavesca) in type 3 Gaucher disease began at two sites, one in the U.K. and the other in the U.S., involving 30 patients. It is hoped that this drug will cross the blood-brain barrier and slow down neurological progression. However the clinical trial has proven to be a stressful time for many patients and their families. The drug must be taken orally up to three times a day and is delivered in a capsule that must be swallowed. This has been a challenge for many of the young children who have had difficulties in swallowing the capsules. The results of the trial are expected in late 2006.

Formation of a Support Service for Neuronopathic Gaucher Disease Families in the U.K.

Gaucher disease is rare and types 2 and 3 Gaucher disease are even rarer. Most patient groups worldwide know of very few patients with type 2 and 3 and when patients are newly diagnosed, usually in their first year of life, it can be difficult for families to meet other families to exchange and provide emotional and practical support. In 1997 the U.K. Gauchers Association recruited Tanya Collin-Histed, whose daughter was diagnosed with type 3 Gaucher disease at 16 months old, to join their Executive Committee to focus on raising awareness of the neurological form of the disease and to support families. Previously the group's attention had mainly focused on type 1. This enabled the group to offer type 3 and in some cases type 2 families someone who understood the issues they faced. Together with Dr. Ashok Vellodi from Great Ormond Street Hospital in London, she set out to raise the profile and understanding of neuronopathic Gaucher disease and bring families together to support each other.

Discovering that a child has neuronopathic Gaucher disease is devastating, but to learn that there are few other sufferers, that there is little information available and little known about the disease, and that there is no effective treatment for the neurological aspects of the disease is at first incomprehensible. The provision of emotional and practical support in this situation is essential and this has included visiting new families in their homes and in hospital to provide emotional support and practical information. Often just meeting an older child with the disease helps families to be more positive about the future.

European Consensus

In 1998, at the 3rd European Working Group for Gaucher Disease, the U.K. patient representative for neuronopathic Gaucher disease made a plea to the Working Group to develop a consensus paper for Europe on the management of neuronopathic Gaucher disease as treatment varied considerably in different countries in terms of dosage of enzyme replacement therapy and clinical management. The European

Working Group took the request forward and a Taskforce led by Dr. Vellodi was formed. Clinicians from the U.K., Sweden, Italy, Germany, and Poland together with the U.K. neuronopathic patients' representative met in the U.K. and developed a European Paper on the "Management of Neuronopathic Gaucher Disease." The Consensus paper was published in *The Journal of Inherited Metabolic Diseases* in 2001 and was revised in 2006.

NEURONOPATHIC GAUCHER DISEASE BOOKLETS

Little or no information was available for families and nonmedical professionals (such as teachers) involved in the day-to-day care of these patients. Consequently in 2001 two booklets were written by Tanya Collin-Histed and Dr. Ashok Vellodi entitled "Neuronopathic Gaucher Disease: A Guide for Parents" and "Neuronopathic Gaucher Disease: Special Educational Needs." Both booklets were produced by Great Ormond Street Hospital Trust and posted on its website. The booklets were distributed to all known families in the U.K. and sent to other European patient groups for information; the Italian Gaucher Association translated the booklets into Italian for their families. The booklets were re-written and updated in 2004 by the authors.

Neuronopathic Family Conferences

Bringing families together has always been a priority and an important aspect of support. In 1997 Dr. Vellodi held a Fun Day in London for all the families attending Great Ormond Street Hospital. The day included activities for the children and talks for the families. This successful event highlighted the value to families of getting together to share their experiences and listen to talks about the disease. In 1998 the U.K. Gaucher Association held separate sessions for type 3 families at their annual conference in London and in 1999 the first type 3 Family Conference was held in Northampton, U.K. A dozen U.K. families attended the conference and a program for the children was provided so that the parents could listen to presentations on topics such as eye movements, substrate reduction therapy, special educational needs, and the auditory pathway.

In 2004 a European Family Conference was held in the U.K. that was attended by families from the U.K., Sweden, Italy, New Zealand, Serbia, Montenegro, Jordan, and Germany. Many of the families from outside the U.K. had never met another family with a child with type 3 Gaucher disease and the event highlighted the need for patient groups throughout Europe and further afield to meet regularly. The European Gaucher Alliance has an important role to play in this.

Families Coping with Neuronopathic Gaucher Disease

The stress of the disease on the patient and their families can be enormous and have devastating effects. Marriages break down and parents may have to give up work or work part time, lowering income levels. Regular trips to hospital for check-ups or re-occurring illnesses mean time off from work or school, which may cause additional problems.

Uncertainty regarding long-term neurological outcome and the lack of a really effective treatment are difficult to deal with. Society's understanding of the condition is also limited and therefore for patients and their families, it is a challenge to cope with aspects of everyday life such as education, employment, and independence.

All patients have an eye movement problem horizontally and in some cases vertically, and in addition the majority of patients have some auditory processing problems and other cognitive issues. The patients look normal, are not blind or deaf and therefore present a real challenge to schools and colleges. The majority will need some degree of support to access the school curriculum. Unfortunately they do not fit neatly into the "boxes" designed by educational authorities for providing support, and there is a lack of understanding of their needs. Parents of other children with type 3 Gaucher disease who have been through this process can be a tremendous support and the two booklets mentioned above offer practical information on how to acquire the support children need. As more is understood, hopefully the assistance available to these patients as they grow up and become adults will improve and enable them to obtain employment and live independently to some degree.

Type 2 Gaucher Disease

For type 2 families, clinical management is focused on symptomatic care. Detailed advice on management should be given as soon as the diagnosis is made. To enable this, such children should be seen as soon as possible at a specialist center. Sadly there is as yet no effective treatment. Their child's frequent visits to hospital and rapidly deteriorating condition mean that parents have little time to adjust to the situation, let alone liaise with other families. For all these reasons it is difficult to support these families. However some type 2 families do make contact for emotional support during their child's illness and following their death. Support should consist of realistic, practical advice on day-to-day management.

Fundraising

Raising money to support the work of patient groups and, if possible, help research is essential. Although the price of the enzyme replacement therapy prohibits fundraising to provide individual patients' treatment, money needs to be raised to enable patients' support groups to carry out their activities. Most personnel work on a voluntary basis, expecting no pay, but the provision of information requires bills to be paid for telephones, printing, conferences, etc.

The National Gaucher Foundation in the U.S. consistently raises money for research and supplies grants to several research centers in the U.S. and internationally. It also raises money for patient education and awareness program and for patient assistance regarding insurance and other needs.

In addition the *Children's Gaucher Research Fund* administered by Greg Macres devotes all the money raised to research on neuronopathic Gaucher disease, as discussed below.

THE CHILDREN'S GAUCHER RESEARCH FUND

The section on the Children's Gaucher Research Fund was contributed by its founder, Greg Macres.

Amazing Progress, But ...

Looking at the history of Gaucher disease, it can be seen that amazing progress has been made over the past 20 years. In the 1980s, a partnership of patients, researchers, and industry was formed that would develop, test, and eventually receive FDA approval in the U.S. for the first ever enzyme replacement therapy. Enzyme replacement therapy, approved by the FDA in 1991, has significantly ameliorated the debilitating systemic symptoms faced by those who suffer from Gaucher disease. The pioneers of this effort should be commended for their impressive accomplishment. However, for children who suffered from neuronopathic Gaucher disease, there was little cause to celebrate because of the inability of the enzyme to cross the blood-brain barrier. Unfortunately, the remaining decade of the 1990s saw a significant reduction in Gaucher disease research. The Gaucher community now found itself with one group of patients with a successful therapeutic intervention, and another group of patients (mostly children) with the potential of neurological progression, and in some cases, certain death.

Enzyme Replacement for Neuronopathic Gaucher Disease

On December 21, 1993, at 11 months of age, our son Gregory Austin Macres was admitted to the hospital. This was the culmination of a 6-month effort to determine the cause of his health problems. We remember thinking, "What would it be like to learn your child has such a serious and potentially fatal disease?" At that point, we still believed Gregory's diagnosis and resolution would be fast and painless. After 6 months of doctor visits and testing, Gregory was diagnosed with type 3 Gaucher disease. As with many other families, obtaining an accurate diagnosis was a long and frustrating process. For some who suffer from the more severe neurological form of the disease (type 2), diagnosis has come within weeks of death.

After learning of the diagnosis, we researched Gaucher disease and found that the National Institutes of Health in Bethesda, Maryland, was conducting studies using enzyme replacement therapy. Gregory was seen at NIH and received weekly infusions of enzyme replacement from 1994 through 1997. Enzyme replacement was a relatively new drug, and it had proven to work well in alleviating the systemic symptoms presented in type 1 Gaucher patients. However, it was made clear by the medical team at NIH that it was yet to be determined if enzyme replacement would help to mitigate or reverse neurological symptoms in type 2 or type 3 Gaucher patients.

Difficult Decisions

Parents of children with type 2 and type 3 Gaucher disease face difficult decisions. Often these decisions are based on limited information, and results are not guaranteed. In December 1996, just prior to Gregory's fourth birthday, it became clear that despite high doses of enzyme replacement therapy, Gregory was suffering from continued neurological progression. With limited options available, it was decided that Gregory would undergo a bone marrow transplant. At the time, a successful bone marrow transplant offered the greatest hope for an optimal clinical outcome. However, so little was known about the molecular mechanism that is responsible for brain dysfunction, that even a successful bone marrow transplant offered no guarantees. Sadly, little Gregory succumbed to complications of the bone marrow transplant and passed away on April 13, 1997.

There are innumerable decisions that families face when caring for a sick child. With the more severe form of neurological disease, parents often must consider performing a tracheotomy. This helps to alleviate breathing difficulties and choking issues in addition to minimizing the risk of pneumonia; however, it does nothing to prevent continued progression of brain disease. A tracheotomy may allow a child to live longer, though this creates the likelihood that those parents and children will endure additional debilitating neurological symptoms as the disease progresses. Families also have the option to donate, via autopsy, brain tissue that is imperative for continued research. It is necessary that families make this decision prior to the passing of their child. For some, this is yet another agonizing decision.

Currently there are clinical trials underway to determine the efficacy of substrate reduction therapy. In the future, families may also have the option of participating in clinical trials for gene replacement or stem cell therapy. The decision to participate may be easy for families with critically ill children. However, there are some type 3 patients who have minimal neurological symptoms and they often move through school and on to college. For these families the decision is balanced between taking a chance on an unproven therapy and taking a chance on the natural progression of the disease.

Frustration

We, along with many other families in the 1990s, endured similar frustrations. Upon diagnosis, families would research Gaucher disease and soon realize that most of the support and medical emphasis was limited to type 1 Gaucher disease. There was a plethora of information pertaining to type 1 Gaucher disease, but information pertaining to the chronic and acute forms of neuronopathic Gaucher disease was sparse at best. Explicit questions relating to how the disease affects the brains in these children were met with a shrug and an apology. It was simply not known how Gaucher disease affected the brain. With further questioning, parents would realize that not only was the medical community unsure as to what molecular mechanism was responsible for brain dysfunction, but that very little research was being conducted to determine the cause. Families were left isolated, with modest support and little hope for a cure.

These frustrations led to the creation of the Children's Gaucher Research Fund, a grass-roots effort of parents across America, whose immediate focus is to fund research that is targeted at finding a cure for brain dysfunction in Gaucher disease. Families who learn of their child's diagnosis progress through the stages of shock and disbelief, to finding the strength to provide the medical care that is required. Upon the death of a child, families transition through grief, and many seek to honor their child's life by joining this effort to find a cure.

The Children's Gaucher Research Fund is Born

The true beginning of the Children's Gaucher Research Fund (CGRF) was on April 19, 1997, when over 400 people met at a small Presbyterian church in Saratogo, California, to say their last goodbyes to our son, Gregory Austin Macres. Due to our frustrations outlined above, and a strong desire to affirm Gregory's life, we expressed our desire to pursue a cure. The fuel for all that has been accomplished thus far began this day, as those in attendance initiated what is now over 8 years of generous giving. In 1999, the CGRF became a legal nonprofit entity and it was at this time that we began contacting families from across America and beyond who had been affected by neuronopathic Gaucher disease.

The goal of the CGRF is to raise funds for medical research and to award those funds to research on the brain, specifically targeted at finding a cure for neuronopathic Gaucher disease (type 2 and type 3). As a by-product of these efforts, the CGRF is also instrumental in providing support for newly diagnosed families and for families enduring the grief that follows the loss of a child. Our commitment to our donor base is simple; 100% of funds donated to the CGRF will be applied to medical research. The founders of the CGRF pay all administrative costs. However, the many individuals who generously donate their time and talent, fulfilling the various administrative needs of the research fund, minimize these costs. The families involved, some of whom are caring for a child, and some of whom have lost a child, have participated in a myriad of fundraising events. This collection of caring parents has now made great strides toward the funding of significant research on neuronopathic Gaucher disease.

Research Funded

In 2002, the Children's Gaucher Research Fund initiated the funding of scientific research by supporting the efforts of Dr. Tony Futerman at the Weizmann Institute of Science in Rehovot, Israel. Over a period of 2 years, the funding of this research answered questions that until now have been a mystery. Those involved in the CGRF understand the importance of basic laboratory research and are prepared to endure the years of effort required to harvest the knowledge that will lead to a cure. Although founded by parents and supported by a generous group of donors, it is the children affected by neuronopathic Gaucher disease that have set the tone for the Children's Gaucher Research Fund. Frustration, isolation, and unimaginable grief can be transformed to commitment, hope, and a cure.

Lysosomal Storage Diseases

Gaucher disease is part of a category of diseases referred to as Lysosomal Storage Diseases (LSD). As a group, LSDs are relatively common, occurring in 1:7000 births. Similar to most LSDs (Tay-Sachs, Niemann Pick, Batten's, Krabbe, Mannosidosis, Fucosidosis, Sialidosis, Mucolipidosis, etc.), type 2 and type 3 Gaucher disease affects the central nervous system. The impact on the brain is what makes these diseases terminal for many of the children who are affected and is also the most difficult part of the disease to understand. Currently, there is no effective treatment for the neurological symptoms of any LSD. Although each of these diseases has a different etiology, it is likely that common pathogenic processes exist. Researchers believe that a better understanding of one of these neuronopathic LSDs will lend itself to a greater understanding of the others. Further, and more exciting, researchers believe that a cure for one of these diseases may help pave the way toward finding a cure for the others.

With this in mind, the CGRF believes it is imperative that collaboration by those involved in neuronopathic LSD research take place on a continuing basis. The CGRF pursued this goal in 2004 by hosting a scientific conference in Bethesda, MD, entitled "Lysosomal Diseases and the Brain." We believe this conference substantiated the need and the desire among researchers to increase collaboration among these disease groups. With over 100 in attendance, M.D.s and Ph.D.s from seven countries participated in two full days of presentations that exclusively focused on brain dysfunction in lysosomal diseases. The importance of this event was captured in a quote when one of the attendees commented, "If the CGRF does nothing more than to bring this group together on a regular basis, you will be fulfilling an essential need within the LSD research community."

CHAPTER **28**

Societal Perspective: Comment

Abby Alpert, Alan M. Garber, and Dana P. Goldman

CONTENTS

Enzyme replacement therapy (ERT) has been a major clinical breakthrough in the treatment of Gaucher disease. Without the protection of the Orphan Drug Act (ODA), it is unlikely that ERT would have become available to treat this extremely rare disease, for which no therapy had previously existed. However, despite its success in improving the quality of life for those who suffer from the disease, the therapy's expense poses a dilemma for payers such as government programs who would like to administer treatment to those who would benefit, but who also cannot ignore the cost to individuals and society. Has the ODA impeded, rather than stimulated, the development of cost-effective technologies? Is the price of Cerezyme too high or will the treatment be provided regardless of the cost? These questions, among others, have been discussed in the chapters of this section.

On the one hand, pharmaceutical firms require substantial assurance of profitability in order to undertake the risk of developing new drugs. The development of drugs for rare diseases is especially risky, because high research and development (R&D) costs must be recovered over a very small patient population. The ODA affects the incentives for development by awarding 7 years of market exclusivity to producers in which to recover fixed research costs. High prices are an inevitable consequence of this government-granted monopoly. Once the drug has been developed, patients, health insurers, and government programs face difficult choices about how to pay for it. They must decide whether the benefits of the new drug are worth the cost. In countries where the government is the primary payer, allocative decisions are often based on formal assessments of costs and benefits. Those countries may limit distribution of the drug to patients who will benefit the most, and some mandate that initial dosages should be low, in order to minimize the cost per patient treated. These countries may also impose price controls. While these cost-controlling measures benefit consumers in the short run, they may adversely impact firms' incentives for drug development, hence harming consumers in the future. In the U.S., in particular, insurers have limited ability to negotiate price reductions for truly unique drugs. They may be required by law, fear of litigation, or contractual agreements to reimburse patients for the use of the drug regardless of cost. Thus can ensure manufacturers of high revenues while enabling patients receiving the treatment to do so at out-of-pocket costs that represent only a fraction of the total payments for the drugs. The costs borne by the insurer are spread across the entire insured population in the form of premium payments.

The case of ERT for Gaucher disease dramatically illustrates the delicate balance that regulatory bodies must consider in devising policies to control health care costs without deterring the development of important new technologies. In this chapter, we consider the necessary economic incentives for drug innovation and, in particular, the challenges that must be overcome to develop treatments for rare diseases. We also discuss the social welfare consequences of the ODA, since excessive rewards to innovation can lead to excessive investments in development of new drugs and can lead to the wrong kind of innovations (e.g., in products that will not be cost-effective). Finally, we consider the appropriate role of regulation following orphan drug approval.

THE DEVELOPMENT OF ENZYME REPLACEMENT THERAPY

The primary justification for government intervention in pharmaceutical research and development is that private markets tend to underinvest in research or supply insufficient quantities of the new drugs they develop. As Meeker and Termeer note in Chapter 26, drug companies must make very large investments in research, clinical testing procedures, and production equipment and facilities before knowing with certainty whether they will be able to develop a product that will gain marketing approval. One widely cited study estimates that approximately $800 million is spent on average to bring a new product to market, including the costs of failed attempts.[1] Since marginal costs (the cost to produce each unit) are typically small (and often negligible) relative to these large fixed research costs, producers must be able to charge a price well above marginal cost in order to recover, or more than recover, their R&D investments. In a competitive setting, with multiple firms vying to treat patients using similar (or identical) therapeutic substitutes, this would not be possible, as the high price would eventually fall to a level approaching the cost of production by firms that did not face fixed costs as high as those faced by the innovator firm. If this were the market that the pharmaceutical innovator expected to face, it could not expect to generate positive returns from its product, and thus would have no incentive to develop the product in the first place. Thus a competitive market would lead to underinvestment in research. Market exclusivity, such as that conferred by the patent system and regulatory restrictions by the U.S. Food and Drug Administration (FDA) is essential in correcting this market failure, as it provides firms with a temporary monopoly and thus the ability to set prices that exceed marginal costs by a substantial margin — hence providing an incentive for innovation.

In addition to these barriers to drug development in general, orphan drugs are designed for a small market, making a low price even less feasible. The ODA has successfully stimulated research in this area by providing 7 years of market exclusivity for orphan drugs. This feature is most important for drugs for which patent protection is not available. Furthermore, the ODA lowers the fixed costs of R&D. A tax credit on clinical trial costs, exemption from user fees, grants to support development, protocol assistance, and less stringent requirements for clinical trial design are among the ways in which the ODA lowers research and development outlays.

In the development of Ceredase, Genzyme did not take advantage of all of the available provisions of the ODA such as the tax credit and protocol assistance.[2] However, the initial discovery and substantial research leading to the development of alglucerase was performed or funded by the federal government. Researchers at the National Institutes of Health (NIH) identified the enzyme and developed a method for harvesting and synthesizing it for patients' use. NIH also sponsored or performed much of the preclinical trial research and, using enzyme supplied by Genzyme, conducted the pivotal trial that led to the drug's approval. In fact, it has been estimated that Genzyme spent less than $30 million on R&D for its initial product.[3] Still, the ODA was essential in bringing this product to market since the substantial federal role in its development would have made the drug unpatentable. (Genzyme later went on to patent its production technique, which had changed significantly following approval). Without the ODA, Genzyme would have risked having its manufacturing process for its large-scale production undertaking replicated by competing firms. Thus, market exclusivity was the primary incentive behind Ceredase's development.

Ceredase developers also benefited from the ODA's flexibility in clinical trial design, since its approval, as discussed in Chapter 25, was awarded on the basis of a study involving only 12 patients. While randomized, controlled trials involving thousands of patients are the FDA's preferred method for testing safety and efficacy, the cost of conducting these larger trials (in this case, by recruiting from a very small population among which many were not even aware of their disease) would have been prohibitively expensive.

The FDA is responsible for bringing to market safe and effective medications and medical devices. In order to do so, it provides firms with an opportunity to make substantial profits. The FDA does not have the legislative authority to base its drug approval decisions on consideration of relative costs and benefits. Though the FDA does not directly provide incentives for developing *cost-effective* medications, its actions have consequences for producer profits (and consumer benefits). For example, the duration of market exclusivity (from approval to patent expiration, or in the case of ODA approvals, 7 years) affects producer profits. The speed with which generics enter the market following patent expiration determines how quickly innovating firms face lower-priced competition, and patients have access to lower-cost versions of drugs. Although some FDA activities therefore lead to price reductions for drugs, the Orphan Drug Act generally works to protect the profits of pharmaceutical firms.

POSTDEVELOPMENT PRICING

Ceredase, and subsequently Cerezyme, have been scrutinized because of their high price, which can easily exceed $300,000 per year for adults on the high dose regimen, and because the federal government bore such a large fraction of the costs of discovering and developing the drug. How did Ceredase and Cerezyme become among the most expensive drugs on the market? The small patient population alone does not explain the high cost of Ceredase and Cerezyme; other orphan drugs that have comparably small patient populations are considerably less expensive. There are some other important factors driving up the price such as consumers' price-insensitivity induced by the presence of health insurance and the high marginal cost of production.

Health Insurance

The high price would not have been possible in the absence of health insurance. As we noted, monopolies are necessary to provide appropriate incentives for pharmaceutical firms to invest in drug innovation. However, market exclusivity can also lead to under-provision of drugs for consumers. Certainly, once the drug has been developed, consumers may find that the benefits of the drug are not commensurate with the cost they bear and decide not to pay. Since some consumers would be willing to pay more than the competitive price (or marginal cost) but less than the monopoly price, a lower, competitive price could increase the quantity of the drug demanded. The relative underconsumption, which results from monopoly pricing — or the difference in benefits between consumption at the competitive price vs. consumption at the monopoly price, is a pure welfare loss to society (known as "deadweight loss"). (By limiting the patent, and thus the time during which a monopoly can exist, this loss can be mitigated over time.)

The presence of insurance has the opposite effect — over-consumption. With insurance, demand is subsidized so that patients face only a fraction of the actual price directly. Few patients with Gaucher disease could afford to pay the full cost of bi-weekly Cerezyme infusions out-of-pocket, but with insurance they may face a co-payment that amounts to a very small percentage of the total price. Under government sponsored insurance, the fee might be an even smaller fraction of expenditures for the drug. Thus, increases in the actual price will not appreciably lower consumer demand. In fact, at a fraction of the price, the drug needs to deliver only a fraction of the benefit for patients to be willing to consume it. Thus, insurance drives up the consumption of drugs beyond the optimal competitive level.

Although insurers directly pay most of the costs of the drug consumed, their price sensitivity is limited; most must pay because they are contractually obligated to cover drugs approved by the FDA. When there are no therapeutically equivalent substitutes, as is the case of Cerezyme, insurers are almost certain to cover the treatment in spite of its high cost. Pressure from patient advocacy groups may also influence the expansion of coverage of certain drugs by third-party payers (as noted in Chapter 27). Thus, when insurance is used to finance the purchase of a drug with no competitors, the pharmaceutical company can extract a higher price than if it were a simple monopoly selling into a typical market in which demand is not subsidized. These effects allow profits to expand beyond the normal monopoly level.

Because monopoly and insurance separately lead to under-consumption and over-consumption, respectively, they may cancel each other out. That is, the quantity of drug consumed might be equal to the quantity consumed in a competitive equilibrium. Of course, the two effects are unlikely to offset each other exactly. The extent to which these two effects lead to either over- or under-consumption depends upon the distribution of treatment benefits across the population and the co-insurance level.[4] In sum, insurance helps to encourage innovation by allowing patients who can benefit from a new drug to gain access to it, despite the high price that has been induced by the government-granted monopoly.

Marginal Cost of Production

Another factor that contributes to the high price of Cerezyme is the marginal cost of production. The marginal costs of small molecule drugs are very small. In the case of a biologic product like Ceredase or Cerezyme, however, marginal costs can be much higher. For Ceredase, the production process, which required thousands of placentae to produce treatment for one patient for one year, was inherently costly. In fact, Genzyme has claimed that production costs accounted for more than half of Ceredase's price — an unusually high share for a pharmaceutical.[5] While it is likely that the recombinant form costs less to produce than the placental form, the production process is still far more complex and expensive than for most nonbiologics. This distinction has important consequences for evaluating the appropriate treatment level, since the cost can be greatly reduced by lowering the dosage. And in the case of such an expensive treatment as Cerezyme, the difference between a high and low dose regimen could mean hundreds of thousands of dollars per patient each year.

Ceredase was approved on the basis of a trial, which, for practical reasons, tested only one dosage, 60 units/kg, at two dosing frequencies for its 12 subjects. Research conducted following Ceredase's approval appeared to demonstrate that equivalent efficacy could be achieved at a lower, fractionated dosage (e.g., 2.3 units/kg, three times a week).[6] However, due to the small sample sizes and, in many instances, short study period, few physicians have embraced the studies as definitively establishing therapeutic equivalence.[7,8] Consequently, several authors recommend initiating therapy at high dosages. In Chapter 25, Waalen and Beutler argue that what they believe to be an unjustified reliance on the high dosage is a consequence of inadequate clinical evaluation under the ODA due to its lax standards for clinical trial design.

If the same therapeutic benefits can be produced at a lower cost, it should make this treatment far more attractive to physicians, patients, and payers. However, testing the efficacy of multiple dosages is very difficult when the potential sample of patients is small. Of course, once the product is on the market, a pharmaceutical firm has little incentive to invest in studies showing that lower dosages provide similar benefits. It is very difficult for a firm to double the price it charges for a drug after it shows that half the dosage is sufficient. Thus, firms with monopoly power do not possess an incentive to search for the cheapest regimen for a desired level of benefits. This effect is even more pronounced when the presence of health insurance makes consumers insensitive to the price of the high dosage regimen. It is interesting to note that countries where the government is payer, such as Israel and the U.K., the lower dosage is mandated (see Chapter 25). Finally, the controversy surrounding the adoption of the low-dose regimen also illustrates how postapproval research conducted outside of the manufacturing firm may not be an immediate "market solution" to the inadequate clinical evaluation of alternative dosages for orphan drugs with high marginal cost. For these reasons, government intervention may be necessary to promote the development of a cost-effective therapeutic strategy.

The uncertainty over the appropriate dosing regimen for Cerezyme also demonstrates the importance of postapproval clinical surveillance. Since the FDA's clinical trial requirements are very lenient for orphan drug innovators (allowing very small samples, open-protocol arrangements, etc.), postclinical studies are particularly important in this context. Genzyme has created the Gaucher Registry to catalogue physician-reported data of over a thousand patients worldwide. But the Registry may not be adequate to address some of the major postapproval issues that have arisen. For example, a 2002 article, referenced by Meeker and Termeer as including the most extensive data from the Registry ever published, examined the effect of Cerezyme on the physical manifestations of the disease without considering treatment effects by dosage or dosing frequency.[9] In order to improve postclinical trial surveillance, it may be necessary for the FDA to require full disclosure of this kind of information and to ensure that the information to be collected is comprehensive. In this case, physicians, researchers, and the government, rather than drug company, bear most of the postapproval cost of evaluating the treatment. This allows for better clinical evaluation without imposing the substantial costs of large clinical trials on the producer, thereby enhancing incentives to innovate.

Welfare Consequences

Finally, while we have considered some of the causes of Cerezyme's high price and the consequences of this price level, we have not addressed whether the price of the drug is too high or, alternatively, whether it is at the level needed to encourage the development of an important treatment. Ultimately, to understand whether the price is *too high*, we must consider not only the costs and benefits faced by consumers and those faced by pharmaceutical firms separately, but the total costs and benefits to society.

The total benefits to society from a given level of drugs supplied can be measured in terms of consumer surplus and producer surplus — together known as social surplus. Consumer surplus is a benefit captured by those patients who pay less than the drug is worth to them. In dollar terms, this benefit is the difference between the price they would be willing to pay and the price they actually pay. On the other hand, producer surplus represents a gain that results when firms sell their product at a price that exceeds the cost of producing the drug, or simply the firm's profits. To determine whether society benefits from the introduction of a new drug, we must consider the social surplus. One approach is to simply consider whether the overall benefits produced by the drug are at least as great as the total costs. While it is relatively straightforward to measure benefits at any single point in time, (i.e., through standard cost-benefit analysis), only a close examination of the entire life cycle of a product can reveal the true benefits that result from a new drug.

Typically, the price of a product represents its value to purchasers. However, when the market is not perfectly competitive, as is the case for Cerezyme and most health care products and services, price is not a good measure of willingness to pay. In this market, the preferred metrics are based on cost-effectiveness or cost-benefit analysis, two techniques that examine the social value of the improvement in mortality and morbidity that result from use of the drug. This may be accomplished by equating an additional life year gained to a dollar value and then examining the difference between the benefits and the cost, or by evaluating the ratio of the cost of the drug to the units of health it produces. Compared to many widely accepted health care interventions, Cerezyme's treatment of a generally nonfatal disease is far from cost effective. In the U.S., treatments that cost less than $100,000 per life-year are typically taken to be cost effective. Given the high cost of treatment in the U.S., it would not meet this standard even if the drug meant the difference between life and death.

The problem with measuring consumers' welfare on the basis of cost-effectiveness analysis is that such an assessment is concerned with the benefits of the drug after it has been developed (ex post) and does not consider the value of the innovation. Pharmaceutical research and development is a uniquely dynamic process, and thus we must compare the research and production costs to the full stream of benefits over a prolonged period of time. The money earned from the treatment has value to society — in fact, this treatment largely built Genzyme into a worldwide biotechnology company with 7000 employees. Future innovations from Genzyme can thus be linked to this product.

Despite these concerns, some countries (in particular, those with a single payer) use cost-benefit or cost-effectiveness analysis to determine whether to supply or reimburse certain drugs. In some cases, their policies result in explicit drug exclusions or price controls that are intended to ensure that the cost of a given benefit is not too high. Whatever the means of controlling costs, too stringent and too widespread a reliance on this method of technology assessment in allocating resources may result in a loss for consumers in the long run by preventing the development of important therapies. In contrast, the U.S., with no single payer, is unable to systematically restrict the use of drugs that are not cost effective and does not directly control prices through regulation. Along with the substantial presence of insurance, this environment allows drug innovation to thrive.

The availability of Cerezyme to treat Gaucher disease *and* its high price are direct consequences of this type of setting. Despite the fact that Cerezyme is not a cost-effective treatment, its supply has not been explicitly restricted. Contrast ERT with, for example, antiretroviral drugs (ARVs), which treat HIV/AIDS. ARVs slow the transition from HIV to AIDS and have dramatically lengthened life expectancy for those living with AIDS. The enormous social benefits conferred by these drugs are reflected in favorable cost-effectiveness ratios. While the price of ARVs is also very high, recent estimates suggest that producers' profits actually represent a very small fraction of the social benefits resulting from the introduction of this intervention.[10] On the other hand, Cerezyme, while providing important benefits, does not meet traditional cost-effectiveness criteria. In this special case, it seems likely that the producer is appropriating all of the social surplus.

CONCLUSION

The primary purpose of the ODA is to stimulate the research and development of treatments for underserved diseases. Although it is largely successful in this respect, it creates challenges at a time of rapidly rising drug expenditures. Its provisions are intended to promote the development of orphan drugs, not to ensure that the drugs brought to market confer benefits that are commensurate with their costs. Moreover, in an effort to lower the costs of the R&D process, drugs are approved with sometimes limited evidence of efficacy and inadequate knowledge of the optimal treatment regimen. In the case of Cerezyme, this leniency may have led to an inappropriately high recommended dosage, which made the therapy more costly than necessary. Of course, these concerns should be balanced against the need to provide companies with sufficient rewards to undertake the costly research and development needed to bring a drug to market.

Surveillance following approval is also imperative, since ODA approval requires only small and relatively short-duration clinical trials. Such surveillance will improve our understanding of a drug's risks and benefits and can lead to improved dosage regimens. Ultimately, in deciding whether to support the development and provision of a new medical treatment, especially for orphan diseases, society must make complex calculations of whether the benefits are worth the cost over many years, an exercise that can only be validated with the information gained from ongoing surveillance.

REFERENCES

1. DiMasi, J.A., Hansen, R.W., and Grabowski, H.G., The price of innovation: new estimates of drug development costs, *J Health Econ*, 22, 151–85, 2003.

2. Goldman, D.P., Clarke, A.E., and Garber, A.M., Creating the costliest orphan. The Orphan Drug Act in the development of Ceredase, *Int J Technol Assess Health Care*, 8, 583–97, 1992.

3. Garber, A.M., No price too high? *N Engl J Med*, 327, 1676–8, 1992.

4. Garber, A., Jones, C.I., and Romer, P., Insurance and incentives for medical innovations, *Forum for Health Economics and Policy*, 2006 (accesed at http://www.bepress.com/fhep/biomedical_research/).

5. Garber, A.M. et al., Federal and Private Roles in the Development and Provision of Alglucerase Therapy for Gaucher Disease, Report No. OTA-BP-H-104, Office of Technology Assessment, Washington, D.C., 1992.

6. Figueroa, M.L. et al., A less costly regimen of alglucerase to treat Gaucher's disease, *N Engl J Med*, 327, 1632–6, 1992.

7. Grabowski, G.A., Treatment of Gaucher's disease, *N Engl J Med*, 328, 1565, 1993.

8. Moscicki, R.A. and Taunton-Rigby, A., Treatment of Gaucher's disease, *N Engl J Med*, 328, 1564; author reply 1567–8, 1993.

9. Weinreb, N.J. et al., Effectiveness of enzyme replacement therapy in 1028 patients with type 1 Gaucher disease after 2 to 5 years of treatment: a report from the Gaucher Registry, *Am J Med*, 113, 112–9, 2002.

10. Philipson, T. and Jena, A., Who benefits from new medical technologies? Estimates of consumer and producer surpluses for HIV/AIDS drugs, *Forum for Health Economics and Policy*, 2005 (accesed at http://www.bepress. com/fhep/biomedical_research/).

CHAPTER 29

Gaucher Associations Around the World

ARGENTINA

Asociación Gaucher Argentina
Avda.Cordoba 5933 – Of. 2
(1414) Ciudad de Buenos Aires
Tel: 011 4776 3283 y 011 4664 0402
Int tel: (0054) 11 4776 3283 y 11 4664 0402
http://www.gaucher.com.ar

AUSTRALIA

Gauchers Association
PO Box 983
Sunbury
Victoria 3429
Tel: (0061) 3 9740 7203

AUSTRIA

OGG — Österreichische Gesellschaft für Gauchererkrankungen
Millergasse 48
A 1060 Wien
Tel: 01 596 5000
Int tel: 0043 1 596 5000

BRAZIL

Associação Brasileira dos Portadores
da Doença de Gaucher
Av. Nilo Pecanha 155
Sobreloja, Centro, CEP 20020-100
Rio de Janeiro
Tel: (0055) 21 220 8633

Associação Paulista dos Portadores
da Doença de Gaucher
Rua Paes de Araujo, 178
Itaim Bibi, São Paulo
SP – 04531-090
Fone: 3167-1988
http://www.appdgaucher.org.br

BULGARIA

National Gaucher Association in Bulgaria
TOMOV@Gaucherbg.org
Tel: (02) 971 5038
Fax: (02) 790 550

CANADA

National Gaucher Foundation
4100 Yonge Street, Suite 310
North York, Ontario, M2P 2B5
Tel: (00) 1 416 250 2850
Fax and messages: (00) 1 416 486 4338
http://www.gaucher.org/

COLOMBIA

Asociación Gaucher de Colombia
Calle 142 No 25-92 Casa 12B
Santafé de Bogotá
Tel: (0057) 1 216 4936
Fax: (0057) 1 274 2695
http://www.gaucher.org.co/

FRANCE

Vaincre les Maladies Lysosomales
9 Place du 19 Mars 1962
91035 Evry Cedex
Tel: (0033) 1 60 91 75 00
http://www.vml-asso.org

GERMANY

Gaucher Gesellschaft Deutschland
An der Ausschacht 9
59556 Lippstadt
Tel/fax: (0049) 02941 18870
http://www.ggd-ev.de

GREECE

Greek Gauchers Association
12 Polyla Street
Gerakas
Athens 15344
Tel: (0030) 010 661 1270

HUNGARY

Kossuth útca 45
H-3916 Bodrogkeresztur
Tel: (0036) 47 396 329
Tel: (0036) 36 428 153

ISRAEL

Israel Gaucher Association
PO Box 33814
Haifa 31338
Tel: 04 950 3403
Int tel: (00972) 4 950 3403
http://www.gaucher.org.il

ITALY

Associazione Italiana Gaucher
Via dell'Arcolaio 33
50133 Firenze
Tel: 055 612 1297
Int tel: (0039) 055 612 1297
http://www.gaucheritalia.org

JAMAICA

Gaucher Patient Contact
Tel: 1 876 933 4597

JAPAN

Gaucher Support Club
41-24, Tajii
Mihara-chou
Minamikawachi gun
Osaka
Tel: 72 362 8184
Int tel: (0081) 72 362 8184
http://www1.odn.ne.jp/mikun/

MEXICO

Asociación Gaucher de Mexico A.C.
Centro Comercial Galerias Reforma
Carretera Mexico — Toluca 1725 Local A-30
Colonia Lomas de Palo Alto
Cuajumalmpa, Mexico D.F. C.P. 05110
Tel: (0052) 53 96 46 82 or 53 42 25 27
Fax: (0052) 55 70 78 88

THE NETHERLANDS

Gaucher Vereniging Nederland
Ruitercamp 155
NL – 3992 BZ Houten
Tel: 030 637 4417
Int tel: (0031) 30 637 4417
http://www.gaucher.nl/

NEW ZEALAND

Gauchers Association of New Zealand
PO Box 74384
Auckland
Tel: (0064) 9 527 1287

PARAGUAY

La Asociación de Pacientes de Gaucher del Paraguay
Cáceres Zorrilla 1076, CP 1771,
Villa Guaraní, Asunción
Tel: (00595) 28 34203

POLAND

Stowarzyszenie Rodzin Osob z Choroba Gauchera
Pawel Oswiecinski
ul. Perzyny 85/3
26-700 Zwolen
Tel: (0048) 48 676 2741

ROMANIA

Romanian Foundation for Lysosomal Diseases
CP 1298 OP 1
3400 Cluj Napoca
Tel: (0040) 94 618 289

RUSSIAN FEDERATION

Moscow
Tel: (007) 095 727 8154

SOUTH AFRICA

Gauchers Society
PO Box 51399
Raedene 2124
Tel: (0027) 11 640 5577

SPAIN

Asociación Española de Enfermos y Familiares de Enfermedad de Gaucher
C/. Pérez del Toro, 41
35004 Las Palmas de Gran Canaria
Tel/fax: (0034) 928 24 26 20

SWEDEN

Morbus Gaucher foreningen
Soldatvägen 19
955 31 Råneå
Tel: (0046) 924 10986

UNITED KINGDOM AND IRELAND

Gauchers Association
19 Downham View
Dursley
Gloucestershire GL11 5GB
Tel: 01453 549 231
Int tel: +44 1453 549 231
http://www.gaucher.org.uk/contents.htm

UKRAINE

Gaucher Association — Charitable Organisation
Tel: 05447 4 85 96
Fax: 05447 4 06 83

UNITED STATES

National Gaucher Foundation
11140 Rockville Pike
Suite 101, Rockville
Maryland 20852-3106
Tel: (00)1 301 816 1515
http://www.gaucherdisease.org/

VENEZUELA

Asociación Gaucher de Venezuela
Urb. Monterrey
Calle B, Qta. Lantepenúltima
Caracas 1050
Tel: (0058) 212 943 3493

Index

A

α-Gal A, deficiency in Fabry disease, 382
ABC transporters, 403, 405
Abdominal distention, 197
  as adverse effect of miglustat, 367
Abducens nerve weakness, 483
  in type 3 Gaucher disease, 429
Ability to pay, *vs.* need, 447–448
Absorptive-mediated transcytosis (AMT), 409
Academic Medical Center of Amsterdam, 275
Accumulating materials
  glucocerebroside, 3–4
  glucopsychosine/glucosylsphingosine, 4
  glucosylceramide, 98
Acid β-galactosidase, deficiency in GM
      gangliosidosis, 386–387
Acid phosphatase, 348
  as biomarker, 260
Acid ß-glucosidase (GCase). *See also* GlcCerase
  cell biology and biochemistry, 40–41
  x-ray structure of, 85–86
Acid sphingomyelinase (ASM), lysosomal
      delivery by ICAM-1-targeted
      nanocarriers, 135–136
Acidic mammalian chitinase (AMCase), 255
Activator molecules, in lysosomes, 126
Active site inhibitors, as pharmacologic
      chaperones, 380
Adherens junctions (AJ), 404
Adult form Gaucher disease, 2
Adverse events, with ERT, 342–343
AL amyloidosis, 257
Alcian blue neurons, 229, 230
Alcian blue/Periodic acid-Schiff (AB/PAS) stain,
      229, 244
Alcian blue staining, 243, 244
Alendronate, use in bone density therapy, 259
Alglucerase, 341

Alpha glucosidase, deficiency in Pompe disease,
      126
Alveolar capillaries, Gaucher cells in, 200, 221,
      222, 520
Amino acid (SLC) transporters, 402
Anemia, 159, 161, 198
  elimination by ERT, 469
  gynecologic concerns, 168–169
Angiotensin-converting enzyme (ACE), as ERT
      response biomarker, 348
Animal models, 141–142
  canine model, 148
  chemically induced, 146
  chimeric mouse model, 146
  current issues, 148
  GCase activator mouse models, 146
  global gene expression, 148
  GM1 gangliosidosis, 378
  mouse models, 140–146
  partial prosaposin deficiency with GCase
      point mutations, 147
  partial saposin deficiency mice, 147
  total saposin deficiency mice, 147
Antiglucosylceramidase immunoreactivity, 234
Aortic calcification, in neuronopathic Gaucher
      disease, 178
Arachnoid membrane
  mechanism of, 399
  role in BBB, 398
Armed Forces Institute of Pathology (AFIP), 240
Aseptic necrosis, 293
Ashkenazi Jews, 156
  carrier screening programs for, 259, 327
  as carriers of Gaucher, 127
  gallbladder involvement in, 161
  hydrolase screening in, 253
  mutant GBA alleles in, 16
  neuroprotective N370S mutation in, 384
  Parkinsonism in, 184
  prevalence of GD among, 322, 323

variations within genotypes, 35
Resting energy expenditure, increased in Gaucher
    disease, 310
Retroviral vectors, in gene therapy for Gaucher
    disease, 7
Ryanodine receptor (RyaR), 100, 101, 234, 236
    increased sensitivity with GlcCer
        concentrations, 237
    release of calcium from, 119

S

Saccadic paresis, 177–178, 180, 181
    in neuronopathic Gaucher disease, 178–179
Safety and efficacy, 452
Sandhoff disease, 252, 360, 371
    mouse model of, 414
    pharmacologic chaperone therapy for,
        387–388
    substrate reduction therapy in mouse model,
        361–362
    subtypes of, 388
Saposin A, 69, 74
Saposin B, 69, 74–75
Saposin C, 67–69, 75, 260
    abnormal juvenile form of Gaucher in
        deficiency, 69
    acid ß-glucosidase requirement for, 85
    biochemical properties, 70–71
    deficiency of, 78
    degradation of glucosylceramide by, 71
    function, 71
    immune system role, 77
    importance to stability of GCase, 51
    mechanism of GCase activation, 56, 94
    membrane destabilization, 74
    patchlike structural domains, 74
    role in lipid antigen presentation, 77
    role in skin, 76
    and stratum corneum integrity, 76
    structure and mechanism of action, 73–74
    threshold levels for variant GCase function,
        147
    triggering of membrane binding by pH
        decrease, 74
Saposin D, 69, 75
Saposin deficiencies
    prosaposin, 77–78
    saposin C, 78
Saposins
    additional functions, 75–77
    biosynthesis of, 70
    general aspects, 70

laboratory features, 251–252
lipid-binding and membrane-perturbing
    properties, 70
precursor of, 70
Saposin C, 70–74
saposins A, B, D, 74–75
Scarce resources
    balancing rare genetic diseases with, 445–446
    ethical issues on allocating, 444
    governmental issues, 449
sCD163, 256
Schindler's disease, enzyme analysis for, 331
Scintigraphic score, dose-response relationship to
    ERT, 315
Scintigraphy, 288, 522
    with autologous ^{99m}Tc-HMPAO-leukocytes,
        298
    with bone-seeking agents, 288
    combined bone marrow and liver-spleen, 302
    correlation with severity of skeletal disease,
        302
    defining regions of interest, 304
    with injection of autologous heat-damaged
        erythrocytes, 292
    long-term follow-up of pattern with ^{99m}Tc-
        Sestamibi, 313, 314
    radiocolloid, 292
    role in differential diagnosis of bone crisis,
        297
    sensitivity and specificity, 288
    three-phase bone, 296
    use in monitoring efficacy of ERT, 313
    whole-body skeletal scans, 295
Screening
    based on ethnicity, 445
    for biochemical markers, 259–260
    defined, 327
    for eligibility and compliance, 452
    moral imperatives issue, 445–446
Screening policy, 321–322, 327
    carrier testing, 327–328
    false positives and, 329
    newborn screening issues, 328–331
    technology for LSD screening, 331–337
Screening technology, 331
    enzyme analysis, 331–332
    metabolite profiling, 332
    protein profiling, 332–333
Seizures, 182
    in neuronopathic Gaucher disease, 179–182
Septic arthritis, of hip joint, 164
Severity score index (SSI), 157, 310
Shapiro, David, 4
Sialidosis, 487

Milton Keynes UK
Ingram Content Group UK Ltd.
UKHW051013071024
449327UK00012B/224